GW00702230

ECOLOGY AND MANAGEMENT OF FOREST SOILS

ECOLOGY AND MANAGEMENT OF FOREST SOILS

THIRD EDITION

Richard F. Fisher
Temple-Inland Inc.

Dan Binkley
Colorado State University

JOHN WILEY & SONS, INC.

New York · Chichester · Weinheim · Brisbane · Singapore · Toronto

0023085
114
FIS
2000

This book is printed on acid-free paper. ⊗

Copyright © 2000 by John Wiley & Sons, Inc. All rights reserved.

Published simultaneously in Canada.

No part of this publication may be reproduced, stored in a retrieval system or transmitted in any form or by any means, electronic, mechanical, photocopying, recording, scanning or otherwise, except as permitted under Sections 107 or 108 of the 1976 United States Copyright Act, without either the prior written permission of the Publisher, or authorization through payment of the appropriate per-copy fee to the Copyright Clearance Center, 222 Rosewood Drive, Danvers, MA 01923, (978) 750-8400, fax (978) 750-4744. Requests to the Publisher for permission should be addressed to the Permissions Department, John Wiley & Sons, Inc., 605 Third Avenue, New York, NY 10158-0012, (212) 850-6011, fax (212) 850-6008, E-Mail: PERMREQ@WILEY.COM.

This publication is designed to provide accurate and authoritive information in regard to the subject matter covered. It is sold with the understanding that the publisher is not engaged in rendering professional services. If professional advice or other expert assistance is required, the services of a competent professional person should be sought.

Library of Congress Cataloging-in-Publication Data:

Fisher, Richard F.
    Ecology and management of forest soils / Richard F. Fisher, Dan Binkley. -- 3rd ed.
       p.   cm.
    Rev. ed. of: Properties and management of forest soils / William L. Pritchett, Richard F. Fisher. 2nd ed. c1987.
     ISBN 0-471-19426-3 (alk. paper)
     1. Forest soils. 2. Soil ecology. 3. Soil management. 4. Forest. management.   I. Binkley, Dan. II. Pritchett, William L. Properties and management of forest soils. III. Title.
SD390.F56   1999
634.9--dc21   99-15960

Printed in the United States of America.

10 9 8 7 6 5 4 3 2 1

*For Karen Dangerfield Fisher
and Jane Austen Higgins*

CONTENTS

◼◼◼◼ PREFACE

Forests can be viewed from a range of perspectives, all of which are based on soils. Forest productivity is a story that centers on photosynthesis and plant growth, but plant biochemistry is supported by nutrient cycles that are essentially a soil story. Within a region, patterns in forest productivity result from spatial variations in soils. The diversity of plant species in forests is largely a soils story as well; across landscapes, the patterns in vegetation are typically modified (or even controlled) by patterns in soils. More than 99 percent of the diversity of life in forest ecosystems resides in soils, where amazingly small, numerous, and important organisms make the rest of the ecosystem (such as trees and mammals) possible.

Forest soils differ in major ways across landscapes, over time, and in response to disturbances and management. Aldo Leopold suggested that the first rule of intelligent tinkering is to save all the pieces. We would recast this idea and say that the first rule of intelligent forest management is to take care of the soil. Taking care of forest soils requires many important insights into the chemistry, physics, and biology of soils, which together comprise soil ecology.

In this book we try to convey the key features of soil ecology that are critical to successful forest management. Our primary audience includes people who are new to this subject, such as undergraduate students, as well as ecologists and forest managers who recognize the value of knowing more about soils. Forest soil scientists may also find some useful ideas in this book, and we hope they will help us improve future editions by pointing out our errors, omissions, and opportunities for clarifying the story of forest soils. We apologize in advance for being unable to include many of our favorite case studies developed by our colleagues in forest soils; the need to tell a clear and coherent story prevents us from going into the detail that most topics deserve. Fortunately, a wide range of more advanced books, journals, and Web pages are available that continue where we stopped.

This book is an amalgamation, update, and substantial expansion of two previously published books (*Forest Soils: Properties and Management* by W. L Pritchett and R. F. Fisher and *Forest Nutrition Management* by D. Binkley). Major changes include a worldwide perspective (rather than a North American focus), more chemistry, greater breadth and depth of case studies, and more synthesis of patterns around the world. The complexity of forest soils goes far beyond current knowledge, and even farther beyond the scope of a single general text such as this one. Forest soil ecology and management includes unavoidable uncertainty that derives from a lack of understanding and site-specific information. Uncertainty is unavoidable, so it needs to be included as a key aspect of management. We present some basic points about hypothesis testing and decision analysis in the hope of stirring creative skepticism among readers about the material we present and about the state of knowledge of forest soil ecology and management.

Acknowledgments

We would like to express our deep appreciation to our colleagues and students for their helpful suggestions and critical comments. We are especially indebted to C. B. Davey, Christian Giardina, Richard Houseman, Jason Kaye, Michele Lee, Andreas Rothe, Yowhan Son, and José Luis Stape for their critical input.

RICHARD F. FISHER
DAN BINKLEY

CHAPTER 1

Introduction

The need for a separate study of forest soils is sometimes questioned on the assumption that a forest soil is no different from a soil supporting other tree crops, such as citrus, pecans, olives, or a soil devoted to agronomic crops. Persons who are not well acquainted with natural ecosystems and who have failed to note even the most obvious properties of soils associated with forests generally make this assumption. Upon close observation of forests, one notices many unique properties of forest soils. The forest cover and its resultant forest floor provide a microclimate and a spectrum of organisms very different from those associated with cultivated soils or horticultural plantations. Such dynamic processes as nutrient cycling among components of the forest community and the formation of soluble organic compounds from decaying debris, with the subsequent eluviation of ions and organic matter, give a distinctive character to soils developed beneath forest cover (Figure 1.1).

In the broadest sense, a forest soil is any soil that has developed under the influence of a forest cover. This view recognizes the unique effects of deep rooting by trees, the role of organisms associated with forest vegetation, and the litter layer and eluviation promoted by the products of its decomposition on soil genesis. By this definition, forest soils can be considered to cover approximately one-half of the earth's land surface. Essentially all soils except those of tundra, marshes, grasslands, and deserts were developed under forest cover and have acquired some distinctive properties as a result. Of course, not all of these soils support forests today. Perhaps as much as one-third of former forest soils are now devoted to agricultural, urban, or industrial use. A better definition of forest soils might be those soils that are presently influenced by a forest cover. Currently, forests of various types cover about one-third of the world's land surface.

When European settlers arrived in what is now the United States, forests covered half of the land area, another two-fifths was grassland, and the remainder was desert or tundra. The eastern seaboard was almost entirely forested, and, largely as a consequence of the difficulties of clearing new land, agricultural settlement was mostly confined to the Atlantic slope until the end of the eighteenth century. Slowly settlement began expanding west of the Appalachian Mountains, and extensive land clearing began in that region. By the middle of the nineteenth century, more than a million hectares of virgin

1

Figure 1.1   Spodosol soil with a thick forest floor underlain by a strongly eluviated horizon and a horizon of illuviation.

forests had been cleared and the land converted to agricultural purposes (Figure 1.2).

This area of cleared forest land gradually increased until well into the twentieth century. As new and better farmlands were opened in the Midwest and West, millions of acres of former croplands were abandoned. Especially in the eastern regions, large areas of degraded farmland reverted to forests, and by 1950 forested areas had increased until they again covered nearly one-third of the total land area of the United States. This forest land base is now slowly diminishing under the relentless pressures of increasing urban and industrial development.

Few truly virgin forests exist today in populated regions of the globe. The conversion of forests to croplands and back to forests has gone through many cycles in sections of central Europe and Asia, as well as eastern North America. Many European forestlands have been managed rather intensively for centuries. At the other extreme, relatively short-term shifts in land use occur in the tropics, where "swidden" agriculture or shifting cultivation, a form of crop rotation involving 1 to 3 years of cultivated crops alternating with 10 to 20 years of forest fallow, is practiced. Such practices alter many properties of the original forest soil.

In recent years, intensively managed plantation forests have been created in several countries of the world. Among these forests are several million hectares of exotic pine and eucalypt forests in the Southern Hemisphere and an even larger area of plantations employing native species in the Northern Hemisphere. The latter includes some 8 million ha of pine plantations in the coastal plains of the southeastern United States and large areas of Douglas-fir plantations in the Pacific Northwest.

Because of the alteration in certain properties of forest soils as a result of intensive management, the distinction between forest soils and agronomic soils has become progressively less evident in some areas. Although some properties

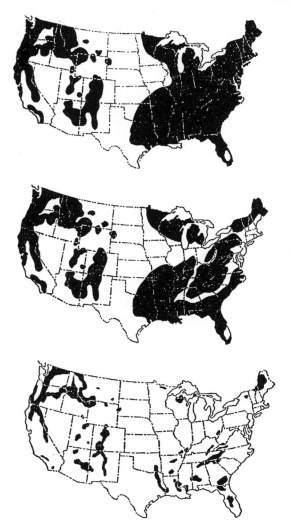

Figure 1.2 Area of primary forest in the United States in 1620, 1850, and 1940.

acquired by soil during its development persist long after the forest cover has been removed and the soil cultivated, other characteristics are drastically modified by practices associated with intensive land use. In this text, we will treat forest soils in the narrow sense as **soils that presently support a forest cover**, but including intensively managed forest soils. Only cursory attention will be given to genesis and classification of forest soils. Emphasis will be placed on understanding various physical, chemical, and biological properties and processes and how they influence forest dynamics and the management of forests.

FOREST SOILS DIFFER IN MANY WAYS FROM CULTIVATED SOILS

The soil is more than just a medium for the growth of land plants and a provider of physical support, moisture, and nutrients. The soil is a dynamic system that serves as a home for myriad organisms, a receptor for nature's wastes, a filter for toxic substances, and a storehouse for scarce nutrient ions. The soil is a product as well as an important component of its environment. Although it is only one of several environmental factors controlling the distribution of vegetation types, it can be the most important one under some conditions. For example, the further removed a tree is from the region of its climatic optimum, the more discriminating it becomes with respect to its soil site. This means that the range of soil conditions favorable to the growth of a species narrows under unfavorable climatic conditions for that species.

Many properties and processes characteristic of forest soils will be discussed in detail in later chapters. At this time, it will suffice to point out a few properties of forest soils that differ from those of cultivated soils. These differences derive, in part, from the fact that often the most "desirable" soils have been selected for agronomic use and the remainder left for native vegetation such as forests and grasslands. Fortunately, soil requirements for forest crops generally differ from those for agronomic crops. It is not unusual to find that productive forest sites are poor for agronomic use. Nevertheless, they may be used for agriculture because of their location with respect to markets or centers of population. Poor drainage, steep slopes, or the presence of large stones are examples of soil conditions that favor forestry over agriculture. However, the choice of land use often results from differences in crop requirements. Good examples are the wet flatlands of many coastal areas around the world. These important forest soils cannot be effectively used for agricultural purposes without considerable investments in water control, lime, and fertilizers.

Not all forest soils are poor. Some soils with excellent productive capacity for both trees and agronomic crops remain in forests today. This is generally because of location, size of holdings, ownership patterns, or landowner objectives. The fact that many forest soils contain a high percentage of stones by volume has a profound effect on both water and nutrient relations (Figures 1.3 and 1.4). Water moves quite differently through stony soil than it does through stone-free soil, and the volume of stones reduces proportionally the volume of water retained per meter of soil depth. Likewise, although stones contain weatherable minerals that release nutrient ions to the soil, the volume of stones reduces proportionally the soil's ability to provide nutrients for plant growth.

Forest trees customarily occupy a site for many years. Their roots may penetrate deeply into subsoil and even into fractured bedrock. During this long period of site occupancy, considerable amounts of organic material are returned to the soil in the form of fallen litter and decaying roots. As a result, a litter layer forms and exerts a profound influence on the physical, chemical, and biological properties of the soil.

Figure 1.3   A colluvial soil with many large stones.

The tree canopy of a forest shades the soil, keeping the soil cooler during the day and warmer during the night than cultivated soils. The presence of forest vegetation and the litter layer also results in more uniform moisture conditions, producing a soil climate nearly maritime in nature.

The physics of both overland and subsurface flow of water in steep forest soils is quite different from that in cultivated soils. Steep slopes under forests have their surfaces protected by the litter layer, their shear strength increased by the presence of roots, and their infiltration capacity enhanced by old root channels.

The more favorable climate of forest soils also promotes more diverse and active soil fauna and flora than are to be found in agronomic soils. The role of these organisms as mixers of the soil and intermediaries in nutrient cycling is of much greater importance in forest soils than in agronomic soils.

The deep-rooted character of trees leads to another unique feature of forest soils. Although the great majority of roots occur at or near the soil surface,

Figure 1.4 A forest soil developed on glacial till with high coarse-fragment content.

deep roots also take up both moisture and nutrients. Thus, deep soil horizons, of little importance to agronomic crops, are of considerable importance in determining forest site productivity (Figure 1.5).

Agronomic soils may be described as products of human activity, in contrast to forest soils, which are natural bodies and exhibit a well-defined succession of natural horizons. This was certainly a valid contrast a few decades ago, and it continues to be valid in most areas today. But the contrast has diminished greatly in the exotic forests of the Southern Hemisphere and in the short-rotation forests of the southeastern United States, the Pacific Northwest, and much of Western Europe. Clear-cut harvesting of trees disturbs the surface litter, resulting in short-term changes in the temperature and moisture regimes of the surface soil. Seedbed preparation by root raking or shearing, disking or plowing, and sometimes bedding incorporates the litter layer with the mineral soil, often enhancing microbial activity. Fertilization increases the nutrient level of the surface soil but may also affect the rate of breakdown of the organic

Figure 1.5 Sandy soils with clay layers at great depth may be poor for agriculture but are often excellent for tree growth.

layer. These practices influence the characteristics of forest soil and render it more like agronomic soil. However, most of these changes are relatively temporary, existing only until a forest cover becomes well established. With the development of the forest canopy and a humus layer, the soil again acquires the properties that distinguish it from cultivated soils.

FOREST SOIL SCIENCE IS AS OLD AS SOIL SCIENCE ITSELF

That trees and other plants are intimately related to the soil on which they grow seems rather obvious, but it took a long time to perceive this and even longer to understand the relationship. It is difficult to say just when perception occurred, but our scientific understanding of soils began with Aristotle (384–322 B.C.) and later Theophrastus (372–297 B.C.), both of whom considered soil in relation to plant nutrition. Cato (234–149 B.C.), Varo (116–27 B.C.), and Virgil (70–19 B.C.) also considered the relationship between plant growth, including trees, and soil properties. However, the knowledge of soils accumulated by Aristotle and the natural philosophers who followed him vanished with the fall of Rome and was unknown to Western scholars for over 1500 years.

Pliny the Elder (A.D. 23–79) gave the most complete account of the ancients' understanding of soil as a medium for plant growth. Pliny spoke not only of crops but also of forests and grasslands. He, like Cato, was a chronicler and recounts mostly folk wisdom and other empirically derived information, much of which is incorrect in light of modern discoveries. Oddly, while the wisdom of the natural philosophers was lost to Western science, much of the traditional knowledge that Pliny chronicled persisted among the agrarian population throughout the Dark and Middle Ages. It was not until the fifteenth century that science (natural philosophy) again turned its attention to the mystery of plant growth. Even then, much of the work in this regard was carried out by the learned nobility, and there was little sharing of knowledge among the various thinkers. In 1563 Bernard de Palissy published his landmark treatise *On Various Salts in Agriculture*, in which he stated that soil is the source of mineral nutrients for plants. However, his work was little known at the time.

The period 1630 to 1750 saw the great search for the principle of vegetation. During this time, any one of five "elements" — fire, water, air, earth, and niter — was considered, from time to time, to be the active ingredient in vegetable matter. It was during this period that Van Helmont (1577–1644) conducted his classic experiment with a willow (*Salix*) tree. He grew 164 lb of willow tree in 200 lb of soil, and only 2 oz of soil were consumed. Since only water was added during the experiment, he concluded that the 164 lb of willow had come from water alone. Van Helmont may not have been the first to conduct this experiment and draw the wrong conclusion — some believe that he was preceded by Nicholas of Cusa (1401–1446) — nor was he the last. Robert Boyle repeated the experiment with *Cucurbita*, obtained similar results, and reached a similar erroneous conclusion. It was not until 1804 that de Saussure successfully explained the experiment when he found that most of the mass of the plant was carbon derived from the carbon dioxide of the air.

Despite the experiments of Van Helmount and Boyle, the quest for the principle of vegetation continued. John Woodward (1699) grew *Mentha* in rainwater, River Thames water, Hyde Park Conduit effluent, and effluent plus garden mold. He found that the plants grew better as the amount of "sediment" increased and concluded that "certain peculiar terrestrial matter" was the principle of growth. Frances Home experimented and reached similar conclusions in the 1750s; however, he noted that "exhausted soils recovered from exposure to the air alone" and therefore concluded that the air must be the ultimate source of the essential materials.

The work of de Saussure and Boussingault in France in the early nineteenth century began a period of rapid scientific advancement. In 1840, Justus von Liebig in Germany published *Chemistry Applied to Agriculture and Physiology*, and modern soil science began. Liebig helped dispel the theory that plants obtained their carbon from the soil and developed the concept that mineral elements from the soil and added manure are essential for plant growth. However, he continued to believe, erroneously, that plants received their nitrogen from the air. It remained for de Saussure to show that plants' source of nitrogen was the soil. Lawes and Gilbert put these European theories to test at the now-famous Rothamsted Experiment Station in England and found them to be generally sound, particularly the concept that has come to be known as "Liebig's Law" or the "Law of the Minimum" (Steward, 1964).

Following the work of these chemists, scientists in many different fields including geology (Dokuchaev, Hilgard, Glinka), microbiology (Beijerinck, Winogradsky), and forestry (Grebe, Ebermayer, Müller, Gedroiz) contributed to the development of what today is termed "soil science." Most of the early research on soils was directed to its use for agricultural purposes because Europe had a critical food shortage; however, the importance of soils to the natural forest ecosystem was recognized by several early scientists.

In 1840, Grebe, a German forester, recognized the importance of soil to forest growth, stating, "In short, almost all of the forest characteristics depend on the soil, and hence, intelligent silviculture can only be based upon a careful study of the site conditions." Pfeil echoed this thought in 1860, and it became a central theme of European forestry (Fernow, 1907). Hilgard (1906) emphasized the relationship of vegetation to soils in North America that had been noted in Europe. The work of Grebe, Pfeil, Hilgard, and others, perhaps unintentionally, laid the foundations of forest ecology. In North America, early ecologists such as Merriam (1898), Cowels (1899), and Clements (1916) knew that soil was important in vegetation dynamics, but none of them understood soils well. Toumey (1916) noted the importance of soils in American silviculture for the first time.

Early research in forest soils was dominated by basic scientific studies. In fact, some very important early soil science research was done on forest soils. Ebermayer's work on forest litter and soil organic matter (1876) had a strong influence on soil science and agriculture. Müller's work on humus forms (1878) marked the beginning of the study of soil biology and biochemistry. Gedroiz (1912) did pioneering work on soil colloids, suggested that soils performed important exchange reactions, and laid the groundwork for modern soil chemistry. These scientists were interested not only in soil properties, but also in the processes that led to the existence of the properties. Much of the early research on forest soils in North America was similarly basic in nature.

As wood production became a pressing problem in Europe late in the nineteenth century and the restoration of degraded forestlands abandoned after agriculture became a necessity in North America in the twentieth century, forest soils research became quite applied. Studies of species selection for reforestation, methods of site preparation for reforestation, methods for estimating the site quality of forestlands that no longer supported forest vegetation, tree nutrition and response to fertilization, and soil changes under intensive forest management dominated forest soils research for much of the twentieth century. However, this situation began to change as concern over the impact of acidic deposition and global change arose, and there is currently a broad spectrum of forest soils research. Today applied research is still necessary, but some of our most pressing environmental problems require more thorough knowledge of the basic processes that take place in soils and their relationship to other ecosystem processes.

H. Cotta in Germany had made the importance of soils to forest production clear as early as 1809. Thus, soils education became an important part of forestry education in Europe. Portions of forestry textbooks were devoted to soil science by Grebe (1852) and by B. Cotta (1852), and in 1893 Raman published a text called *Forest Soil Science* in German. In 1908, Henry published the first forest soils text in French.

The science of forest soils was slow to develop in North America because of the lack of any compelling need for soils information during the early period of forest exploitation. Only after World War I did the ideas of managing selected forests for sustained yields, reforestation of abandoned farmlands, and the establishment of shelterbelts in the Midwest begin to take hold. Perhaps the greatest impetus to the scientific study of forest soils was the publication of textbooks on the subject by Wilde (1946 and 1958) and Lutz and Chandler (1946). These books were widely used by students throughout the United States for many years, and North American forest soil scientists today are largely the "academic offspring" of these intrepid scholars.

SUMMARY

In this text, forest soils are considered to be soils that presently support forest cover. These soils differ in many ways from agronomic soils. They have forest floors, organic layers that cover the mineral soil. They have diverse fauna and flora that play major roles in their structure and function. They are often wet or steep, shallow to bedrock, or have a high stone content. Soil layers that occur at great depth are important to forests.

The influence of soils on forests and other vegetation was known to the ancients, although our current understanding of soils did not begin to develop until the nineteenth century. The study of forest soils is as old as soil science itself. Researchers working on forest soils made many of the early discoveries that form the foundation of modern soil science.

CHAPTER 2

Forest Soils and Vegetation Development

The development of soil and associated forest vegetation is a complex and continuing process. It has taken place over thousands of years through a complicated sequence of interrelated events. Although a number of factors are involved in the development of both soils and forests, none is more important than climate. Climate, vegetation, and soil form an interdependent, dynamic complex: when one member of this complex undergoes change, the others also change, and a new equilibrium is established.

Soils, like climate, play vital roles in the development of forests. They provide water, nutrients, and support for trees and other forest vegetation. Soils are derived from parent materials of different mineral composition, and these differences result in soil properties that influence both the composition of the forest and its rate of growth.

As parent rocks weather, a soil develops, influenced not only by the mineralogy of the rocks and the physical factors of the environment, but also by the biota of the area that contributes to mineral weathering and the buildup of organic material in the soil. Eventually a series of horizons form in a typical well-developed forest soil: an **organic** layer (**O**), an organically enriched mineral layer (**A**), an **eluviated** layer (**E**), an **illuviated** layer or zone of accumulation (**B**), and a mineral layer little altered by soil-forming processes (**C**). The formation of these distinctive soil horizons results from a series of complex reactions and simple rearrangements of matter termed **pedogenic** processes (Buol et al., 1997).

PEDOGENIC PROCESSES OPERATE SIMULTANEOUSLY AT VARYING RATES

All pedogenic processes appear to operate to some degree in all soils, however, they operate at different rates at different times, and the dominant processes in any one soil body cause it to develop distinctive properties. All pedogenic processes involve some phase of (1) addition of organic and mineral materials to the soil as solids, liquids, and gases; (2) losses of these materials from a given horizon or from the soil; (3) translocation of materials from one point to another in the soil; or (4) transformations of mineral and organic matter within the soil.

Additions to the soil occur in both organic and inorganic forms. Wind and rain act as agents of transport for dust particles, aerosols, and organic compounds washed from the forest canopy. These additions are generally not large, but they can be substantial as a result of catastrophic storms or volcanic action. The most significant addition to forest soil is organic matter. The death and decay of live roots contribute a significant amount of organic matter directly to the soil in each growing season (McClaughetty et al., 1982). The surface litter layer contributes to the accumulation of soil organic matter and exerts considerable influence on the underlying mineral soil, as well as on the associated populations of microorganisms and soil animals. The process by which organic matter is added to the A horizon, thus darkening it, is called **melinization**.

Processes involving losses include leaching of mobile ions such as calcium, magnesium, potassium, and sulfate, as well as surface erosion by water or wind, solifluction, creep, and other forms of mass wasting. The latter are generally not problems in forests except on steep slopes and in mountainous areas. The dominant processes in forest soils are those related to translocations or transformations within the soil body. **Eluviation** is the movement of material out of a portion of a soil profile. It is a major process in the development of the A horizon and is the dominant process in the development of the E horizon. **Illuviation** is the movement of material into a portion of a soil profile. It is the dominant process in the development of most forest soil B horizons. More specifically, **decalcification** is a series of reactions that removes calcium carbonate from a soil horizon, and **lessivage** is the downward migration and accumulation of clay-sized particles, producing an **argillic** or clay-enriched horizon.

Probably the two most important soil-forming processes in forest areas are **podzolization** and **desilication**. Podzolization is a complex series of processes that brings about the chemical migration of iron, aluminum, and/or organic matter, resulting in a concentration of silica in the eluviated albic (E) horizon (Figure 2.1) and an accumulation of organic matter and/or sesquioxides in the illuviated spodic (Bh or Bs) horizon (Stobbe and Wright, 1959; Ponomareva, 1964; Brown, 1995).

Podzolization is the dominant soil-forming process in the boreal climatic zone, but it is not restricted to a specific climatic regime. Extensive areas of Aquods (Groundwater Podzols) occur on the coastal plain of the southeastern United States, and "giant podzols" with extremely thick E and Bs horizons occur in the tropics (Eyk, 1957; Klinge, 1976). The processes leading to podzolization intensify certain types of vegetation, especially hemlocks (*Tsuga*), spruces (*Picea*), pines (*Pinus*), and heath (*Calluna vulgaris*) in northern latitudes and kauri (*Agathis australis*) in New Zealand. Organic compounds leached from the foliage and litter (Bloomfield, 1954) or produced during litter and soil organic matter decomposition (Fisher, 1972) lead to rapid mineral

Figure 2.1   Podzolized profile of a spodosol in a spruce-fir stand in the boreal forest zone of Quebec.

weathering (Herbauts, 1982) and translocation of organic complexes downward in the soil profile (Ugolini et al., 1977).

Desilication (ferrilization) is the dominant soil-forming process in forested areas of the intertropical zone, where high temperatures and high rainfall favor rapid silica loss and the concentration of iron immobilized as ferric oxides under oxidizing conditions. In a well-drained environment, this process leads to the formation of an oxic horizon, a zone of low-activity clay that lacks weatherable primary minerals and is resistant to further alteration. In a soil with a fluctuating water table, this process leads to the formation of plinthite, which may occur as scattered cells of material or may form a continuous zone in the soil. When exposed to repeated wetting and drying, plinthite becomes indurated to ironstone, which has also been termed **laterite** (Alexander and Cady, 1962; Soil Survey Staff, 1975).

Pedoturbation is a third process that often plays a significant role in the development of forest soils. **Pedoturbation** is the biological or physical (freeze-thaw or wet-dry cycles) churning and cycling of soil materials, thereby mixing or even inverting soil layers (Bockheim et al., 1997). In many forests, wind-throw of mature trees causes significant mixing of soil horizons. Discontinuous horizons are often developed in this process, and soil horizons may even be inverted in some cases. A characteristic "pit and mound" surface condition is also often developed (Lyford and MacLean, 1966).

EXTERNAL FACTORS GUIDE SOIL FORMATION

The pedogenic processes involved in the genesis of soils are conditioned by a number of external factors. Simply stated, these factors are the forces of weather and of living organisms, modified by topography, acting over time on the parent rock (Jenny, 1980). These factors were first outlined by Dokuchaev, the Russian scientist who laid the foundation for our understanding of soil genesis in 1883. At the turn of the twentieth century, the Russian forester Morozov (1904, quoted by Remezov and Progrebnyak, 1969) wrote:

> From its inception, forestry showed interest in the manner of influence of tree stands on the soil; such effects of the canopy and of the litter were regarded by the forester as a means of changing the soil in order to conserve its fertility, facilitating the growth of new forests, etc.... As early as the time of Müller, foresters began using such expressions as "beech soil," "oak soil," etc., not merely in the sense of a soil suitable for the given species, but with emphasis on the idea that the soils are actually being influenced by the tree stand.

Müller realized that descriptions of naturally occurring patterns could not separate cause from effect, and he was less certain than Morozov that vegetation modified the soil (Handley, 1954). The degree to which vegetation modifies soils (rather than reflects preexisting conditions) is still a topic of debate.

In 1941, Jenny captured people's attention by writing these ideas in the form of an equation that he hoped could someday be solved quantitatively. Although no one ever solved this equation, it focused thinking on how to deal quantitatively with soil formation. Systems thinking and simulation modeling developed in the 1950s and 1960s, offering better ways to conceptualize and quantify the formation of soils. This view allowed reciprocal interactions to be included, whereas Jenny's approach assumed that soil-forming factors worked independently. In a systems context, climate, parent material, and the effects of topography are viewed as driving variables, variables that affect the state of a system and may change over time but are not affected by the system. Animals and vegetation are viewed as state variables, features of the system of interest, affected by driving variables and by each other, and time becomes the dimension over which soil development occurs (Figure 2.2). This approach is not as simple as Jenny's equation appeared to be, but it provides a more realistic framework for the way soils really develop (Bryant and Arnold, 1994).

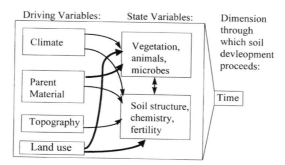

Figure 2.2 Diagrammatic representation of a systems model of soil development.

PARENT MATERIAL

Parent materials consist of mineral material, organic matter, or a mixture of both. The organic portion is generally dead and decaying plant remains and is the major source of nitrogen in soil development. Mineral matter is the predominant type of parent material, and it generally contains a large number of different rock-forming minerals in either a consolidated state (granite, conglomerate, mica-schist, etc.) or an unconsolidated state (glacial till, marine sand, loss, etc.). Knowledge of the parent material of youthful soils is essential for understanding their properties and management. As soils age, the influence of parent material on their properties and management decreases but never disappears.

Although a thorough discussion of rock-forming minerals is impossible in this book, we will outline the structure and chemical composition of the main types in order to better explain the behavior of parent material as it is subjected to soil-forming processes (Table 2.1).

Nonsilicates

Oxides, hydrous oxides, and **hydroxides** are among the most abundant and important soil-forming minerals. Hematite hydrates rather easily to form limonite and turgite, which are important in many forest soils as coatings on particles and as cementing agents in hardpans. Magnitite is a heavy mineral resistant to weathering and often occurs in soils as a black sand with strong magnetism

Goethite, gibbsite, and boehmite are common minerals in soils. They occur as coatings on other mineral or rock particles, and gibbsite and boehmite often accumulate in the spodic horizons. In tropical forest soils, limonite, hematite, gibbsite, boehmite, and other iron and aluminum oxides and hydroxides accumulate in the B horizon and may form plinthite.

Sulfates, sulfides, and **chlorides** are uncommon in forest soils. Pyrite is the

TABLE 2.1 Soil-Forming Minerals

Crystal Mineral Structure	Group	Mineral Species	Chemical Formula
Nonsilicates	Oxides	Hematite	Fe_2O_3
		Magnetite	Fe_3O_4
	Hydrous oxides	Limonite	$2Fe_2O_3 \cdot H_2O$
		Turgite	$2Fe_2O_3 \cdot H_2O$
	Hydroxides	Gibbsite	$Al(OH)_3$
		Boehmite	$AlO(OH)$
		Goethite	$FeO(OH)$
	Sulfates	Gypsum	$CaSO_4 \cdot H_2O$
	Sulfides	Pyrite	FeS_2
	Chlorides	Halite	$NaCl$
	Carbonates	Calcite	$CaCO_3$
		Dolomite	$(Ca,Mg)CO_3$
		Siderite	$FeCO_3$
	Phosphates	Apatite	$Ca(CaF \text{ or } CaCl)(PO_4)_3$
		Variscite	$AlPO_4 \cdot 2H_2O$
		Sytrengite	$FePO_4 \cdot 2H_2O$
		Barrandite	$(Al,Fe)PO_4 \cdot 2H_2O$
Ortho- and ring silicates	Epidote	Epidote	$4CaO \cdot 3(Al,Fe)_2O_3 \cdot 6SiO_2 \cdot H_2O$
	Garnet	Almandite	$Fe_3Al_2Si_3O_{12}$
	Zircon	Zircon	$ZrO_2 \cdot SiO_2$
	Olivine	Olivine	$2(Mg,Fe)O \cdot SiO_2$
Chain silicates	Amphibole	Hornblende	$Ca_3Na_2(Mg,Fe)_8(Al,Fe)_4Si_{14}O_{44}(OH)_4$
		Tremolite	$2CaO\ 5MgO\ 8SiO_2 \cdot H_2O$
		Actinolite	$2CaO_5(Mg,Fe)O$
	Pyroxene	Enstatite	$MgO \cdot SiO_2$
		Hypersthene	$(Mg,Fe)O \cdot SiO_2$
		Diopside	$CaO \cdot MgO \cdot 2SiO_2$
		Augite	$CaO \cdot 2(Mg,Fe)O \cdot (Al,Fe)_2O_3 \cdot 3SiO_2$
Sheet silicates	Mica	Muscovite	$K_2Al_6Si_6O_{20}(OH)_4$
		Biotite	$K_2Al_2Si_6(Fe,Mg)_6O_{20}(OH)_4$
	Serpentine	Serpentine	$Mg_3Si_2O_5(OH)_4$
	Clay minerals	Hydrous mica	$KAL_4(Si_7Al)O_{20}(OH)_4 \cdot nH_2O$
		Kaolinite	$Si_4Al_4O_{10}(OH)_8$
		Vermiculite	$Mg_{0.55}(AL_{0.5}Fe_{0.7}Mg_{0.48})O_{22}(OH)_n \cdot nH_2O$
		Montniorillonite	$Ca_{0.4}(AL_{0.3}Si_{7.7})Al_{2.6}(Fe_{0.9}Mg_{0.5})O_{20}(OH)_4 \cdot nH_2O$
Block silicates	Feldspar	Orthoclase	$(Na,K)_2OAl_2O_36SiO_2$
		Plagioclase	$[Na_2 \text{ or } Ca]\ OAl_2O_36SiO_2$
	Quartz	Quartz	SiO_2

most common and important of these minerals. It occurs in igneous and metamorphic rocks and is often abundant in mine spoil materials. Pyrite alters to a variety of iron oxides or hydroxides and releases sulfur as sulfuric acid. Thus, soil or spoil material high in pyrite is highly acidic in reaction and may become more acids as weathering progresses.

Carbonates are the major source of calcium and magnesium in soils. Calcite and dolomite occur in igneous, sedimentary, and metamorphic rock and are commonly added to agricultural soils to alter the soil's acidity; however, they are uncommon constituents and are generally ineffective as amendments in acid forest soils. Siderite is common in some forest soils and is an important cementing agent in some hardpans. Carbonates are generally present only in the lower horizons of humid zone forest soils. In more arid forests, secondary and even primary carbonate minerals may occur at or near the soil surface.

Phosphorus is one of the mineral nutrients that is commonly deficient in forest soils. Phosphorus occurs as apatite in all classes of rocks, but it is much less abundant in highly siliceous rocks. Apatite does not persist long, particularly in acid soils. Weathering leads to increasing amounts of phosphorus bonding to iron and aluminum ions through complex ion-exchange reactions. The chemical properties of phosphate in these forms resemble those of variscite, strengite, fluoroapatite, hydroxyapatite, and octacalcium phosphate. The high stability of these minerals and their very low secondary minerals such as solubility at low pH lead to the low phosphorus status of many forest soils.

Silicates

Silicates account for 70 to 90 percent of the soil-forming minerals. These minerals have complex structures in which the fundamental unit is the silicon-oxygen tetrahedron. This is a pyramidal structure with a base of three oxygen ions (O^{2-}), an internal silicon ion (Si^{4+}), and an apex oxygen ion (Figure 2.3). The four positive charges of the silicon ion are balanced by four negative charges from the oxygen ions, and the tetrahedral unit (SiO_4^{4-}) has four excess negative charges. In nature these tetrahedra are linked in a number of different patterns by silicon ions sharing oxygen ions in adjoining tetrahedra. This leads to the average composition of silica being SiO_2.

Varying patterns of tetrahedral linkage lead to a variety of distinctive and characteristic mineral forms and are the basis for classification. In addition, the type of linkage is an important factor in determining the mineral's resistance

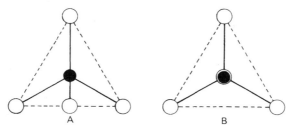

A   B

Figure 2.3 Diagram of the silicon-oxygen tetrahedron: (A) side view; (B) top view. Open circles are O_2 and solid circles are Si.

to weathering. Isomorphous (ion-for-ion) substitution of aluminum for silicon in the tetrahedral structure leads to structural and electrostatic imbalances; the inclusion of cations such as sodium, potassium, calcium, and magnesium in the mineral; and a change in the resistance to weathering. The silicate minerals are commonly divided into four broad classes based on crystalline structure: ortho- and ring silicates, chain silicates, sheet silicates, and block silicates.

Ortho- and **ring silicates** are the most variable silicate minerals. Olivines have the simplest structure, with separate silicon-oxygen tetrahedra arranged in sheets and linked by magnesium and/or iron ions. Olivines are common and are a frequent constituent of many basic and ultrabasic igneous rocks; however, olivines are the most easily weathered silicates and do not persist in soils. Zircon is another important mineral in which a cation provides linkage for silica tetrahedra. Each zirconium ion is surrounded by eight oxygens, resulting in a very strong structure and great resistance to weathering. Because of its persistent nature, zircon is often used as a marker in soil genesis studies. Garnets are complex silicates with alumina octahedra and iron linkages. Many species of this group are found in soils, where they are fairly persistent.

Chain silicates are divided into the amphiboles and the pyroxenes. The latter are composed of tetrahedra linked together by sharing two of the three basal oxygens to form continuous chains (Figure 2.4). These chains are then variously linked by cations such as calcium, magnesium, iron, sodium, and aluminum to form a variety of mineral species. The pyroxenes are fairly easily weathered by cleavage at the cation linkage.

Amphiboles are composed of double chains of silicon-oxygen tetrahedra (Figure 2.5) linked together by calcium, magnesium, iron, sodium, or aluminum. These minerals are only slightly resistant to weathering. The chain silicates are in reality ferromagnesian metasilicates and occur in both igneous and metamorphic rocks. They are primary sources of calcium, magnesium, and iron in soils.

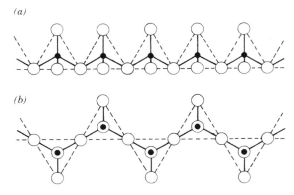

Figure 2.4   Diagram of the pyroxene chain: (a) side view; (b) top view. Open circles are oxygen and solid circles are silicon.

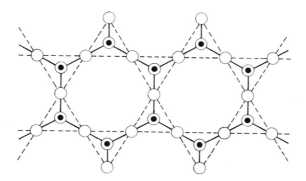

Figure 2.5 Diagram of the amphibole chain. Open circles are oxygen and solid circles are silicon.

Sheet silicates are represented by primary minerals such as mica and the secondary clay minerals. Both of these groups are composed of various combinations of three basic types of sheet structures: the silicon tetrahedral sheet (Figure 2.6), the aluminum hydroxide sheet (Figure 2.7), and the magnesium hydroxide sheet (Figure 2.8). The silicon tetrahedral sheet is composed of silicon-oxygen tetrahedra linked in a hexagonal arrangement with the three basal oxygens in one plane and the apical oxygens in another. The aluminum hydroxide sheet is made up of aluminum-hydroxyl octahedra in which the hydroxyl ions are in two planes and the aluminum ions are sandwiched in a third plane between them. To satisfy all valences in this structure, only two-thirds of the positions in the middle plane are occupied by aluminum ions. The magnesium hydroxide sheet is similar in structure, but because magnesium is divalent, all of the sites in the middle plane are filled.

The micas are abundant and important soil-forming minerals that occur in a wide variety of igneous and metamorphic rocks. The basic structure is two silicon tetrahedral sheets with either an aluminum hydroxide (muscovite) sheet or a magnesium hydroxide (biotite) sheet between them (Figure 2.9). One-quarter of the silicon ions have been replaced by aluminum ions, causing the sheet to have a net negative charge. This charge imbalance is satisfied by potassium, which bonds the sheets together. In biotite, about one-third of the magnesium in the middle sheet is replaced by ferrous iron. Mica is altered in soil to form hydrous mica (illite), kaolinite, chlorite, or even serpentine.

Serpentine is a secondary mineral that results from the alteration of olivine, hornblendes, biotite, and other magnesium minerals. Soils derived primarily from serpentine rock are infertile due to their high magnesium and low calcium content (Bohn et al., 1985). They also often have peculiar physical properties and are sometimes referred to as serpentine barrens.

Clay minerals are secondary alumino-silicates that are of great importance in soils. Six groups of these minerals are important in soils: hydrous mica, kaolinite, vermiculite, montmorillonite, chlorite, and allophane. The first five

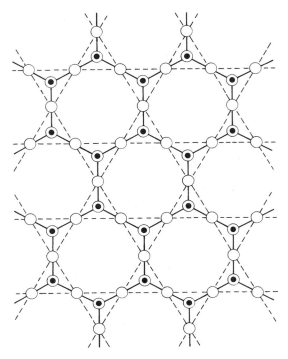

Figure 2.6   Diagram of the silicon-oxygen tetrahedral chain. Open circles are oxygen and solid circles are silicon.

are crystalline and are composed of silicon tetrahedral sheets and aluminum hydroxide and magnesium hydroxide sheets in various combinations. Allophane is an amorphous hydrous aluminum silicate or a mixture of aluminum hydroxide and silica gel with a variable ratio of silica to aluminum. It is prevalent in soils of volcanic origin but is also a common constituent of many other soils.

Hydrous mica, as its name implies, consists of hydrated muscovite and biotite particles of clay size. There are fewer potassium linkages in this mineral than in primary mica, but there are still sufficient linkages to prevent the lattice from expanding on hydration.

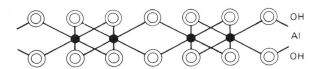

Figure 2.7   Diagram of an aluminum hydroxide sheet.

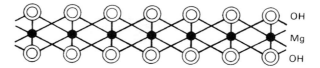

Figure 2.8 Diagram of a magnesium hydroxide sheet.

Kaolinite is the simplest of the true clay minerals. It is composed of one aluminum hydroxide sheet and one silicon tetrahedral sheet in which each apical oxygen replaces one hydroxyl group of the aluminum hydroxide sheet to form a 1:1-type clay structure (Figure 2.10). Vermiculite is hydrated mica with the potassium ions replaced by calcium and magnesium and with increased isomorphic substitution of aluminum for silicon. It is made up of one hydroxyl sheet and two silicon tetrahedral sheets and is said to have a 2:1-type clay structure (Figure 2.11). This structure is capable of expansion on hydration because calcium and magnesium do not reform the broken potassium linkages. Therefore, the intersheet spacing varies with the degree of hydration.

Montmorillonite is also a 2:1 clay (Figure 2.12). The sheets that make up the structure exhibit considerable substitution of aluminum for silicon, as well as iron and magnesium for aluminum. The mineral has high affinity for water and expands and contracts a great deal in response to hydration and dehydration.

Chlorite exists as both a primary and a secondary mineral. In its primary form it is a 2:1 layer silicate with a magnesium hydroxide sheet between the mica layers in place of potassium. Secondary chlorite, as it is formed in soils,

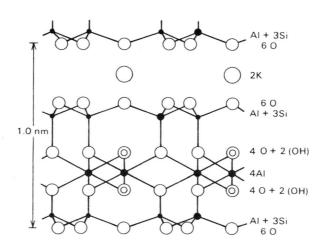

Figure 2.9 Diagram of muscovite mica.

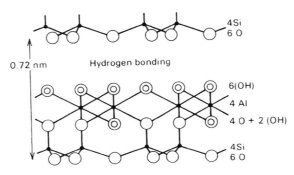

Figure 2.10 Diagram of the structure of kaolinite. Adjoining sheets of the 1:1 lattice are held rigidly together by hydrogen bonding. The intersheet spacing is a constant 0.72 nm.

differs from the primary form by having an aluminum hydroxide interlayer sheet.

Clay minerals have several properties that are quite influential in soil behavior: their cation-exchange capacity and ionic double layer, their ability to absorb water, and their ability to flocculate and disperse, to name but a few.

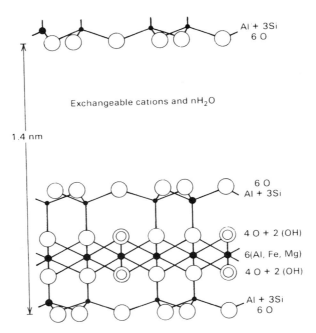

Figure 2.11 Diagram of one possible structure of vermiculite.

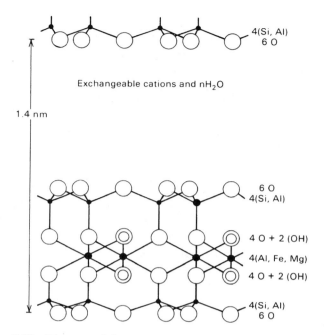

Figure 2.12   Diagram of the generalized structure of montmorillonite.

Minerals such as kaolinite, with its rigid structure and absence of isomorphic substitution, have low cation-exchange capacity because their exchange sites are confined to the broken edges of crystals; where bonds are satisfied by hydrogen or other cations. Montmorillonite and vermiculite, on the other hand, have a great deal of isomorphic substitution of aluminum for silicon, resulting in a large number of exchange sites and a high cation exchange capacity. Allophane, with its intermediate structure, has a high but variable cation exchange capacity that is strongly influenced by external conditions, even the techniques used to measure it.

When clay particles are moist, each particle is covered with negative charges that are satisfied by positive charges of ions in the surrounding solution. The innermost ions are most tightly held, but the outer ionic shell is also attached to some degree to the particle. This unit, the clay particle and its surrounding ionic double layer, is called a micelle.

Although clay minerals have a characteristic basal spacing — that is, the fixed distance from any point in one layer to the same point in the adjacent layer — many have the ability to expand and contract as they undergo hydration and dehydration. This ability, along with the small size of clay particles and the resulting large surface area, gives clays the capacity to retain a great deal of water against the force of gravity.

Flocculation and dispersion are also important properties of clays. **Flocculation** is the process of coagulation of individual particles into aggregates, and **dispersion** is the separation of the individual particles from one another. Flocculation in soils is brought about by a thin double layer such as the one that occurs when clays are calcium saturated. Dispersion results from a thick double layer such as the one that occurs when clays are sodium saturated.

Block silicates are minerals composed of silicon-oxygen tetrahedra linked at their corners into a continuous three-dimensional structure. The simplest of these is quartz, which is composed entirely of silicon-oxygen tetrahedra. Quartz, with its very regular and continuous structure, is extremely hard and resistant to weathering.

The feldspars are also block silicates, but they display a considerable amount of isomorphic substitution and contain a high proportion of basic cations. This substitution stresses even the three-dimensional structure and makes feldspars more easily weathered than quartz. There are three principal types of feldspar: sodium, potassium, and calcium. Members of the sodium to potassium series are known as orthoclase or alkali feldspars, and members of the sodium to calcium series are called plagioclase feldspars.

Stability series based on resistance to weathering have been developed, particularly for the silicate minerals. Figure 2.13 presents a comparison

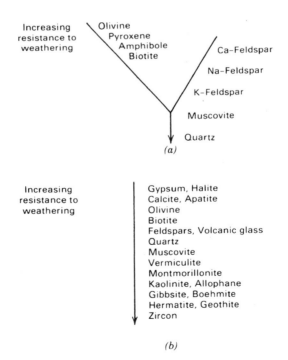

Figure 2.13 Mineral stability series for sand- and silt-sized particles (a) and for clay-sized particles (b) (adapted from Bohn et al., 1985).

between stability series for sand- and silt-sized mineral particles and for clay-sized mineral particles.

PARENT MATERIAL CLASSIFICATION

Parent materials are made up of consolidated or unconsolidated mineral material that has undergone some degree of physical or chemical weathering. In order to organize our thinking, a simple system of classification based on parent material properties important in soil formation is useful.

A generalized classification of the rock types that commonly constitute consolidated soil parent materials is given in Table 2.2. Unconsolidated parent materials are usually classified according to the size and uniformity of their constituent particles, as well as their mode of origin. Thus, we often speak of alluvial deposits, loess, drift, till, marine clay, and so on. Such a system is not as accurate as that used for consolidated materials. Since a wide variety of materials can occur as alluvium or till, it is necessary to modify these terms not only with an indication of their texture and uniformity, but also, when possible, with an indication of the type of rock from which they were derived.

Probably the most important characteristic of parent materials is their wide variability in composition within a short distance. Unconsolidated materials can vary greatly in texture, degree of sorting, and petrography in the space of a few meters, either horizontally or vertically. Even consolidated materials may have as much as a fivefold variation in one or more constituents within the space of a few centimeters. This is particularly true of the accessory constituents, which, although they make up only a small portion of the material, may markedly influence the chemical and physical properties of the rock.

CLIMATE AND WEATHERING

Climate is generally considered the most important single factor influencing soil formation and it tends to dominate the soil-forming processes in most forested areas. It directly affects weathering through the influence of temperature and moisture on the rate of physical and chemical processes.

Physical weathering plays an especially important role during the early stages of soil development, particularly in the degradation of parent materials. Changes in temperature result in rock disintegration by producing differential expansion and contraction. Changes in temperature also may convert water to ice. Since water reaches its maximum density at about $4°C$, its volume increases as the temperature falls below that point. The maximum expansion is about 9 percent, and forces as great as 2.1×10^5 kPa may result. These forces can split rocks apart, wedge rocks upward in the soil, and heave and churn soil materials (Figure 2.14).

TABLE 2.2   Generalized Classification of the Major Soil-Forming Rock Types

Rock Type	Petrography
	Ultrabasic
Peridotite	Coarse-grained igneous: oilvine, augite, hornblende, plagioclase, biotite, magnetite
Serpentine	Coarse to medium-grained igneous: serpentine with accessory magnetite
	Basic
Basalt	Fine-grained, sometimes porphyritic, igneous: plagioclase, augite, olivine, magnetite, apatite
Dolerite	Medium-grained igneous: plagioclase (labradorite), augite, olivine, hornblende, biotite, magnetite, apatite, quartz
Gabbro	Medium- to coarse-grained igneous: plagioclase, augite, olivine, accessory hornblende, biotite, magnetite
	Intermediate
Andesite	Glassy to fine-grained igneous: plagioclase with feldspar phenocrysts, augite, hornblende, biotile
Diorite	Medium-grained igneous: plagioclase (labradorite), augite, olivine, hornblende, biotite, accessory magnetite, apatite, quartz
Syenite	Coarse- to medium grained igneous: orthoclase, hornblende, augite, biotite, accessory quartz, zircon, apatite, magnetite
Trachytes	Fine-grained igneous: orthoclase with phenocrysts of biotite, augite, hornblende, olivine, accessory apatite, magnetite, zircon
	Acidic
Rhyolite	Glassy to fine-grained igneous: orthoclase, quartz, orthoclase phenocrysts, accessory biotite, hornblende, apatite, zircon
Granite	Coarse-grained igneous: quartz, orthoclase, plagioclase, muscovite, biotite, accessory epidote, augite, apatite, zircon, magnetite
Aplite	Fine-grained granite variety
Quartz porphyry	Macrocrystalline igneous: quartz and feldspar with phenocrysts of quartz, feldspar, muscovite, biotite, augite
Pegmatite	Very-coarse-grained igneous: orthoclase (microcline), quartz, muscovite, garnet
Obsidian	Volcanic glass of granitic composition
Gneiss	Coarse-grained metamorphic: orthoclase, quartz, biotite, muscovite, hornblende
Arkose	Coarse-grained sedimentary: sandstone derived from granite and high in feldspar
Siliceous	Medium- to coarse-grained sedimentary: quartz, feldspar,

TABLE 2.2 Continued

Rock Type	Petrography
	Acidic
sandstone	*accessory biotite, muscovite, hornblende, magnetite, zircon*
Acidic conglomerate	*Very-coarse-grained sedimentary: quartz and feldspar in a siliceous matrix*
	Extremely Acidic
Quartzite	Medium- to fine-grained metamorphic: quartz, accessory muscovite, magnetite
Ferruginous sandstone	Medium- to fine-grained sedimentary: quartz, accessory mica, hornblende, magnetite, zircon
Schist	Medium-grained metamorphic: quartz, feldspar, mica, garnet
Mica-schist	Medium- to fine-grained foliated metamorphic: biotite, quartz, muscovite, accessory epidote, hornblende, garnet
Shale	Fine-grained sedimentary: quartz, other highly variable constituents
Slate	Metamorphosed shale: quartz, accessory muscovite, chlorite
	Carbonaceous
Limestone	Crystalline, oolitic, or earthy sedimentary: calcium carbonate, other highly variable constituents
Marble	Metamorphosed limestone
Chalk	Soft white limestone of foraminifera remains
Dolomite	Medium to coarse crystalline sedimentary: equal parts calcium and magnesium carbonate
Calcareous sandstone	Medium- to fine-grained sedimentary: quartz, calcium carbonate, many accessories
Calcareous conglomerate	Very coarse sedimentary: quartz, feldspar, many others in a calcareous matrix

Chemical weathering can manifest itself in many ways. Initially, the principal agent is percolating rainwater charged with carbon dioxide dissolved from the atmosphere. Calcium and magnesium carbonates and other rock minerals such as feldspars and micas are affected. These latter minerals are hydrolyzed by the acid solution to produce clay minerals and to release cations, some of which become attached to the clay particles. Resistant materials such as quartz are scarcely affected and tend to accumulate in the soil as sand and silt particles, while the water-soluble carbonates tend to be dissolved and removed.

Leaching of the readily soluble components continually depletes the surface soil of chlorides, sulfates, carbonates, and basic cations. On balance, this produces more and more acidic conditions. The degree of this impoverishment is controlled by the intensity of leaching from water passing through the soil,

Figure 2.14 Cobbles in a Eutrocrept (Brown forest) soil in a Finnish esker. (Note that there is only slight soil podzolization in this low-rainfall area.)

the speed of the return of materials to the soil surface via organic remains, and the weathering of rock fragments. In most forested regions there is a positive rainfall balance (after subtraction of transpiration, evaporation, and runoff losses), resulting in percolation, the formation of acid surface soils, and the absence of any accumulated carbonates in a subsoil horizon.

BIOLOGICAL WEATHERING

Silicate minerals can be altered by biological systems. For example, tree roots and microorganisms can alter mica by causing the release of potassium and other ions from the mineral lattices (Boyle and Voigt, 1973). This process is a combination of the replacement of the interlayer K^+ and structural multivalent ions in the mica by H^+ ions from organic acids and the subsequent chelation of these released ions by organic acids (Boyle et al., 1974).

Many large organic molecules form polydentate ligands and are called chelates, from the Greek word for "claw." Such molecules absorb or chelate cations and are important not only in mineral weathering but also in the translocation of multivalent ions, including transition metals such as copper and iron, through the soil.

TOPOGRAPHY, TIME, AND BIOTA

Topography can have a profound local influence on soil development, over-shadowing that of climate (Daniels and Hammer, 1992). Soil relief affects development mainly through its influence on soil moisture and temperature regimes and on leaching and erosion. For example, in most coastal plains, small variations in elevation can have a pronounced effect on drainage. Water in soils with restricted drainage often becomes stagnant, and microorganisms and plant roots use dissolved oxygen faster than it can be restored. Under the ensuing anaerobic conditions, iron, aluminum, manganese, and some other heavy metals are reduced to a more soluble condition. Under reducing conditions, these metals move in the soil solution until they eventually oxidize and precipitate. Such precipitation produces gleyed conditions that are characterized by a gray to grayish-brown matrix sprinkled with yellow, brown, or red mottles or concretions in the B horizon. Subsoil colors are generally reliable guides to drainage conditions, ranging from dull bluish-gray for very poorly drained, reduced locations to bright yellows and reds for better-drained areas.

The genesis of a soil begins when a catastrophic event initiates a new cycle of soil development. The length of time required for a soil to develop, or to reach equilibrium with its environment, depends on its **parent material, climatic conditions, living organisms**, and **topography**. Weakly podzolized layers developed under hemlock within 100 years after fields in New England were removed from cultivation. These layers have also formed within relatively short periods in local depressions resulting from tree uprootings in hemlock forests. Recognizable mineral soil development under black spruce, and perhaps red spruce and balsam fir, can also be fairly rapid (Lyford, 1952). Buol et al. (1997) point out instances in which forest soils on glacial moraines in Alaska developed a litter layer in 15 years, a brownish A horizon in silt loam in 250 years, and a thick Spodosol (Podzol) profile in 1000 years.

Soil development from bare rock to muskeg took about 2000 years at one location in Alaska. The time required to develop modal Hapludalf (gray-brown podzolic) soil was estimated to be more than 1000 years in the northeastern United States. Development of distinctive profile characteristics can be extremely slow in arid regions and in some local areas where drainage or other conditions may slow the weathering processes. In short, there is no absolute time scale for soil development.

The roles of **biota** in soil development can hardly be overemphasized. Tree roots grow into fissures and aid in the breakdown of bedrock. They penetrate some compacted layers and improve aeration, soil structure, water infiltration and retention, and nutrient-supplying capacity. In addition to tree roots, a multitude of living organisms in the soil are responsible for organic matter transformations, translocations, decomposition, accretion of nitrogen, and structural stability. Bacteria and fungi are intermediaries in many of the chemical reactions in the soil, and many animals play a vital role by physically mixing soil constituents. Furthermore, the soil may be protected from erosion by a vegetative ground cover. A forest cover can significantly modify the temperature and moisture conditions of the soil below by its influence on the amounts of light and water that reach the soil surface, by a reduction in runoff and an increase in percolation, and by an increase in water loss as a result of evapotranspiration. Roots penetrate deep into the profile, extract bases, and return them to the surface in litter fall. The litter of many deciduous species that normally occur on base-rich soils decomposes rather rapidly, resulting in base-enriched humus. On the other hand, the litter of most conifers that normally occur on base-poor soils is rather resistant to decay and may encourage the process of podzolization. The composition of leachates produced beneath various plant covers can have a profound effect on soil formation. The presence of a thin, continuous Bh horizon below the E horizon on some windthrow mounds but not on others in a disturbed forest probably is caused by the particular vegetation growing on or near those mounds (Lyford and MacLean, 1966).

Pyatt (1970) outlined examples of changes in soil processes that appear to have resulted from changes in vegetation brought about by human activity. One example is the increase in soil wetness resulting from deforestation during Bronze Age times in areas of Great Britain. Increased soil moisture apparently was responsible for the development of gleying within formerly well-drained soils on the north York moors. A similar process commenced during medieval times with the spread of heath vegetation at the expense of woodland on some sites in south Wales.

Because of the close relationship between soils and plant communities and the influence of each on the formation of the other, an examination of the process of vegetation development should promote understanding of the influence of the soil on these processes.

VEGETATION AND SOIL DEVELOP TOGETHER BUT AT QUITE DIFFERENT RATES

It is generally believed that the Earth was formed more than 4.5 billion years ago and that the land area consisted of a single supercontinent. Some 200 million years ago, this supercontinent broke into fragments that largely define today's continents, and the fragments began their slow, ponderous voyages across the planet (Matthews, 1973). Although North America and Europe are reported to be moving apart by a distance equal to a man's height during his lifetime, the two continents may have been joined in the north as recently as 65 million years ago, with the Appalachian and Scandia mountains forming a continuous range.

Giant glaciers scoured the northern part of North America and the whole of northwest Europe as recently as 10,000 to 20,000 yr ago. The vegetation of many of the warmer areas of the world has apparently had a rather constant composition for millennia. Revegetation of the abandoned ice fields and coastal areas inundated during intraglacial periods has taken place relatively recently.

Tree species that evolved and simultaneously colonized the European and North American continents while these two land masses were joined in the pre-Cretaceous period were largely eliminated by the repeated advances of ice sheets during the Pleistocene. It appears that while most species could continue to retreat to the south in advance of the North American glaciers, they were not so fortunate in Europe and Asia. Trapped between the arctic ice mass in the north and the glaciers advancing from the several mountain ranges in the south, many species were apparently eliminated. These events help explain why the flora of much of Europe and Asia is simpler in species composition than that of North America and why Eurasia and North America now have few indigenous plant species in common.

We often view plant communities as static entities when in fact they are dynamic and ever-changing. The type of vegetation change that is most closely linked with soil change is termed succession. F. E. Clements originally defined succession in 1916 as continuous, directional change in the species composition of a community by extinction and colonization, leading to a single ultimate community. Today the belief in a single ultimate community is uncommon, and we define **succession** as sequential change in the relative abundance of the dominant species in a community following a disturbance (Huston and Smith, 1987).

Forest succession generally proceeds from less stable, relatively short-lived communities, often termed **pioneer** communities, to more stable, long-lived communities, sometimes termed **climax** communities. The most stable community for a given area in equilibrium with the climate and soils is considered the natural vegetation for that area. The basic premise is that if a naturally well-drained surface is left completely undisturbed for a protracted period, with no major climatic changes or other disturbances, a time sequence of plant communities will occupy it, and ultimately a community will establish itself that is stable enough to persist relatively unchanged for a protracted period of time.

This succession of communities alters the chemical, physical, and biological properties of the soil through their occupancy of the area, and it is likely that these alterations contribute to the relative change in abundance of the dominant species that characterizes succession. The development of plant communities on virgin sites such as new lava flows, newly exposed glacial material, or new sand bars is termed **primary succession**, while succession after a disturbance of an existing biotic community is termed **secondary succession**.

Certain developmental trends that characterize forest successions were outlined by Whittaker and Woodwell (1972) as (1) progressive increases in the height of the dominant tree species; (2) increasing diversity of growth forms and stratal differentiation within communities; (3) increasing species diversity; (4) progressive soil development with increasing depth of organic matters accumulation and horizon differentiation; (5) increasing community productivity and respiration, with progress toward balance of these activities; (6) increasing biomass and increasing stock of the nutrients held in the biomass; and (7) increasing relative stability of species populations. The authors felt that a climax community is defined not by maximum productivity and species diversity, but rather by maximum biomass accumulation and by a steady state of species populations, productivity (approximate balance of gross photosynthesis and total respiration), and nutrient circulation.

A classic example of plant and soil succession is the development of hardwood forest on granite bedrock in the northeastern United States. Initially, lichens colonize the bare rock. These organisms are symbiotic associations of algae and fungi. The algal component is often capable of fixing atmospheric nitrogen. The fungal component produces a series or organic acids, called lichen acids, that aid in the chemical breakdown of the minerals that make up the rock. These actions, plus the entrapment of dust and debris by the lichen surface, lead to the development of a very thin rime of soil, which is invaded by mosses and ephemeral herbs. As weathering action in the rock and the accumulation of dust and debris continue, the herbs become more robust and grasses may invade. After several decades, shrubs become established along with pioneering tree species. These species must be able to grow in full sunlight and withstand the periodic droughts that occur in the still shallow soil. As weathering continues, the solum deepens and tree species with more demanding nutrient and moisture requirements are able to invade the site. As the soil matures, the pioneering *Pinus* species are displaced by *Betula*, *Fagus*, and *Acer* species.

Another example of soil development and community succession occurs in the seaside mangrove forests of the tropics. Red mangrove (*Rhiophora mangle*) invades tidal mud flats and coastal shoals. It can tolerate the periodic inundations of salt water that occur as the tide changes. In time a mature forest of red mangrove develops, and the dense mass of stilt roots is effective in accumulating sediment and raising the soil surface. Other mangroves (*Avicennia*, *Cinocarpes*, and *Laguncularia*) gradually invade the system, and finally black mangrove (*Avicennia nitida*) dominates the stand. The water table may eventually be lowered sufficiently by transpiration to allow invasion by other forest species.

Parent material may dictate the plants that initially colonize an area, and it can affect the species composition of stable communities. This is quite evident in glaciated boreal regions. Finer-textured deposits are often imperfectly drained and are invaded by larch and alders. Over time, internal drainage improves as soil structure develops, and these areas become stable spruce-fir stands. In the same region, coarser deposits are initially occupied by open stands of aspen and birch or pine and shrubs. Eventually, all but the coarsest of these deposits develop a finer-textured B horizon and gradually become stable spruce-fir forests. However, the coarsest sand may remain in a series of pine forests.

Numerous examples of secondary succession in abandoned agricultural fields exist in the eastern United States from New England to Florida, in southwestern France, and in areas of shifting cultivation in the tropics. An interesting case is that of New England, where agriculture flourished until the middle of the nineteenth century. Agricultural areas began to decline with the opening of farmlands of the Midwest via the Erie Canal. Abandoned upland fields and pastures were invaded by red spruce and white pine mixed with such hardwood pioneers as birch, cherry, and aspen. Once the coniferous forests were well established, they were themselves invaded by more tolerant hardwoods: white ash, red oak, sugar maple, and black birch. Because the overstory conifers seldom reproduce under their own canopy, they were eventually displaced by these hardwoods plus the more tolerant hemlock, beech, and basswood (Figure 2.15). Griffith et al. (1930) described the soil profile beneath the virgin white pine-hemlock forest as "podzolic" (Spodosols).

Cultivation of the cleared land removed all traces of horizons; thus, the invading old-field pines took root in undifferentiated soil profiles. The authors investigated the changes that were brought about in the soil by the white pine and the succeeding hardwoods. The pine developed a layer of felted needles on

Figure 2.15 An example of vegetation succession in New England. Note the relic stone fences, the invasion of junipers into more recently abandoned fields and pastures, and the stable hardwood forest on the older abandoned agricultural land in the background.

the forest floor. The arrested decomposition of organic debris limited the thickness of the dark brown mineral surface horizon because of a lack of infiltration of humus colloids. The authors concluded that this degenerative process was reversed with the invasion of hardwoods. The rapid decomposition and incorporation of the base-rich hardwood forest litter quickly increased the thickness of the dark brown mineral surface horizon.

More recent evidence indicates that dramatic changes take place in soil nutrient dynamics during secondary succession (Fisher and Stone, 1969; Fisher and Eastburn, 1974). Abandoned fields have low levels of extractable nitrogen and phosphorus. Pines tolerate this poor nutritional status, while the new microflora and fauna associated with the conifers rapidly decompose soil organic matter and increase the levels of extractable nitrogen and phosphorus. The more demanding hardwoods then invade these more hospitable sites. Another change in microflora and fauna occurs, the temporarily degraded soils again accumulate organic matter, and the darkened surface soil thickens.

The study of succession has concentrated primarily on plant communities and only rarely on animal communities. It is clear that soil undergoes successive changes as great as if not greater than those that occur above-ground, although these changes may occur on a somewhat different time scale. The changes are not as easily observed as those in vegetation and have received little attention. Adding these soil dynamics to our consideration of succession might help us to understand this complex phenomenon better and give new direction to the continuing debate on succession.

SOIL PROPERTIES INFLUENCE VEGETATION DEVELOPMENT

Five easily observable properties of soil (**texture, structure, color, depth** and **stoniness**) can be used to infer a great deal about how a particular soil influences plant growth. Texture, whether the soil is a sand or a clay, in large measure determines the moisture and nutrient-holding capacity of the soil. Structure, whether the soil is massive, single grain, or blocky, modifies the influence of texture on water-holding capacity and its corollary, oxygen availability. Color indicates the amount of organic matter in the soil, which is generally well correlated with fertility. Depth tells us how much water can be held and what volume of soil is exploitable for nutrients, and stoniness tells us how much nonsoil objects dilute soil volume. If we know something about the climate, these properties can go a long way toward allowing us to determine the type and vigor of the plant community that any given soil might support. Many of these properties are strongly influenced by parent material but they change during soil development, and the ability of an area to support vegetation changes as the soil matures.

The type of **parent material** from which a soil is derived can influence its base status and nutrient level. Changes in vegetation type have been observed to coincide with changes in underlying parent rock. For example, soils derived from sandstone are generally acid and coarse-textured, producing conditions under which pines and some other conifers have a great competitive advantage. On the other hand, limestone may weather under the same climatic conditions into finer-textured, fertile soils that support demanding hardwoods and such conifers as redcedar. In the Appalachian Mountains, it is not unusual to find a band of pine growing on sandstone-derived soils parallel to a strip of hardwoods growing on limestone soils. Although not all limestone-derived soils are productive, this parent material exerts an important influence on the distribution and growth of forest trees. Differential chemical weathering of the same parent material can also influence the distribution and growth of forest vegetation because of the changes in acidity, base status, and nutrient availability associated with the intensity of weathering.

Vegetation development on soils derived from transported materials may differ from that on adjacent soils derived from bedrock. In fact, the mode of transport may greatly influence the particle-size distribution (texture) of the soil. Wind-deposited soils are likely to be of finer texture than those deposited by running water. Outwash sands laid down by flowing water from melting glaciers are generally coarser-textured than the glacial till soils formed by the grinding action of advancing glaciers. The latter soils are generally more fertile and support hardwoods, while the former often support pine forests. Soils with high silt and clay content may offer more resistance to root penetration than coarser soils, thus excluding trees intolerant of drought or of low oxygen conditions in the root zone. On the other hand, sandy soils hold less water and nutrients but more air than clays. The rate of movement through a soil varies inversely with soil texture because the minute interstitial spaces in fine-textured soils offer resistance to water movement. While finer-textured soils hold more water per unit volume of soil, they also lose more water by direct evaporation than do sandy soils.

SUMMARY

The development of soil and associated forest vegetation is a complex and continuing process. Soils play vital roles in the development of forests, and forests likewise play vital roles in the development of soils. All pedogenic processes appear to operate to some degree in all soils; however, they operate at different rates at different times, and the dominant processes in any one soil body cause it to develop distinctive properties. These processes involve some phase of (1) addition of organic and mineral materials to the soil as solids, liquids, and gases; (2) loss of these materials from a given horizon or from the soil; (3) bioturbation and translocation of materials from one point to another in the soil; or (4) transformations of mineral and organic matter within the soil.

The mineralogical makeup of the soil's parent material plays a dominant role in the character of soil, and an understanding of the soil's parent material is nearly always necessary to understand the soil. For example, soils that develop from acidic and basic parent materials are nearly always quite different. As a succession of differing plant communities occupy a soil, they have different effects on the soil's character. In forests, bioturbation, or the lack of it, plays an important role in soil development. In a systems context, climate, parent material, and the effects of topography are viewed as driving variables, which affect the state of a system, and may change over time, but are not affected by the system. Animals and vegetation are viewed as state variables, features of the system of interest, affected by driving variables and by each other, and time becomes the dimension over which soil development occurs.

CHAPTER 3

Soils of the Major Forest Biomes

Forests cover about one-third of the earth's land surface, and trees play a major role in all ecosystems except the tundra, deserts, some grasslands, and wetlands. Forests are assemblages of biotic and abiotic features, with trees dominating the interactions between the environment and the biota. Trees profoundly affect the soils in which they grow, including the microclimate of the soil. In turn, the growth and survival of trees are affected by the soil, climate, and other environmental factors.

Differences in temperature, precipitation, and physiography produce great diversity in productivity and species composition throughout the earth. Adequate nutrients, along with a proper balance between temperature and precipitation, are of prime importance in determining the suitability of an area for living organisms. The composition of a plant community is strongly influenced by climate and soil. The boreal forests of North America are similar in vegetation, soils, and climate to the boreal (taiga) forests of northern Europe and Asia. Such broad, common characteristics have led to grouping of vegetation and soils into major ecosystem types or biomes.

Forest soils are shaped by the influence of plants and animals on parent materials for varying lengths of time, with strong influence from environmental features such as temperature and water supply. Broad patterns of soil development and climate can lead to some useful generalizations, but local variations in soil-forming factors can result in an astonishing variety of soils within relatively small areas. For example, 9 of the 12 orders in the U.S. soil classification system occur on the island of Puerto Rico (too dry for Aridisols, no recent volcanic activity for Andisols, and no permafrost for Gelisols). Ten of the 12 orders occur on the island of Hawaii (no Spodosols or Gelisols). Soil formation is mostly a local story that scales up to a global story only in vague generalities. For example, Oxisols occur only in warm and wet environments, but most soils in such environments are *not* Oxisols.

SOILS COMMONLY DIFFER AS MUCH WITHIN REGIONS AS AROUND THE GLOBE

Many early ideas about patterns in soil properties were developed before much information was at hand (Richter and Babbar, 1991), and unfortunately, some disproven ideas live on. For example, it is still reported that tropical eco-systems have most of their nutrients tied up in biomass, with relatively little stored in mineral soils because rapid decomposition prevents accumulation of soil organic matter. This description is true for some tropical sites, just as it is for some temperate sites, but many tropical soils are quite rich in nutrients and organic matter (Table 3.1). No general relationship has been found between overall soil nutrient stores and latitude. Low-nutrient soils are as common at high latitudes as in the tropics.

The variations in soil properties across relatively small regions is illustrated by comparing the total phosphorus and extractable calcium in a set of representative subtropical soils from New South Wales in Australia and from the temperate Sierra Occidental region of Chile. These soils were chosen by the authors of review chapters to illustrate the common suite of soils in each region (Turner and Lambert, 1996; Fölster and Khanna, 1997). On average, the Chilean soils had about two-thirds more exchangeable calcium than the Australian soils (Figure 3.1), and about two and a half times more phosphorus (because of the presence of high-phosphorus volcanic ash soils in Chile). However, the calcium content of the Australian soils spanned more than an eightfold range and phosphorus more than a threefold range. The Chilean soils were similarly variable, spanning a 7-fold range for calcium, and an 18-fold range for phosphorous. One may accurately state that the Chilean soils on average had higher nutrient levels than the Australian soils, but this would not be a very useful statement given the large overlapping ranges of values for individual sites.

TABLE 3.1   Soil Chemical Pools for Representative Soils Throughout the Tropics

Feature	Young Alluvial Soil, Venezuela, 1.7 m/Yr Rain	Old, Weathered Terrace in Colombia, 3.0 m/Yr Rain	Weathered Ash, Hawaiian Soil, 4.0 m/Yr Rain	Weathered Metamorphic Rock, Nigeria, 1.4 m/Yr Rain	Unusually Poor Pleistocene Sediments, Venezuela, 2.4 m/Yr Rain
Forest floor mass (Mg/ha)	2	15–50	9	5–15	50–120
Soil C (Mg/ha)	58	81	80	63	47
Soil N (Mg/ha)	7	5	7	6	2
Total P (Mg/ha)	1.4	0.7	—	0.8	0.6
Exch. K (kg/ha)	460	120		650	70
Exch. Ca (kg/ha)	7100	31		2600	7
Exch. Mg (kg/ha)	960	40		290	16
Acid saturation (%)	3	90		25	93

Source: Fölster and Khanna (1997) and Binkley and Resh (1999).

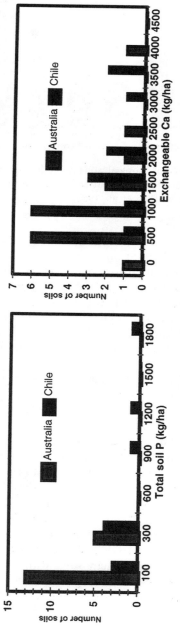

Figure 3.1 Total soil phosphorus and extractable calcium (estimated to a depth of 20 cm) for 17 representative soils of the Sierra Occidental in Chile (from Fölster and Khanna, 1997) and from New South Wales in Australia (from Turner and Lambert, 1996). On average, the old tropical Australian soils had lower quantities of these nutrients than the younger temperate soils from Chile, but the overlap among regions is too large to provide much insight into individual sites between the regions.

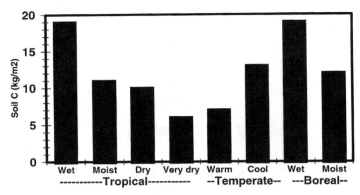

Figure 3.2 The carbon content of tropical soils is higher on wetter sites, and the carbon content of soils shows no particular pattern with latitude (from over 1500 soil profiles, from data summarized by Post et al., 1982).

Soil variability is typically high at all scales: from region to region, within regions, and even within single stands or compartments (Fölster and Khanna, 1997). At the scale of 5000 ha of a forestry operation in East Kalimantan, Indonesia, exchangeable soil calcium (to a depth of 0.5 m) ranges from 60 to 3240 kg/ha of calcium across individual stand compartments. At the level of a single compartment in the Jari plantations in the Amazon basin of Brazil, exchangeable soil calcium (to 0.6 m in depth) varied from 40 to 290 kg/ha.

The long-term fertility of soils may depend most strongly on the accumulation and turnover of soil organic carbon and nitrogen. Tropical soils tend to cycle carbon and nitrogen faster than temperate soils, but the faster turnover does not result in lower total pool sizes. The storage of soil carbon and nitrogen among 1500 soils around the world is related somewhat to soil moisture regime (wetter soils have more carbon and nitrogen), but not to latitude (Figure 3.2). Despite the efforts of scientists to dispel the myth of "poor soils in the tropics" (Sanchez et al., 1982; Duxbury et al., 1989; Richter and Babbar, 1991), it has taken on a life of its own.

Although soil properties do not relate well to latitude, these properties do vary consistently among soil taxonomic groups (such as soil orders in the U.S. classification system). Motavalli et al. (1995) found that allophanic (amorphous clay) soils had more total nitrogen than soils with smectitic clays, which exceeded the nitrogen in kaolinitic or oxic (amorphous sesquioxide) clays. Interestingly, the mineralizability of nitrogen from each of these different soil types was essentially constant; soils with more nitrogen mineralized proportionally more nitrogen.

SOIL TYPES ARE DISTRIBUTED UNEVENLY AROUND THE GLOBE

The U.S. system of soil classification places all the soils of the world into 12 classes called orders. The soil order with the greatest coverage of the globe is Aridisols (Table 3.2), but these soils are unimportant for forests because they are too dry to support trees. Together, Entisols and Inceptisols provide much of the forestland of the temperate and boreal regions. Spodosols commonly form under conifers (and sometimes under hardwoods too), typically on coarser materials under high-rainfall regimes. Andisols are volcanic soils restricted primarily to the margins of continents. These soils tend to be high in amorphous clays and soil organic matter and often support very productive forests. Alfisols are common under hardwood forests in areas with well-developed soil profiles. Ultisols take tens of thousands of years to form, and they dominate many tropical areas, as well as some old, unglaciated areas with

TABLE 3.2   Global Distribution of Soil Orders

Soil Order	Similar Categories in Some Other Soil Classification Systems	Land Area (% of Globe)
Aridisols, very dry areas	Xerosols, Solonetz, Solochaks, Yermosols	23.4
Vertisols with shrink-swell clays	Grumusols, Vertisols	2.4
Entisols, undeveloped horizons	Regosols, Lithosols, Arenosols	11.0
Inceptisols, B horizon developing	Acid brown soils, Gleysols, Cambisols	16.0
Alfisols, B horizons with clay accumulation carbon	Grey wooded soils, Luvisols, Planosols	13.5
Mollisols, high carbon in top horizon	Prairie soils, Chernozems, Rendzinas	4.1
Spodosols, B horizon with iron and aluminum	Podsols	3.6
Gelisols, soils underlain by permafrost	Gelic members of Gleysols, Cambisols, and others	6.0
Ultisols, acid soil with clay accumulation	Red-yellow podzolics, Acrisols	8.3
Oxisols, extremely weathered	Latosols, Ferralsols	8.7
Andisols, developed in volcanic material	Andosols	1.8
Histosols, organic soils	Bog soils, Histosols	1.2

Source: USDA Natural Resource Conservation Service, modified from Brady and Weil (1996).

high precipitation such at the U.S. Pacific Northwest and western Canada. Oxisols are highly weathered soils with high concentrations of clays that have low exchange capacity; they occur primarily on ancient landforms in the tropics. Histosols comprise major forest soils in areas of poor drainage in temperate and boreal areas; forest growth is commonly improved by draining Histosols.

MAJOR FOREST TYPES OCCUR ON A VARIETY OF SOILS

The forests of the world span a vast range of climatic conditions and soil properties, and the variations across relatively small landscapes may be almost as large as the variations across hemispheres. This wide range of situations is reflected in the major soil types that are found within major types of forests, and we conclude our discussion of soil and forest associations with a table that summarizes the major orders that are found most commonly among forest types (Table 3.3).

TROPICAL FORESTS ARE DIVERSE AND OCCUR ON A DIVERSE RANGE OF SOILS

Tropical forests are arbitrarily defined as all forests between the Tropic of Cancer and the Tropic of Capricorn, an area about one-third of the Earth's land surface (4.6 billion ha) that contains about 45 percent of the world's population. It is the region of most rapid population growth and, in recent years, an area of rapid changes in land-use patterns. Trade winds, monsoons, and mountain ranges provide enormous variations in the amount and distribution of rainfall, producing a wide range of vegetation, from permanently humid evergreen rain forests through various types of semideciduous and deciduous forests to savanna, semidesert, and desert.

Some 42 percent (1.9 billion ha) of tropical land areas contain significant forest cover. Of this amount, 58 percent (1.1 billion ha) is closed forest in which the forest canopy is more or less continuous (Figure 3.3). The remaining 42

TABLE 3.3   Major Soil Types in Forest Regions

Major Forest Types	Most Common Soil Orders
Boreal forests	Gelisols, Spodosols, Histosols, Inceptisols
Temperate conifer, mixed, hardwood, and montane forests	Alfisols, Inceptisols, Ultisols, Spodosols, Entisols, Mollisols
Tropical rain forests, monsoon forests, dry forests	Ultisols, Inceptisols, Oxisols, Andisols

Figure 3.3 Neotropical wet forest with a wide diversity of overstory trees and understory plants growing on a highly weathered soil.

percent (0.8 billion ha) is open forest or woodland in which the canopy is not continuous and the trees may be widely scattered.

Tropical forests are among the Earth's most diverse ecosystems. These forests are supplying an increasing proportion of the world's demand for wood products. Native tropical forests are being harvested and landscapes converted to second-growth forests, agricultural fields, and pastures. Intense pressure from people seeking food, fuel wood, forage, and timber (Figures 3.4, 3.5) is changing tropical forests on a global scale. In some areas, tropical forests are recovering on former cropland and pastureland, with trees reestablishing on soils altered by previous land management.

Figure 3.4 Wet montane forest that has been extensively cleared for agriculture.

Figure 3.5   Landless farmers preparing to colonize a recent clear cut in a tropical rain forest life zone.

TROPICAL CLIMATES ARE WARM YEAR ROUND, AND PRECIPITATION VARIES AMONG REGIONS

The equatorial belt between the Tropics of Cancer and Capricorn possesses great diversity in climate and soil conditions, ranging from cold highlands to hot lowlands and from marshes to deserts. The most common features in this portion of the Earth are relatively uniform air temperature between the warmest and coolest months of the year and little variation in soil temperatures. These constants apply to all tropical areas, regardless of location. Although mean annual air temperatures decrease, on average, by about 0.6°C for every 100-m increase in elevation, they remain fairly constant from month to month at any specific location. The mean monthly air temperature variation is 5°C or less between the three coolest and the three warmest months, with the least variation closest to the equator.

Soil temperatures are **isothermic**, with less than a 5°C difference between mean summer and mean winter temperatures at 50-cm depth (Soil Survey Staff, 1975). Surface soil temperatures beneath forest canopies approximate those of the air, but with the removal of the canopy, surface soil temperatures rise dramatically. For example, a daily maximum temperature of 25°C at the soil surface beneath a canopy may increase to 50°C or more after canopy removal, especially during dry periods (Chilicke, 1980).

Rates of tree growth are high in many areas of the tropics because of extended growing seasons that are limited (if at all) by periods of drought rather than by cold seasons. Wood growth rates of 10 Mg/ha annually are common in tree plantations of the tropics, rates that are matched only by those of the most productive temperate plantations (Binkley et al., 1997).

Annual rainfall can vary from near zero to more than 10 m in tropical areas, generally decreasing with increasing latitude but greatly influenced by local relief and seasonal wind patterns. The seasons in the tropics are nominally classified as **wet** and **dry** rather than winter and summer. The length of the dry season increases with the increase in latitude, being essentially zero at the equator. The period of heaviest rainfall usually comes during periods of greatest day length (or when the sun is directly overhead). On average, there is essentially no moisture limitation to tree growth in approximately one-third of the tropics, but there are slight to severe moisture limitations during 2 to 6 months or more in the remaining areas.

Precipitation interacts strongly with soil properties in determining forest productivity. Along the central Atlantic coast of Brazil, differences in geology have led to development of a variety of soils in a local (< 100 km) landscape, from sandy Entisols to sandy or clayey Ultisols to sandy Oxisols (Figure 3.6). This diversity of soil types is overlain by a gradient in precipitation, with areas near the coast receiving about twice the annual precipitation of inland sites. At the dry end of the precipitation gradient, the productivity of *Eucalyptus* plantations is similar among soil types (Figure 3.7). Productivity increases with increasing precipitation, but the gains from extra rainfall are more pronounced for the Ultisols than for the Oxisols.

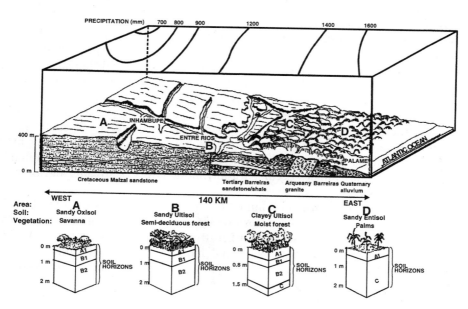

Figure 3.6 Soils near the Atlantic coast of Bahia, Brazil, range from marine-derived Entisols to Oxisols formed in Cretaceous sandstones over a distance of about 100 km. Precipitation drops by half over the same distance. "Tropical soils" cannot be represented by simple generalizations any better than temperate forests can (modified from Pessotii et al., 1983, used by permission of Copener Florestal, S.A.).

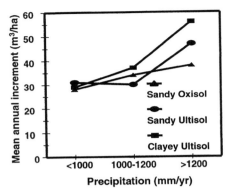

Figure 3.7 Interaction between soil type and precipitation in determining productivity of *Eucalyptus* (clones of *E. grandis × urophylla*) in Bahia, Brazil (data from Stape et al., 1997).

RAIN FORESTS HAVE NO PROLONGED DRY SEASON

Tropical rain forests have developed where year-round rainfall and temperatures are favorable for a high level of biological activity. Three distinct formations of tropical rain forests are (1) American, (2) African, and (3) Indo-Malaysian. These three formations are remarkably similar in structure and general appearance. Trees in these formations are largely evergreen and have a wide variety of leaf forms, and the dominant species often develop large flank buttresses on the lower trunk. Trees support a variety of plants that can survive without soil contact (epiphytes), such as orchids, bromeliads, ferns, mosses, and lichens. A dense, mature rain forest has a compact, multistoried canopy that allows little light to penetrate to the ground. The understory may be rather open in these dense forests. Rain forests that have been opened up, either by nature or by humans, quickly grow into dense, second-growth forests. Similar dense, impenetrable jungles occur naturally along edges of streams and other natural clearings.

The high temperatures and humidity of rain forests ensure that litter reaching the forest floor decomposes rapidly, sustaining rapid nutrient cycles. The soils of the rain forests and other tropical formations are extremely variable (Richter and Babbar, 1991) and have been less well studied than temperate soils. Furthermore, the dissemination of information concerning their properties and management has been impeded by the large variety of systems used in their classification. Sanchez (1976) contrasted the, 1938 USDA, French (ORSTROM), Belgian (INEAC), Brazilian, and United Nations FAO systems of classifying tropical soils with the U.S. Soil Taxonomy (Soil Survey Stall, 1975). He summarized the distribution of tropical soils at the suborder level from a generalized map (Aubert and Tavernier, 1972) based on the U.S. taxonomy.

Parent material weathering in tropical Oxisols is deeper than in any other group of soils (Figures 3.7, 3.8). In many areas, weathering and leaching gradually remove a large part of the silica from the silicate minerals, forming a thick subsurface oxic horizon that is high in hydrous oxides of iron and aluminum (Figure 3.9). Less intensive weathering produces Ultisols. Oxisols and Ultisols are the most commonly occurring soils in rain forests, and both types have high acidity and frequently low supplies of phosphorus, potassium, calcium, nitrogen, magnesium, and various micronutrients. These soils are generally high in exchangeable aluminum but low in effective cation-exchange capacity, resulting in a moderate to high leaching potential. The capacity to immobilize phosphorus fertilizer is often high, but phosphorus fertilization is still very effective.

Floodplain areas are often occupied by Inceptisols that may be rich or poor in nutrients, depending on the source of the sediment from which they are formed. Andisols also occur in rain forests, as well as in isotemperate montane forests in the tropics. On less well drained sandy soils, Spodosols (Aquods) may

Figure 3.8 Many Oxisols and Ultisols in tropical and temperate forests are weathered to tremendous depths.

Figure 3.9   Deep profile of an Oxisol in a tropical rain forest in French Guiana.

develop. Areas of impeded drainage are not unusual in tropical rain forests, but deep peat accumulations appear to be rather rare. On the other hand, the organic matter content of mineral soils is generally higher than might be expected in an area of such high biological activity. This underscores the importance of distinguishing at least two general pools of soil organic matter: relatively young, labile pools with high throughput, and large, stabilized pools of organic matter with very slow rates of turnover.

MONSOON FORESTS HAVE SEASONAL PERIODS OF DROUGHT AND RAIN

In tropical areas where significant seasonal droughts are experienced, the rain forest gives way to "rain green" or monsoon deciduous forests. The formation is not as distinct as that of the evergreen rain forest and, in fact, consists of a transitional zone of semideciduous seasonal forest as well as deciduous seasonal forest. The transitional forest is similar in structure and appearance to the rain forest, except that the trees are not as tall and the forest may have only two canopy tiers. Monsoon forests are found in areas with relatively mild drought periods, usually no more than 5 months, with less than 9 cm of rain. In areas with more distinctive drought periods, the forest becomes predominantly seasonal deciduous. Deciduous trees such as teak (*Tectona grandis*), with seasonal opening of the canopy and a subsequent show of flowers in the lower tier, are common to both the transitional semideciduous and seasonal deciduous communities with this formation. Lianas and epiphytes also become less abundant in areas with distinctive dry periods. Formations of monsoon forests have been described in Central and South America, Indo-Malaysia, and Africa (Eyre, 1963). The American formation is typically found in a belt skirting both the northern and southern fringes of the Amazon Basin, along the east coast of Central America, and some West Indies islands. The Indo-Malaysian formation stretches from northeast India and Indochina through Indonesia to northern Australia. In the wetter portions of the areas, teak and a deciduous legume, *Xylia xylocarpa*, are dominant; in drier sections, *Dipterocarpus tuberculatus* (eng tree), bamboo, *Acaela*, *Butea*, and other leguminous species are abundant. The African formation is rather distinct. Except for a discontinuous belt along the northern edge of the west African tropical rain forest, the other African communities with this formation merge rapidly into savanna woodlands, perhaps kept in check by frequent burning. Soils with these formations may be less highly weathered than those of the tropical rain forests. They generally include various groups of Ultisols, Oxisols, and Vertisols.

MANY TROPICAL FORESTS ARE DRY

The deciduous seasonal forests of Asia and America often grade into drier regions with no sharp change in vegetation. In these drier areas, many dominants disappear and those that persist have a lower, spreading canopy. Many of these low-growing, bushy trees and shrubs possess thorns, and where the woody plants form a closed canopy, the formation is called dry tropical forest. Dry tropical forests are found in northern Venezuela and Colombia, much of northeast Brazil, the West Indies, and Mexico. They also show up in India, Myanmar (formerly Burma), and Thailand in Asia and in vast areas of Africa and Australia. Both deciduous and evergreen species are found in this formation. The leaves of both types are usually small and cutinized and possess other xeromorphic adaptations. Although great attention has been focused on rates of loss of tropical rain forests through conversion to other land uses, a far larger proportion of the dry tropical forests have been lost.

Soils developed in these dry tropical forests are often light in color, low in organic matter, and not severely leached. They may even have an accumulation of calcium salts in the subsurface horizon. These soils are mostly classified as Inceptsols, and Alfisols.

Adjacent to the tropical thorn forests but in drier areas, semidesert shrub and savanna woodlands grade into true desert areas. In these formations, trees occupy a small percentage of the area but their effects may be profound (Rhoades, 1997), especially for nitrogen-fixing trees (Rhoades et al., 1998). Soils beneath these scattered trees are frequently Argids and generally have a darker-colored surface layer than adjoining soils. This darkening results from the organic matter produced by the trees and from the entrapment of wind-blown organic debris beneath the trees.

TROPICAL MONTANE FORESTS HAVE MODERATE, UNIFORM CLIMATES

At higher altitudes in the tropics, montane forests occur in isotemperate climates. Tropical montane forests are species rich and contain many unique species, including species from common temperate-forests families such as Pinaceae, Cupressaceae, Betulaceae, Cornaceae, Fagaceae, Magnoliaceae, and Ulmaceae. Tropical montane forests, in the Southern Hemisphere are often dominated by species of Araucariaceae and Podocarpaceae. Common soils are Andisols, Inceptisols, Alfisols, and Ultisols.

PLANTATION FORESTRY IS EXTENSIVE THROUGHOUT THE TROPICS

Intensively managed plantations of one or more tree species are attractive development projects for the tropics, and the area of tropical plantations increased from less than 1 million ha in 1950 to 45 million ha in the 1990s (Brown et al., 1997). Plantations offer opportunities to relieve pressures on the mixed natural tropical forests while providing a source of wood products including pulp, charcoal, fuel, and others. Much of this expansion of tropical forest plantations has involved reforestation of agricultural fields (Figure 3.10), and some has followed clearing of native forests.

The intensive management of introduced tree species may provide more rapid growth and a greater economic return than would native species. Successful plantations in tropical areas produce 4 to 10 times the amount of usable wood produced by a natural forest. However, introduced species must be carefully matched to local soil, climate, and other environmental conditions. *Pinus caribaea*, *P. oocarpa*, *P. elliotti*, *P. taeda*, and *Gmelina arborea* are used operationally in forestry in many tropical countries. Short rotations of fast-growing and coppicing hardwoods, such as *Tectona grandis* and species of *Eucalyptus* (Figure 3.11), *Acacia*, and *Leucaena* have been established for the production of pulpwood, charcoal, fuelwood, and other forest products (Nam-

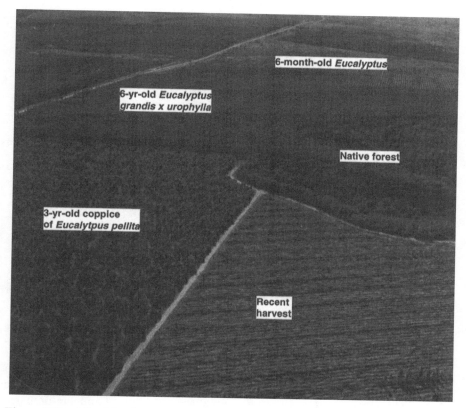

Figure 3.10 Mixed landscape with intensively managed *Eucalyptus* (stand size averages about 25 ha) and unmanaged native forest in Bahia, Brazil. Current laws require at least 20 percent of company lands to be retained in native vegetation (photo by Nilton Souza, courtesy of Copener Florestal, S.A.).

biar and Brown, 1997). Plantation forestry also can be labor intensive, an important consideration in low-employment areas of the tropics (Evans, 1992).

A number of concerns have arisen about the establishment of plantation monocultures in the humid tropics. For example, intensive plantation management often involves methods of harvesting, site preparation, and planting requiring heavy machinery. These operations may be suited to large blocks of relatively level land with sandy to sandy loam soils, but may damage more fragile or sloping soils. The use of wheeled vehicles can compact fine-textured soils, reduce infiltration rates, increase runoff, and promote erosion on sloping lands. Bulldozers with cutting and clearing blades are sometimes used to eliminate harvest debris and residual vegetation from areas to be planted (Figures 3.12 and 3.13). If the harvest debris is burned prior to blading, a portion of the nutrients contained in the vegetation may be left on the site as

Figure 3.11 A 30-month-old *Eucalyptus saligna* plantation in Hawaii reaches 10 m in height.

ashes instead of being concentrated into windrows. However, a significant part of the ash, along with some of the nutrient-rich topsoil, is often pushed into the slash piles even when conscientious efforts are made to keep the blade above the soil surface. Losses of nutrients during harvesting and site preparation operations may be sufficient to create the need for mineral fertilizer additions to sustain rapid growth of plantations.

Figure 3.12 A bulldozer with a shearing blade being used to prepare a site for conversion to pine.

Figure 3.13 Windrows such as these contain excessive amounts of topsoil and will reduce the average quality of the site.

SOME TEMPERATE BROAD-LEAVED FORESTS ARE EVERGREEN

Evergreen broad-leaved forests extend outside the tropics in some areas of Japan, southern China, parts of Australia, and the southeastern United States. Evergreen oaks make up most of the species of Asian broad-leaved formations. Vast forests of broad-leaved evergreen trees once covered southwestern and southeastern Australia before European colonization. These forests were composed almost entirely of *Eucalyptus* species. The cut-over forests are still widespread, with substantially altered species composition. The eucalypts are quite different from the broad-leaved evergreens of the northern midlatitudes in both appearance and habitat. Many soils of Australia are very old, with low precipitation. Widely varying properties of these soils result from some distinctive geomorphic characteristics of the country. Australia has suffered only limited uplift, dissection has been modest, basement rocks have been subjected to little juvenile weathering, and only minor glaciation occurred on the continent. Despite a wide range in parent materials, climatic environment, and age, most of the forested soils are very old Ultisols (Udults), with some Alfisols (Ustalfs) and Vertisols. Interestingly, Oxisols are absent from Australia.

TEMPERATE RAIN FORESTS ARE AMONG THE LARGEST IN THE WORLD

Northwestern coastal North America is an area of high rainfall, well distributed throughout the year, and mild temperatures. This marine climate favors the development of dense stands of unusually tall and massive conifers, with aboveground biomass of more than 1500 Mg/ha. The dominant species include

Sitka spruce (*Picea sitchensis*), which ranges from Alaska through British Columbia and southward along the coasts of Washington and Oregon. These northern coastal rain forests also include western redcedar (*Thuja plicata*), western hemlock (*Tsuga heterophylla*), and Douglas-fir (*Pseudotsuga menziesii*). In northern coastal California and extreme southern coastal Oregon, the giant coastal redwoods (*Sequoia sempervirens*) are a common forest dominant (Figure 3.14). The landscapes that contain redwood forests are less affected by summer droughts as a result of frequent fog. Western species of hemlock, pine, spruce, firs, and false cypress are also found in this coastal region.

Other temperate rain forest formations are found along the western side of the South Island of New Zealand, in the Valdevian forests of southern Chile, and in southeastern Australia. These Southern Hemisphere formations are largely broad-leaved evergreens. Apart from this fact, the beech (*Nothofagus*) forests of New Zealand and the *Eucalyptus regnans* forests of Australia resemble in size and general appearance the temperate rain forests of other regions and contain some of the tallest trees in the world.

Soils developed under temperate rain forests commonly accumulate large amounts of organic matter, and have a moderate abundance of layer lattice clay and argillic horizons. Common soils include some Sposodols (especially on coarse-textured parent materials), extensive Alfisols (Udalfs and Ustalfs) and some Inceptisols (particularly Umbrepts) and Ultisols (including Humults).

Figure 3.14   Massive stand of mature redwood in a temperate rain forest of northern California.

LOWLAND CONIFEROUS FORESTS ARE MAJOR SOURCES OF TIMBER

These forests are well represented in North America, where they make up several distinct formations, all of which are largely the result of human activity. Extending across the lake states into southern Ontario and northern New England, a belt of white spruce, white pine, red pine, and hemlock flourished until the middle of the nineteenth century. After this midcontinent forest was cut over for its valuable timber, much of the land was diverted to agricultural use. Pines occur in pure stands on dry sands, but they are usually mixed with oaks, aspen, and other hardwoods.

The pine forest of the coastal plain and piedmont of the southeastern United States is composed of several species collectively called southern pines. The dominant species are loblolly (*Pinus taeda*), shortleaf (*P. echinata*), longleaf (*P. palustris*), and slash (*P. elliottii*) pines. Pine forests in this region often have frequent fires that limit the development of hardwood trees. In recent years, southern pines have been widely used in intensively managed plantations for pulp, veneer, and sawn-log products. These forests occur on highly weathered low-base status soils, mostly Ultisols and the Aquod group of Spodosols. Aquods have developed on sandy soils in humid areas under a wide range of temperature conditions, from temperate to tropical zones.

TEMPERATE ZONE MIXED FORESTS INCLUDE CONIFERS AND HARDWOODS

Except on the west coast of North America, high-altitude coniferous forests of the Northern Hemisphere generally grade into broad-leaved deciduous forests at lower elevations. This change is usually achieved through a transition zone where the two forest formations exist side by side. These zones, with diverse mixtures of species, vary from a few miles to several hundred miles in width and are found in Europe and Asia as well as in North America. Species of birch, beech, maple, and oak are commonly found growing with the conifers. Mollisols, Inceptisols, and Alfisols are commonly found under these mixed forests. Since these productive soils are well suited to agriculture, most have been cleared, plowed, and cropped, and in some cases abandoned and reforested. In several lowland areas of the Southern Hemisphere there are mixed forests of broad-leaved evergreens and conifers. They occur in several climatic regions in Chile, southern Brazil, Tasmania, northern New Zealand, and South Africa. The species in this formation vary rather widely among the continents, as might be expected, but they are quite similar in life forms and general appearance. Such genera such as *Araucaria*, *Podocarpus*, and *Agathis* are found in many of these mixed evergreen forests. The predominant soils in these areas are Alfisols and Ultisols.

The broad-leaved, winter-deciduous forests of eastern and north-central North America, western and central Europe, and northern China generally lie south of the mixed coniferous-deciduous forests. These areas are heavily populated and intensively farmed; few relatively undisturbed forests remain.

In France and the British lowlands, *Quercus robur*, *Fagus sylvatica*, and *Fraxinus excelsior* are the dominant species, with other species of oak, birch, and elm becoming important in local areas. The tree associations gradually change when moving to southern Europe, and *Quercus lusitanica*, *Q. pubescens*, *Acer platenoides*, *Castanea sativa*, and *Fraxinus ornus* become more plentiful. Some of the best soils from the northern Ukraine across to the southern Urals once carried deciduous forests dominated by various oak species. Except for shelterbelts and windbreaks, these forests have been converted to agricultural uses.

A similar but more diverse formation of hardwoods is found in the United States, extending from the Appalachian Mountains west beyond the Mississippi River, from the mixed forests of New England and the lake states southward to the southern coniferous and mixed forests of the coastal plain. It reaches westward into the prairies along the valleys of several rivers. Although noted for its diversity, the American formation is dominated by species of oak and hickory (*Carya*), with basswood (*Tilia americana*), maple, beech (*Fagus sylvatica*), and ash (*Fraxinus*) species found in the eastern regions. Along the lower slopes of the southern Appalachians, several other species of oaks, sweetgum (*Liquidambar styracflua*), and yellow poplar (*Linodendron tulipifera*) appear, often with an understory of wild cherry (*Prunus serotina*), and dogwood (*Cornus florida*).

The East Asian formation of deciduous hardwoods is very similar to the formations of Europe and North America. However, northern and central Chile have a formation that is more closely related to the nearby evergreen forest than to the deciduous genera so common in the Northern Hemisphere formations. In southern Argentina and in the Valdevian forests of southern Chile, hardwood forests are common on Andisols.

Precipitation is relatively heavy and well distributed throughout the year in many areas supporting temperate forests of deciduous hardwoods. Summers are generally warm and humid, and winters are cool to cold, with heavy snowfall in the coldest regions. Differences in nutrient demands of the various species, plus the diversity of parent material on which they grow, have resulted in a wide range of soil properties in these forests. Spodosols have developed on parent materials of granites and sands in many northern hardwood forests in North America, as well as in some oak-birch forests in France and other parts of Europe. Mollisols are common on the calcium-rich clays of central and southern Europe, and Alfisols are sometimes found in a narrow zone along the margins of the wooded steppes in the central European basins and in the Ukraine.

Most Alfisols of the northern part of the United States, as well as the Ultisols in the South, were initially fertile and easily cultivated. They were exploited early in the settlement of the country, then often abandoned when market conditions no longer allowed row crops to be profitable. After abandonment, these areas were often planted to pine or reverted to pine-hardwood mixtures.

TEMPERATE MONTANE CONIFER FORESTS

Alpine ecosystems occupy high-elevation territories where trees cannot grow, so by this definition, the highest-elevation forests are called subalpine forests. Subalpine forests around the Northern Hemisphere are commonly composed of spruce species, sometimes alone or mixed with fir, birch, and occasionally pines. Below these forests are found a great diversity of montane forests, with conifers in many cooler or drier regions, and hardwoods in warmer and moister regions. Although these patterns are typical, they often do not apply to specific situations.

Pine, spruce, larch, and fir comprise the subalpine forests of western and central Europe. Spruce and fir are found at 1000 to 2000 m in the Alps and at 1500 to 2500 m in the Pyrenees. Similar associations are found at subalpine heights on the mountains of central and eastern Asia, extending southward to the Himalayas. Subalpine forests of Europe extend southwest into the Mediterranean peninsulas and even as far as the Atlas Mountains in North Africa and the mountains of Turkey and Spain. Pines (*Pinus sylvestris* and *P. nigra*) and firs are common dominant species, with cedars (*Cedrus* species) being of local importance.

In North America there is an almost continuous belt of subalpine and montane forest extending the whole length of the Cascade and Sierra Nevada Mountains from Canada to California in the West and along the length of the Rocky Mountains as far south as New Mexico. The altitude of the lower limit of the subalpine forest varies from 600 m in southern Alaska to 1000 m in British Columbia to 2700 m in New Mexico. Engelmann spruce (*Picea engelmannii*), subalpine fir (*Abies lasiocarpa*), and lodgepole (*Pinus contorta*), bristlecone (*P. aristata*), and limber (*P. flexilis*) pines are widely distributed in these western areas. The montane forests of the Pacific Northwest occur on the middle slopes of mountains from the Rockies westward, usually interposed between the subalpine forest and the temperate rain forest along the coast. At lower elevations in the Southwest they give way to scrub communities. The dominant species are ponderosa (*Pinus ponderosa*), western white (*P. monticola*), and lodgepole pines; Douglas-fir (*Pseudotsuga menziesii*); and grand fir (*Abies grandis*).

The Appalachian province consists of a series of mountain ranges and high plateaus extending from Newfoundland in the north to Georgia and Alabama in the south. The Appalachian subalpine forests are dominated by black and white spruce, balsam fir, and tamarack (*Larix laricina*) in the north and by red spruce (*Picea rubens*), fraser fir (*Abies fraseri*), and eastern white pine (*Pinus strobus*) farther south. The altitudes of these forests rise southward, from sea level in Newfoundland to 200 m in the Adirondacks, 900 m in West Virginia, and nearly 1800 m in Georgia.

The soils of subalpine forests are often of glacial or colluvial origin and generally are weakly developed. Spodosols and Inceptisols predominate in the higher latitudes. In the lower latitudes Spodosols are infrequent, and Alfisols and Ultisols become more common. In the Rocky Mountains, Inceptisol, Alfisol, Spodosol, and Mollisol soils occur beneath subalpine and montane forests.

BOREAL FORESTS COVER VAST AREAS

This vast northern forest, composed largely of conifers, extends southward from the tundra treeline across Canada, Alaska, Siberia, and Europe and is referred to as boreal forest or simply taiga. Although boreal forests are characterized by conifers, hardy species of *Populus*, *Betula*, and *Salix* extend throughout the forested boreal landscape.

Within the boreal forest, species are distributed according to landforms and landscape position, with soil drainage being the key factor. In North America, a gradual transition in species spans the continent. Red pine (*Pinus resinosa*), eastern hemlock (*Tsuga canadensis*), yellow birch (*Betula alleghaniensis*), and maple (*Acer*) of southern and eastern Canada are gradually replaced by a predominantly forested belt across the middle of Canada composed chiefly of white spruce (*Picea glauca*), black spruce (*P. mariana*), balsam fir (*Abies balsamea*), jack pine (*Pinus banksiana*), white birch (*Betula papyrifera*), and aspen (*Populus tremuloides*) (Figure 3.15). These species give way to white spruce, black spruce, and tamarack (*Larix laricino*) near the tundra. In a somewhat analogous fashion, pine (*Pinus sylvestris*), spruce (*Picea abies*), and European larch (*Larix decidua*) predominate in the European boreal forests, and to the east in Siberia the dominant species become *Abies siberica*, *Larix siberica*, and *Picea siberica*.

The mean annual rainfall of most boreal forests is less than 900 mm/yr, with many forests receiving less than half of this maximum. Evapotranspiration rates are low owing to the short growing season, and boreal forest soils tend to be moist (and often saturated) except on well-drained landforms or in very dry summers. Some boreal forest soils are underlain by permanently frozen parent material called permafrost; the zone that freezes in winter and thaws in summer is called the active layer. The depth of the active layer may increase following fires that remove canopies and insulating forest floors, although this is not a universal pattern (Swanson, 1996). The forest floors are often thick, with well-developed layers of humus overlain by carpets of moss or lichen. The organisms that decay the litter produce organic acids that aid in leaching minerals important for plant growth from the upper soil (A_2 or E) horizon. The sesquioxides of iron and aluminum are removed from this eluviated

Figure 3.15   Spruce-fir stand in the Canadian boreal forest.

horizon during the podzolization process, and the resultant sandy horizon is light gray in color (as shown in Figure 2.1). The iron and aluminum compounds that are removed from the upper horizons are mostly redeposited in the illuviated horizons lower in the profile. Soils of the boreal forest are predominantly classified as Spodosols and Inceptisols to the south and Gelisols to the north, with Histosol soils found in bogs or fens.

The positive water balance of the boreal results in a high water table in large expanses of nearly level terrain. Where the water table remains within about 0.5 m of the surface for a large part of the year, trees are stunted or absent. Extensive areas of the boreal zone are covered by organic soils (Histosols), with various mosses (*Sphagnum* and *Polytrichum* species), flowering plants (*Eriophorum* and *Trichophorum* species), and dwarf shrubs (*Vaccinium* and *Calluna* species). Various names are used for these mires in different countries: bogs, muskegs, and fens.

Windthrow mounds are numerous on some Spodosols of the boreal zone, occupying up to 30 percent of the forest floor area in parts of the Great Lakes region of North America (Buol et al., 1980). They are formed by the uprooting of trees in storms, and they typically have relief of as much as 0.5 to 1 m and a diameter of 1 to 3 m. This disturbance creates a characteristic pit and mound surface morphology and complicates the characterization of boreal soils (Fisher and Miller, 1980).

The major effect of humans on most boreal forests has been through the ignition of fire, but increasingly large portions of the boreal forests are being harvested for wood products. Very little of the boreal forest realm has been cleared and plowed for agriculture.

SUMMARY

Forest soils are variable at all spatial scales—from beneath one tree to another, across slopes and landforms, and across major gradients in climate. Some generalizations are possible, but not many; on average, tropical soils are no poorer or richer than temperate soils. The major types of forests around the world have some correspondence to soil types, but again, the variations in soil types within a forest type are striking. Major challenges and opportunities in understanding and managing forest soils depend on local details rather than on regional generalizations.

CHAPTER 4

Physical Properties of Forest Soils

Soil physical properties profoundly influence the growth and distribution of trees through their effects on soil moisture regimes, aeration, temperature profiles, soil chemistry, and even the accumulation of organic matter. Some physical properties can be altered intentionally by management, including draining wet soils and plowing subsoil to break up hardpans. The most widespread result of management on the physical properties of soils may be inadvertent soil compaction during harvesting operations.

The physical properties of soils constrain soil water properties, chemical reactions, and especially biologic activity. The importance of overall physical soil features is clear in a 4-year-old plantations of *Eucalyptus grandis*, where the volume was twice as great on ridgetop soils as on toeslope soils (Figure 4.1), primarily as result of differences in soil depth. The major physical properties of soils include texture (size distribution of particles), structure (aggregation of particles), porosity, stoniness, and depth.

In this chapter we examine the most important physical features of soils, discussing the major ranges of conditions that are common in forests and the effects of management activities.

SOIL TEXTURE IS FUNDAMENTAL

Soil can be conveniently divided into three phases: solid (including mineral and exchange phases), liquid, and gas. The solid phase makes up approximately 50 percent of the volume of most surface soils and consists of a mixture of inorganic and organic particles varying greatly in size and shape. The size distribution of the mineral particles determines the texture of a given soil. Schemes for classifying soil particle sizes have been developed in a number of countries. The classification used by the USDA, based on diameter limits in millimeters, is outlined in Table 4.1.

Mineral soils are usually grouped into three broad textural classes—sands, silts, and clays—and combinations of these class names are used to indicate mixtures (Figure 4.2).

The most important differences in soil texture relate to the surface areas of particles of different sizes. Medium-sized sand has a diameter of 0.25 to

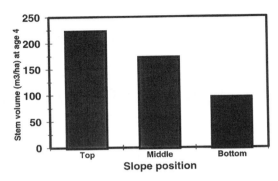

Figure 4.1 The stem volume of *Eucalyptus grandis* at age 4 years was twice as great on deep Oxisol (lateritic) soils at the top of a slope in Argentina as on shallower soils at the bottom of the slope (data from Dalla-Tea and Marco, 1996).

0.50 mm, about 7000 grains per gram, and a surface area of 0.013 m^2/g (Barber, 1995). Silt has a diameter of 0.002 to 0.050 mm, about 20 million particles per gram, and a surface area of 0.09 m^2/g. Clay particles are less than 0.002 mm in size, with 400 billion particles per gram, and a surface area of 1 to 10 m^2/g. The surface area per gram spans a range of three orders of magnitude, with dramatic effects on water potential, organic matter binding, cation exchange, and overall biotic activity.

Cobbles and **gravels** are fragments larger than 2 mm in diameter. They are not included in the particle-size designations because they normally play a minor role in agricultural soils. However, coarse particles are common in forest soils (up to 80 percent or more of some mountain soils). Major properties of coarse fragments are recognized by adjectives that indicate modification of the texture name (Table 4.2).

A moderate amount of rock in a fine-textured soil may favor tree growth. Coarse fragments may increase penetration of air and water, as well as the rate

TABLE 4.1   USDA Classification of Soil Particle Sizes

Name of Soil	Diameter Limits, mm
Sand	0.05–2.0
Very coarse	1.0–2.0
Coarse	0.5–1.0
Medium	0.25–0.5
Fine	0.10–0.25
Very fine	0.05–0.10
Silt	0.002–0.05
Clay	Less than 0.002

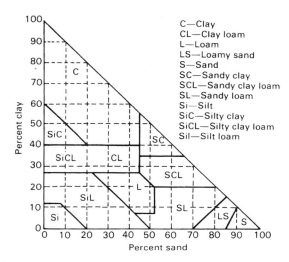

C—Clay
CL—Clay loam
L—Loam
LS—Loamy sand
S—Sand
SC—Sandy clay
SCL—Sandy clay loam
SL—Sandy loam
Si—Silt
SiC—Silty clay
SiCL—Silty clay loam
Sil—Silt loam

Figure 4.2 Soil textural triangle: percentage of clay and sand in the main textural classes of soils; the remainder of each class is silt.

of soil warming in spring. Nevertheless, a coarse skeleton dilutes the soil and can be detrimental to tree growth if it occupies a large volume in sandy soils, because of low water-holding capacity and exchange capacity. Coarse-textured materials contribute little to plant nutrition.

TEXTURE INFLUENCES TREE GROWTH

Soil texture has major effects on forest growth, but these effects are indirect, manifested through the effect of texture on features such as water-holding

TABLE 4.2   USDA Classification Scheme for Coarse Fragments

Shape and Kind of Fragment	Size and Name of Fragment		
Rounded fragments	2 mm to 8 cm in diameter	8–25 cm in diameter	>25 cm in diameter
	Gravelly	Cobbly	Stony or bouldery
Thin, flat fragments	2 mm to 15 cm in length	6–38 cm in length	>38 cm in length
	Channery	Flaggy	Stony or bouldery

capacity, aeration, and organic matter retention. For example, deep, coarse, sandy soils often support low-productivity stands of pines, cedar, scrub oak, and other species that cope well with moisture and nutrient stress. Some sands have lenses of finer-textured materials, and variations in texture strongly influence overall soil properties (such as water retention).

SOIL STRUCTURE MODERATES EFFECTS OF SOIL TEXTURE

Soil structure refers to the aggregation of individual mineral particles and organic matter into larger, coarser units. This aggregation generally reduces bulk density (megagrams of soil per cubic meter), and increases water movement and aeration. Field descriptions of soil structure usually give the type or shape, class or size, and degree of distinctness in each horizon of the soil profile. The following types of structure are recognized by the U.S. Natural Resource Conservation Service: platy, prismatic, columnar, angular blocky, subangular blocky, granular, crumb, single grain, and massive. The size class of aggregates ranges from very fine to fine, medium, coarse, and very coarse. Structural classes are determined by comparing a representative group of peds with a set of standardized diagrams. Grade is determined by the relative stability or durability of the aggregates and by the ease of separating one from another. Grade varies with moisture content and is usually determined on nearly dry soil and designated by the terms weak, moderate, and strong. The complete description of soil structure, therefore, consists of a combination of the three variables, in reverse order of that given above, to form a type, such as moderate fine crumb.

Important drivers of soil aggregation include mineral chemistry, salts, clay skins, oxide coatings, growth and decay of fungal hyphae and roots, freezing and thawing, wetting and drying, and the activity of soil organisms (especially earthworms). Soil texture has considerable influence on the development of aggregates. The absence of aggregation in sandy soils gives rise to a single-grain structure, whereas loams and clays exhibit a wide variety of structures. Soil animals, such as earthworms and millipedes, favor the formation of crumb structure in the surface soil by ingesting mineral matter along with organic materials, producing casts that provide structure to the soil. The intermediate products of microbial synthesis and decay are effective stabilizers, and the cementing action of the more resistant humus components that form complexes with soil clays gives the highest stability.

Soil aggregation is also influenced by different tree species. In an experiment with 35-year-old plots with different tree species, the average size of aggregates ranged from 1.5 mm under white pine to 2.1 mm under Norway spruce (Scott, 1996). Across the species, average aggregate size increased with increasing fungal mass ($r^2 = 0.66$) and declining bacterial biomass ($r^2 = 0.72$; bacterial biomass declined as fungal mass increased, $r^2 = 0.87$).

The influence of tree species on soil aggregation was also apparent from a "lysimeter" experiment in California in which 50 m^3 chambers were filled with soil and planted to Coulter pine or oak (Graham and Wood, 1991). After 40 years, soils under the influence of pine lacked earthworms, had developed a clay-depleted A horizon, and had accumulated enough clay in the B horizon to qualify as an argillic horizon. Soils under oak (*Quercus dumosa*, a shrub) developed a 7-cm A horizon (90 percent of which was earthworm casts) enriched in humus and clay relative to the underlying C horizon. In this experiment, the plant species affected earthworm activity, which dominated structural development of the soil.

BULK DENSITY ACCOUNTS FOR THE COMPOSITION OF MINERALS, ORGANICS, AND PORE SPACE

Bulk density is the dry mass (of <2mm material) of a given volume of intact soil in Megagrams per cubic meter (which also equals kilograms per liter). Well-developed soil structure increases pore volume and decreases bulk density. The particle density of most mineral soils varies between the narrow limits of 2.60 and 2.75, but the bulk density of forest soils varies from 0.2 in some organic layers to almost 1.9 in coarse sands. Soils high in organic matter have lower bulk densities than soils low in this component. Soils that are loose and porous have low mass per unit volume (bulk density), while those that are compacted have high values. Bulk density can be increased by excessive trampling by grazing animals, inappropriate use of logging machinery, and intensive recreational use, particularly in fine-textured soils.

Increases in soil bulk density are generally harmful to tree growth for the same reasons that structure affects soil properties. Compacted soils have higher strength and can restrict penetration by roots. Reduced aeration in compacted soils can depress the activities of roots, aerobic microbes, and animals. Reductions in water infiltration rates are common when soils become compacted, and anaerobic conditions may develop in puddled areas.

The importance of differences in soil bulk density (and soil strength) for tree growth is illustrated by an extensive characterization of variation in soil bulk density across a stand of *Eucalyptus camaldulensis* on an Inceptisol in the savanna region of central Brazil (Figure 4.3). Where the bulk density of the 0- to 20-cm soil averaged 1.25 kg/L, stem volume of the 4-year-old trees was 25 m^3/ha (Gonçalves et al., 1997). Where the bulk density averaged 1.06 kg/L, stem volume was more than three times greater (90 m^3/ha).

Pore volume refers to that part of the soil volume filled by water or air. The proportions of water and air change over time, and soil water content drastically affects soil aeration. Gas molecules diffuse about 10,000 times faster through air than through water. Coarse-textured soils have large pores, but their total pore space is less than that of fine-textured soils (although puddled clays may have even less porosity than sands). Because clay soils have greater total pore space than sands, they are normally lighter per unit volume (have

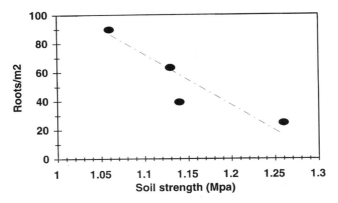

Figure 4.3 Variations across a stand in bulk density of an Inceptisol in the Brazilian savanna strongly affected the volume of wood accumulated by *Eucalyptus camaldulensis* at age 4 years (data from Pereira, 1990, cited in Gonçalves et al., 1997).

lower bulk density), but differences in soil structure can override the basic influence of texture on pore volume and bulk density.

Pore volume is conveniently divided into **capillary** and **noncapillary pores.** Soils with a high proportion of capillary (small-diameter) pores generally have high moisture-holding capacity, slow infiltration of water, and perhaps a tendency to waterlog. By contrast, soils with a large proportion of noncapillary (large-diameter) pores generally are well aerated, and have rapid infiltration and low moisture-retaining capacity. The pore size distribution of a soil may be broad, with large portions of capillary and noncapillary pores, particularly where soil animals and old root cavities provide large-radius pores in clayey soils.

Sandy surface soils have a range in pore volume of approximately 35 to 50 percent, compared to 40 to 60 percent or higher for medium- to fine-textured soils. The amount and nature of soil organic matter and the activity of soil flora and fauna influence pore volume and soil structure. Pore space is reduced by compaction and generally varies with depth. Some compact subsoils may have no more than 25 to 30 percent pore space. Tree species can also change the distribution of pore sizes in soils; soils under Norway spruce have been reported to have more pore space than adjacent soils under beech (Nihlgård, 1971).

The pore volume of forested soils is normally greater than that of similar soil used for agricultural purposes because continuous cropping results in a reduction in organic matter and macropore spaces (Figure 4.4). Porosity of most forest soils varies from 30 to 65 percent.

Soil aggregates are generally more stable under forested conditions than under cultivated conditions. Continued cultivation tends to reduce aggregation in most soils through mechanical rupturing of aggregates and by a reduction

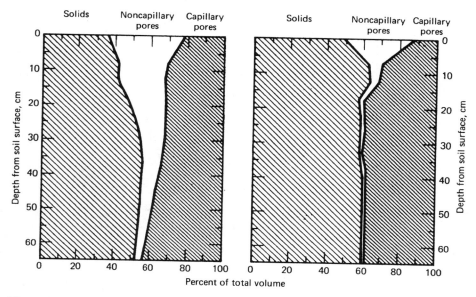

Figure 4.4 Porosity as measured in the surface 60 cm of Vance Soil (Typic Hapludult) in the South Carolina Piedmont: (left) in an undisturbed mixed hardwood forest and (right) in abandoned farmland (Hoover, 1949).

in organic matter content and associated cementing action of microbial exudates and fungal hyphae (Figure 4.5).

LIFE IN THE SOIL DEPENDS ON THE SOIL ATMOSPHERE

Soil air is important primarily as a source of oxygen for aerobic organisms, including tree roots. Soil air composition, like air volume, is constantly changing in a well-aerated soil. Oxygen is used by plant roots and soil microorganisms, and carbon dioxide is liberated in root respiration and by aerobic decomposition of organic matter (Figure 4.6).

Gaseous exchange between the soil and the atmosphere above it takes place primarily through **diffusion**. Consumption of oxygen by respiration in the soil leads to a gradient from high carbon dioxide in the ambient air into the low-oxygen soil air. The oxygen content of air in well-drained surface soils seldom falls much below the 20 percent found in the atmosphere, but oxygen deficits are common in poorly drained, fine-textured soils. Under these conditions, gas exchange is very slow because of the small, water-filled pore spaces. In very wet soils, carbon dioxide concentrations may rise to 5 or 6 percent and oxygen levels may drop to 1 or 2 percent by volume (Romell, 1922). However, oxygen is not necessarily deficient in all wet soils in spite of the absence of

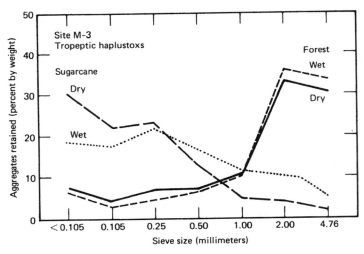

Figure 4.5   Stability of aggregates of an Oxisol used for forestry and for sugarcane in Hawaii (Wood, 1977). The forest soil has much larger aggregates that do not pass through a 1-mm sieve, whereas most of the sugarcane soil aggregates are much smaller. Reproduced from *Soil Science Society of America Journal* 41, no.1 (1977):135, with permission of the Soil Science Society of America.

voids. If the soil water is moving, it may have a reasonably high content of oxygen brought in through **mass flow** of water. Soils saturated with stagnant water are low in oxygen, and they are very poor media for the growth of most higher plants. Soil air usually is much higher in water vapor than is atmospheric air, and it may also contain a higher concentration of such gases as methane and hydrogen sulfide, formed during organic matter decomposition.

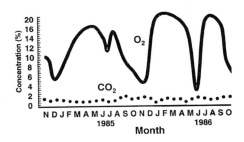

Figure 4.6   Trends in soil oxygen (solid line) and carbon dioxide (dotted line) concentrations at 20-cm depth in gleyic podzol (Spodosol) developed in glacial till in northern Sweden (data from Magnusson, 1992). During saturated periods, oxygen concentration dropped below 5 percent, and carbon dioxide concentration varied between 1 and 2.5 percent.

Oxygen concentrations in soil air as low as 2 percent are generally not harmful to most trees for short periods. Some *Alnus*, *Taxodium*, *Nyssa*, and *Picea* species can thrive at low levels of soil oxygen. Seedling root growth of many species is reduced by an oxygen content of 10 percent or lower. Any restriction in gas exchange between the tree roots and the atmosphere will eventually result in the accumulation of carbon dioxide.

SOIL STRUCTURE CAN BE HARMED BY INAPPROPRIATE MANAGEMENT ACTIVITIES

Compaction by heavy equipment or repeated passages of light equipment compresses the soil mass and breaks down surface aggregates, decreasing the macropore volume and increasing the volume proportion of solids. Reductions in air diffusion and water infiltration often combine with increases in soil strength in compacted soils. Compaction occurs more frequently on moist soils than on dry soils (because water lowers soil strength, "lubricating" soil particles) and more often on loamy-clay soils than on sandy soils. The effects of compaction may be less permanent in fine-textured soils (especially those containing considerable shrink-swell clays) than in some coarse-textured soils because of swelling and shrinking as a result of wetting and drying.

The use of large machines in forest harvesting is the primary driver of increases in soil bulk density and strength and decreases in pore volume. For example, 10 passes with a rubber-tired skidder substantially increased soil strength in a harvesting unit in Australia (Figure 4.7), although in this case, the amount of soil compaction may not have been great enough to affect growth.

A careful greenhouse study illustrates the pieces of the soil compaction puzzle (Simmons and Pope, 1988). When soils at -0.3 MPa were compacted to increase bulk density from 1.25 to 1.55 mg/m^3, the air-filled porosity declined from 36 percent to 20 percent, and soil strength increased from 2.3 to 4.1 MPa. When these soils were moister (with soil water potential near field capacity — 0.01 MPa), air-filled porosity declined from 22 percent in the 1.25 mg/m^3 soil to 6 percent in the 1.55 mg/m^3 soil. The wetter soils had much lower soil strength (0.4 to 0.8 MPa, too low to impede root growth), but the most compacted soil had poor root growth because of anaerobic conditions that resulted from the low air-filled pore volume.

Under severe compaction, soils may puddle. Puddling results from the dispersal of soil particles in water and the differential rate of settling, which permits the orientation of clay particles so that they lie parallel to each other. The destruction of soil structure by this method may result in a dense crust that has the same effect on soil conditions as a thin, compacted layer. The crust is most common on soil surfaces where the litter has been removed by burning or by mechanical means. Reduced germination and increased mortality rates of loblolly pine seedlings have been observed on soils compacted or puddled by logging equipment.

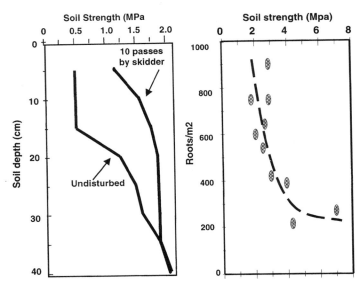

Figure 4.7 Soil strength increased substantially after 10 passes with a rubber-tired skidder in Australia, and root penetration declined as root strength increased. However, the domain over which these effects occurred did not overlap; the skidder did not increase soil strength enough to reduce root growth (modified from Gracen and Sands, 1980).

SOIL COLOR INCORPORATES EFFECTS OF ORGANIC MATTER AND OTHER FACTORS

Color is an obvious characteristic of soils that is used to differentiate soil horizons and classify soils. In many parts of the world, soils may be described as Red and Yellow Podzolics, Brown Earths, Brown Forest soils, and Chernozens (black soils).

Soil color depends on pedogenic processes and the parent material from which the soil was derived. Most soil minerals, such as quartz and feldspars, are light in color. Color is generally imparted by small amounts of colored materials, such as iron, manganese, and organic matter. Red colors generally indicate ferric (oxidized iron) compounds associated with well-aerated soils. Yellow colors may signify intermediate aeration. Ferrous (reduced iron) compounds of blue and gray colors are often found under reduced conditions associated with poorly aerated soils. Mottling (marbling of colors) often indicates a zone of alternately good and poor aeration. Manganese compounds and organic matter produce dark colors in soils. Color intensity is often used to estimate organic matter content. It is not a foolproof system, however, because the pigmentation of humus is less intense in humid zones than in arid regions. Brown colors predominate in slightly decomposed plant materials, but more thoroughly decomposed amorphous material is nearly black.

Color itself is of no importance to tree growth, but color may indicate several important characteristics of soil. These characteristics include geologic origin and degree of weathering of the soil material, degree of oxidation and reduction, content of organic material, and leaching or accumulation of such chemical compounds as iron, which may greatly influence soil quality.

Dark-colored surface soils absorb heat more readily than light-colored soils but because of their generally higher content of organic matter, they often have higher moisture content. Therefore, dark soils may warm less rapidly than well-drained, light-colored soils. Soil color influences the temperature of bare soils, but it has less effect on the temperature of soils beneath forest canopies. Soil color becomes important after fires, when removal canopy combines with blackening of the soil surface to increase temperature (see Chapter 10).

SOIL TEMPERATURE INFLUENCES BIOTIC AND ABIOTIC PROCESS RATES

Soil temperature is a balance between heat gains and losses. Solar radiation is the principal source of heat, and losses are due to radiation, conduction, and convection (and evaporation of water in some cases). When well-developed canopies are present, the temperature of upper soil layers varies more or less according to the temperature of the air immediately above it. Temperature fluctuations in deeper soil layers are moderated. In the absence of a forest canopy, topsoil temperature may be much higher than air temperature because of direct absorption of solar radiation.

Soil temperatures generally decrease with increases in elevation, although postharvest soil temperatures may be extreme at high elevations on sunny days. Soil temperatures in winter are commonly warmer beneath snowpacks than are air temperature (Figure 4.8). Conifer forests may also have cooler soils than hardwood deciduous species (Figure 4.8) for several reasons. The higher leaf area of conifers may intercept more light and reduce solar heating of the soil, and may reduce convective heat losses as well. Conifer canopies also intercept more snow in winter, providing shallower snow layers on the ground to insulate the soil from frigid winter air. Temperatures also tend to decrease with latitude, and seasonal swings become more pronounced (Figure 4.8).

Aspect (direction of slope) influences soil temperature. Soils receive more solar radiation on south-facing slopes in the Northern Hemisphere and on north-facing slopes in the Southern Hemisphere. Higher temperatures typically lead to greater rates of evapotranspiration, so the heating effect on the soil may be amplified by drying.

The tree canopy and forest floor moderate extremes of mineral soil temperatures. They protect the soil from excessively high summer temperatures by intercepting solar radiation, and they reduce the rate of heat loss from the soil

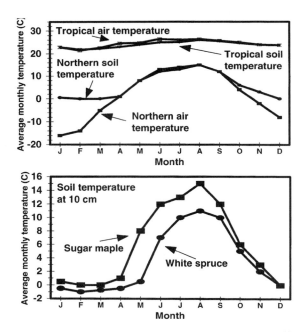

Figure 4.8 Seasonal traces of temperatures for air and upper soil at the Luquillo Experimental Forest in Puerto Rico (tropical site) and at Isle Royale, Michigan, for the northern forests of sugar maple and white spruce. The upper graph provides air temperature and soil temperature for the tropical, closed-canopy forest, which shows little seasonal variation and no difference between air and soil temperatures. The soil in the sugar maple stand remains warmer throughout the winter than the air because of the insulating snow layer. The lower graph contrasts soil temperatures under sugar maple and white spruce; spruce soils are colder because of less snow cover in winter and perhaps because of lower light penetration of the canopy (Luquillo data from X. Zou, personal communication; Isle Royale data from Stottlemyer et al., 1998).

during the winter. Forest cover may influence the persistence of frost in soils in cold climates; however, freezing generally occurs earlier and penetrates deeper in bare soil than in soil under a forest cover.

The effect of forest canopies on reducing frosts has long been attributed to canopies' absorbing longwave radiation emitted by the soil and then reradiating some of the energy back to the soil in a form of the greenhouse effect. Direct measurements of radiation budgets have not supported this story (Löfvenius, 1993); it appears that the beneficial effects of canopies on soil temperatures at night derive from the effects of canopies on air movement. Air at the surface of bare soils may become very cold as a result of radiative heat loss to the cold night sky. The presence of even a few trees per hectare can reduce the development of such cold air layers by intercepting some of the air currents higher above the ground and creating turbulent airflow that prevents stratification.

Part of the ameliorating effect of the forest cover on soil temperature is due to the forest floor (O horizon), as shown for a clay soil in Figure 4.9. Organic layers have low thermal conductivity, and they lower maximum summer temperatures and raise minimum winter temperatures. Diurnal fluctuations are also dampened.

Snow also insulates soils because of its low thermal conductivity. A porous snow covering of about 45 cm was sufficient to prevent soil freezing in frigid northern Sweden (Beskow, 1935), and a covering of 20 to 30 cm was sufficient to prevent freezing in southern Sweden. Winter soil temperatures may often be higher at northern locations and high elevations, where the snow cover is thick. Sartz (1973) reported that aspect influences soil freezing in Wisconsin because the direction of the slope affects the amount of snow accumulation. Therefore, depending on the conditions, frost may be deeper on southern slopes than on northern slopes (Table 4.3).

Specific heat and **conductance** of heat are inherent properties of soils that influence swings in daily temperature. Specific heat refers to the number of Joules needed to raise a unit mass of water by a defined number of degrees; raising the temperature of 1 kg (or 1 L) of water from 15° to 16°C requires an input of 4.2 MJ of energy. Conductance refers to the movement or penetration of thermal energy into the soil profile. Both specific heat and conductance are influenced somewhat by texture, and especially by soil water content and organic matter content. Water has high specific heat and high conductance.

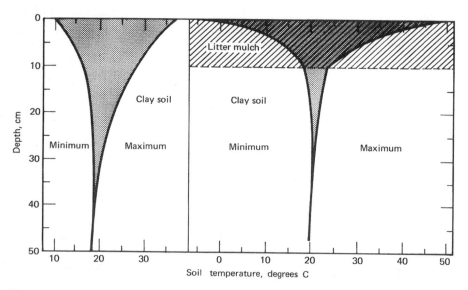

Figure 4.9 Daily temperature variations with depth for (left) an unmulched and (right) a mulched clay soil where the litter mulch has a lower thermal conductivity (Cochran, 1969).

TABLE 4.3 Snow and Frost Depths on North and South Slopes in Wisconsin (in cm)

	Snow		Frost	
Year and Date	North	South	North	South
1970 (February 25)	25	0	8	11
1971 (February 25)	55	25	0	0
1972 (March 10)	45	25	16	23
1973 (March 01)	5	0	21	12

Source: Sartz (1973).

Wet soils are slower to change their temperature for two reasons: it takes a lot of energy to warm the water, and meanwhile the water transmits heat deeper into the soil. Organic matter has low conductance and impedes the movement of thermal energy.

Favorable soil temperature is essential for the germination of seeds and the survival and growth of seedlings. In some cold climates, a dense canopy may keep soil temperature so low that it delays germination and slows up seedling development. On the other hand, removal of the forest cover may result in increases in soil temperature to lethal levels for germinating seedlings. Soil surface temperatures vary among substrates, with high temperatures in dark residues (especially charred residues) and lower temperatures in moist, decayed logs (Figure 4.10). Clearcut areas have different radiation and wind regimes, which tend to accentuate daily swings between low nighttime and high daytime temperatures (Figure 4.11).

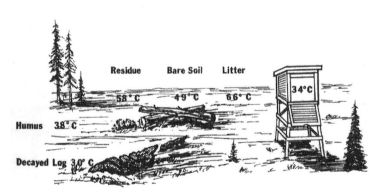

Figure 4.10 Surface temperatures of substrates in a clearcut area in the mountains of Montana (Hungerford, 1980).

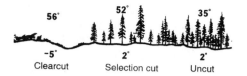

Figure 4.11 Average surface temperatures in uncut, partially cut, and clearcut units in the mountains of Montana (Hungerford, 1980). Night temperatures in the clearcut area fall below 0°C as a result of reduced turbulence without trees, which allows air at the soil surface to supercool; daytime temperatures reach extreme levels under high-altitude sunshine.

Root growth is also affected by soil temperatures, as shown by Lyford and Wilson (1966). They found that day-to-day variations in growth rates of red maple root tips closely paralleled variations in surrounding soil temperatures.

SOIL WATER IS PART OF THE HYDROLOGICAL CYCLE

The supply of moisture in soils largely controls the types of tree that can grow, and the distribution of forests around the world relates to patterns in precipitation and soil moisture. Water is essential to the proper functioning of most soil and plant processes. Carbon uptake by plants requires the moist interiors of leaves to be exposed to relatively dry air. Given the concentration gradient between wet leaves and dry air, about 1000 molecules of water are lost for every molecule of carbon dioxide acquired. In addition to serving the metabolic needs of the plant, water is critical for many functions in the soil. Water is a solvent and a medium of transport of plant nutrients, the medium for the action of the exoenzymes produced by soil microbes, and an important influence on soil temperature and aeration.

The retention and movement of water in soils and plants involves energy transfers. Water molecules are attracted to each other in a polymer-like grouping, forming an open tetrahedral lattice structure. This asymmetrical arrangement results from the dipolar nature of the water molecule. Although water molecules have no net charge, the hydrogen atoms sit to one side of the oxygen atom, giving the molecule a partial positive charge on one side and a partial negative charge on the other. As a result, the hydrogen atoms of one molecule attract the oxygen atom of an adjacent molecule. This hydrogen bonding (along with partial covalent bonding) accounts for the forces of adhesion, cohesion, and surface tension that largely regulate the retention and movement of water in soils. **Adhesion** refers to the attractive forces between soil surfaces and water molecules. At the water-air interface, **surface tension** may be the only force retaining water in soils. It results from the greater attraction of water molecules for each other (**cohesion**) than for the air.

Water with high free energy tends to move toward a zone of low free energy—from a wet soil to a dry soil and from the upper soil to the lower soil. The amount of movement depends on the differences in the energy states between the two zones; these differences are referred to as differences in potential. The potential of water in soil has three components. **Matric potential** is the attraction of water to soil surfaces; small pores hold water more tightly than larger pores do. Water that is low in dissolved ions tends to move into areas with more concentrated salts, which represents the **osmotic potential**. Water also tends to move toward the center of the Earth because of the **gravitational potential**. The total soil water potential is the combined effects of matric, osmotic, and gravitational potentials. In physics, matter tends to move from zones of higher potential to zones of lower potential, and the unit for potential is the Pascal ($= 1$ Newton/m^2; 25 years ago, the units commonly used in the United States were bars or atmospheres, which equal about 100 kPa, or 0.1 MPa). Common soil water potentials range from near zero for very wet soils to -3.0 MPa (MegaPascal) or lower. The water potential at the bottom of a lake may have a positive potential because the mass of water above it has a positive pressure (or head).

Investigators have long attempted to devise useful equilibrium points or constants for describing soil moisture. Such terms as **field capacity** and **permanent wilting point** have found their way into soils literature over the years. Most of these terms deal with hypothetical concepts and do not apply equally well to all soil conditions. Nevertheless, they are employed commonly and may be of use.

Field capacity describes the amount of water held in the soil after gravity has drained most of the water that is easily drained. Field capacity is difficult to determine accurately. Factors such as soil texture, structure, and organic matter content influence measurements. Soil layers of different pore size markedly influence water flow through a profile, greatly affecting field capacity. Field capacity may be defined as the water content and water potential after 1 or 2 days of draining following full rewetting of the soil, or 0.03 MPa for silt loam soils and 0.01 to 0.001 MPa for sandier soils.

Permanent wilting point is a classic term for the soil moisture potential at which plants remain permanently wilted, even when water was added to the soil. Just as field capacity has been widely used to refer to the upper limit of soil water storage for plant growth, the permanent wilting point is used to define the lower limit. The use of a test plant, such as a sunflower, is the most widely accepted method for determining this soil water potential, but it has no particular relevance to trees.

Soil water-holding capacity is a useful term that refers to the quantity of water held within soils between the freely drained level of field capacity and some arbitrary potential beyond which plant uptake becomes minimal. For example, the Commerce silt loam holds about 45 percent moisture (water mass as a percentage of equivalent soil dry mass) at field capacity and about 10 percent moisture at -2 MPa, for a water-holding capacity of 35 percent of the

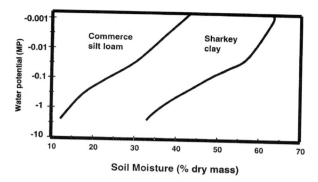

Figure 4.12 Relations of soil water potential to soil moisture content in Sharkey clay and Commerce silt loam (modified from Bonner, 1968).

soil's dry mass (Figure 4.12). Clays hold more water than silt loams, but they hold it more tightly. The Sharkey clay holds about 65 percent water at field capacity and 35 percent at −2 MPa, for a water-holding capacity of 30 percent of the dry soil mass. At 35 percent moisture, roots obtain water easily from the silt loam (0.01 MPa) but not from the clay (−2.0 MPa).

Soil water potential also affects the activities of microbes. A variety of studies have shown that carbon dioxide formation (an integrated measure of microbial activity) is reduced by half as soil water potential declines from saturated to −0.2 MPa (Sommers et al., 1980). Nitrogen mineralization also declines as soil moisture declines. One study in Kenya found that gross Nitrogen mineralization dropped from 2.2 to 0.08 µg/g daily as soil water potential fell from −0.06 to −5.9 MPa (Pilbeam et al., 1993).

WATER FLOW IN UNSATURATED SOILS DEPENDS ON WATER CONTENT AND HYDRAULIC CONDUCTIVITY

The availability of soil water to plants depends on its potential and on the hydraulic conductivity of the soil. In saturated soils, water uptake by trees is not limited by the rate of water movement through soils. As water drains from the soil, macropores empty and water is present only in capillary pores, which hold water with strong negative potential and also retard the flow of water. The rate at which water can move through a soil, **hydraulic conductivity**, is related to water-filled pore size and water potential. For example, conductivity at 0.02 Mpa would be about 10,000 times greater than at −1.0 MPa. At very negative potentials, water in sands is held only at points of contact between the relatively large particles. Under these conditions, there is no continuous water film and no opportunity for liquid movement. Layers of sand in a profile of fine-textured soil, therefore, can inhibit downward movement of water in a fashion similar to that of compact clay or silt pans.

Layering or stratification of materials of different texture is common in soils as a result of differences in the original parent material or pedogenic development. These factors can result in silt or clay pans or lenses of sand or gravel. Layers of fine-textured materials over coarse sands and gravel, as well as the reverse situation, are common in glacial outwash sands in glaciated terranes. Many Ultisols (highly weathered soils with low base saturation) have sandy A horizons underlain by clay-rich B horizons. Changes in texture throughout a profile tend to slow water movement whether the change down through the profile is from coarse to fine (the fine horizon has lower saturated conductivity) or from fine to coarse (water cannot leave the fine layer and enter the coarse layer except near saturation).

FACTORS AFFECTING INFILTRATION AND LOSSES OF WATER

The term infiltration is generally applied to the entry of water into the soil from the surface. Infiltration rates depend mostly on the rate of water input to the soil surface, the initial soil water content, and internal characteristics of the soil (such as pore space, degree of swelling of soil colloids, and organic matter content). Overland flow happens only when the rate of rainfall exceeds the infiltration capacity of the soil. Because of the sponge-like action of most forest floors and the high infiltration rate of the mineral soil below, there is little surface runoff of water in mature forests. Overland flow is rarely a problem in undisturbed forests even in steep mountain areas.

When rainfall does exceed infiltration capacity, the excess water accumulates on the surface and then flows overland toward stream channels. This surface flow concentrates in defined stream channels and causes greater peak flows in a shorter time than water that infiltrates and passes through the soil before reappearing as streamflow. High-velocity overland flow can erode soils. Soil compaction by harvesting equipment, disturbances by site preparation, and practices that reduce infiltration capacity and cause water to begin moving over the soil surface are of great concern to forest managers. Special care is warranted on shallow soils and those that have inherently low infiltration rates. Soils that are wet because of perched water tables or because of their position along drainage ways are unusually susceptible to reduction in porosity and permeability by compaction.

The litter layer beneath a forest cover is especially important in maintaining rapid infiltration rates. This layer not only absorbs several times its own mass of water, but also breaks the impact of raindrops, prevents agitation of the mineral soil particles, and discourages the formation of surface crusts.

The incorporation of organic matter into mineral soils, artificially or by natural means, increases their permeability to water as a result of increased porosity. Forest soils have a high percentage of macropores through which large quantities of water can move—sometimes without appreciable wetting of the soil mass. Most macropores develop from old root channels or from

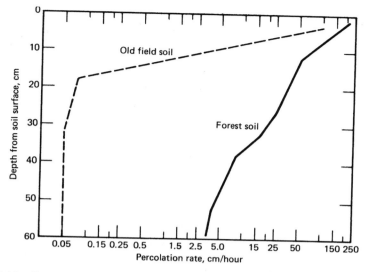

Figure 4.13 Comparative percolation rates by soil depth for a forest soil and an adjacent old-field soil in the South Carolina Piedmont (Hoover, 1949).

burrows and tunnels made by insects, worms, and other animals. Some macropores result from structural pores and cracks in the soil. As a consequence of the better structure, higher organic matter content, and presence of channels, infiltration rates in forest soils are considerably greater than in similar soils subjected to continued cultivation (Figure 4.13). Wood (1977) found that water infiltration rates were higher on 14 of 15 forest sites than on adjacent sites used for pastures or for pineapple or sugarcane crops in Hawaii. In this study, lower bulk densities and greater porosities were found in forest-covered soils than in nonforested soils.

The presence of stones increases the rate of water infiltration in soils because the differences in expansion and contraction between stones and the soil produce channels and macropores. However, stones reduced the retention capacity of soils for 40 Oregon sites (Figure 4.14).

Burning of watersheds supporting certain types of vegetation may result in a well-defined hydrophobic (water-repellent) layer at the soil surface (see Chapter 10), which may increase erosion rates. This heat-induced water repellency results from the vaporization of organic hydrophobic substances at the soil surface during a fire, with subsequent condensation in the cooler soil below.

Evaporation from surface soils increases with increases in the percentage of fine-textured materials, moisture content, and soil temperature and decreases with increases in atmospheric humidity. Evaporation losses are also influenced by wind velocity. Heyward and Barnette (1934) found that the upper soil layers of an unburned longleaf pine forest contained significantly more moisture than

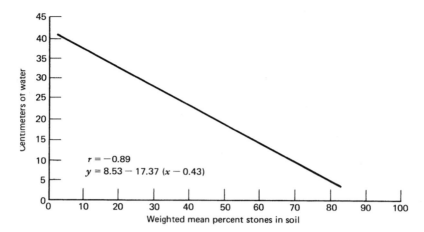

Figure 4.14 Retention storage capacity in the surface 120 cm of 40 soil profiles representing four soil series in Oregon as a function of average stone content (Dyrness, 1969).

did similar soils in a burned forest because of the mulching effect of the O horizon in the unburned areas.

Fine-textured soils have higher retention capacity for water than sands, and can store larger amounts of water following storm events. The effective depth of the soil (rooting depth) and the initial moisture content are factors that also influence the amount of rainwater that can be retained in soils for later use by plants.

The condition of a forest has a strong influence on soil water as a result of differences in water use and water input. Clearcutting a mixed conifer forest in Montana led to substantially wetter soils throughout the year (Figure 4.15). The difference was related to lower water use in the absence of trees but also to greater snow input (Figure 4.16). Snow that accumulates in conifer canopies

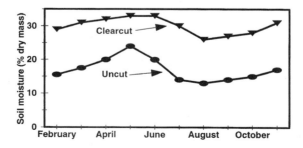

Figure 4.15 Seasonal course of soil moisture in a clearcut and an uncut forest in the mountains of Montana (data from Newman and Schmidt, 1980).

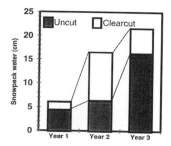

Figure 4.16   The water content of the snowpack across 3 years was substantial higher in clearcut areas than in adjacent mature conifer forests in the mountains of Montana (data from Newman and Schmidt, 1980).

may sublimate without reaching the soil surface, lowering the input of water to the soil relative to clearcut areas, with less sublimation (or interception loss).

Differences in interception loss can also be important among tree species that support substantially different leaf areas. For example, Helvey and Patric (1988) contrasted the interception loss for a watershed dominated by a mixed-age, mixed-species hardwood forest with a watershed planted with a single-age stand of white pine. Average annual interception loss from the mixed-hardwood stand (250 mm/yr) was less than half of the loss from the white pine stand (530 mm/yr). Soil leachates were more concentrated under the pine stand, due in part to the reduced quantity of water leaching through the soils (data in Johnson and Lindberg, 1992).

TREES MAY GET WATER FROM THE CAPILLARY FRINGE

The upper surface of the zone of saturation in a soil is called the groundwater table. Extending upward from the water table is a zone of moist soil known as the **capillary fringe**. In fine-textured soils, this zone of moisture may approach a height of 1 m or more, but in sandy soils it seldom exceeds 25 to 30 cm. In some soils, the height of the water table fluctuates considerably between wet and dry periods. In a forested soil in New England (Lyford, 1964), water tables were highest during late autumn, winter, and early spring; gradually lower in late spring and early summer; and higher again in the autumn. In the flatwoods of the southeastern coastal plain of the United States, the water table may be above the soil surface during wet seasons and fall to a depth of 1 m or more during dry seasons.

Soil depth often reflects the volume of growing space for tree roots above some restricting layer, and in wet areas this restricting layer may be a high water table. The water table may be a temporary (perched) zone of saturation above an impervious layer following periods of high rainfall.

Large seasonal fluctuations in the depth to the water table are harmful to root development for many species. Roots that develop during dry periods in the portion of the profile that is later flooded may be killed under the induced anaerobic conditions. A reasonably high water table is not detrimental to tree growth as long as there is little fluctuation in its level. In an experiment in the southeastern coastal plain of the United States, slash pines grown for 5 years in plots where the water table was maintained at 45 cm by a system of subsurface irrigation and tile drains were 11 percent taller than trees in plots where the water table was maintained at 90 cm, but they were 60 percent taller than trees in plots where the water table was allowed to fluctuate normally (White and Pritchett, 1970).

The primary benefits derived from drainage of wet soils may result more from an increase in the nutrient supply than from an increase in the soil oxygen supply. The improved nutrition is brought about by an increase in the volume of soil available for root exploitation, and by faster mineralization of organic compounds. For example, fertilizer experiments in nutrient-poor wet soils of the coastal plain indicated that tree growth was increased as much by draining as by fertilization (Pritchett and Smith, 1974).

Trees obtain moisture from the water table or the capillary fringe within reach of their deep roots even when the water table is at a considerable depth. Thus the water tables of some wet soils are lowered to a greater extent where trees are grown than where grasses and other shallow rooted plants are grown. Wilde (1958) reviewed a number of reports of general rises in the soil water table following removal of a forest cover from level areas in temperate and cold regions. The increase in the water table level results from a reduction in water losses from evaporation on leaf surfaces (interception loss) and from transpiration of water from within leaves.

TREES REQUIRE GREAT QUANTITIES OF WATER

Tree roots absorb vast quantities of water to replace that lost by transpiration and that used in metabolic activities. Under favorable conditions, this loss can be as much as 6 mm/day (60,000 L/ha) during summer. Trees use 500 to 1000 kg of water for each kilogram of dry matter produced. Tree roots can effectively exploit soil moisture even when the soil moisture content is relatively low. The mycelia of mycorrhizal fungi may play an important role in the extraction of soil, water and nutrients.

In spite of the efficiency of most tree root systems, the distribution of trees is controlled to a great extent by the water supply. Wherever trees grow, their development is limited to some degree by either too little or too much water. Even in relatively humid areas of the temperate zone, forest soils may be recharged to field capacity only during the dormant season or for a brief time after very wet growing periods.

PHYSICS IN A LANDSCAPE CONTEXT CAN DETERMINE SOIL CHEMISTRY

The physical properties described above need to be placed in a landscape context. Most forest soils are affected by their context on landscapes. Ridgetop soils may be excessively well drained. Midslope sites receive water and nutrients from upslope sites but lose some water to downslope sites. Lower-slope positions may receive more water and nutrients than leach away, and in some cases may become saturated and experience periods of low redox potential. These landscape issues are sometimes lumped together under the term topography, and many generalizations may be offered. We stress that landscape position (or topography) is very important in forest soils but that broad generalizations may not be helpful. For example, the sequence of soil moisture supply, nutrient supply, and overall fertility commonly increases from ridges, to midslopes, to toeslopes in relatively young soils on the sides of hills, mountains, and valley sides. Other situations, such as the *Eucalyptus* plantation on an Oxisol in Argentina, show exactly the reverse pattern (Figure 4.1).

Rather than try to establish generalizations about the role of landscapes in soil physics, we present some of the biogeochemistry of a single landscape sequence (catena) for a boreal forest in Sweden. The same biogeochemical factors are integrated in different ways at other sites, but this case study illustrates the sorts of features that emerge in landscape contexts.

The Betsele transect is located in northern Sweden in the valley of the Umeå River (Table 4.4, Figure 4.17). The transect is 100 m long, descending a gentle

TABLE 4.4 Stand and Soil Characteristics Along a 90-m Transect of Soils Developed in Sandy Till in Sweden

Feature	Upper End	Lower End
Dominant tree	Scots pine	Norway spruce
Site index (m at 100 years)	17	28
Basal area (m^2/ha)	22	32
Forest floor morphology	Thin Oa (L), indicating little soil faunal mixing Thick Oe (F) Thin Oi (H)	Thick Oi (H) only, indicating major soil fauna activity
Forest floor solution		
pH	3.8	6.4
NO_3 (μmol/L)	10	160
PO_4 (μmol/L)	300	8
Forest floor Extractable Fe + Al (mg/g)	2	35

Source: Data from Giesler et al. (1998).

Figure 4.17 Endpoints of a landscape sequence from well-drained (upper) soils with Scots pine, Norway spruce, and *Vaccinium* to a discharge area (that receives groundwater inputs) with Norway spruce and a diverse, tall herbaceous layer (photos from Peter Högberg).

(2 percent) slope of soils formed from sandy till deposits. The upper end of the transect is dominated by a 125-year-old stand of Scots pine with an understory of dwarf shrubs (mostly *Vaccinium*). The lower end is dominated by a Norway spruce stand of the same age, with a variety of tall herbs in the understory. The upper end is much less productive (site index of 17 m at 100 years) than the lower end (site index 28 m at 100 years).

The key driver of the differences in soil chemistry and overall soil fertility is soil water. The upper end of the transect is excessively drained; water leaches downward out of the soil. The lower end is in a position to receive the "discharge water" from the upslope ecosystems, providing abundant water supply and changing nutrient cycles. The major biogeochemical implications are that the upper ecosystems acidify and are nitrogen limited, and the lower ecosystems receive abundant alkalinity (acid-neutralizing capacity) from the discharge water and have high nitrogen availability. One surprising feature is the low (and limiting) concentrations of phosphorus at the lower slope position. Here, mixing of the organic litter with mineral soil brings phosphorus into contact with iron and aluminum, allowing sorption and removal of phosphorus from the soil solution (similar to the earthworm story in Paré and Bernier, 1989; see Chapter 9). Alternatively, Giesler et al. (1998) suggest that iron and aluminum enrichment of the forest floor at the toeslope may have resulted from dissolved inputs, particularly iron, which may have high solubility during anaerobic periods under saturated conditions. Overall, the authors conclude that the 3-unit gradient in soil pH was driven primarily by the differences in base saturation (caused by input of alkalinity into discharge water) (see Chapter 5) rather than by differences in acid quantity or strength.

SUMMARY

The physics of soils is fundamental to soil temperature, water relations, chemistry, and the life that depends on the soil. Differences in soil physics among soils, or within a soil over time, have major influences on tree growth. Soil texture refers to the size of mineral particles in the soil, and soil structure concerns the three-dimensional conglomerations of mineral particles and organic matter. The pore space within soils is important in influencing water infiltration, soil atmosphere composition, and ease of penetration by roots. Forest management operations may affect soil structure, especially by altering soil pore space and bulk density. Soil temperature regimes differ by regions and management activities, driving differences in the rates of soil processes (such as decomposition) and in the species composition and growth of forests. Water is held under negative potential within freely draining soils, and the characteristics of water retention and release as a function of water potential determine the availability of water to plants. Forest harvest reduces water losses through interception/evaporation and transpiration, increasing water yield from forest lands and the average soil water content. The physics of soils is often influenced by adjacent areas where upslope ecosystems supply water, chemicals, and sometimes material that affect both the biota and long-term development of soil structure and other properties.

CHAPTER 5

Soil Chemistry and Nutrient Uptake

The chemistry of soils is fascinatingly complex, involving inorganic reactions between solid phases (including minerals, mineral surfaces, and organic matter), the liquid phase (near surfaces and in the bulk soil solution), and an incredible diversity of soil organisms. In this chapter, we focus on the chemistry of the soil solution as the pool of chemicals available for plant use. The soil solution is a dilute soup of dozens of chemicals, with the soup supplied with chemicals from atmospheric inputs and from the solid phases, and depleted by processes of plant uptake and movement into the solid phases (Wolt, 1994). The broader aspects of biogeochemistry determine the amount of flow (or flux) of chemicals through the soil solution on an annual scale. In this chapter, we focus on shorter-term issues of what comprises the soil solution, how the soil solution is buffered by relatively rapid interactions with the solid phases, and how the soil solution provides nutrients for plant uptake.

MAJOR SOIL ANIONS INCLUDE CHLORIDE, SULFATE, BICARBONATE, AND SOMETIMES NITRATE

Soil solutions contain dissolved chemicals, and many of these chemicals carry negative charges (anions) or positive charges (cations). The total charge of anions in forest soils commonly ranges between 100 and 500 $\mu mol_c/L$ (micromoles of charge per liter, or 10^{-6} moles of charge per liter), and maintenance of electroneutrality requires an equivalent concentration of cations. Figure 5.1 illustrates the anion composition of soil solutions from A horizons in six temperate forests. The Norway spruce forest in Norway has a soil solution dominated by chloride and sulfate. The chloride shows a moderate influence of inputs of salt from the ocean, and the sulfate concentrations indicate substantial deposition of sulfur from polluted air. The loblolly pine soil in North Carolina, in the United States, has some chloride, a large amount of sulfate, and some bicarbonate or alkalinity. As described later, soils contain high partial pressures of CO_2 as a by-product of respiration, and the CO_2 dissolves into the soil solution to form carbonic acid. Depending on the pH of the soil solution, some of the carbonic acid may dissociate to form H^+ and bicarbonate (HCO_3^-). Both the Norway spruce and loblolly pine soils have carbonic acid

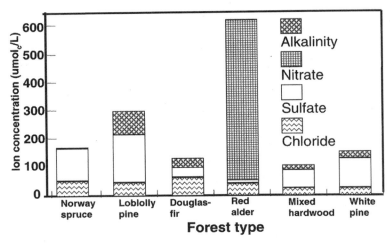

Figure 5.1 Concentrations of anions in solutions collected by lysimeters placed in upper mineral soil from representative temperate forests (data from Johnson and Lindberg, 1992).

in the soil solution, but only the pine soil has a pH level high enough for substantial dissociation to form HCO_3^- anions. Some of the alkalinity of soil solutions comes from soluble anions of organic acids rather than from inorganic HCO_3^-. The Douglas-fir soil solution in Washington State in the United States has almost no nitrate, in contrast to extremely high concentrations of nitrate in the soil solution of an adjacent stand of nitrogen-fixing red alder. The mixed hardwood and white pine soil solutions come from watershed experiments in North Carolina and demonstrate greater deposition of sulfate in the larger conifer canopy. Other minor anions in soil solutions include phosphate and fluoride.

MAJOR SOIL CATIONS INCLUDE SODIUM, POTASSIUM, CALCIUM, MAGNESIUM, AND SOMETIMES ALUMINUM

The cations that balance the charge of the anions interact strongly with the solid phase of the soil, particularly with the cation exchange complex (see later). Sodium is important in some soils (Figure 5.2), particularly near coastlines, where inputs of marine salts are important. Potassium is a key nutrient for plants and is very mobile in soils; most soils have substantial quantities of potassium in solution, but potassium never dominates the total cation suite. Calcium is the dominant cation in most forest soil solutions; the major exceptions are found in soils with pH values below about 4.5, where dissolved aluminum ions become important (such as the Norway spruce soil in Figure 5.2). The increase in nitrate concentrations under red alder results in

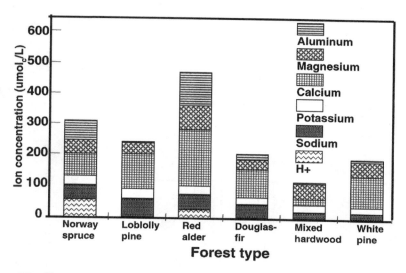

Figure 5.2 Concentrations of cations in solutions collected by lysimeters placed in upper mineral soil from representative temperate forests (data from Johnson and Lindberg, 1992).

increases in cation concentrations, and aluminum responds more dramatically than calcium (see Figure 5.6). Magnesium concentrations are commonly 20 to 50 percent of the calcium concentrations, although some soil types may have more magnesium than calcium (such as the mixed hardwood site in Figure 5.2). Other minor cations in soil solutions include ammonium and a suite of metals including iron, manganese, zinc, and copper.

Soil solutions change in composition over time and with depth in the profile. The white pine soil from Figures 5.1 and 5.2 showed different patterns of change in solution concentrations with depth in the profile. Concentrations of sulfate reached a peak in the upper mineral soil (Figure 5.3), followed by strong reductions in the BC horizon after biologic uptake and adsorption on iron and aluminum sequioxides (see later). Nitrate concentrations were high at the bottom of the forest floor but then dropped to near zero as this limiting nutrient was absorbed from the solution by microbes and plants. Chloride concentrations showed little change with depth, demonstrating the unreactivity of chloride with minerals and biota. Some biological cycling of chloride is indicated by the match in upper soil and lower soil concentrations because the water percolating to the lower depth was reduced almost by half as a result of plant water uptake from the upper soil. Alkalinity showed the reverse pattern, generally increasing with depth. Soil solution cations generally decreased with depth, indicating that geochemical and biological processes remove these ions from the soil solution before it percolates very far into the soil.

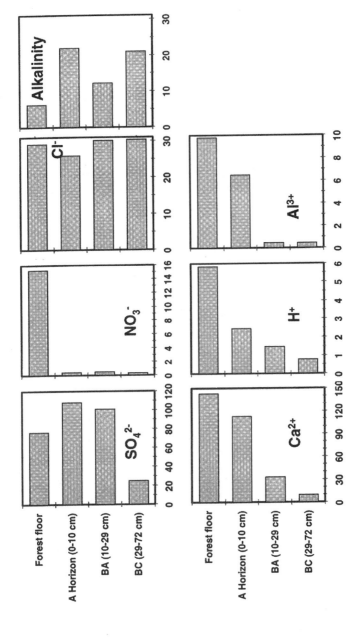

Figure 5.3 Soil solution chemistry profiles (all units are $\mu mol_c/L$) for a white pine soil at the Coweeta Hydrologic Laboratory in North Carolina (data from Johnson and Lindberg, 1992).

SOIL SOLUTIONS ALSO CONTAIN SILICIC ACID, DISSOLVED ORGANIC CHEMICALS, AND GASES

Most forest soil solutions also contain nonionized chemicals. Silicic acid (sometimes called silica) is a by-product of weathering of many primary minerals, and concentrations of silicic acid in forest soil solutions are commonly on the order of 1 to 10 mg/L. Silicic acid does not dissociate at normal pH levels in forest soils, and it is relatively soluble. Therefore, silicic acid formed in soils mostly leaches out of the soil profile.

Dissolved organic compounds in soil solutions range from simple sugars to large, complex fulvic acids. The six soil solutions illustrated in Figures 5.1 and 5.2 ranged from about 300 to 2000 µmol/L of dissolved organic carbon (assuming a carbon:nitrogen ratio of 20; Johnson and Lindberg, 1992). The organic carbon concentration of soil solutions is commonly about the same as that of the inorganic ions (on the order of 1 mole of carbon for 1 mole of charge of ions).

Soil solutions exchange gases with the soil atmosphere, and the soil solution commonly contains <1 mg/L of CO_2 and 10 mg/L of oxygen. Depletion of oxygen leads to anaerobic conditions, changing many biogeochemical reactions of the soil system.

The importance of low-molecular-weight, soluble organic acids in soil solution chemistry has not been explored thoroughly. Plant roots and microbes excrete organic acids into soils, and organic acids leach from forest litter. Major acids include acetic, formic, oxalic, malic, citric, and shikkimic acids (Fox, 1995; Kryszowska et al., 1996). Some mat-forming mycorrhizae accumulate over 800 kg/ha of calcium oxalate (Sollins et al., 1981). The concentration of malic, succinic, and lactic acids in soil solutions in an oak/pine forest in Alabama was 0.5 mmol/L (Hue et al., 1986), and the concentration of oxalic acid in the rhizosphere of slash pine exceeded 1 g/kg soil (Fox and Comerford, 1990). Leachates from spruce forest floor in Maine averaged 10 µmol/L of acetic acid, 0.7 µmol/L of formic acid, and 3.3 µmol/L of oxalic acid (Kryszowska et al., 1996). The sum of these three acids comprised a little more than 2 percent of the dissolved carbon in the soil solution.

Some of these acids chelate metals, increasing the solubility of micronutrients such as iron and zinc and reducing the toxicity of monomeric Al^{3+}. The acids may also increase phosphorus availability by chelating the iron and aluminum that was bound to phosphorus (see later). The phosphorus sorption sites may also be reduced through the accumulation of low-molecular-weight organic acids.

Aluminum biogeochemistry is heavily influenced by soluble organic matter. The classic model of soil podzolization involves binding of aluminum (and iron) by soluble organic matter in the upper soil, followed by precipitation in the B horizon. For example, Driscoll et al. (1985) found that the concentration of monomeric aluminum in three Spodosols ranged from about 5 to 30 µmol/L, and 65 to 100 percent of the aluminum was bound with soluble organic matter.

This binding of aluminum with organic matter may be particularly important in studies that examine potential aluminum toxicity in relation to acid deposition; organically bound aluminum is nontoxic, and assays of total monomeric aluminum in soil solutions may greatly overestimate the concentration of toxic inorganic, monomeric aluminum.

SOLUTION CHEMISTRY IS REGULATED BY INPUTS, OUTPUTS, AND REVERSIBLE REACTIONS

The variety of patterns in soil solution chemistry derives from chemical reactions and interactions between geochemical and biological components of soils (Figure 5.4). This chapter deals primarily with the reversible (sometimes called equilibrium) reactions that influence the soil solution (the bottom half of Figure 5.4). The fluxes into and out of the soil solution are covered primarily in Chapter 9.

THE SOLID PHASE OF THE SOIL HAS FOUR MAJOR COMPONENTS

The solid phase of the soil can be viewed as consisting of four functional components: the mineral phase, the readily reversible exchange complex, the specific ion adsorption complex, and organic matter. The release of ions from primary or secondary minerals is called weathering, and this process accounts for the initial sources of most of the calcium, magnesium, potassium, and

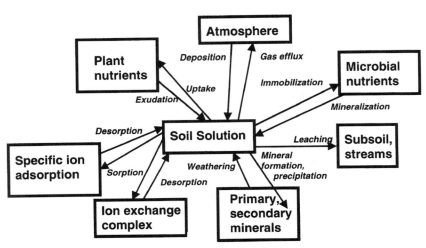

Figure 5.4   Major pools of soil chemicals and processes that transfer chemicals to and from the soil solution.

phosphate cycling in most forests. Weathering processes are strongly influenced by biotic activity, including production of H^+, low-molecular-weight organic acids that complex with ions from soil minerals, and complex organic acids in leachates of decomposition organic matter. Rates of mineral weathering are discussed in Chapter 9; here we focus on the processes.

The two primary processes that degrade soil minerals are:

1. Acid hydrolysis, in which H^+ displaces other ions from minerals, often forming silicic acid and bicarbonate salts with sodium, potassium, calcium, and magnesium.

2. Organic complexation, in which complexing anions (such as oxalate) have a very strong affinity for atoms that comprise the mineral (such as calcium).

Rock surfaces weather very slowly, on the order of 1 mm in 200,000 years for basalt (Colman and Dethier, 1986). In forest soils, interest in mineral weathering focuses on small particles of primary and secondary minerals that weather much faster than rocks. The dissolution and removal of soil minerals across the landscape of the U.S. Pacific Northwest averages about 0.3 Mg/ha annually (Dethier, 1986). In situ weathering of minerals to form sapprolite proceeds at a rate of about 1 m per 30,000 years of soil development (Colman and Dethier, 1986).

The weatherability of major classes of igneous minerals follows a reverse sequence of the Bowen reaction series of progressive crystallization as magma cools (Goldich, 1938). Olivene is one of the first minerals to crystallize as magma cools; it has a high energy of formation and weathers easily. Moving down the sequence of crystal formation in the direction of lower weatherability, the sequence progresses from olivene to augite to biotite mica to potassium feldspar, muscovite mica, and quartz. The sequence of decreasing weatherability for feldspars goes from calcium plagioclase to calcium-sodium plagioclase to sodium plagioclase to potassium orthoclase, mica, and quartz.

In addition to high energies of formation, the feldspars that crystallize at higher temperatures have tetrahedra that are half aluminum and half silicon. At temperatures on the Earth's surface, aluminum is more stable in octahedral structures; the aluminum charge in feldspar tetrahedra is balanced in part by Ca^{2+} ions between the tetrahedra, weakening the mineral structure (Bohn et al., 1985). More resistant feldspars contain only one-quarter aluminum, and the single available charge is met with monovalent Na^+ or K^+. Potassium feldspars are more stable than sodium feldspars because K^+ fits better between the tetrahedra. Finally, the degree of linkage of tetrahedra also influences weatherability; a high degree of linkage is more stable.

Another weathering sequence lists the minerals that characterize soils in various stages of weathering. The Jackson-Sherman stages start with gypsum, carbonates, olivine/pyroxene/amphibole, and feldspars in relatively unweathered soils (Sposito, 1989). Intermediate weathering stages characteristically have quartz, illite, vermicullite, or smectites. Advanced stages have kaolinite, gibbsite, and iron and titanium oxides.

The weathering of silica minerals includes alteration of the original structure (such as loss of potassium and gain of silica by muscovite to become mica) or complete dissolution and recrystallization (such as formation of kaolinite from dissolved silica and aluminum). If all the products of weathering remain in solution, the weathering reaction is congruent; if some of the products are not soluble, the reaction is incongruent. For example, the weathering of albite is congruous when aluminum ions and silicic acid are formed:

$$NaAlSi_3O_8 + 6H_2O + 2H^+ \rightarrow Na^+ + Al(OH)_2^+ + 3Si(OH)_4 + 2H_2O$$

Alternatively, precipitation of gibbsite from the weathering of albite is an incongruent reaction:

$$NaAlSi_3O_8 + 6H_2O + H^+ \rightarrow Al(OH)_3 + Na^+ + 3Si(OH)_4 + H_2O$$

Biotic effects on mineral weathering are most apparent in rhizosphere soils, where organic complexation and simple reductions in soil pH may drive mineral dissolution. In a Norway spruce plantation in Sweden, rhizosphere soil was sampled by shaking off the soil that adhered to fine roots (Courchesne and Gobran, 1997). X-ray diffraction showed that this rhizosphere soil had substantially less amphibole and expandable phyllosilicates than were present in the bulk soil; no differences were found between the soil types for potassium feldspars. The rhizosphere soil also had higher concentrations of oxalate-extractable aluminum and iron. These differences in mineral weathering were associated with much greater organic matter concentrations in the rhizosphere soil and with a fivefold greater concentration of exchangeable base cations (Gobran and Clegg, 1996). A great deal of research is needed to provide a better understanding of the differences between bulk soils and those that are intimately affected by tree roots ad mycorrhizae.

The solubility of minerals may regulate the concentrations of ions in solution. For example, the concentration of phosphate in soil solutions depends on the solubility of calcium phosphates, iron phosphates, or aluminum phosphates. Above pH 6.5, calcium-phosphate compounds are the least soluble form in most soils (Figure 5.5), and phosphorus solubility increases with further increases in pH. At lower pH levels, aluminum-phosphate compounds dominate, and phosphorus solubility drops by one order of magnitude for each one unit decline in pH. If calcium phosphate (from apatite minerals or fertilizers) is present in low pH soils, phosphorus tends to be released from calcium into the soil solution, followed by precipitation (or specific adsorption; see below) by aluminum.

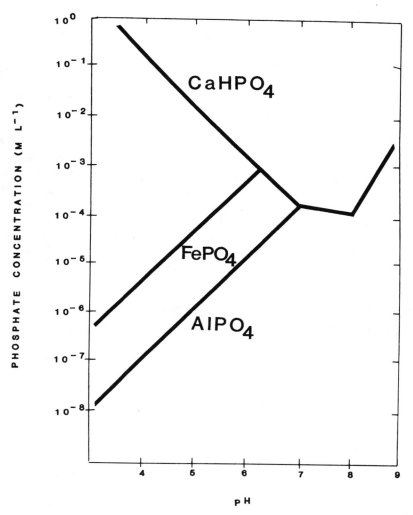

Figure 5.5 Solubility of phosphate depends on the availability of calcium, iron, aluminum, and soil pH (after Lindsay and Vlek, 1977; reproduced from *Minerals in Soil Environments*, 1977, page 658; used by permission of the Soil Science Society of America).

CATIONS EXCHANGE IS A REVERSIBLE, ELECTROSTATIC SORPTION

Soil textbooks typically describe two types of positive soil charges that can retain cations: permanent charge that derives from isomorphic substitution of ions within minerals and pH-dependent charge that increases as pH rises. The key difference between these two types of soil charge results from the inability of soil clays to retain adsorbed H^+. If a clay is saturated with H^+ in a laboratory, most of the H^+ is rapidly replaced by aluminum ions in a matter of hours (Bohn et al., 1985). Therefore, the ions on permanent charge sites can be displaced by concentrated salt solutions. Concentrated salt solutions cannot displace H^+ from soil organic matter, so the proportion of these sites available for cation exchange varies with soil pH.

Some positive charges can develop at low soil pH, providing some electrostatic anion exchange capacity for anions. If the pH is low enough, the size of the salt-exchangeable positive and negative charges may be equal. When the anion exchange capacity and cation exchange capacity are equal, the pH level is called the point of zero charge (PZC). At first glance, this might suggest that no ions will be retained on the soil surfaces because the charges cancel one another. However, the charges on soil colloids are not adjacent to each other and cannot dissolve and move, so the PZC point is important only for issues of flocculation (soils with a pH at the PZC flocculate and behave as uncharged colloids; Sposito, 1989), not for ion exchange properties.

CATION VALENCE AND HYDRATED RADIUS EXPLAIN THE SELECTIVITY OF THE EXCHANGE COMPLEX

In electrostatic cation exchange, smaller cations with higher valence (charge per atom) are held more tightly to exchanges sites than are larger cations with lower valence. Accounting for hydrated radius and valence leads to the lyotropic series of cations in order of ease of replacement by ions further along in the series (Bohn et al., 1985):

$$Na^+ > K^+ \sim NH_4^+ > Mg^{2+} > Ca^{2+} > Al^{3+}$$

Given an exchange complex with a distribution of adsorbed cations, what determines the relative concentrations of ions supported in soil solutions? Two features are important. The total concentration of anions is matched by the total concentration of cations, so changes in the ionic strength of solutions lead to different quantities of cations in solution. The distribution of individual cation species is commonly described by selectivity coefficients that are analogous to the equilibrium constant of chemical reactions.

A variety of algebraic details have been used to characterize the selectivity coefficients of soil exchange reactions. These selectivity coefficients are at the heart of most simulation models that attempt to model the changes in soil solution as a function of the total ionic strength of the soil solution (see Reuss and Johnson, 1986, for a helpful overview).

Consider the replacement of Ca^{2+} on the exchange complex (represented CaX) by Al^{3+}:

$$CaX + 2/3\,Al^{3+} \rightarrow 2/3\,AlX + Ca^{2+}$$

The equilibrium expression for this reaction (using the Kerr approach to selectivity coefficients) is:

$$K_s = \frac{[AlX]^{2/3}[Ca^{2+}]}{[CaX][Al^{3+}]^{2/3}}$$

The form of equilibrium constants is the concentration of the products multiplied in the numerator (with each species raised to the power of the number of molecules in the reaction) and the concentrations of the reactants multiplied (again raised to the power of the number of molecules) in the denominator.

This equation can be rearranged to describe the ratios of the concentrations on the exchanger and in solution:

$$\frac{[AlX]^2}{[CaX]^3} = K_s \frac{[Al^{3+}]^2}{[Ca^{2+}]^3}$$

(both sides of the equation were cubed to eliminate the fractional powers). The ratio of exchangeable aluminum to calcium will not change substantially over short periods of time, giving

$$[Al^{3+}] = Kt[Ca^{2+}]^{3/2}$$

(where Kt is the square root of the right side of the equation above). The significance of these equations is that the changes in the concentration of Al^{3+} should be to the 3/2 power of the changes in concentration of Ca^{2+}.

How well does this description work in real soils? Seasonal changes in nitrate concentrations in soil solutions varied from 10 to 2500 $\mu mol_c/L$ in a red alder soil solution in Washington (Reuss, 1989). This wide range of nitrate concentrations led to huge variations in soil solution cations. The equations above predict that the change in the concentration of Al^{3+} would be the 3/2 power ($= 1.5$ power) of the change in the concentration of Ca^{2+}. The actual relationship is shown in Figure 5.6, where the Al^{3+} concentrations graphed as a function of the Ca^{2+} concentrations showed that the best fit was with an exponent of 1.76, a bit steeper than the predicted 1.5.

THE RATIO OF ALUMINUM TO BASE CATIONS MAY (OR MAY NOT) BE IMPORTANT

Many studies have shown that some agricultural crops are sensitive not only to the total concentration of aluminum in soil solution, but also to the ratio of aluminum to base cations. Higher concentrations of aluminum may be tolerable if the concentrations of base cations are also high (Marschner, 1995). This concept has been extended to forest soils in an attempt to understand the

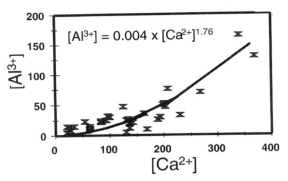

Figure 5.6 Changes in solution nitrate concentrations provide two-order-of-magnitude changes in concentrations of aluminum and calcium. Theory predicted that aluminum would increase to the 3/2 power of calcium; the actual relationship was a bit steeper (1.76, $r^2 = 0.83$; reworked from data of Reuss, 1989).

possible impacts of soil acidification from acid deposition. As discussed in Chapter 16, any value of the concept of aluminum:base cations has been discussed but not clearly established in forest soils.

SOIL ACIDITY INVOLVES ACID-BASE REACTIONS BETWEEN SOLID AND SOLUTION PHASES

The dynamics of soil cations on the exchange complex and in the soil solution also depend on the acidity of the soil. The hydrogen ions in solution influence exchange reactions, and the hydrogen ion concentrations, in turn, are influenced by the distribution of cations on the exchange complex.

Hydrogen ions are free, reactive protons that are involved in many chemical reactions, particularly hydrolysis (the gain or release of H^+) and redox reactions. Hydrogen ion concentrations span several orders of magnitude in soils and biological systems. A logarithmic scale is used to describe H^+ activities, expressed as pH:

$$pH = -\log A_{H^+}$$

where A_{H^+} is the activity of H^+. Under oxidized soil conditions, common soil ranges for pH are 3.5 (very acidic) to 8 (mildly alkaline; Lindsay, 1979). Soils that resists changes in pH as H^+ is added or removed are referred to as well-buffered soils. In forest soils, pH tends to be <4.0 only in organic horizons that lack aluminum (or iron). In soils where aluminum dominates the exchange complex, pH values commonly fall between 4.0 and 4.5. Exchange complexes dominated by so-called base cations (such as calcium, magnesium, and potassium) maintain pH levels of 5.0 to 6.5, and soils with high carbonate contents (such as those derived from limestones) may have pH values higher than 6.5.

Despite all the attention that soil pH has received in forest soils (including our discussion here), pH seems to have limited practical importance in relation to tree nutrition and growth. Some severe pH conditions can be very problematic for trees. Drainage of wetland soils can lead to oxidation of reduced sulfur compounds, driving the pH below 3.0 and making tree growth (and survival) low. High pH is almost never a problem in forests except on dry sites, where high pH and high salt concentrations combine to impair tree growth. Finally, some tree species are well adapted to soils with near neutral pH (pH 5.5 to 7.0); others grow well only in acidic soils (pH < 5.5), and most trees do well across the full range of common pH values. A decline in pH of 1 unit in response to natural processes or acid deposition has never been shown to affect tree growth. Increases in pH following forest harvesting or applications of lime commonly have no effect on growth, sometimes decreasing it and sometimes increasing it.

If pH is rarely of great relevance to tree growth, why has so much effort been invested in measuring and investigating it? Part of the answer comes from agricultural practices because food plants are often more sensitive to soil pH than in forests. Soil pH is also very easy to measure, and many researchers hope that an easy measure of a factor that is affected by a variety of soil processes might have more power of interpretation. Finally, some scientists expect that more insight can be obtained by understanding pH more thoroughly than at present.

THE pH OF SOIL SOLUTIONS IS BUFFERED BY THE SOLID PHASES

The pH of soils is important for a variety of reasons, including the solubility of aluminum (which is toxic to many plants and organisms), the weathering of minerals, and the distribution of cations on the exchange complex.

The concentration of H^+ in a soil with pH 5 is about 10^{-5} mol/L or 10 μmol/L. At pH 4 the concentration would increase to about 100 μmol/L, and at pH 6 it would decline to just 1 μmol/L. As noted above, the sum of the concentrations of all cations in forest soil solutions tends to be 100 to 500 μmol_c of charge per liter (μmol_c/L). Hydrogen ions typically comprise a small fraction of most forest soil solutions, but these low fractions are matched by strong effects.

What determines the pH of forest soils? Soil pH depends on the equilibrium between the soil solution and the solid phase of the soil. This buffering can be viewed as cation exchange, but the chemistry is more akin to acid/base reactions. The solid phase is a mosaic of weak acids, including clays (smectite, illite, kaolinite) and organic matter. Processes that remove H^+ from the soil, such as mineral weathering, essentially "titrate" the weak acids and make them more dissociated. Highly dissociated acids maintain higher pH values in solutions than undissociated acids. In soils terminology, the degree of acid dissociation is gauged as the base saturation, which is the proportion of the exchange complex dominated by base cations. A soil that has a base saturation of 50 percent would have half of the exchange complex sorbing calcium, magnesium, and potassium and the other half sorbing aluminum and H^+.

Aluminum is included as an acid cation in soils terminology because each monomeric aluminum atom is associated with six water molecules. Depending on the pH of the soil solution, some of these water molecules dissociate and provide H^+ to the soil solution. Under acidic conditions below pH 5, all six water molecules retain their H^+. Between pH 5 and 9, one water molecule loses a H^+, and above pH 9 an additional H^+ is lost. This ability of hydrated aluminum ions to denote and accept H^+ would make it eligible to be an acid or a base, depending on the situation. A further reason may be that exchangeable aluminum is present in forest soils only under acidic conditions; at higher pH values, aluminum oxides are too insoluble to participate in the soil solution in any substantial way.

The total quantity of acid present in the soil solid phase is commonly on the order of $1000\,kmol_c/ha$ or more, vastly exceeding the minute quantities present in soil solution at any one time (commonly $<0.1\,kmol_c/ha$). The balance between the solid and solution phases depends on:

1. The salt concentration of the soil solution
2. The total quantity of acids present
3. The degree of dissocation of the soil acids
4. The acid strength of the solid phase

The acid-base behavior of forest soils can be described in terms that are used to describe acid-base reactions in laboratories. It is important to note that the actual suite of reactions in soils encompasses a broad array of processes, including protonation/deprotonation of organic acids, titration of hydro-aluminum complexes, and even mineral weathering. The rates of some of these processes may be slow enough that equilibrium conditions do not apply. However, the general context of laboratory acid-base systems and forest soils is similar enough to provide a solid picture of how soil pH differs among soils and across time.

Factor 1: Increases in Salt Concentration Decrease pH

If a soil solution had no ions in solution, the pH (7) would be determined simply by the tendency of water to dissociate into H^+ and OH^-. Forest soil solutions contain substantial amounts of ions, commonly 100 to 500 $\mu mol_c/L$. These ions equilibrate with readily exchangeable cations, and the final mixture includes H^+ (at least in acidic soils). If the total concentration of anions in the soil solution increases, the increase in cation concentration will include some extra H^+, and the pH will decrease. This effect of salt concentration is sometimes called the salt effect in the forest soils literature because additions of any salt (such as NaCl or KNO_3^-) will lower the pH of a soil solution even though the solid phase itself has not been altered substantially. The salt concentration effect is commonly on the order of 0.1 to 0.4 pH unit, with the greatest variations occurring in areas where inputs of sea salt are variable, and in soils where accumulations of nitrate are large (and variable). For example, Richter et al. (1988) found that very dilute soil solutions (conductivity of soil solution of 2 mS/m) declined by about 0.3 units in pH when measured in a stronger salt solution ($0.01 M$ $CaCl_2$), whereas soils with higher initial salt concentration (conductivity > 5 mS/m) showed no further effect of added salt on pH.

The same effect of salt concentration is apparent from field studies in areas where roofs have been built over small forest areas to block acidic deposition. The ambient deposition in the Solling area of Germany contained about 500 μmol_c/L of cations and of anions. Under the rainfall exclusion roof, very clean water (about 1 μmol_c/L) was used to replace rainwater (Bredemeier et al., 1995). Within 6 months, the electrical conductivity of the soil solution (a measure of the total salt concentration) dropped from 26 mS/m to 6 mS/m and the pH of the soil rose from 3.7 to 4.1. The rise in pH resulted from the lower salt concentration in the soil under the roof, not from any major long-term change in the soil exchange complex.

Factor 2: Soil pH Decreases as Total Soil Acidity Increases

The quantity of acid that constitutes the solid phase commonly differs by severalfold among soil types and may change by 10 to 50 percent within a soil as a result of changes in soil organic matter. Clays commonly have a charge of 50 $mmol_c$/kg (kaolinite) to 1000 $mmol_c$/kg (smectite) and even 3000 $mmol_c$/kg (allophane). Soil organic matter contains 1000 to 5000 $mmol_c$/kg. Given that soils also contain sand and silt with very little charge, the average total charge density in soils is commonly 20 to 100 $mmol_c$/kg.

How does acid quantity affect the pH of soils? For soluble acids such as acetic acid, the pH is directly determined by the equilibrium constant for the acid dissociation:

$$1.8 \times 10^{-5} = \frac{[H^+][CH_3COO]}{[CH_3COOH]}$$

When no other chemicals affect this equilibrium, the equilibrium concentration of H^+ will be 1.3×10^{-4} mol/L, which equals pH 3.9. If the total concentration of acetic acid were only 0.1 mmol/L, the pH would be 4.4. This simple picture does not apply to soils, where the acid anion is part of the solid phase rather than free to dissociate in solution. Therefore, the effect of acid quantity on soil pH derives from the combined effect of total acid quantity on influencing the balance between acid neutralizing capacity (ANC) and base neutralizing capacity (BNC), which is factor 3 below.

Factor 3: The Distributions of Cations Is Often Called Base Saturation

The degree of dissociation of a weak acid depends on the nature of the acid, but also on the extent to which it has been titrated by addition of H^+ (or OH^-). The pH maintained by a weak acid will increase as the acid is titrated to become more dissociated (Figure 5.7). The equivalent of this phenomenon in soils is the degree to which the solid phase acid (the exchange complex) is protonated (acid saturated) or deprotonated (base saturated). An exchange complex of a soil is dominated by H^+ (or Al^{3+}), maintains a low soil pH, and progressive replacement of the H^+ by so-called base cations leads to the maintenance of a high pH.

In soil science, the degree of dissociation of the exchange complex is commonly referred to as base saturation:

$$\text{Base saturation} = \frac{[Ca^{2+}] + [Mg^{2+}] + [K^+]}{[H^+] + [Al^{x+}] + [Ca^{2+}] + [Mg^{2+}] + [K^+]}$$

This is roughly equivalent to the dissociation of the solid phase exchange complex. A methodological complication arises in equating base saturation with the degree of dissociation of the exchange complex. Much of the H^+ is not removed from the exchange complex when solutions of strong salts are added; these cations are removed only by titration of the soil with a strong base. Various methods and definitions have been developed to deal the salt-exchangeable acid:

1. Soil pH in water: H^+ in soil solution.
2. Soil pH in $0.01 M$ $CaCl_2$: H^+ in a soil solution with total salt concentration buffered at $0.01\ M$.

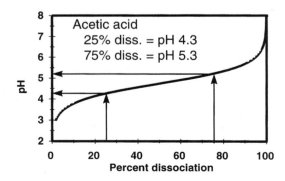

Figure 5.7 Titration of acetic acid with a base (such as OH^-) dissociates or deprotonates the acid, and the dissociated acid maintains a higher pH.

3. Exchangeable acidity: H^+ displaced from exchange complex in 1 M salt solution.

4. Titratable acidity: H^+ that can be consumed by addition of a strong base up to a defined endpoint (often pH 8.2 based on carbonate equilibrium). Titratable acidity is also called base-neutralizing capacity (BNC). Soils also have an acid neutralizing capacity (ANC), where H^+ can replace so-called base cations on the exchange complex.

The classic use of base saturation deals with exchangeable acidity, which is only a subset of the total acidity of the solid phase.

As noted above, the pH of a solution containing a weak acid depends on the degree of dissociation of the acid. Similarly, the pH of a soil is often related to the soil base saturation, as base saturation is a measure of acid dissociation. Soil pH in 16 stands (on a single soil type) in Hawaii ranged from 4.3 to 4.9 (Rhoades and Binkley, 1996). The relationship between base saturation and pH was strong, with an r^2 of 0.90. When the dissociation of the soil acids was expressed in terms of BNC (titratable acidity), the relationship improved significantly to 0.99.

The degree of base saturation of the soil acids can be reduced by removing base cations, which is a scenario for the effects of acid deposition on soil acidity. Added sulfate and nitrate may leach away base cations such as potassium and calcium, lowering the base saturation of the exchange complex and lowering the pH of the soil solution. A second approach to changing the degree of dissociation of soil acids involves adding (or removing) soil organic matter. In acidic soils, most of the charge capacity of organic matter is protonated. Addition of soil organic matter to a soil without commensurate additions of base cations will lead to an overall decrease in acid dissociation and a reduction in soil solution pH (as happened in the alder and Douglas-fir case discussed below).

Factor 4: The Strength of Soil Acids also Influences pH

The strength of weak acids is characterized by equilibrium constants (K) representing the tendency of the acid to dissociate. The equilibrium constants are often converted to pK values (the negative of the logarithm of K). An acid that is half dissociated will have a pH that matches its pK. Acids that dissociate easily have a low pK and maintain low solution pH levels. For example, carbonic acid is weaker than acetic acid, maintaining lower solution pH for the same concentration and level of dissociation (Figure 5.8).

In soils, the acid strength varies greatly among the solid-phase acids, with kaolinite and vermiculite being weaker than smectites. In other words, a kaolinite clay that is half saturated with base cations and half with aluminum (with some H^+) will sustain a solution pH value that is higher than that maintained by a smectite clay under the same conditions.

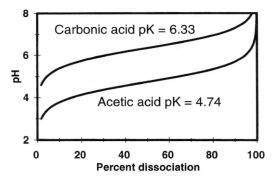

Figure 5.8 Carbonic acid is weaker than acetic acid, and a 1 mmol/L solution maintains a pH about 1.5 units higher for the same acid quantity and degree of dissociation.

Acid strength appears to be a key feature in determining the effects of some tree species on soil pH. In a replicated set of plantations of Norway spruce, white pine, and green ash, the pH was 3.8 under spruce, 4.2 under pine, and 4.6 under ash (Binkley and Valentine, 1991). The primary cause of the lower pH under spruce was the greater acid strength of the soil. Lower base saturation (22 percent under spruce, 42 percent under pine, 52 percent under ash) played a secondary role.

COMPARISONS OF TITRATION CURVES CAN EXPLAIN WHY pH DIFFERS AMONG SOILS

If soil pH differs among sites, or within a site over time, which factors account for the differences? An empirical approach uses titrations of soils to determine the relative importance of factors that determine differences in soil pH among sites or over time. Comparisons of pH in water and in dilute salt (such as 0.01 M $CaCl_2$) identify any role of differences in the ionic strength of the soil solutions. Titrations are then used to characterize the relationship between pH and the degree of dissociation of the solid phase acids. Additions of base identify the BNC (titratable acidity), and additions of strong acid identify the ANC. The titration curves may differ among soils as a result of greater acid quantity (ANC + BNC). The curves may show the same acid quantity, but the degree of dissociation [position along the curve, ANC/(ANC + BNC)] may differ. Finally, the quantity of acid and the degree of dissociation may be the same, but the strength of the solid-phase acids may differ, giving different pH levels in the soil suspension (see Valentine and Binkley, 1992, for methodological details). This empirical approach identifies the major factor(s) responsible for differences in pH, but the estimated magnitude of the contribution of each factor is only approximate. Variations in the titration (such as length of time

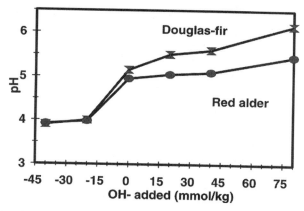

Figure 5.9   Titration curves for soils from adjacent stands of Douglas-fir and red alder (data from Van Miegroet et al., 1989). Both species had similar ANC to pH 4, representing the same level of exchangeable base cations. Alder soil had twice the BNC to pH 5.5, indicating much larger pools of soil acids.

and which ion saturates the exchange complex) alter the specific calculations.

For example, soil pH (0–7 cm in depth) was 5.0 in a 50-year-old stand of red alder compared with 5.2 for a matched adjacent plantation of Douglas-fir (Van Miegroet et al., 1989). Titrations of the soils showed the same ANC (about the same quantity of base cations) but much larger BNC (acid quantity) under alder (Figure 5.9). The dissociation of the acid complex was 41 percent in Douglas-fir soil but only 29 percent in red alder soil.

If the alder soil had the same total quantify of acid as the Douglas-fir soil, its pH would have been higher (5.2). If the acid strength of the alder soil were adjusted to match that of the Douglas-fir soil, the pH would also have risen to about 5.2. If the acid quantity in the Douglas-fir soil were increased to match that of the alder soil, the pH would have dropped from 5.2 to about 4.8. If the acid quantity in the Douglas-fir remained the same but the strength matched that of the red alder soil, the pH would have dropped to 5.0. These calculations indicate that the most important difference between these soils was the quantity of acid in the soil (which altered the degree of dissociation), with the greater acid strength playing an additional role. The studies that have examined the factors that account for differences in pH among soils have found that acid quantity, dissociation, and strength all play a role in different cases (Table 5.1).

SOIL pH INCREASES FROM UPPER-SLOPE TO LOWER-SLOPE SITES

Most studies that have examined soil pH from ridges to toeslopes have found that pH increases. The major factor driving this pattern is the downslope

TABLE 5.1 Contributions of Soil Acid Factors in Determining Differences in pH Among Species and Over Time

Comparison	Difference in pH	Relative Ranking of Factors	Reference
Loblolly pine soil across 20 years	pH declined from 4.5 to 3.9	Dissociation > strength	Binkley et al. (1989)
50-year-old Norway spruce vs. white pine	Spruce pH 3.8 Pine pH 4.2	Strength > dissociation	Binkley and Valentine
50-year-old Norway spruce vs. green ash	Spruce pH 3.8 Ash pH 4.6	Strength > dissociation	(1991)
50-year-old white pine vs. green ash	Pine pH 4.2 Ash pH 4.6	Dissociation > strength	
8-year-old *Eucalyptus* vs. *Albizia*	*Eucalyptus* pH 4.5 *Albizia* pH 4.3	Dissociation > strength	Rhoades and Binkley (1995)
55-year-old conifer–alder vs. conifer, rich site	Conifer pH 3.9 Alder-conifer pH 3.6	Dissociation > strength	Binkley and Sollins (1992)
55-year-old Douglas-fir–alder vs. Douglas-fir	Douglas-fir pH 4.0 Alder–Douglas-fir pH 4.0	Increased dissociation under alder offset by increased acid strength	
50-year-old Douglas-fir vs. red alder	Douglas-fir pH 5.2 Red alder pH 5.0	Dissociation (owing to increased quantity) > strength	Calculated from Van Miegroet et al. (1989)

movement of bicarbonate (HCO_3^-). High concentrations of CO_2 dissolve in the water in forest soils to form carbonic acid (H_2CO_3), which dissociates to H^+ and HCO_3^-. The H^+ tends to react within the profile, whereas the HCO_3^- leaches downslope. The incoming HCO_3^- may then consume H^+ in the downslope soils, leading to higher soil pH levels.

PHOSPHATE AND SULFATE CONCENTRATIONS DEPEND ON SPECIFIC ADSORPTION

Phosphate and sulfate may be adsorbed electrostatically on positively charged sites that develop on the edges of some clay lattices or organic molecules. A more important type of sorption for these anions is called specific adsorption on iron and aluminum oxides. This ligand exchange involves substitution of phosphate or sulfate for one or more molecule of water (or OH^-; Figure 5.10).

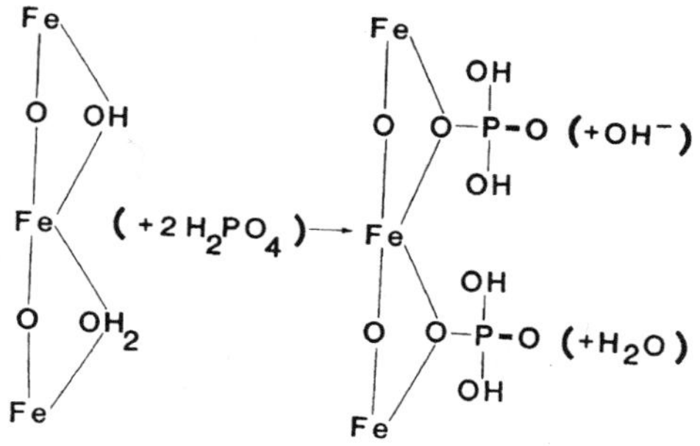

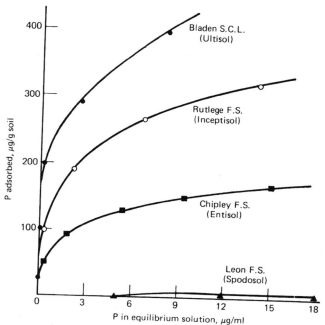

Figure 5.10 Specific sorption of phosphate by iron sesquioxides (sesqui- means 3/2, referring to the number of oxygen atoms per iron atom) may release OH^- or H_2O to the soil solution.

The overall sorption of phosphate or sulfate by soils may be gauged from sorption isotherms. These curves are developed by equilibrating soil samples (for a specified time) with various levels of phosphate or sulfate in solution. The disappearance of the ion from solution is the amount adsorbed by the soil (Figure 5.11). Some of the phosphate or sulfate may actually be immobilized by the soil biota in nonsterile soils.

Figure 5.11 Phosphate adsorption isotherms for four soils from the southeastern Coastal Plain of the United States, representing four soil orders (S.C.L. = sandy clay loam; F.S. = fine sand). Reproduced from *Soil Science Society of America Proceedings* 38, #2 (1974): 252, by permission of the Soil Science Society of America.

Sorption of phosphate and sulfate is often related positively to sesquioxide concentrations and negatively to organic matter concentrations. For example, in the 0- to 15-cm depth, 14 Oxisols in Queensland had a range of sesquioxides from 14 to 30 percent by soil weight, and phosphate sorption capacity ranged from 0.05 to 0.2 mg/g (Oades et al., 1989). The organic matter concentration declined below the 15-cm depth. The increase in phosphorus sorption capacity with increasing sesquioxide concentration was much greater, ranging from 0.05 to 1.2 mg/g soil for the same range of sesquioxide concentrations.

Sulfate adsorption also relates to soil properties such as sesquioxides and pH. Harrison and Johnson (1992) examined soils from 10 temperate forests and found that sulfate sorption related well to oxalate-extractable aluminum in Spodosols and Inceptisols ($r^2 = 0.6$). Ultisols adsorbed large quantities of sulfate, regardless of the concentration of soil aluminum.

How reversible is the specific adsorption of phosphate and sulfate? This is a key question in studies of the availability of phosphate fertilizers and the changes in soil chemistry that follow increases or decreases in atmospheric deposition of sulfate. Some evidence of reversibility has been found in short-term laboratory studies. For example, Harrison and Johnson (1992) found that about two-thirds of the sulfate adsorbed in the soils from 10 temperate forests was readily released in an extraction with phosphate, and one-third appeared to be irreversibly retained.

Soil organic matter may play a role in the capacity of soils to sorb phosphate and sulfate. Organic matter may sorb or coat the active surfaces of sesquioxides, reducing the available surfaces for phosphate and sulfate adsorption. Soil animals modify soil organic matter, and therefore may play an important role in modifying adsorption of phosphate and sulfate. In Quebec, the presence of earthworms in sugar maple forests greatly reduced the phosphorus supply to trees by mixing phosphorus from surficial organic layers with mineral soil; aluminum bound the phosphorus (Pare and Bernier, 1989a, b). The activity of worms may have a somewhat opposite effect on phosphorus sorption within mineral horizons. In a microcosm experiment, pine needles were added to soils with and without the presence of earthworms. Where the worms had mixed the needles into the mineral soil, the soil sorption capacity for phosphorus was substantially reduced (Figure 5.12). Much more work is warranted on the overall effects of animals on soil chemistry and nutrient dynamics (see Chapter 6).

NUTRIENT SUPPLY AND UPTAKE

Nutrient supply can be viewed in three ways:

1. The mass of an element that is readily available in the soil solution;
2. The mass of an element in "labile" pools that may be accessed through the activity of microbes and plants; or
3. The mass of an element actually acquired by plants.

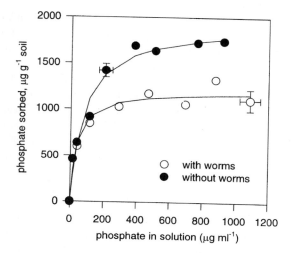

Figure 5.12   One-month laboratory incubation in microcosms with and without worms showed substantially lower phosphorus sorption capacity with worms than without worms (error bars = 1 standard deviation; Winsome, Horwath, and Powers, unpublished data), used by permission of Thais Winsome.

The nutrients present in soil solution are typically a small fraction of the quantity cycled among the soil, microbes, and plants each year. The turnover rate for the nutrients contained in soil solution is typically on the order of hours to weeks; over these periods, the flux into and out of the soil solution will match the size of the pool contained in the soil solution. Soil solution concentrations are viewed as "intensity" factors capable of driving chemical reactions.

The size of labile pools is difficult to assess, as they include a diverse array of compounds, and depend in part on the activity of the microbes and plants. The labile pool of soil nitrogen typically equals 1 to 3 percent of the total soil nitrogen pool, and includes the release of atoms contained in microbial biomass, plant detritus, and humified soil organic matter (see Chapter 9). The labile pool of soil phosphorus is even more complicated, including major contributions from minerally and organically bound pools. The nutrients released from labile pools represent "capacity" or "buffer" factors, which determine the rate of resupply of the soil solution in response to nutrient uptake.

The actual quantity of nutrients available for use within plants depends on these intensity and capacity factors, but also on the acquisition by the plants from these pools. The investment by trees in the production of roots and mycorrhizal fungi is highly variable, both among sites and among species, and this belowground production can be a major determinant of nutrient acquisition by trees.

How large is the pool of nutrients dissolved in soil solution? The size of the soil solution pool differs across scales of time, both small and large. As a general illustration, consider a solution in a fertile soil that has 1 mg of nitrate-nitrogen per liter, and a soil moisture level of 30 percent (water:soil by weight in a soil with a bulk density of 1.0) to a depth of 20 cm. The total quantity of soil water (to this depth) would be $60 L/m^2$, or $600 m^3/ha$, giving $60 mg/m^2$ (0.6 kg/ha) of nitrogen. A typical forest might use 100 kg/ha of nitrogen annually, so the amount of nitrate in the soil solution at one time would be quite small relative to the annual uptake. Microbial processes replenish the soil solution pool of nitrate, and this buffering of the soil solution pool provides the sustained source of nitrogen (and other elements) for tree growth.

The buffering of soil solution concentrations of nutrients depends on several processes, with some key differences among nutrients. Forest soils commonly contain 5 to 50 kg/ha of nitrogen as exchangeable ammonium, and the ammonium concentrations in soil solution may be partially buffered by sorption/desorption reactions. The overall picture of dissolved soil nitrogen is dominated by microbially mediated processes. Microbes commonly absorb more nitrogen daily than trees do; fortunately for the trees, microbial biomass turns over relatively rapidly (on a time scale of days to weeks), and the microbial pool of nitrogen may be tapped by trees. Laboratory studies with ^{15}N show that microbial uptake (immobilization) of nitrogen averages 75 to 95 percent of the nitrogen released through decomposition (gross mineralization), leaving only a fraction for tree uptake (net mineralization). The balance between microbial uptake and tree uptake remains unclear, although the importance of this competition has been well established (reviewed by Kaye and Hart, 1997). For example, Schimel et al. (1989) added ^{15}N to soils and found that 24 hours later, microbial biomass contained about half of the added nitrogen, and trees only 10 percent. The key issue, of course, is, what happens to the microbial nitrogen when the microbes die? How much goes into recalcitrant organic pools in soils and how much is obtained by trees? The interacting processes need much more attention before answers to this question can be generalized.

The soil solution concentrations of cation nutrients are well buffered by the soil exchange complex; removal of K^+ from the soil solution tends to lead to displacement of K^+ from the exchange complex. The balance between adsorbed cations and cations in the soil solution depends on the nature of the exchange complex, the relative distribution of ions in the exchange complex, and the ionic strength of the soil solution.

The concentrations of phosphate and sulfate in soil solutions are buffered both by microbially mediated process (as for nitrogen), and by geochemical sorption and desorption. Unlike the sorption and desorption of cations, phosphate and sulfate may be "specifically" adsorbed to sesquioxides, which leads to limited exchange between the solid and solution phases.

NUTRIENT UPTAKE DEPENDS ON SOIL CHEMISTRY AND THE ABSORBING SURFACE AREA OF TREES

The complex processes involved in nutrient acquisition by plants can be summarized by considering a few key features (after Barber, 1995): the concentration of the ion in solution, the buffering of the solution concentration (discussed above), the absorbing area of the roots and mycorrhizae, and the kinetics of uptake.

A key distinction between roots and mycorrhizae involves the surface area per mass of root or hyphae. A typical plant root hair might have a radius of 0.15 mm compared with a radius of 0.005 mm for hyphae. A kilogram of hyphal biomass would yield 30 times more absorbing area than the same mass of root hair biomass.

The importance of diameter in roots versus hyphae is even more important than the relative surface area would indicate. Large-diameter roots act as large cylinders, with zones of depletion extending outward into the soil as a result of nutrient uptake; each square millimeter of root surface area draws on a zone of soil perpendicular to the root surface. Hyphae act as very tiny cylinders, and each square millimeter of hyphal area draws on a large area of "curved" soil around it. This geometrical effect is illustrated in Figure 5.13, where plant roots deplete the concentrations of phosphorus in soil solution to a distance of 0.8 cm. Fungal hyphae in the same situation can absorb more phosphorus (per unit of surface area) because the depletion zone is far smaller. In the simulation in Figure 5.13, the surface areas of the roots and hyphae are the same, but the root mass weighs 30 times more than the hyphae. Because of the smaller radius

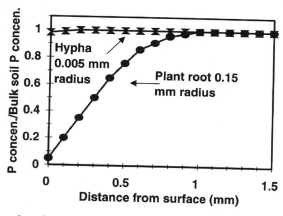

Figure 5.13 The phosphorus concentration near the surface of a root is much lower than that in the bulk soil solution, but the smaller diameter of mycorrhizal hyphae permits phosphorus uptake without local depletion of soil solution phosphorus (modified from Barber, 1995).

of the hyphae, the fungus can absorb twice as much phosphorus as the root hairs, even when the surface areas are equal. Nutrient uptake by plants has reciprocal effects on soil solution chemistry. Depletion of nutrient concentrations can promote rates of ion diffusion and mineral dissolution.

NUTRIENT UPTAKE ALSO DEPENDS ON UPTAKE KINETICS

The absorption of nutrients by roots and mycorrhizae tends to show Michaelis-Menten-type kinetics for enzymatic reactions:

$$V = \frac{V_{max}^* C}{K_m + C}$$

where V = the rate of uptake, V_{max} = the maximum rate of uptake when the nutrient concentration is not limiting, C = the concentration of the nutrient in the soil solution, and K_m is the concentration where $V = \frac{1}{2}V_{max}$. In the example in Figure 5.14, the maximum rate of uptake would be 40 nmol/m² of phosphorus absorbing area per second, and half of this value would be realized when the solution concentration of phosphorus was 5 µmol/L. As described below, trees that are limited by a particular element (such as phosphorus) tend to show higher V_{max} (when excised roots are exposed to saturating levels of phosphorus in solution) than trees with adequate nutrient supplies.

Putting all the factors together (soil chemisty and root characteristics), how sensitive is root uptake to changes in any single parameter? This is difficult to determine by direct measurement because of the number of variables involved, but Silberbush and Barber (1983) simulated the changes in phosphorus uptake when each factor was doubled (Figure 5.15). Doubling the phosphorus concentration in soil solution increased the simulated uptake by almost threefold. Buffering of the phosphorus concentration and diffusivity of phosphorus in the soil showed linear effects on phosphorus uptake, but doubling of the V_{max} for

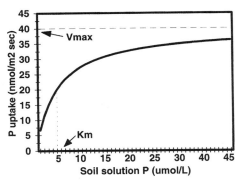

Figure 5.14 Hypothetical pattern of phosphorus uptake as a function of soil solution phosphorus concentration.

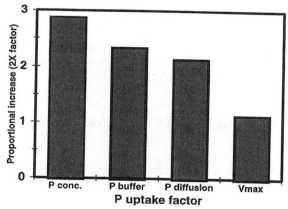

Figure 5.15 Simulated effect of changing factors that influence phosphorus uptake by twofold (data from Barber, 1995).

uptake increased uptake only marginally. As mentioned above, substituting an equivalent mass of mycorrhizal hyphae for roots would double the rate of phosphorus uptake by reducing the absorbing radius.

Uptake kinetics at the root surface depend on the active uptake capability of the roots. Roots that are deficient in a particular nutrient tend to have greater capacity for taking up that nutrient. This uptake capacity is measured in excised roots that are exposed in laboratories to adequate concentrations of labeled nutrients (such as ^{32}P or ^{15}N).

Bioassays of excised roots from trees in control plots of Norway spruce showed high rates of uptake of both phosphorus and nitrogen, indicating possible deficiencies (Figure 5.16). In roots from plots that had been fertilized with nitrogen alone, nitrogen uptake rates dropped but phosphorus uptake

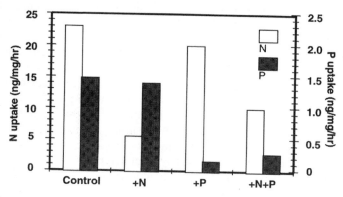

Figure 5.16 Bioassay of excised Norway spruce roots show the root surface uptake kinetics in response to fertilization treatments (data from Jones et al., 1994).

rates remained high, indicating an alleviation of nitrogen deficiency but not of phosphorus deficiency. Roots from plots fertilized with phosphorus alone showed low phosphorus uptake kinetics and high nitrogen uptake kinetics, and roots from plots fertilized with both nutrients showed reduced uptake kinetics for both nutrients (Jones et al., 1994).

SOIL SOLUTIONS ARE SAMPLED IN SEVERAL WAYS, AND NO WAY IS PERFECT

Soil solutions are sampled most commonly with tension lysimeters. A lysimeter is a gauge for sampling internal fluids, and tension lysimeters are made of porous materials that can sustain a negative potential (or tension), sucking in water from the surrounding soil. Other porous materials used in tension lysimeters include fritted glass and Teflon. Tension lysimeters are small cup-shaped devices on the ends of plastic tubes or, in some cases, flat plate-like devices. Tubes attached to the lysimeters create negative potentials through the use of a hanging column of water or a vacuum pump. The negative potential may be relatively slight, such as 0.01 to 0.03 MPa, sampling soil water that is held only loosely by the soil matrix. In other situations, a stronger negative potential is applied to pull out larger volumes of water in a shorter time period.

An alternative design uses plates of acrylic or other material to funnel water that flows by gravity into a tube and collection bottle. These zero-tension lysimeters are good at collecting water during high flow periods, but they undersample the soil after free-draining water has percolated out of the soil.

Do different lysimeter designs yield soil water with similar chemistry? The short answer is "no." Silkworth and Grigal (1981) compared the solutions obtained by four types of tension lysimeters: large ceramic plates, small ceramic plates, fritted glass, and cellulose fiber. The concentrations of calcium in collected solutions varied by a factor of 2, and the concentration of magnesium varied by more than sixfold. The ratio of calcium to magnesium ranged from 1:1 to 3.6:1, indicating major differences among the lysimeters. Haines et al. (1982) compared the ability of two types of lysimeters to measure the effects of clearcutting on soil solution chemistry. Stainless steel troughs covered with fiberglass netting served as zero-tension lysimeters; water trickled through the netting and was collected in a tube that led to a collection bottle. The zero-tension lysimeters collected about seven times more water just below the forest floor than ceramic-plate tension lysimeters. At 30 cm into the mineral soil, the pattern was reversed and the tension lysimeters collected twice as much water as the zero-tension lysimeters. Again, the concentrations of nutrients differed substantially among lysimeter types.

A different approach to sampling soil solution involves taking a soil sample and centrifuging it (at more than 10,000 rpm for an hour or more; Giesler et al., 1996). This centrifugation compacts the soil tightly at the bottom of the tube, with the water sitting on top, ready to be decanted. Zabowski and Cole

(1990) compared the chemistry of solutions obtained from centrifuging with those collected from low-tension lysimeters. They found that the centrifuge solutions generally had higher concentrations of ions and showed greater variations throughout a year of monthly sampling than the lysimeter solutions. The centrifugate may illustrate the chemistry as affected by biological processes, whereas the lysimeter water may resemble the water that is lost through leaching deeper into the soil.

A third approach is to simply place soil (and vegetation) in a container and collect the water leaching from the container — a monolith lysimeter approach. An example of this type of lysimeter was mentioned in Chapter 4. Graham and Wood (1991) characterized the effects of 40 years of development of different types of vegetation in large monolith lysimeters (50 m^3).

SOIL SOLUTIONS AND PLANT UPTAKE ARE LINKED TO ECOSYSTEM BIOGEOCHEMISTRY

These processes and patterns of soil solution chemistry and plant uptake interact with many other processes and patterns in forest ecosystems; this larger picture is called biogeochemistry. Biogeochemistry examines the interactions among geology, chemistry, and life, focusing on ecosystem productivity and nutrient cycling. A key feature of biogeochemistry is the rates at which nutrients are taken up by trees, and how these rates compare to rates of nutrient inputs and the overall store of nutrients in soils. For the soils with solution data in Figures 5.1 and 5.2, the annual uptake of nutrients is higher than the annual inputs from the atmosphere for all elements except for sulfur in all forests, magnesium in Norway, and nitrogen in the nitrogen-fixing red alder (Table 5.2). In all cases, the annual uptake of nutrients represents a very small fraction (1 percent or less) of the total quantities of these elements stored in soils. The controls and interactions among these ecosystem processes are the subject of Chapter 9.

SUMMARY

The chemistry of soil solutions includes many different elements and inorganic compounds, as well as organic compounds and gases. The soil solution is the immediate source of most nutrients used by plants, and the composition and dynamics of the soil solution depend on interactions with the solid phases of the soil, as well as on the overall ecosystem biogeochemistry. Cation exchange reactions are important for elements such as calcium, potassium, and aluminum, and specific ion exchange is important for phosphate and sulfate. The total concentration of ions in solutions influences the acidity of the soil solution and the exchange reactions between the soil solution and the solid phases. Soil acidity is similar to the acid-base properties of any weak acids and

TABLE 5.2 Rates of Nutrient Uptake (kg/ha Annually to Meet Requirements for Aboveground Growth) in Relation to Annual Inputs from the Atmosphere and Total Soil Nutrient Contents

		Element					
Forest Type		Nitrogen	Phosphorus	Potassium	Calcium	Magnesium	Sulfur
Norway spruce	Uptake	27.6	4.0	21.8	21.3	2.9	1.9
	Uptake/input	2.5	—	8.7	9.8	0.40	0.09
	Uptake/soil total (to 50 cm)	0.007	—	0.0003	0.0007	0.0006	—
Loblolly pine	Uptake	30.2	7.3	12.9	26.4	5.6	4.3
	Uptake/input	2.2	240	2.6	3.8	6.2	0.12
	Uptake/soil total (to 80 cm)	0.02	0.006	0.0004	0.005	0.0005	0.002
Douglas-fir	Uptake	13.2	2.3	9.9	17.2	1.6	1.5
	Uptake/input	2.7	74	3.9	4.9	1.9	0.14
	Uptake/soil total	0.003	0.0006	0.0004	0.0004	0.00006	0.002
Red alder	Uptake	80.3	3.3	31.7	51.9	7.2	4.6
	Uptake/input	1.0	110	13	15	8.8	0.42
	Uptake/soil total	0.01	0.002	0.002	0.002	0.0003	0.007
Mixed hardwoods	Uptake	34.5	2.2	41.3	47.7	12.2	5.6
	Uptake/input	4.7	8.9	30	17	24	0.30
	Uptake/soil total	0.009	0.0008	0.0002	0.004	0.0001	0.004
White pine	Uptake	37.0	4.1	27.3	32.9	8.1	6.1
	Uptake/input	5.1	17	20	12	16	0.33
	Uptake/soil total	0.01	0.002	0.0002	0.009	0.0001	0.008

Source: From data in Johnson and Lindberg. (1992).

bases, except that the anion portion of the soil acidity is solid. Soil pH differs across landscapes and under the influence of species and management. Differences in pH depend on differences in the types of acids present, the degree of dissociation of the soil acids (often referred to as base saturation), and the total salt concentration in the soil. Nutrient uptake by plants results in local depletion of the soil solution, and characteristics of the roots and mycorrhizae have reciprocal influences on the composition of the soil solution. Soil solutions are commonly sampled with lysimeters or with centrifugation of soil samples. No single method provides a satisfactory representation of the diverse aspects of soil solution chemistry.

CHAPTER 6

Biology of Forest Soils

The variety, number, and activity of organisms found in forest soils are generally much greater than those in agricultural soils. In all but a few cloud forest formations, the majority of the biological diversity in forests is found below the surface of the litter layer. These organisms are not charismatic, and most are even difficult to see, but they are crucial to the proper functioning of forest ecosystems. Since forest soils are seldom tilled or mixed by human beings, this great activity of a wide variety of decomposing and mixing organisms is of paramount importance. Soil biota are the tillers of uncultivated soils (Hole, 1981). The favorable environment of the forest floor encourages the proliferation of myriad organisms that perform many complex tasks relating to soil formation, slash and litter disposal, nutrient availability and recycling, and tree metabolism and growth.

The soil biota form an intricate food web through which nutrients and energy pass. The study of this food web has lagged far behind the study of soil phenomena (DeAngelis, 1992; Hall and Raffaelli, 1993; Coleman and Crossley, 1996; Benckiser, 1997). The importance of the processes within this web to the maintenance of the soil's productive potential cannot be overestimated (Fisher, 1995). Food webs have been constructed for few soils, but it is clear that cultural practices applied to the soil can dramatically alter their structure.

The majority of soil organisms are ubiquitous, but the inoculation or introduction of organisms to a new environment is largely a haphazard occurrence. The presence or absence of one specific soil organism can have tremendous positive or negative effects on the vegetative assemblage at a particular location. How well the organisms thrive in new surroundings depends on the organisms and on soil factors such as moisture, temperature, aeration, acidity, and nutrient and energy supplies. The same factors greatly influence the spatial distribution of organisms in the soil. Conditions favorable for most organisms can be found at some level within the litter layer and underlying soil horizons.

Most soil animals make their homes on or near the surface litter or humus layers, where space and light conditions fit their particular needs. They are rather mobile, but because their distribution is dependent on an organic substrate, they seldom venture deeply into the mineral soil (Hole, 1981). Microflora, on the other hand, are found throughout the soil profile, with each type occupying its own niche. The photosynthetic organisms such as blue-green algae are found only on surface materials where light is not limiting for photosynthesis. Some microorganisms get their carbon from atmospheric carbon dioxide and their energy from the oxidation of inorganic substances; therefore, they are less restricted in their spatial distribution. The vast majority of organisms get their carbon from complex organic materials, and their abundance is directly dependent on environmental conditions and the presence of a proper organic substrate. Some of the smallest of these microorganisms are not free in the soil but are held to clay or organic colloids by cation exchange forces. Organic exudates stimulate the growth of other microorganisms at the root surface. They are often 10 to 100 times more numerous in the rhizosphere, that portion of the soil in the immediate vicinity of roots, than in other nearby areas of mineral soil. Generally, the number of organisms is greatest in the forest floor and the rhizosphere, with a decreasing gradient with soil depth (Gray and Williams, 1971).

SOIL ORGANISMS PERFORM A WIDE VARIETY OF FUNCTIONS

Forest soils contain a multitude of organic and mineral substances available as carbon and energy sources, as well as the physical environment suitable for a vast array of plant and animal populations ranging in size from microscopic bacteria to fairly large animals. Perhaps the only feature that many of these organisms have in common is that they spend all or a major part of their lives in the soil. Because of the diverse and complex nature of these populations, several classification schemes have been devised. The most commonly used system divides the organisms into two broad groups: the plant and animal kingdoms. They may be further grouped arbitrarily on a taxonomic, morphological, or size basis or, in the case of microorganisms, on some physiological basis using a variety of metabolic characteristics or oxygen requirements. Grouping according to the functions that organisms perform in the environment is of particular interest in forest soils because of its relevance to soil formation, properties, and management.

Size divisions are useful for soil animals. Macrofauna (>2 mm) are species large enough to disrupt soil structure through their activity and include small mammals, earthworms, termites, ants, and other arthropods (Dindal, 1990; Stork and Eggleton, 1992; Berry, 1994). These organisms create macropores that influence infiltration rate and gas exchange. They are responsible for a great deal of soil mixing and can profoundly influence horizonation and rooting pattern. They often consume very high carbon:nitrogen ratio organic material and produce lower carbon:nitrogen ratio detritus.

Mesofauna (0.1–2 mm) are those soil-dwelling organisms large enough to escape the surface tension of water and move freely in the soil but small enough not to disrupt soil structure through their movement. They include some mollusks, springtails, mites, wood lice, and other arthropods. These organisms are extremely important in decomposition and mineralization processes (Crossley et al., 1992).

Microfauna (<0.1 mm) are usually confined to water films on soil particles. This group includes protozoa, turbellaria worms, rotifers, nematodes, and gastrotricha worms. These organisms are important predators on fungi and bacteria. They regulate microbial populations and are important in mineralization processes.

SOIL FAUNA PLAY A VITAL ROLE IN NATURAL SOIL ECOSYSTEMS

Essentially all fauna that inhabit the forest environment influence soil properties in some way that eventually affects tree growth. Fauna range in size from wild beasts (and sometimes domestic animals) to simple one-celled protozoa. Their importance to the soil, however, is essentially inversely proportional to their size. The effects of large animals on forest soils are minimal except on overgrazed lands, where a reduction in ground-cover vegetation or an alteration in species composition, plus soil compaction and reduced water infiltration, result in soil erosion (Lavelle, 1997).

MACROFAUNA ARE MOVERS AND SHAKERS

Vertebrates

Vertebrates, consisting largely of four-legged animals, influence the soil through fertilization, trampling, scarification, and a form of cultivation. Many such animals burrow in the soil and aid in the breakdown of its organic constituents. They assist in mixing this organic material with the inorganic surface soil and with the transport of the former into the soil profile. Decomposition of organic debris is hastened as a result of its use as food by animals and its mixing with soil mineral materials. Woodchucks or marmots, moles, gophers, mice, shrews, and ground squirrels are particularly important in soil development. **Moles** are especially active in European forests, supporting a mull humus layer, and it has been suggested (Lutz and Chandler, 1946) that the names mole and mull are of common origin. While large animals may trample and compact the soil surface, penetration of water and air into the soil is greatly facilitated by the actions of smaller animals. These small quadrupeds are chiefly carnivorous, devouring worms, larvae, and insects. However, a few rodents destroy seedlings and young trees, and their beneficial actions as soil cultivators may be overshadowed by this destructive behavior (Crouch, 1982).

Mice, shrews, and other small animals are abundant in many forest soils. Hamilton and Cook (1940) found that in mixed hardwood forests of the northeastern United States there was an average of 4814 g/ha of small mammals, or about twice the carrying capacity for whitetailed deer. Small mammals have a pronounced influence on microrelief and other soil properties. Their labyrinths of interconnecting tunnels allow for ready penetration of air and water, while their nests, dung, stored food, and carcasses all enhance the organic matter content and fertility of the soil (Troedsson and Lyford, 1973). It has been suggested that the absence of an eluviated horizon in some Eutrochrepts of the Northeast may be caused largely by the mixing of surface and subsurface materials by fauna (Lyford, 1963). Burrowing animals have been reported to bring as much as several milligrams per hectare per year to the surface in forests (Paton et al., 1995). Burrowing rodents may perform functions in arid and semiarid areas similar to those of the earthworm in humid forests.

Arthropods

Arthropods comprise a broad group of soil animals with articulated bodies and limbs that populate both the forest floor and the surface soil. Macrofauna-sized arthropods are largely important as soil mixers; however, some are important decomposer organisms. **Ants** occur in essentially all forest soils, and although they are predominantly carnivores, they are very important soil mixers. Ants may transport large amounts of subsoil to the surface. As much as 10 Mg/ha/yr has been reported in temperate forests, and amounts closer to 100 Mg/ha/yr have been observed in subtropical and tropical forests (Paton et al., 1995) Lyford (1963) reported that the return of subsoil material to the forest floor in small litter-concealed mounds by ants is an important process in the development of some Spodosols of New England and possibly in other temperate zone soils. He considered that the mineral material of the entire A horizon of some, if not most, of the forest soils of that region consists of material returned by ants to the surface from the B horizon over a period of many years. Fine-textured materials returned to the surface of coarse-textured soils provide increased cation exchange capacity, and available moisture and may be important in burial of seeds, roots, and charcoal. In the wet soils of subtropical and tropical areas, **crayfish** are important in soil mixing, aeration, and drainage. They also aid in the decomposition process. **Centipedes** are predatory animals feeding on other soil fauna and play only a minor part in soil formation.

Millipedes are largely saprophagous, feeding on dead organic matter. They are generally confined to soils with mull humus layers, especially those supporting stands of deciduous trees. Lyford (1943) noted that millipedes favored leaves of certain trees, particularly leaves with high calcium content. Millipedes are considered important in the formation of mull humus layers, although possibly not as important as earthworms.

Termites are important in tropical and subtropical wet and dry climates. They may be divided into three groups on the basis of food habits: those that feed on wood, those that consume litter and humus, and those that cultivate fungi for food. Termites do not produce their own cellulase; rather, they rely on a rich gut fauna of flagellates. Not all termites build mounds, but all tunnel in the soil, and they are important in transporting soil, decomposition, and mixing organic material with mineral soil.

Annelids

The ordinary **earthworm** is probably the most important component of soil macrofauna (Edwards, 1998). Although there are many species of earthworms and earthworm taxonomy is in a state of flux, all large and middle-sized worms have been classified as *Lumbricidae*, while the smaller, light-colored "pot-worms" have been considered *Enchytraeidae* (Kevan, 1962).

The large, reddish *Lumbnicus terrestris*, a species introduced from Europe, occurs widely in the Americas. A high population of earthworms is generally associated with mull humus formation, and this is particularly true of *L. terrestris*, which may make up as much as 80 percent of the total soil fauna by weight. These earthworms feed on fallen leaves and organic debris and pass it, together with fine mineral particles, through their bodies. Each year earthworms have been estimated to pass from 30 to 1100 mg/ha/yr of soil material through their bodies, where it is subjected to digestive enzymes and to a grinding action (Paton et al., 1995). The rate of casting by earthworms has been estimated to range from 10 to as much as 200Mg/ha/yr (Paton et al., 1995). Earthworm casts are higher in total and nitrate nitrogen, available phosphorus, potassium, calcium, magnesium, and cation exchange capacity and lower in acidity than is the soil proper. Earthworms mix bits of organic materials into the mineral soils, and they promote good soil structure and aeration through their burrowing action. As a result of this transporting and mixing action, the upper layer of certain forest floors takes on the crumbly structure of the so-called earthworm mull. The concentration of earthworms in forest soils has been estimated to be from 500,000 to more than 2.5 million per hectare, the actual numbers depending on several climatic and soil factors. Although earthworms may fluorish in acid soils, highly acid soils support fewer earthworms than less acid soils, with the optimum range being about pH 6.0 to 8.0. Sandy soils and soils that dry excessively are not favorable habitats for earthworms. *Lumbnicus rubellus* and *L. festivus* appear to be more acid tolerant than *L. terrestris* and may be more common than the latter in coniferous forests and in mor humus.

Allobaphora species are approximately the same size as the *Lumbricus* species but are much lighter in color. They are active in forest soils of Europe and North America and are particularly important in the development of hardwood mull humus. The smaller worms, such as *Octolaseum* and *Dendrobaena* species, also devour organic debris, thereby improving the physical and chemical properties of the surface soil. Because of their smaller size, generally no more than a few centimeters long, they have not been considered as important as the other worms in forest soil formation. However, since these Enchytraeidae have less stringent environmental requirements than their larger relatives, they may be approximately as numerous in mor as in mull humus layers (Hendrix, 1995).

MESOFAUNA FRAGMENT DEBRIS AND PROMOTE SOIL STRUCTURE

Arthropod dominate the mesofauna of forest soils. Springtails, woodlice or sow bugs, spiders, ticks, and mites are abundant in essentially all forest soils and are particularly important in forest litter decomposition. The primary consumers chew and move plant parts on the surface and into the soil. Some, such as sow bugs, are active feeders on dead leaves and wood and are important in the disintegration of freshly fallen leaves.

Saprophagous **mites** constitute one of the most important groups of Arachnida, and by virtue of their great numbers, often exceeding 1 million per square meter, they play a major role in producing a crumb structure in some surface organic layers. They feed on decaying leaves, wood, fungal hyphae, and feces of other animals (Wallwork, 1970). Mites are important at several nodes in the soil food web (Coleman and Crossley, 1996; Benckiser, 1997). The **springtails** and **bristletails** are small, wingless, saprophagous insects that feed on decaying materials of the forest floor. Adults and larvae of many **beetles** and **flies** contribute to the breakdown of organic debris and improve the structure of surface soil (Kevan, 1962).

Fires generally reduce the number of arthropods in the forest floor, but the reduction is largely temporary and not all genera are reduced equally. Carabids are the most numerous arthropods in protected forests of northern Idaho, but *Acarina, Chilopoda, Thysanoptera, Protura,* and *Thysanura* are most numerous in areas that had received recent prescribed burns (Fellin and Kennedy, 1972).

MICROFAUNA THRIVE IN SOIL WATER FILMS

Nematodes

These nearly microscopic, nonsegmented **roundworms** are most common in mull and grassland soils. They usually inhabit only the surface 5 to 10 cm, where population densities exceeding $10^6/m^2$ are not uncommon. Nematodes have a high reproductive capacity, and their populations can expand rapidly as soil conditions change. Much of the decline in agricultural production on newly cleared forest plots is associated with the increase in specific plant parasitic nematode populations. Only about a tenth of the 10,000 or so known nematodes are soil inhabitants. Although the populations of nematodes are always highest in the vicinity of plant roots, only a few appear to be root parasites. Most prey on bacteria, algae, fungi, protozoa, rotifers, or other nematodes.

Although some nematodes are parasites of trees, they are important as population regulators and nutrient concentrators in the soil ecosystem. The impact of their parasitism versus their role in decomposition is poorly understood (Benckiser, 1997). In a survey of 3-year-old slash pine plantations in the lower coastal plains of the United States, plant parasitic nematodes were found in all 34 sampled sites. Spiral (*Helicotylenchus*) and ring (*Criconemoides*) nematodes were found most frequently, although 11 other plant parasitic genera were also identified. The actual damage to tree vigor and growth is not known, but several of the genera found in these plantations have been reported to cause damage to pine. In addition to the two genera mentioned above, others included sheath (*Hemicycliophora*), lance (*Hoplolaimus*), stunt (*Tylenchorhynchus*), and dagger (*Xiphinema*) nematodes.

Rather spectacular increases in growth have been reported for tree seedlings planted in some fumigated soils. This response is presumed to result from a reduction in the population of nematodes that attack pine roots, and tree responses have been particularly notable on dry, sandy sites. However, neither the actual causes of the increased growth nor the longevity of the effect of fumigation are well understood. The total concentration of nematodes in the soil may return to normal in a year or two following fumigation, but the spectrum of genera may be altered for longer periods.

Rotifers

These aquatic animals are active in soil when moisture is sufficient; at other times, they may form protective shells or simply enter a state of anabiosis. The soil rotifers are about equally divided between those that feed on organic debris and those that feed on algae or protozoa. They generally inhabit the litter and humus layers, and their populations can be quite high ($10^5/m^2$) on favorable sites.

Protozoa

Protozoa are the most abundant soil fauna; Waksman (1952) reported that their number varied from 1500 to 10,000 per grain of forest soil. These one-celled organisms may exist in either an active or a cyst stage, but they are generally aerobic and occur in the upper horizons. Their diet consists largely of decomposing organic materials and bacteria. Soil conditions that favor their development are similar to those that favor bacteria. They are found in soils supporting both hardwood and coniferous forests. In a strict sense, they are the only major group of soil fauna classified as microorganisms.

SOIL MICROFLORA ARE SMALL IN SIZE, GREAT IN IMPORTANCE

Soils contain three major groups of microflora: bacteria (actinomycetes), fungi, and algae (Sylvia et al., 1998). These microflora can also be conveniently divided into two broad classes with respect to their energy and carbon sources: **heterotrophic** forms, which require preformed organic compounds to serve as sources of energy and carbon, and **autotrophic** forms, which obtain their energy from sunlight or by the oxidation of inorganic compounds and their carbon by the assimilation of CO_2. Most types of bacteria and all fungi, protozoa, and larger animals are heterotrophs. Only algae and a few types of bacteria share with higher plants the capacity to use sunlight as a source of energy.

Bacteria

Although bacteria are small—rarely more than several micrometers in length—they are especially prominent in soils because of their large numbers. Bacteria and fungi dominate in well-aerated soils, but bacteria alone account for most biological and chemical changes in anaerobic environments. In fact, the ability to grow in the absence of O_2 is the basis for further grouping of bacteria into three distinct categories: **aerobes** that live only in the presence of O_2; **anaerobes** that grow only in the absence of O_2; and **facultative anaerobes** that can develop in either the presence or the absence of O_2. The more meaningful grouping for understanding soil processes, however, is that based on energy and carbon source requirements.

Autotrophic bacteria are of two general types: **photoautotrophs**, whose energy is derived from sunlight, and **chemoautotrophs**, whose energy comes from the oxidation of inorganic materials. The chemoautotrophs are limited to a relatively few bacterial species, but their importance to soils greatly exceeds their numbers. Bacteria using **nitrogen** compounds as energy sources include those that oxidize ammonium to nitrite (*Nitrosomonas* and *Nitrosococcus*) and those that oxidize nitrite to nitrate (*Nitrobacter*). The energy-yielding reactions involving these organisms are the following:

$$2NH_4^+ + 3O_2 \rightarrow 2NO_2^- + 4H^+ + 2H_2O \quad (Nitrosomonas)$$
$$2NO_2^- + O_2 \rightarrow 2NO_3^- \quad (Nitrobacter)$$

The number of these organisms in forest soils is often quite low, and nitrification proceeds at a very slow pace under most acid forest soil conditions. The population of these bacteria expands rapidly when acidity is corrected or when a source of nitrogen is available. The composition of organic materials in the soil also affects the rate of nitrification. Because litter layers generally have wide carbon:nitrogen ratios, a large part of the nitrogen compounds mineralized during decomposition is used by the nitrifying organisms to maintain their own population. This disappearance of inorganic nitrogen following additions of nitrogen-poor litter is termed nitrogen immobilization. Only after the carbon content of certain materials has been reduced to a ratio of below about 20:1 does any significant amount of nitrogen become available to higher plants (Waksman, 1952). Since materials with such ratios are uncommon in acid forest soils, organic compounds and ammonium are apparently the principal sources of soil nitrogen available to trees (McFee and Stone, 1968). A low level of nitrate is found in forest soil at any given time.

However, this does not necessarily mean an absence of nitrification because nitrates can be rapidly lost by leaching, plant uptake, and denitrification.

Nitrification has long been assumed to be solely a chemoautotrophic process, but there is increasing evidence of heterotrophic nitrification. A large number of heterotrophic bacteria and actinomycetes as well as some fungi, are capable of producing at least small amounts of nitrate from ammonium. The large number of nitrifying heterotrophs may compensate for their inefficiency, and they may exert significant influence on the rate of nitrate synthesis under some conditions. This may be particularly true in acid forest soils that commonly have carbon:nitrogen ratios in excess of 20:1 where competition for NH_4^- nitrogen between heterotrophs and autotrophs begins to reduce autotroph activity.

Of the chemoautotrophic mineral oxidizers, bacteria involved in the oxidation of inorganic **sulfur** are perhaps the most important in forest soils. Sulfur exists as sulfide in several primary minerals, and it is added to forest soils as plant and animal residues and to rainwater. Elemental sulfur is sometimes added to nursery soils to increase acidity and to control certain plant pathogens. The major part of the sulfur in soils exists in organic combinations and, like nitrogen, it must be mineralized to be useful to trees. Tree roots absorb sulfur largely as sulfate. While the initial decomposition of these organic materials and their conversion to inorganic sulfur compounds is accomplished by heterotrophic organisms, the oxidation of sulfides and elemental sulfur to sulfates can be carried out by both heterotrophs and chemoautotrophs. Bacteria of the genus *Thiobacillus* are the principal autotrophic oxidizers in well-aerated soils. This genus contains eight species with widely varying habitat requirements, but the acid-loving aerobe *T. thiooxidans* is probably the most prevalent in forest soils. The reaction involved in sulfur oxidation by this organism is generally expressed as follows:

$$2S + 3O_2 + 2H_2O \rightarrow 2H_2SO_4$$

The oxidation of elemental sulfur can result in the mobilization of some slowly soluble soil minerals as the result of the sulfuric acid formed. The solubility of phosphorus, potassium, calcium, and several micronutrients may be increased by the acidification resulting from this reaction.

Oxidation of inorganic sulfur compounds can also be accomplished by heterotrophic bacteria, actinomycetes, and fungi under some conditions and, in the case of *T. denitrificans*, oxidation can take place anaerobically. This chemoautotroph can use nitrate as a terminal electron acceptor and convert nitrate to gaseous nitrogen while oxidizing sulfur compounds. Under some anaerobic conditions in water-saturated soils, however, inorganic sulfur compounds are reduced to sulfides instead of being oxidized. The anaerobic *Desulfovibrio desulfuricans* is largely responsible for the reduction of sulfates, but it has a narrow acidity range. Because it is usually limited to pH 6 and above, the formation of sulfides in acid forest soils is not common even in wet areas. However, hydrogen sulfide may accumulate to toxic levels for tree roots in stagnant marshes where soil acidity is near neutral. *Desulfovibrio* bacteria use sulfates as electron acceptors, but they are not considered autotrophs since they use carbohydrates as electron donors (energy sources).

Several types of bacteria are involved in the transformations of **iron**, **manganese**, and other heavy metal compounds. The most important chemoautotroph for iron oxidation is *T. ferrooxidans*. As the name of the genus indicates, the bacterium may also derive energy by oxidizing inorganic sulfur when such compounds are present. Because iron is readily available to higher plants only in the reduced state, the action of this organism can result in iron deficiency in some well-aerated soils. It is in the ferrous state that iron is leached in soils, and a kind of iron pan may form when iron is oxidized or complexed with organic molecules in a lower horizon. The reduction of ferric iron by such aerobes and by facultative anaerobes such as *Bacillus*, *Clostridium*, and *Pseudomonas* is heterotrophic.

Heterotrophic bacteria, which require preformed organic compounds as sources of energy and carbon, comprise the largest group of soil bacteria. This diverse group includes free-living and symbiotic nitrogen fixers and bacteria that decompose fats, proteins, cellulose, and other carbohydrates. They include both aerobic and anaerobic forms.

Biological nitrogen fixation is accomplished by free-living bacteria (or blue-green algae) and by symbiotic associations composed of a microorganism and a higher plant or a fungus. It is primarily through the action of these organisms that part of the huge reservoir of atmospheric nitrogen is rendered available to higher plants. The free-living bacteria capable of utilizing nitrogen gas (N_2) are primarily aerobic species of *Azomonas*, *Azotobacter*, *Beijerinckia*, and *Azospirillum* and anaerobic species of *Clostridium* and *Desulfovibrio*, although strains of several other genera are also capable of the transformation under some conditions. Most are apparently not obligate because they can obtain nitrogen from organic and inorganic nitrogenous compounds as well as from the atmosphere.

Free-living nitrogen fixers can maintain themselves over a wide range of soil conditions, but the conditions under which they fix atmospheric nitrogen are limited. For example, *Azotobacter*, an obligate aerobe, prefers temperatures around 30°C and generally fails to fix N_2 below pH 6. *Beijerinckia* species are also aerobic; they grow well in acid conditions (perhaps as low as pH 3), but they are apparently confined largely to tropical soils. The dominant anaerobes are of the genus *Clostridium*. They are like the blue-green algae in that they are most numerous in wet and flooded soils and they are active over a range of pH 5 to 9. The wide carbon:nitrogen ratios of forest floor materials appear to favor N_2 fixation by nonsymbiotic organisms, but this type of fixation is not believed to be of much practical significance in forest soils. The amount of nitrogen fixed annually probably averages less than 2–5 kg/ha.

Symbiotic nitrogen fixation can be of considerable importance in many forest soils. The classical example of such a symbiosis is that between plants of the Leguminosae and bacteria of the genus *Rhizobium*. This relationship forms root nodules within which N_2 is converted to organic forms. There are more than 10,000 species of legumes, many of which grow wild in forests, although not all are symbiotic N_2 fixers. They are sometimes planted in young forests as a source of nitrogen under special silvicultural conditions (Rehfuess and Schmidt, 1971). Lupines have been planted in conjunction with grasses and pine to stabilize sand dunes in New Zealand (Gadgil, 1971a, b). Other leguminous plants may have a place in intensively managed young plantations on infertile soils such as glacial outwash and coastal sands of many parts of the world. Some reseeding or perennial plants may be maintained for 5 to 7 years and can fix 50 to 200 kg/ha/yr of nitrogen in fertilized areas or in plantations where soil phosphorus and potassium are not deficient. Since forest soils are often very acid, tolerant species must be selected for planting in unamended sites. The expense usually involved in establishing an effective stand of leguminous plants in forests has probably been the greatest deterrent to their use.

Nitrogen-fixing trees are widely used to provide nitrogen in agroforestry systems. *Leucaena lucocephala* has been the tree most widely studied and used (National Academy of Sciences, 1977), but it is not well adapted to acid soils (Evans and Szott, 1995). Many nitrogen-fixing trees are adapted to acid soils (Tilki and Fisher, 1998), and greater efforts to find acid soil-adapted multi-purpose tree species are underway. Both the tree and the microbe must be able to tolerate acid conditions, which are often accompanied by high aluminum and manganese levels, in order for successful nitrogen fixation to take place.

Several genera of nonleguminous plants (*Alnns, Myrica, Hippopliac, Elaragnus. Shepherdia, Casuarina, Coriaria,* and *Ceanothus,* to name a few) possess root nodules that contain actinomycete symbionts of the genus *Frankia* and are capable of nitrogen fixation (Torrey, 1978). These plants, mostly trees and shrubs, are more abundant than legumes in many forests and may have considerable ecological significance on some sites. Annual fixation of nitrogen varies widely among these genera from a few kilograms by *Myrica gale* to more than 100 kg/ha by pure stands of alder (Youngberg and Wollum, 1970; Silvester, 1977). A few gymnosperms such as *Podocarpus* and *Cycas* possess nodule-like structures, but proof of nitrogen fixation is confirmed only for the latter. *Cycas* can apparently accomplish nitrogen fixation in association with blue-green algae under some conditions.

Organic matter-decomposing bacteria play a major role in the degradation of vast amounts of forest litter, plant roots, animal tissue, excretory products, and cells of other microorganisms. These materials are both physically and chemically heterogeneous and include such constituents as cellulose, hemicellulose, lignin, starch, waxes, fats, oils, resins, and proteins. With such a diversity of organic materials, it is not surprising that a complex population of heterotrophic bacteria, as well as fungi and actinomycetes, is involved. The bacteria involved include both aerobic and anaerobic forms and the mechanisms of decomposition vary, depending on environmental conditions and the participating organisms. In all cases, the organic matter provides the microflora with energy for growth and carbon for cell formation. The end products are carbon dioxide, methane, organic acids, and alcohols, in addition to bacterial cells and resistant organic materials.

The aerobic cellulose-decomposing bacteria are intolerant of poor aeration and high soil acidity. Since their activity ceases below about pH 5.5, they are found chiefly in hardwood mull humus and in humus developed under mixed pine-hardwood stands (Waksman, 1952). The anaerobic bacteria tolerate strongly acid and poorly drained soil, but organic matter breakdown by these organisms is slow; consequently, organic matter accumulates in many poorly drained soils. Although the subsidence of organic soils after drainage may be due primarily to shrinkage in cold climates, subsidence of drained peatlands in warm climates results mostly from biological oxidation.

The rate of decomposition of plant materials depends largely on their nitrogen content, with protein-rich substrates being metabolized most readily. Forest litter tends to be decomposed slowly because of its wide carbon:nitrogen ratio, resulting in litter accumulations on the forest floor. Because extra nitrogen is needed for the expanding microbial population necessary to decompose carbonaceous materials, the addition of materials with a wide carbon:nitrogen ratio may result in temporary nitrogen deficiency. For example, when raw organic materials such as sawdust are applied to forest nurseries, a deficiency of nitrogen for seedling growth results unless extra nitrogen fertilizer is applied to replace that immobilized in the tissue of cellulose-decomposing bacteria.

In some wet soils with a pH near neutral, the activity of **denitrifying organisms** may be quite high. Under anaerobic conditions, certain bacteria derive their oxygen supply from the oxides of nitrogen (anaerobic respiration), reducing nitrates to nitrite and then to nitrous oxide or elemental nitrogen. True denitrification is largely limited to the genera *Pseudomonas*, *Achromobacter*, *Bacillus*, and *Micrococcus*. Because these organisms are facultative anaerobes, they can survive under a wide range of soil conditions, and the presence of a large population gives most soils a significant denitrifying potential. However, conditions must be favorable for the organisms to change from aerobic respiration to a denitrifying type of metabolism. This normally occurs in the presence of nitrates and a source of readily available carbohydrates when the demand for oxygen by the microflora exceeds the supply. Anaerobic conditions that favor denitrification are found in flooded soils and even in microsites on well-drained soils, but very little volatile loss of nitrogen is normally expected in acid forest soils because of the scarcity of nitrates in such soils.

Actinomycetes

These heterotrophic organisms are in actuality bacteria. They are unicellular microorganisms that produce a slender, branched mycelium that may undergo fragmentation or may subdivide to form asexual spores. Numerically, the actinomycetes are second only to bacteria in most soils. They are not only taxonomically similar to bacteria; many of their environmental requirements are also similar. They are typically aerobic organisms. Peats, waterlogged areas, and soils whose pH is greater than about 5 are unfavorable habitats for most actinomycetes. They are more numerous in warm climates than in cooler ones. Because of these environmental influences, actinomycetes are believed to be less important in cellulose decomposition in forest soils than in prairie or pasture soils.

The activities of actinomycetes in soil transformations are not well understood, but in many respects their activities are similar to those of fungi. They are active in the decomposition of cellulose and a range of other organic materials. They are not good competitors and appear to play their role in decomposing resistant components of plant and animal tissue. Species of the genus *Streptomyces* are also capable of chitin hydrolysis, while *Nocardia* species metabolize paraffins, phenols, steroids, and pyrimidines. The end products of their activity are complex molecules assumed to be important in the humus fraction of mineral soils. Their absence in strongly acid soils may account for the development of layers that are high in lignin resistant to decomposition (Wilde, 1958).

Certain actinomycetes are capable of synthesizing antibiotics, but the significance of such compounds under field conditions is not clear. The ability to excrete antibiotics or the capacity to produce enzymes that are responsible for lysis (killing) of bacteria and fungi may play an important role in microbial antagonism and in regulating the composition of the soil community. Actinomycetes also cause certain soil-borne diseases of plants, such as potato scab.

Fungi

The microbial biomass within the decomposing litter of forest soils is predominantly fungal, and fungi are probably the major agents of decay in all acidic environments. The large variety and number of mushrooms (fruiting bodies) seen in the forest during wet periods attests to the wide distribution of fungi. Fungi possess a filamentous network of hyphal strands in the soil. The fungal mycelium may be subdivided into individual cells by cross walls or septae, but many species are nonseptate. Fungal mycelia permeate most of the forest floor and are readily seen in mor and moder humus types. Taxonomically, most soil fungi are placed in one of two broad classes: Hyphomycetes and Zygomycetes. Species of Hyphomycetes produce spores only asexually; the mycelium is septate; and conidia of asexual spores are borne on special structures known as conidiospores. Zygomycetes and other fungi, in contrast, produce spores both sexually and asexually. At least six other classes of fungi are found less frequently in soils, but the biodiversity of soil fungi is very poorly understood (Coleman and Crossley, 1996). Fungi are so diverse that it is difficult to classify them on the basis of either morphology or source of carbon since the dominant soil genera can utilize a variety of carbonaceous substrates. They can be most rationally divided into functional groups important in forest soils:

1. **Decomposition of cellulose** and related compounds is one of the most important activities of fungi. They are active in the early stages of aerobic decomposition of wood and other organic debris on the forest floor. These materials include hemicelluloses, pectins, starches, fats, and the lignin compounds particularly resistant to bacterial attack. By degrading plant and animal remains, the fungi participate in the formation of humus from raw residues and aid in nutrient cycling and soil aggregate stabilization.

2. Proteinaceous materials are utilized for both nitrogen and carbon by fungi; as a consequence of proteolysis, they are sources of ammonium and simple nitrogen compounds in the soil. Under some conditions, **competition** between fungi and higher plants for nitrate and ammonium leads to a reduction in available nitrogen for the higher plants.

3. Some fungi are **predators** on such soil fauna as protozoa, nematodes, and certain rhizopods and may thereby contribute to the microbiological balance in soil.

4. **Pathogenicity** is another attribute of some soil fungi. There are both obligate and facultative parasites among this group. Members of the genera *Rhioctonia, Pyritium,* and *Phytophthora* cause damping-off disease among nursery seedlings. *Fusarium* species may cause root rot in the nursery and in older plants. Representatives of the genera *Armillaria, Verticillium, Phymatotrichurn,* and *Endoconidiophora,* among others, may also invade roots of older trees.

5. Fungi form **symbiotic associations** called **mycorrhizae**, or fungus root, with roots of higher plants. The adaptation to root tissue may be associated with the complex nutrient demand of the microorganism, and many fungi have been cultivated in artificial media. The mycorrhizal fungi are very important to the nutrition and growth of trees, and because of their critical relationship with higher plants, they will be discussed in a separate chapter.

6. Mycelia of fungi may somehow be responsible for the development of a hydrophobic property in some forest soils. These soils, mostly sands, are slow to wet once they become air-dry; consequently, their capacity for water retention is greatly impaired. The mechanism of this water **repellency** is not well understood.

Algae

Soil algae are commonly unicellular but may also occur in short filaments or colonies. Taxonomically, they are divided into the green, blue-green, and yellow-green algae and the diatoms. Typically, they possess chlorophyll, which

Figure 6.1   Lichen-shrouded spruce limb in a cool, damp region of Norway.

enables them to use light as an energy source for fixation of carbon dioxide (photosynthesis), thus giving them photoautotrophic nutrition. Algae are most commonly found in fertile soils that are well supplied with bases, available nitrogen, and phosphorus. They tend to be sparse in infertile, acid sands. In temperate regions, green algae and diatoms may be most common, while blue-green algae are most common in tropical areas.

Algae aid in solubilization of soil minerals and thus hasten the process of weathering. They generate organic matter from inorganic substances and increase the humus content of soils. Strains of blue-green algae can assimilate or fix atmospheric nitrogen, thus adding to the nitrogen supply of some soils. These organisms are particularly active in wet and flooded soils and in surface soils whose alkalinity has been increased following burning. Since they do not depend on organic matter as an energy source, they are early colonizers in barren areas, preparing the way for later invasion by higher plants.

Lichens are the result of a symbiotic association between fungi and algae. They usually form crust-like colonies and are often pioneer forms on freshly exposed mineral soil and bare rock, providing organic matter for succeeding higher forms of plant life. They also colonize trees, as illustrated by the encrusted branches of spruce in Norway (Figure 6.1). Green algae are apparently the most common symbiont in lichen formation, but blue-green algae (cyanobacteria) may be predominant in some temperate forests. Denison (1973) concluded that the lichen *Lobaria oregona* contributed 2 to 10 kg/ha of nitrogen annually in Douglas-fir forests of Oregon.

HIGHER PLANT RHIZOSPHERES ARE HOTBEDS OF MICROBIAL ACTIVITY

The roots of higher plants exert a profound influence on the development and activity of soil microorganisms (Lynch, 1990). Roots grow and die in the soil and supply soil fauna and microflora with food and energy. More important, live roots create a unique niche for soil microorganisms, resulting in a population distinctly different from the characteristic soil community. This difference is due primarily to the liberation by the root of organic and inorganic substances that are readily consumed by the organisms in its vicinity. The rhizosphere effect has been demonstrated with a wide variety of forest trees and other plants (Katznelson et al., 1962). Many factors such as the kind and stage of development of the trees, the physical and chemical properties and moisture content of the soil, and other environmental conditions such as light and temperature influence the rhizosphere effect. These factors may act directly on the soil microflora or indirectly by influencing plant growth.

The most important influence of the growing plant on the rhizosphere flora results from the root excretion products and sloughed tissue that serve as sources of energy, nitrogen, or growth factors. The plant root absorbs inorganic nutrients from the rhizosphere, thereby lowering the concentration available for both plant and microbial development. On the other hand, root respiration may increase rhizosphere acidity and hasten the solubilization of less-soluble inorganic compounds. By this means, the amount of available phosphorus, potassium, magnesium, and calcium may increase (Youssef and Chino, 1987). Because the microflora are strong competitors for these nutrients, there can be a temporary deficiency due to immobilization in spite of increased solubility.

There is a high population of ammonifying bacteria in the rhizosphere of many plants that is apparently stimulated by the presence of organic nitrogen compounds. Although these bacteria contribute to increased mineralization in the rhizosphere, net mineralization (and plant-available nitrogen) may be low due to the large amount of immobilization by microorganisms themselves. Fisher and Stone (1969) suggested that in secondary plant succession, conifer rhizospheres mineralized or otherwise extracted some fraction of the soil nitrogen and phosphorus that had been resistant to microbial action under the previous vegetation (Figure 6.2). Fox and Comerford (1992) showed that low-molecular-weight organic acids produced by pine roots increase phosporus solubility in the Spodosols.

There is good evidence that nitrogen fixation by free-living bacteria is greater in the rhizosphere of nonleguminous plant roots than in adjacent soil (Richards and Voigt, 1965). The excretions that stimulate strains of *Azotobacter*, *Beijerinckia*, and *Azospirillum* to prosper in some rhizospheres, but not in others, have not been identified. Coinciding with the increase in nitrogen in the rhizosphere is a greater density of denitrifying organisms, but since the population of autotrophic nitrifiers is not stimulated by root exudates, the level

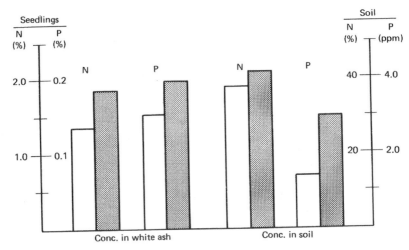

Figure 6.2 Concentrations of nitrogen and phosphorus in white ash seedlings planted in the open (open bars) and at the edge of a conifer canopy (shaded bars), and of hydrofluoric acid-extractable organic nitrogen (as a percentage of total nitrogen) and Morgan-extractable phosphorus in soil (average of four locations) (Fisher and Stone, 1969). (Adapted from *Soil Science Society of America Proceedings* 33:995–961 by permission of the Soil Science Society of America.)

of nitrates in the rhizosphere may not be sufficient to result in significant denitrification,

Root excretions affect the germination of the resting structures of several fungi, perhaps as a result of the energy sources in the rhizosphere. The stimulation can be deleterious to the host plant if the fungus is pathogenic. There are **allelopathic** relationships, or chemical interferences, between some tree species. In some instances, this apparently results from root exudates of one species inhibiting the root development and growth of the other species. Allelopathy may result from rhizosphere microorganisms altering exudates to form materials that are toxic to some species (Tubbs, 1973; Fisher, 1979). On the other hand, there is evidence that the rhizosphere flora protects the root from some soil-borne pathogens. Some antibiotic-producing microorganisms are found in abundance in the rhizosphere, but what effect these organisms have on pathogens is not known. It appears certain that exudates from some rhizosphere organisms form a kind of buffer zone, protecting roots against the attack of soil pathogens. An example is the protection afforded pine roots against *Phytophthora cinnamomi* by mycorrhizal fungi (Marx, 1977). Rhizosphere microflora also produce considerable amounts of growth substances such as indoleacetic acid, gibberellins, and cytokinins that may influence the growth of the host plant. Root exudates strongly influence microbial actions in the root zone, which in turn influence nutrient mobilization and uptake. Bowen and Rovira (1969) reviewed the early work in this area. Recent work by Grayson et al. (1997) and Dinkelaker et al (1997) has increased our knowledge of these phenomena, but our understanding of the complex relationships in the rhizosphere remains vague.

Beare et al. (1995) proposed that there are other arenas of intense microbial activity in the soil. In addition to the rhizosphere, they identify the Drilosphere, that portion of the soil influenced by earthworm secretions; the Detritusphere, areas of concentration of organic detritus such as the litter layer; and the Aggregatusphere, complex mineral-organic aggregates as "hot spots" of microbial activity.

SOIL CONDITIONS STRONGLY INFLUENCE SOIL ORGANISM ACTIVITY

Many factors affect the population density and species diversity of soil organisms. Among the most important factors are the supplies of oxygen and moisture, soil temperature, levels of inorganic nutrients, and the amount and character of soil organic matter.

Soil animals are less affected by soil conditions than are the microflora. Animals often occupy the transition zone between the forest floor and the mineral soil. Some of the larger animals spend a significant part of their lives outside the confines of the soil; consequently, they may be more affected by such environmental factors as weather, flooding, fire, and site disturbances than by soil conditions. Adverse temperature and site perturbations may destroy many animals or drive the more mobile animals from their normal habitats.

Since soil animals include both primary consumers and predators, they obtain their energy by consuming plant parts or preying on other organisms. Soil acidity and the level of inorganic nutrients indirectly affect animal numbers and activity by affecting the food supply. For example, soils rich in calcium and other bases produce plants with higher concentrations of these elements, rendering them more palatable to such animals as the common earthworm.

Animal populations in general are large in fertile soils capable of producing abundant food supplies, but soil chemical properties normally have less direct influence on animals than do physical properties. Some fine-textured and compacted or gravelly soils discourage burrowing animals. High water tables and impervious layers may greatly restrict animal activity. Soil temperature also significantly influences animal populations.

Moisture affects microbial activity because, as a component of protoplasm, it must be available for vegetative development. An overabundance of water, however, restricts gaseous exchange, lowers the available oxygen supply, and creates anaerobic conditions. Aerobes, anaerobes, and facultative anaerobes may all function in the soil at the same time. Because microbial populations are sensitive to soil moisture conditions, community size and composition in a given soil vary with fluctuations in moisture. Aerobes predominate in well-aerated soils, but a largely anaerobic population is found under waterlogged conditions. The maximum density of microorganisms, however, is usually found at a soil water-holding capacity of 50 to 75 percent.

Temperature affects the activity of all soil organisms but not to the same extent. Each microorganism has an optimum temperature for growth and a range outside of which development ceases. Most soil organisms grow best between 25° and 35°C, but they can survive and develop at both higher and lower temperatures. Some organisms grow readily at temperatures up to 65°C and have temperature optima at 35° to 45°C. Temperature affects population size and the rate of biochemical processes carried out by the microflora up to the optimum temperature for the transformation. Ordinary soil temperatures seldom kill bacteria.

The addition of carbonaceous materials directly affects the numbers and activities of all heterotrophic organisms and indirectly affects autotrophic organisms. The application of sawdust, the turning under of a green manure crop in a nursery, or the incorporation of forest floor material during site preparation all stimulate population numbers and activities, sometimes resulting in nutrient immobilization.

Highly acid conditions inhibit activities of many common bacteria, algae, and actinomycetes, but most fungi are able to function over a wider pH range; consequently, fungi commonly dominate the microbial population of acid forest soils. This is not necessarily because fungi respond better to acidic conditions; rather, it is a consequence of the lack of microbiological competition for the available food supply.

Soil microorganisms require inorganic nutrients, but the addition of fertilizers may affect the activity of these organisms only when it stimulates plant growth. The exceptions are those instances when the supply of nutrients in the soil does not meet microbiological demands.

The season of the year influences temperature, moisture, and food supplies and indirectly regulates microbial activity. Microorganisms are usually most active in spring and fall; activity declines markedly during hot, dry summers and cold winters. Numbers of organisms fluctuate closely with seasonal changes in temperature and moisture (Table 6.1).

Another secondary ecological variable that influences soil microorganisms is soil depth. The greatest concentration of organisms is in the top few centimeters of forest soils, with a rapid decline in the number of most organisms with depth. The decline with depth is probably due to the decrease in organic matter and oxygen.

SOIL ORGANISMS MAY BE VALUABLE INDICATORS OF SOIL HEALTH

Sustainable development has become a byword for ecologically sensitive treatment of the industrialized environment. Whether in forestry or in agriculture, maintenance of soil productivity or "soil health" is essential to sustainable

TABLE 6.1   Summer and Winter Microbial Populations in Two Swamp Soils in South Carolina

Microorganism	Summer		Winter	
	Mucky Clay	Loam	Mucky Clay	Loam
	(Millions per gram of oven-dry soil)			
Aerobic bacteria	77.8	67.5	26.1	20.8
Anaerobic bacteria	3.4	2.8	2.6	1.8
Actinomycetes	6.2	4.4	1.6	1.1
Fungi	0.6	0.6	0.6	0.5
Total[a]	88.0	75.3	30.9	24.2

[a]Difference among soils and between seasons are significant at the 1 percent level.
Source: Priester and Harms (1971).

development. The concept of soil quality is not new, but efforts to find indicators of soil quality or soil health are rather recent (Doran et al., 1994).

There is no doubt that soil organisms contribute greatly to soil health, but finding biological indicators of soil health has been a difficult task. The conceptual difficulties in selecting and using biological indicators are many: the absence of clear baseline data, high temporal and spatial variability, the lack of a consistent response to perturbation, and the lack of consistency of indictors across soil types and climatic zones. Nonetheless, many biological indicators of soil health are being tested, and probably some of them will eventually be adopted for assessing the sustainability of soil use (Pankhurst et al., 1997); Powers et al., 1998).

SUMMARY

The variety, number, and activity of organisms found in forest soils are generally much greater than those in agricultural soils. The majority of the biological diversity in forests is found below the surface of the litter layer. The soil biota form an intricate food web through which nutrients and energy pass. The majority of soil organisms are ubiquitous, but the inoculation or introduction of organisms to a new environment is largely a haphazard occurrence.

Essentially all fauna that inhabit the forest environment influence soil properties in some way that eventually affects tree growth. Vertebrates influence the soil through fertilization, trampling, scarification, and a form of cultivation. Arthropods are important largely as soil mixers; however, some are important decomposer organisms. The ordinary earthworm is probably the most important component of soil macrofauna. Arthropods dominate the mesofauna of forest soils. Springtails, woodlice or sow bugs, spiders, ticks, and mites are abundant in essentially all forest soils and are particularly important in forest litter decomposition.

Soils contain three major groups of microflora: bacteria (actinomycetes), fungi, and algae. Bacteria and fungi dominate in well-aerated soils, but bacteria alone account for most biological and chemical changes in anaerobic environments. Autotrophic bacteria are important in soils in the transformation of compounds of nitrogen, sulfur, iron, and so on. Heterotrophic bacteria are important in decomposition and nitrogen fixation. The microbial biomass within the decomposing litter of forest soils is predominantly fungal, and fungi are probably the major agents of decay in all acidic environments.

The roots of higher plants exert a profound influence on the development and activity of soil microorganisms. Many factors affect the population density and species diversity of soil organisms. Among the most important factors are the supplies of oxygen and moisture, soil temperature, levels of inorganic nutrients, and amount and character of soil organic matter.

CHAPTER 7

Soil Organic Matter

ORGANIC MATTER FUELS THE SOIL'S ENGINE

Soil organic matter, although it forms only a small fraction of most forest soils ($\sim 1-12$ percent), has a profound impact on the physical and chemical properties of soils. Its impact on soil biology is even more enormous. Most soil organisms are heterotrophic and gain their energy by decomposing soil organic matter. The activities of these organisms drive the majority of the transformations that take place in the soil. The quality and quantity of soil organic matter determine the performance of the soil's engine, just as does the fuel you place in your car or your body.

Soil organic matter is of great scientific interest for two somewhat different reasons. The first is its role in maintaining soil productivity (Dyck and Skinner, 1990; Powers et al., 1990; Turner and Gessel, 1990); the second is its role as a source or sink for atmospheric carbon (Johnson, 1992a; Lal et al., 1998a,b). Forest soils are unique in that organic carbon resides both within the mineral soil and on the surface of the mineral soil as organic horizons termed the forest floor.

The forest floor is undoubtedly the most distinctive feature of a forest soil. The term **forest floor** is generally used to designate all organic matter including litter and decomposing organic layers resting on the mineral soil surface. These organic matter layers and their characteristic microflora and fauna are the most dynamic portion of the forest environment and the most important criterion distinguishing forest soils from agricultural (cultivated) soils. Forest scientists sometimes refer to the forest floor as humus or the humus form. These terms are confusing because humus is generally defined as a portion of the forest floor and because humus or humic materials exist within the mineral soil (Stevenson, 1994; Piccolo, 1996).

The forest floor is the zone in which vast quantities of plant and animal remains disintegrate above the surface of the mineral soil (Figure 7.1). Some of this material gradually becomes mixed into the mineral soil and, together with the subterranean portions of the plants, forms the soil organic matter fraction. During the decomposition process, the soil organic matter, including cells of dead microorganisms, serves as a source of carbon for successive generations

Figure 7.1   Forest floor under an old-growth fir-birch stand in Quebec. Note the large amount of fallen logs and debris in this boreal forest.

of organisms. The forest floor provides a source of food and a habitat for myriad microflora and fauna. Their activity is essential to the maintenance of nutrient cycles, particularly those of nitrogen, phosphorus, and sulfur. The removal of forest litter for use as animal bedding in Germany prompted concern over site degradation and led to Ebermayer's (1876) classical work on litter production.

Forest litter layers physically insulate the soil surface from extremes of temperature and moisture content, offer mechanical protection from raindrop impact and erosional forces, and improve water infiltration rates. Foresters, as well as soil microbiologists, generally use the term **humus** rather broadly to mean any organic portion of the soil profile. However, they have long known that there are differences in the humus beneath various forest types and perhaps even beneath the same forest types growing on different soils (Wollum, 1973).

FOREST FLOORS HAVE A DISTINCTIVE STRUCTURE

Three layers, or strata, of the forest floor are customarily designated by forest soil scientists, although they do not all occur on all soils (Hesselman, 1926; Green et al., 1993).

The **L** or **litter layer** consists of unaltered dead remains of plants and animals whose origin can be readily identified. Waksman (1936) defined humus as "all the plant and animal residues brought upon or into the soil and undergoing decomposition." It must be recognized that, while the litter is largely unaltered, it is in some stage of decomposition from the moment it lands on the ground and therefore should be considered part of the humus layer.

The **F** or **fragmentation layer** is a zone immediately below the litter, consisting of fragmented, partly decomposed organic materials that are sufficiently well preserved to permit identification of their origin.

The **H** or **humus layer** consists largely of well-decomposed, amorphous organic matter. It is largely coprogenic, whereas the F layer has not yet passed through the bodies of soil fauna. The humified H layer is often not recognized as such because it has a friable crumb structure and contains considerable mineral materials. It may be mistakenly included in the **A1** horizon of the mineral soil.

For a time, the U.S. Soil Conservation Service (Soil Survey Staff, 1975) compromised on a relatively simple procedure, dividing surface organic horizons of mineral soils into two subdivisions, O_1 and O_2. The O_1 corresponded to the L layer and most of the F layer. The Soil Conservation Service has also called this layer **Aoo**. The O_2 corresponded to the H layer and had been previously designated as **Ao**. The most recent terminology (Soil Survey Staff, 1993) defines **Oi** as the slightly decomposed organic layer, **Oe** as the organic layer of intermediate decomposition, and **Oa** as the highly decomposed organic layer. These new designations closely parallel the older L, F, and H designations still in common use.

HUMUS LAYERS COME IN A VARIETY OF TYPES

There is little unanimity among soil scientists on classification systems for the organic layers of forest soils. The term forest floor refers to all organic materials resting on, but not mixed with, the mineral soil. Muller (1879) proposed the first generally acceptable classification of forest floors based on field experiences in Denmark. His classification scheme was largely based on morphological characteristics, but he recognized that variations in properties were due primarily to differences in biological activity in the forest floor. He divided the forest floor into two broad groups and described their relation to forest growth and their effects on soil development. He termed the superficial deposit of organic remains **mor humus** and the intricate mixture of amorphous humus and mineral soil muld or **mull humus**. Muller described mor humus as compacted and sharply delineated from the mineral soil below, while mull humus possessed a diffuse lower boundary and a crumb-like or granular structure.

Heiberg and Chandler (1941), Handley (1954), and Remezov and Pogrebnyak (1969) have given historical accounts of the evolution of terminology and classification of forest floor types and have pointed out some of the variations in nomenclature that have been used in various parts of the world. For example, the terms raw humus, peat, acid humus, duff, and mor have been used to describe one type of forest floor, while mild humus, leaf mold, and mull have all been used to designate another type.

Some general conclusions can be drawn concerning the origin of the distinctive features of mor and mull forest floor, but probably none of them holds for all soil conditions and plant communities. For example, mull forest floors are generally formed under hardwood forests and under forests growing on soils well supplied with bases. Mor forest floors are most often found under coniferous forests and heath plants often growing on spodic (podzolic) soils. Mulls and mors, however, are by no means found exclusively under these forest types or on these soils (Romell, 1935). In fact, other factors in addition to the type of litter also influence the development of the forest floor. Climate has an influence apart from its effect on soil and vegetation development. Furthermore, the fragmentation, decomposition, and mixing brought about by organisms associated with the litter layer have a powerful effect on humus development.

SEVERAL FOREST FLOOR CLASSIFICATIONS ARE USED IN NORTH AMERICA

It is evident that not all forest floors or humus layers fit easily into one or the other of the two broad groups, mor and mull. The heterogeneous nature of soils in the United States and elsewhere has been recognized, and various subdivisions of the two forest floor types have been proposed. For example, Romell and Heiberg (1931), as revised by Heiberg and Chandler (1941), described humus layer types for northeastern United States upland soils. They defined mull as a humus layer consisting of mixed organic and mineral matter in which the transition to a lower horizon is not sharp. They divided mull humus into (1) coarse, (2) medium, (3) fine, (4) firm, and (5) twin types. They described mor humus as a "layer of unincorporated organic material usually matted or compacted or both, distinctly delimited from the mineral soil unless the latter has been blackened by the washing in of organic matter." They divided mor humus into (1) matted, (2) laminated, (3) granular, (4) greasy, and (5) fibrous types on the basis of morphological properties. An example of mor humus is shown in Figure 7.2. However, while this classification was adequate for the northeastern United States, some forest floors in other regions were not clearly identified under this system.

Figure 7.2 Thick mor humus in a spruce stand in central Norway. Note the gray, leached E layer over a spodic horizon.

In a report for the Soil Science Society of America, Hoover and Lunt (1952) proposed a more comprehensive system based on the presence or absence of a humified layer; the degree of incorporation of organic matter into the upper mineral soil layer; and the structure, thickness, and organic matter content of the humus layer and/or Ah horizon. They proposed three major categories: **mull, duff mull,** and **mor.** They described duff mull as having a humified layer with an underlying Ah horizon essentially similar to that of a true mull and possessing characteristics of both mulls and mors. This type of humus has also been termed **moder** (Kubiena, 1953; Edwards et al., 1970). Lyford (1963) believed that moder humus results from continued mixing of mineral soil into the lower part of the forest floor by ants or rodents without destroying the organic horizons.

Green et al. (1993) proposed a comprehensive classification of organic and organic enriched mineral horizons. They defined three taxa — mor, moder, and mull — at the order level that are differentiated by the type of F horizon and the relative prominence of the Ah horizon. Sixteen taxa that reflect differences in the nature and rate of the decomposition process are differentiated at the group level.

While these and other detailed systems of forest humus classification (Romell and Heiberg, 1931; Mader, 1953; Wilde, 1958, 1966) may have merit for technical reporting, they have only minor value in predicting site produc- tivity or in facilitating other management decisions and are probably too complex for general field use. For field purposes, grouping of forest floors into three broad types—mor, moder, and mull—is adequate; however, the classification of Green et al. (1993) has yet to be widely tested.

Mulls are generally less acid than mors; consequently, bacteria are more abundant in mulls, while fungi are the most important microorganisms in the mor forest floor. Mulls generally support nitrification rather well, but mors do not. Large earthworm populations are typically associated with mull humus and, in fact, earthworms and arthropods are believed to be essential in its formation. As a consequence, mull humus often contains significant amounts of mineral material, while mor humus is essentially all organic matter.

Mor forest floors are generally found under the spruce-fir forests of the boreal region of northern and eastern Canada and under much of the coniferous forests of Scandinavia and Siberia. Moders exist under most natural hardwood and mixed pine-hardwood stands such as those of New England and the northern hardwood region. Mulls are often found in the north-central United States and in central Europe. It should be emphasized, however, that local site conditions may alter the type of humus layer developed. Forest floors beneath pines in the southeastern United States are intermediate between mor and mull (Heywood and Barnette, 1936; Metz et al., 1970) but do not fit the classic definition of moder. This may well be a historical anomaly since many of these stands are on abandoned agricultural land and may be too young to have fully developed forest floors.

Quesnel and Lavkulich (1981) suggested that all morphological classification systems of the forest floor are incompatible with soil or ecological classifications because they do not truly reflect forest floor genesis. They propose a system of classification based on physical and chemical properties that would truly reflect the genesis of the forest floor portion of the soil. The system of Green et al. (1993) may fulfill this requirement.

Although **large fallen wood** is not commonly considered to be part of the forest floor, it definitely should be. Often, the average annual input of wood is equal in weight to the input from litterfall (Sollins, 1982). This wood serves an important role as a nutrient sink, a carbon source for microorganisms, and a home for animals. There may be intricate and essential relationships between the presence of large fallen wood, microorganisms, and site productivity, although these relationships have yet to be clearly demonstrated. We do know that this relatively stable component of the forest floor accumulates nutrients, protecting them from loss due to leaching, and serves as a site for biological nitrogen fixation (Grier, 1978; Jorgensen et al., 1984).

FOREST FLOOR DECOMPOSITION VARIES AMONG HUMUS TYPES

The amount and character of the forest floor depend to a large extent on the decomposition rate of the organic debris. The rate of breakdown of the floor material is influenced by the physical and chemical nature of the fresh tissue:

the aeration, temperature, and moisture conditions of the floor and the kinds and numbers of microflora and fauna present. Because the decomposition processes are largely biological, these rates are influenced by the same factors that govern the activity of the organisms. Phosphorus and base concentrations, as well as the ratios of carbon to nitrogen and carbon to phosphorus in the debris, affect the activity of microorganisms.

The rate of breakdown of fresh litter can be quite rapid, with a turnover rate varying from 1 to 3 years in temperate and cool climates to a few months in the tropics. The percent loss of dry leaf weight of four tree species in Tennessee during one year was: mulberry, 90 percent; redbud, 77 percent; white oak, 58 percent; and pine, 40 percent (Edwards et al., 1970). In a similar study, the authors reported that beech leaves lost 64 percent of their weight, oak lost 80 percent, and elm, birch, and ash leaves broke down completely after 1 year.

Litter decomposition rates can be quite slow on cold, dry, or nutrient-poor sites. Fahey (1983) measured a rate of dry weight loss of only 10 percent in nutrient-poor, high-altitude lodgepole pine in Wyoming. Gholz et al. (1985a) found that on nutrient-poor sites in the warm, humid coastal plain of Florida, decomposition rates were as low as 15 percent during the first year. The abundance of various soil biota (Swift et al., 1979) and lignin content (Meetemeyer, 1978) also strongly influence litter decomposition.

As noted earlier, the process of decomposition often begins even before the plant debris is added to the forest floor. Leaf exudates promote the invasion of pathogens while the leaves are still on the trees, and further invasion by fungi occurs during the first few weeks of weathering after the litter reaches the forest floor. The leaves darken, and many of the water-soluble sugars, organic acids, and polyphenols are leached out during that period. As the water-soluble polyphenols are removed by weathering, the litter becomes more palatable to arthropods and earthworms. In these early stages of litter breakdown a large microbial population may be present, but it is generally inactive. Apparently, without preliminary fragmentation by soil animals, the ubiquitous microbes can only slowly decompose the constituents of many leaf species. In temperate regions, earthworms, rotifers, and arthropods including arachnids and crustaceans are responsible for most fragmentation, and if, for any reason, this fragmentation is retarded, the entire process of decomposition is slowed.

Estimates of annual amounts of deciduous leaf litter converted to animal feces range from 20 to 100 percent, according to a review of the subject by Edwards et al. (1970). A succession of species of microflora was identified on fecal pellets, but only minor chemical changes occurred after passage of litter through the guts of soil arthropods. The increase in surface area resulting from ingestion of the litter by arthropods is considered their most important contribution to ultimate breakdown of litter. While some soil fauna are capable of breaking down cellulose with the aid of enzymes in their alimentary tracts, most evidence indicates that the chemical processes of humification are caused more by microbes than by soil fauna (Piccolo, 1996).

In some regions, termites are particularly important in the reduction and decomposition of the large amounts of wood that reach the forest floor. Fungal mycelia ramify through cracks in the wood and soften the tissue, and many insects and larvae then invade the moist tissue. The feces of these animals are a rich substratum for microorganisms. The spectrum of organisms involved depends to a large extent on the nature of the organic debris, but they all work toward the gradual transformation of complex compounds into such simple materials as carbon dioxide, water, nitrogen, other mineral elements, and complex amorphous humic material.

Materials accumulate on the forest floor and, even under aerobic conditions, complete oxidation seldom takes place. While carbohydrates, proteins, and pectins disappear rather rapidly, waxes, resins, and lignins persist for years. As a result, considerable cell substance is synthesized and, together with the modified lignins, constitutes the bulk of the soil humus. The latter is a dark, structureless material composed of complex polymers of phenolic substances (Stevenson, 1994).

MANY FACTORS CONTROL FOREST FLOOR ACCUMULATION

The accumulation of organic materials on the forest floor is largely a function of the annual amount of litterfall minus the annual rate of decomposition. Athough many environmental factors affect the rate of litter decomposition, the rate of litter fall is remarkably uniform among tree species growing under similar soil and climatic conditions.

Periodic **litterfall** is determined by collecting all debris falling to the forest floor in a litter trap and, after drying and weighing, converting to an area basis. A litter trap generally consists of a square frame with low sides and a mesh bottom to allow for drainage. An **annual return** of 2 to 6 Mg/ha appears to hold for most conifers and hardwoods in cool, temperate regions; however, up to 12 ton/ha/yr are produced in tropical rain forests (Bray and Gorham, 1964). The relationship of litter production to latitude is shown in Figure 7.3.

The annual litterfall in a mature northern hardwood stand in New Hampshire averaged 5.7 ton/ha, with leaves, branches, stems, and bark contributing 49.1, 22.2, 14.1, and 1.7 percent, respectively (Gosz et al., 1972). Other structures (bud scales, fruit, and flowers), insect frass, and miscellaneous tissue contributed 10.9 percent. Overstory trees contributed 98 percent of the litterfall. Bray and Gorham (1964) noted that nonleaf litter made up an average of 30 percent of the total litter in forests worldwide. They also reported that an average of 9 percent of total litter was derived from understory vegetation, but they pointed out that the amount of this litter varied with the density of the forest canopy.

The current mass of a forest floor is influenced not only by the annual rates of litterfall and decomposition, but also by the age of the floor or the elapsed time since the last fire or other disturbance. The increase in forest floor thickness is rather rapid in the early stages of stand development and in the

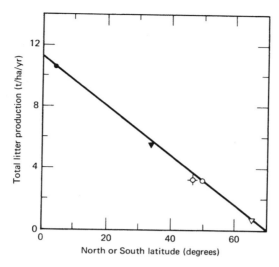

Figure 7.3 Annual production of litter in relation to latitude. Line fitted to means for climatic zones: ●, equatorial; ▼, warm temperate; ◌, cool temperate (North America); ○, cool temperate (Europe); ▽, arctic-alpine (Bray and Gorham, 1964) Used with permission. *Advances in Ecological Research*, Vol 2. Copyright Academic Press, Inc. (New York).

first decade or so following burning. Eventually, a condition of near equilibrium is reached in which the rate of decomposition is about equal to the annual input of organic materials (Bray and Gorham, 1964; Wells, 1971). Hayward and Barnette (1936) thought that a balance between input and output from the forest floor was probably reached in about 10 years in a mature stand of southern pine protected from fire. However, Gholz and Fisher (1982) found that in slash pine plantations on nutrient-poor soils in the same region, the forest floor was still accumulating after 35 years.

The progress toward an equilibrium is much more rapid under favorable growing conditions than under extremes of temperature or moisture. McFee and Stone (1965) reported that the weight of the organic matter in forest floors under mature yellow birch–red spruce stands growing on well-drained outwash sands in the Adirondack Mountains of New York increased from a mean of 131 tons at 90 years of age to 265 tons at 325 years.

Available soil moisture influences the accumulation of forest floor material as a result of its effect on tree growth. Wollum (1973) reported that dry weights of forest floors increased from 9.4 to 80.8 ton/ha along a moisture gradient from dry (piñon-juniper) sites to the wet end of the gradient (white fir) in New Mexico. To be sure, the accumulation of forest floor organic matter, as well as the development of peat soils, may result from reduced microbial activity because of site wetness, especially where there is a plentiful supply of organic litter but a limited period with temperatures favorable for decay. The slow rate of decomposition of organic debris is a major reason for the chronic nitrogen deficiencies noted in some mature stands in the Canadian boreal forest (Weetman and Webber, 1972).

Generally, there is a greater accumulation of mor humus than of mull humus. This apparently results from the slow rate of litter decomposition under mor humus-forming conditions. Both mor and duff mull forest floors were found under old-growth conifer stands in the Cascades (Gessel and Balci, 1965), with the mor possessing a total weight of 158 ton/ha and the duff mull 103 ton/ha. These investigators also found that the weights of forest floors under immature Douglas-fir in drier eastern and wetter western Washington averaged 28 and 14 ton/ha, respectively. Similar values have been found for unincorporated organic debris in forest soils in Minnesota. In mature stands of hardwood in Minnesota, dry matter averaged about 45 ton/ha, while in conifer stands the dry weight of forest floors averaged about 112 ton/ha.

PHYSICAL PROPERTIES OF FOREST FLOORS

Forest floors have wide-ranging physical and chemical properties. Taken together, these properties make the forest floor unique among soil horizons. All the layers of the forest floor have low bulk densities (<1 g/cc). Bulk densities as low as 0.12 g/cc have been reported (Gessel and Balci, 1965; McFee and Stone, 1965). The L layer usually has the lowest density and the H layer the highest. The proportion of large pores is great in both the L and F Layers. This gives forest floors with these layers great hydrologic conductivity, which generally increases the infiltration capacity of the whole soil. The H layer in some forest floors is very compact and may not encourage higher infiltration rates, particularly when dry.

The forest floor in general has a high water-holding capacity, approximately 0.15–0.2 cm per centimeter of depth (Gessel and Balci, 1965; Remezov and Pogrebnyak, 1969; Wooldridge, 1970). This capacity is highest in the H layer and lowest in the L layer. The higher percentage of large pores in the L and F layers leads to increased aeration on wet sites, but it causes forest floors with these layers to dry quickly, and plant roots can frequently occupy the forest floor only temporarily. Mors generally have higher moisture-holding capacities than moders or mulls.

The forest floor has poor heat-conducting properties. It thus insulates the soil, reducing the rate of heat exchange to and from the mineral soil. Mor and moder forest floors are also poor at conducting moisture upward and reduce evaporation from the mineral soil. These forest floor types also protect the mineral soil from the force of raindrop impact and reduce erosion. Mull forest floors do not provide as much protection against either evaporation or erosion.

CHEMICAL PROPERTIES OF FOREST FLOORS

Fresh litter consists of rich, complex organic compounds whose composition may vary widely; however, some general statements can be made about the composition of the raw material from which the forest floor develops. The relative ash content of plant parts is lowest for bole wood and highest for leaves. The phosphorus concentrations in components of slash pine average 0.07, 0.016, 0.018, and 0.006 percent in foliage, bark, branches, and bole wood, respectively. The total ash content of bole wood usually ranges from 0.2 to 1.0 percent, and sapwood normally has a higher content than heartwood.

Bray and Gorham (1964) reported that the ash content of gymnosperm litter ranged from 2 to 6 percent, while that of angiosperm litter ranged from 4 to 14 percent. The leaves of hardwood species generally contain higher concentrations of nitrogen, phosphorus, potassium, calcium, and magnesium than do the leaves of conifers (Table 7.1). Leaves of oak or beech growing on soils low in bases may contain lower concentrations of calcium than do needles of spruce or fir growing on fertile soils. Ash contents of litter of species that usually pioneer in forest development and that often occur on the more infertile soils are generally lower than ash content of species in later successional communities and on more fertile sites (Bray and Gorham, 1964). The age of the leaves at the time they reach the forest floor also influences their composition. In most species, the percentages of nitrogen, phosphorus, and potassium decrease as the growing season progresses. However, a decrease in concentration does not

TABLE 7.1 Nitrogen Concentrations in Leaf and Weight Loss, Nitrogen Deficit, and pH Following 3 Months of Decomposition in the Laboratory

Species	N (%)	After Incubation		
		Weight Loss (% of Original)	N Deficit	Acidity (pH)
Hemlock	1.65	36.8	12.9	6.3
Juniper	1.40	31.5	13.1	7.3
Red pine	1.50	18.2	12.7	5.7
Alder	2.18	34.2	24.3	5.5
Dogwood	1.63	44.6	17.8	7.2
Tulip poplar	1.87	48.8	36.8	7.1
		Average Values		
Conifers	1.52	28.8	12.9	5.9
Hardwoods	1.89	42.5	26.3	6.5

Source: Voight (1965). Reproduced from *Soil Science society of America Proceedings* 29 (1965): 757, by permission from the Soil Science society of America.

always mean a decrease in content because the dry weight of individual leaves often increases throughout the season.

The chemical composition of the forest floor has a significant effect on the rate of litter decomposition, nutrient release, soil organism population, and tree growth. The soil and the vegetation from which the forest floor was developed both influence the acidity, carbon:nitrogen ratio, carbon:phosphorus ratio, and concentration of mineral constituents in the forest floor. These properties influence soil development and stand composition.

As a general rule, mor humus is more acid than mull humus, but because these two types of humus often develop under wide ranges of tree species and soil types, the reaction (pH) of their humus layers also varies within wide limits. Wilde (1958) suggested that the pH of mull humus could vary from 3.0 to 8.0. Heyward and Barnette (1936) found the pH of the F layer under longleaf pine to range from 3.4 to 5.0. These values were 0.25 to 1.0 pH units lower than those of the Ah horizon of the corresponding soil. Lutz and Chandler (1946) reported average values of pH 4.3, 4.5, and 4.9 for the L, F, and H layers, respectively, of mor humus under jack pine and 4.5, 5.9, and 6.5 for the same layers in mull humus under a maple-basswood stand. The reaction of the L, F, and H layers under spruce averaged pH 5.1, 4.9, and 4.7, respectively, while that of the corresponding layers under birch averaged 5.9, 5.7, and 5.7, respectively, in Russian forests (Remezov and Pogrebnyak, 1969). Nitrogen concentration in the humus layers, which varies from approximately 1.5 to 2.0 percent, appears to be correlated with acidity (Voigt, 1965).

The carbon:nitrogen ratio gives an indication of the availability of nitrogen in floor material and of the forest floor's rate of decay. The carbon:nitrogen ratio of forest floors ranges from approximately 20 to 150 or more. In general, this ratio is large in young stands and decreases with age. The ratio is greater in mor humus than in mull humus and does not appear to approach that of humus in agricultural soils. Ratios of the latter humus materials may be as low as 12, a value at which nitrogen mineralization proceeds at a rapid rate.

The chemical composition of the principal layers of the forest floors of some conifer and hardwood stands is given in Table 7.2. All of these stands had been protected from fire for extended periods. Note that the humus from birch stands was generally higher in all nutrients than that from spruce stands in Russia. The nutrient concentration of humus from stands of northern coniferous species was generally higher than that of humus from three southern pine stands. The low concentration of nutrients in the southern pine forest floor probably reflects the low fertility of the soils of the lower coastal plain and may explain the relatively large accumulation of litter in these stands. There is a slower rate of decomposition than might be expected from the favorable temperature and moisture conditions in this area, apparently due to the low nutrient status of the litter.

The concentrations of potassium, calcium, and magnesium generally decreased from the surface litter to the lower humus layers under some stands, while aluminum concentrations increased with depth. This indicates that the bases are eluviated to a greater extent than some other elements. On the other hand, the increases in aluminum concentration, as well as those of iron and manganese in the more decomposed layers, reflect a concentration of these elements and perhaps some contamination from the mineral soil.

Relatively large quantities of nutrients are stored in the forest floor (Cole and Rapp, 1981). The amount and composition of the forest floor, which are influenced by the forest vegetation, climate, mineral soil, and the accumulation period following a major disturbance, dictate the total content of nutrients. In some forest soils, such as glacial outwash sands or sandy soils of coastal regions, the forest floor represents the major reserve of nutrients for tree growth. Although the total nutrient content of forest floors in warm regions may be only a fraction of that of forest floors in cooler areas, the rates of decomposition and nutrient turnover are much more rapid in the warm temperate forests than in cooler forests. It is reported that organic matter accumulation due to the slow rate of decomposition in soils of the boreal forest and other cool-climate areas sometimes results in nitrogen deficiencies (Weetman and Webber, 1972). Regardless of whether the forest floor is developed under a cool or a warm climate, it is the home of most soil organisms, the reservoir of most nutrients involved in the cycling process, and the very life of the soil itself.

NATURAL AND MANAGED PROCESSES ALTER THE FOREST FLOOR

The usual explanation of factors influencing the accumulation and decomposition of the organic debris comprising the forest floor is, understandably, oversimplified. Both natural and human forces act independently and in concert to disrupt an otherwise orderly process.

Wildfires are perhaps the most dramatic of nature's forces of change. A fire can reduce the surface organic layers of the floor to a thin coating of ash in only a few minutes. Not only is the insulating blanket of the floor destroyed by such fires, but the resulting ash may significantly affect the nutrient status of the underlying mineral soils (Grier, 1975). In contrast, **prescribed fires** can and should be controlled so that little more than the litter layer is disturbed and no permanent damage is done to the mineral soil (Wells, 1971; Kodama and Van Lear, 1980).

Lyford (1973) considered the uprooting of trees by **windthrow** and other natural forces a feature of all forested areas and a common disturbance of the organic layers. The pit and mound microrelief resulting from uprooted trees may persist for several decades, particularly in fine-textured soils. Thin or discontinuous mineral soil horizons occur on the mounds, and deep layers of organic matter form in the pits left by such uprootings.

Among nature's other agents operating to alter the forest floor are fossorial mammals, such as gophers, moles, and shrews, and crustaceans (Figure 7.4).

TABLE 7.2 Mass and Elemental Content of Some Forest Floors

Forest Type	Layer	Oven-Dry Mass (Mg/ha)	N	P	K	S	Ca	Mg	Al
						(kg/ha)			
Spruce Russia[a]	L	2.9	34	4	7	4	36	9	9
	F	8.2	119	9	11	7	97	25	48
	H	10.1	133	9	9	8	97	30	110
Birch Russia[a]	L	1.1	15	2	3	1	15	4	5
	F	6.1	101	13	11	13	84	20	29
	H	10.9	129	19	5	46	136	29	137
Longleaf-slash pine USA[b]	L	10.2	53	5	6	—	45	12	—
	F	22.8	123	14	9	—	95	21	—
Slash pine USA[c]	L	15.0	67	4	6	—	99	10	—
	F	24.5	137	4	9	—	282	12	—
Conifer Mor USA[d]	L	14.4	162	15	14	—	—	—	—
	F	22.4	313	24	24	—	—	—	—
	H	121.0	1565	102	89	—	—	—	—
Conifer Moder USA[d]	L	13.6	171	15	15	—	—	—	—
	F	18.0	266	22	21	—	—	—	—
	H	71.7	956	77	67	—	—	—	—
Red pine USA[e]	L	5.3	33	6	7	—	29	3	—
	F	31.7	453	25	25	—	86	13	—
	H	31.2	353	25	25	—	44	22	—
Hemlock-maple USA[e]	L	5.8	39	3	4	—	22	1	—
	F	45	669	45	45	—	94	13	—
	H	31.4	367	28	188	—	31	13	—
Southern pine USA[b]	L	—	53	5	6	—	45	11	—
	F	—	123	13	8	—	96	20	—
Virginia pine USA[f]	L	—	18	1	3	—	11	2	—
	F	—	217	13	10	—	77	9	—
Shortleaf pine	L	—	18	2	3	—	12	2	—

Species	Component									
USA[f]	F	—	217	13	10	—	77	—	9	—
	L	—	28	2	4	—	17	—	4	—
Loblolly pine USA[f]	F	—	178	13	10	—	65	—	9	—
	H	—	59	4	4	—	20	—	3	—
Eastern white pine USA[f]	L	—	23	2	3	—	28	—	3	—
	F	—	124	10	7	—	82	—	8	—
Douglas-fir Eastern Washington USA[d]	All	—	327	29	42	—	—	—	—	—
Douglas-fir Western Washington USA[d]	All	—	193	16	15	—	—	—	—	—
Black spruce Canada[g]	All	—	1214	213	382	—	102	—	430	—
Red spruce Canada[g]	All	—	1465	100	1952	—	253	—	154	—
Birch-spruce USA[h]	All	—	3187	152	91	—	617	—	—	—
Piñon-juniper USA[i]	All	—	80	12	21	—	216	—	41	—
Ponderosa pine USA[i]	All	25.1	191	15	80	—	432	—	83	—
White fir USA[i]	All	80.8	883	85	235	—	2674	—	339	—
Slash pine USA[c]	All	—	183	8	14	—	341	—	20	—

[a] Remezov and Pogrebnyak (1969).
[b] Heyward and Barnette (1936).
[c] Pritchett and Smith (1974).
[d] Gessel and Balci (1965).
[e] Lyford, Harvard Forest.
[f] Metz et al. (1970).
[g] Weetman and Webber (1972).
[h] McFee and Stone (1965).
[i] Wollum (1973).

Figure 7.4 Mineral soil transported into forest floor material by a crayfish on the coastal plain.

These animals often pile soil around burrow entrances and move and mix soil when making tunnels (Troedsson and Lyford, 1973). Abandoned runways eventually collapse or become filled with soil material from above. Transport of mineral soil into the organic horizons may also occur as a result of the activity of ants, termites, earthworms, rodents, and other small animals. Where the mineral soil materials are deposited by these fauna, there are likely to be sudden changes in soil properties. The reverse process, transport of organic matter into the mineral soil, also occurs as a result of animal activity.

Human influence on the forest floor has often been more dramatic than that of nature. In their efforts to increase site productivity by intensive management, human beings have altered the equilibrium that is normally established under mature forests. Timber harvesting, slash burning, site preparation, prescribed burning, fertilization, and even thinning can dramatically alter forest floor properties.

Burning can be considered an acceleration of the natural process of oxidation that the forest floor continuously undergoes. It is a rather drastic operation but, when properly controlled, burning does little long-term damage and may produce some benefits by reducing understory competition and improving conditions for both symbiotic and nonsymbiotic nitrogen fixation. Burning has also been used to reduce excessive buildup of mor humus and to increase mineralization in these acid organic layers. The residual ash results in a decrease in acidity and an increase in the base content of the surface layers (see Chapter 10).

Increasing the amount of sunlight and precipitation reaching the forest floor can sometimes alter organic layers. **Thinning** or removal of forest stands by **harvesting** frequently results in higher soil temperature and moisture, as well as increased decomposition and mineralization of the organic layers. Presumably, the increase in decomposition rate results from increased microbial activity in general, with bacterial activity assuming a more important role in the latter stages of decomposition.

Most forms of **site preparation** for intensive forest management temporarily destroys the forest floor by mixing the organic layers with the mineral soil. Such manipulation may concentrate the humus and increase the aeration and oxidation of organic matter. Schultz and Wilhite (1974) found that the organic matter content of the 0- to 15-cm layer of a flatwoods soil in north Florida was not significantly affected by shallow disking but was increased by 33 percent in low beds after 4 yr. Bedding of these soils also significantly increased the levels of available nitrogen, potassium, calcium, and magnesium in the tree rooting zone during the first few years after planting (Haines and Pritchett, 1965).

Attempts have been made to improve the nutrient status and rate of cycling from the forest floor by altering the stand composition. It is generally believed that nutrients in humus under mixed stands are more readily cycled than nutrients under pore stands. European research has indicated that soil productivity can be improved by introducing hardwood species into coniferous stands. The resultant forest floor is higher in bases, lower in acidity, and more rapidly decomposed. However, interplantings of some hardwood species such as beech are not particularly beneficial to the productivity of spruce stands. It is doubtful that the interplanting of hardwood species on sandy soils or on other sites not favorable for hardwoods will substantially improve site conditions.

Chemical fertilizers often have a direct effect on the forest floor by increasing litter fall. The dry weight of the forest floor under a 15-year-old slash pine stand on a somewhat poorly drained soil fertilized with 45 kg/ha of nitrogen and 267 kg/ha each of phosphorus and potassium was 39.6 ton/ha, compared to 8.2 ton/ha on unfertilized plots. Nutrients in the forest floor were also increased by fertilization. Nitrogen was increased from 53 to 205 kg/ha even though only 45 kg/ha was added as fertilizer (Prichett and Smith, 1974). Fertilizer, wastewater, and sewage sludge application may also drastically decrease the dry weight of the forest floor (Harrison et al., 1995).

ORGANIC CARBON IS A VITAL CONSTITUENT OF THE SOIL

Soil organic matter in the principal source and sink of plant nutrients in both natural and managed forests. Ninety-five percent of nitrogen and sulfur and 25 percent of the available phosphorus in surface soils are found in the soil organic matter. The carbon:nitrogen:sulfur ratio of forest soils is approximately 200:10:1, although it varies somewhat with the type of parent material and the level of atmospheric pollution (Stevenson, 1994). The carbon:potassium and nitrogen:potassium ratios of forest soil vary, depending on the parent material, degree of weathering, and vegetation type. Tropical soils usually contain less total phosphorus and a higher proportion of organic phosphorus than do temperate soils (Coleman et al., 1989). Sanchez et al. (1982) found that tropical forests had higher soil organic matter contents and lower carbon:nitrogen ratios than grasslands, while in temperate areas, grasslands had higher soil organic matter contents and lower carbon:nitrogen ratios than forests.

Nitrogen is held in soils primarily in the organic form. In soils where inorganic sulfur does not accumulate, soil organic matter is the principal source of sulfur. Phosphorus is present in soils in both organic and inorganic forms, but organic phosphorus is the major source of labile phosphorus in soils. The availability of micronutrients (iron, manganese, copper, boron, molybdenum, zinc) is often controlled by soil organic matter turnover. Constituents of the soil organic matter also complex and lower the toxicity of aluminum and manganese species found in many soils.

SOIL ORGANIC CARBON OCCURS IN A WIDE VARIETY OF FORMS

Soil organic matter consists of all of the carbon-containing substances in the soil, except carbonates. It is a mixture of plant and animal residues in various stages of decomposition, the bodies of living and dead microorganisms, and substances synthesized from breakdown products of the above (Schnitzer, 1991). Soil organic matter occurs in solid, colloidal, and soluble states. Solid detritus makes up only a small part of soil organic matter. The largest and most important fractions of soil organic matter are the colloidal and soluble fractions (McColl and Gressel, 1995). The majority of soil organic matter is insoluble and is bound as macromolecular complexes with calcium, iron, or aluminum or in organomineral complexes (Stevenson, 1994).

The organic portion of these complexes is made up of diverse complex compounds called humus. **Humic acid** is that fraction of complex humic substances that is soluble in water under acid conditions (pH < 2) but insoluble under less acid conditions. **Fulvic acid** is that fraction of complex humic substances that is soluble in water of nearly any acidity. The fraction of humic substances that is completely insoluble in water is termed **humin** (MacCarthy et al., 1990). A greater proportion of the more soluble fulvic acids characterizes the humus of forest soils, whereas the humus of peats, grassland, and agricultural soils contains a greater proportion of the less soluble humic acids. The truly soluble soil organic matter is a small proportion of the total soil organic matter. It is composed of amino, fatty, and nucleic acids, carbohydrates, polyphenols, and organic acids. In order of increasing resistance to decay, the organic constituents of the soil are carbohydrates (sugars, starches, hemicellulose, cellulose), proteins and amino acids, nucleic acids, lipids (fatty acids, waxes, oils), lignins, and humus. Lignins are complex, highly aromatic polymers that are difficult to describe and are an important source of humus. Along with lipids, they are insoluble in water and are resistant to microbial decomposition (Tan, 1993). Humus is more or less stable and may remain in the soil for hundreds or thousands of years (Paul and Clark, 1996).

Soil organic matter is frequently divided into heavy and light fractions or into active, passive, and slow fractions. The latter breakdown is more useful in a dynamic functional sense. The active fraction consists of materials with relatively low carbon:nitrogen ratios and short half-lives. This fraction includes microbial biomass, some particulate organic matter, polysaccharides, and other labile organic compounds. The turnover time of the active fraction is rapid and can be measured in months.

The passive fraction of soil organic matter is made up of very stable compounds with extremely long half-lives. This fraction includes organo-mineral complexes, most of the humin, and much of the humic acid portion of the humus. This fraction accounts for 80 to 90 percent of the organic matter in most forest soils. It lends stability to the soil and accounts for most of the cation exchange capacity and water-holding capacity contributed to the soil by organic matter.

The slow fraction of soil organic matter is intermediate between the active and passive fractions. This fraction includes some particulate organic matter with high carbon:nitrogen ratios and high levels of lignin or polyphenolics, as well as other chemically resistant compounds. The half-lives of the components in this fraction may be measured in decades, but this fraction is still an important source of substrate for soil microbes and mineralizable nitrogen.

More recent approaches to examining the nature of humified materials may not separate the material; instead, they may simply characterize the relative frequencies of major types of C bonds. Fourier transform infrared (FTIR) spectroscopy uses infrared light to affect the vibrational frequencies of chemical bonds, and different types of bonds resonate differently. Oak litter from a calcareous site in Israel showed increasing absorbance with depth of soil sampling for phenolic OH- groups, indicating greater humification of deeper material (McColl and Gressel 1995).

Nuclear magnetic resonance (NMR) is a type of spectroscopy that depends on characteristic spin frequencies in an oscillating magnetic fields. The terminology (and technology) are complex; the current state of the art is referred to as nondestructive solid state ^{13}CNMR with cross-polarization and magic-angle-spinning (CPMAS ^{13}CNMR), perhaps even with dipolar dephasing. The spectra produced by CPMAS ^{13}CNMR identify the major types of carbon bonds. For example, humus derived from woody litter was comprised of about 70 percent lignin-carbon, and 5 percent carbohydrate-carbon, and litter derived from leaf, twig, and root litter was 53 percent lignin-carbon and 13 percent carbohydrate-carbon (deMontigny et al. 1993). The spectra for wood-derived humus lacked any peak for tannin-carbon, compared with a strong peak for tannin-carbon in the litter-derived humus (Figure 7.5).

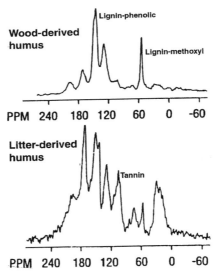

Figure 7.5   Dipolar-dephased ^{13}C CPMAS NMR spectra for humus (>80 % amorphous material) derived from wood and litter from a western redcedar-western hemlock forest. Differences in the origin of humic materials are evident from the high lignin-methoxyl peak for the wood-derived humus and for the tannin peak for the litter-derived humus (after deMontigny et al. 1993).

SOIL ORGANIC MATTER PERFORMS MANY FUNCTIONS

As mentioned at the beginning of this chapter, soil organic matter is the fuel that runs the soil's engine. The life of the soil is carried out largely using soil organic matter as an energy source. The actions of soil fauna and flora enabled by soil organic matter are important in nutrient cycling; in maintenance of soil porosity, hydraulic conductivity, and bulk density; and in soil detoxification processes.

Soil organic matter is important for the formation of soil aggregates (Kay, 1997). Organic compounds, along with fine clays and some amorphous and crystalline inorganic compounds, form the cement between mineral grains in soil aggregates. Soil organic matter contributes to aggregation and to aggregate stability. Since soil structure is important for the soil's ability to receive, store, and transmit water, to support root growth, to the diffusion of gases, and to the cycling of nutrients, soil organic matter is important in determining a soil's productive potential.

Soil organic matter is also an important contributor to the soil's chemical characteristics. Humic substances can have a net negative charge resulting from the dissociation of H^+ from hydroxyl (—OH), carboxylic (—COOH), or phenolic (C_6H_{12}—OH) groups. This dissociation is pH dependent, and at high pH the cation exchange capacity of humus (1500–3000 mmol/kg) may exceed that of silicate clays. Add to this the high nitrogen, phosphorus, and sulfur content of soil organic matter and its importance to soil productivity is clear.

The concern for soil organic matter and its role in soil productivity arises from the fact that soil organic matter is in dynamic equilibrium between inputs and outputs and accumulates to an amount that is controlled by climate, topography, vegetation, and disturbance regime. We generally measure soil organic matter by a one-time determination of soil organic carbon, but this is akin to determining a person's wealth by looking at his assets and disregarding his liabilities. Ecosystems produce new organic matter and decompose old organic matter every year. When these inputs and outputs stabilize, soil organic matter also stabilizes, but if we decrease inputs or increase outputs through cultural activities, we alter the balance and soil organic matter may change dramatically. This could, in turn, decrease productive potential. This is a complex and poorly understood chain of events, and the exact role of soil organic matter in soil productivity declines is unknown.

The soil is both an important sink and source of carbon. The land has been a net source of carbon to the atmosphere since about 1860, when rapid expansion of agriculture began. Schlesinger (1995) estimated that the net annual release of carbon from agricultural lands was about 0.8 Pg/yr, or about 14 percent of the fossil fuel emissions. Forests store about 330 mg/ha of carbon in the vegetation and about 300 mg/ha in the soil worldwide. Roughly 46 percent of the vegetative carbon and 33 percent of the soil carbon are stored in tropical rain forests. Currently, these forests are a net source of atmospheric carbon, but only a small fraction of that amount comes from soil carbon. Temperate forests are in general a net sink for carbon, as forests reclaim land from agricultural uses. However, the source-sink relationships of soil carbon are dynamic (Lal et al., 1998a,b).

The amount of carbon stored in forests and the fluxes of carbon through them are dependent not only on the age of the forest but also on the cultural practices applied to it. Young forests accumulate carbon both above- and belowground, while older forests may be at equilibrium. Harvests temporarily increase carbon evolution, but new growth quickly causes the system to become a net sink. Harvesting followed by burning and site preparation usually causes some loss of soil carbon to the atmosphere. Repeated harvests and intensive culture of short-rotation plantations may not only release carbon to the atmosphere bit also reduce carbon storage to obtain an equilibrium. It is clear that management can alter carbon storage in forests and their soils. How best to manage forests so as to optimize carbon storage has yet to be determined.

SUMMARY

Soil organic matter, although it forms only a small fraction of most forest soils, has a profound impact on the physical and chemical properties of soil. Soil organic matter plays important roles in maintaining site productivity and acts as a sink for atmospheric carbon. The forest floor is undoubtedly the most distinctive feature of a forest soil, and it often contains the major portion of the soil's organic matter. Forest floors have distinctive structure and are generally classified by their structure. They contribute significantly to the hydrologic and nutrient cycles of the site. Soil organic matter is the fuel that drives the biological engine that is at the heart of most soil processes. Although the forest floor is an important feature of forest soils, the organic portion of the mineral soil is probably more influential in most soil processes. It is a principal source of nutrients; is vitally important in determining soil structure, bulk density, and hydraulic conductivity; and contributes the majority of the ion exchange capacity in many forest soils. The amount of carbon stored belowground as soil organic matter may be nearly equal to that stored aboveground in the forest vegetation.

CHAPTER 8

Soil and Roots

Dr. Hans Jenny, a well-known soil scientist, once lamented that "trees and flowers excite poets and painters, but no one serenades the humble root, the hidden half of plants." The paucity of information on roots is particularly acute for trees because of the great difficulties in extracting these large underground organs without destroying or modifying them. The extent and gross morphology of root systems of a number of tree species have been described (Kramer, 1969; Bilan, 1971; Fayle, 1975; Sutton, 1980). Many studies have dealt with growth responses to variations in soil and site properties (Sutton, 1991). However, studies on physiological processes have largely been confined to roots grown in nutrient solutions and to excised roots. Such approaches can be rewarding, but care has to be exercised in extrapolating the results of these studies to field conditions.

Plant roots supply the connecting link between the plant and the soil, and studies of tree root systems are especially pertinent to forest soil science. Roots provide anchorage for the tree and serve the vital functions of absorption and translocation of water and nutrients. They exert a significant influence on soil profile development, and on dying, roots contribute to soil organic matter content (McClaugherty et al., 1982; Fahey et al., 1988). It should not be surprising that the growth and distribution of roots are influenced by essentially the same environmental factors that affect growth of the aboveground portion of the tree. Not only do variations in the chemical, physical, and biological properties of the soil have profound effects on tree roots, but the influence of such factors as light intensity, air temperature, and wind may be reflected as much in root growth as in shoot growth. This is illustrated (Figure 8.1) by the close relationship between root weight and branch weight in radiata pine in New Zealand (Will, 1966).

ROOT SYSTEMS HAVE A CHARACTERISTIC FORM AND AN ENORMOUS EXTENT

When grown under favorable soil conditions, each tree species tends to develop a distinctive root system. This characteristic pattern of root development often persists throughout the life of the tree, but it may become appreciably modified in later years by unfavorable site conditions. Soil texture, compaction, available moisture, impeding layers, and nutrition are some of the factors that can influence the pattern, depth, and extent of root development (Sutton, 1991). Some species hold rigidly to a genetically predetermined root system form and simply do not occur on sites where conditions preclude the development of their normal root system. Many other species have a rather plastic root system form and adapt it to site conditions.

Factors such as stand density, competition among individuals, soil micro-relief, and the percentage of coarse fragments in a soil have a significant effect on the extension of lateral roots. Under the most favorable conditions, it is not uncommon for trees to possess some lateral roots that extend two to three times the radius of the crown (Zimmerman and Brown, 1971; Stone and Kalisz, 1991).

FORM MAY BE OBLIGATORY OR FACULATIVE

Root systems can be characterized on the basis of (I) rooting habit, which relates to the form, direction, and distribution of the larger framework roots; and (2) root intensity, which pertains to the form, distribution, and number of small roots. Although the habit, or form, of a root system is influenced by local site conditions, it tends to be under some degree of genetic control. Most tree root systems can be classified as having a tap, heart, or flat root habit (Figure 8.2). Taprooted trees are characterized by a strong downward-growing main root, which may branch to some degree. Such root systems occur in species of *Carva, Juglans, Quercus, Pinus,* and *Abies.* Numerous strong roots radiating diagonally from the base of the tree without a strong taproot are characteristic of the heart root form. Such systems are found in species of *Larix, Betula, Carpinus,* and *Tilia.* Strong laterals from which vertical sinkers grow downward, as in *Populus, Fraxinus,* and some species of *Picea,* characterize the flat root habit.

The rooting habit of a tree has considerable influence on the type of habitat in which it will thrive. Root form may determine whether or not a species is capable of fully exploiting a site and competing successfully with neighboring species. Root systems of species that are under strong genetic control, such as longleaf pine, tend to retain their rooting habit regardless of the soil conditions to which they are subjected; consequently, they grow well only in a limited range of site conditions. On the other hand, a few species, such as red maple, can adapt their juvenile root systems to a variety of environments, and the species can become established and grow on both wet and dry sites.

Trees that develop strong taproots are capable of penetrating the soil to great depths for support and moisture. For example, many pines, such as longleaf pine, have well-developed taproots and are capable of surviving on deep, relatively dry, sandy sites. This well-developed taproot form is also shared by a number of deciduous trees, exemplified by burr oak and black

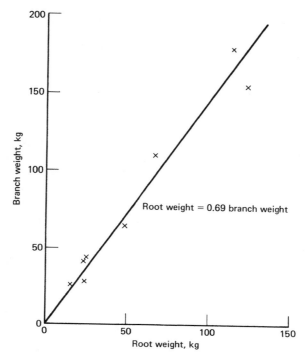

Figure 8.1 Relationship between root weight and branch weight in an 18-year-old radiata pine (Will, 1966). Used with permission of the New Zealand Forest Research Institute.

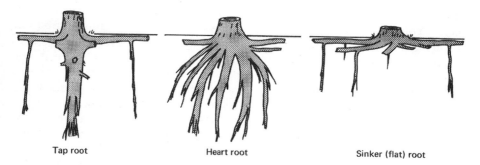

Figure 8.2 Schematic presentation of tree rooting habits.

walnut. Many taprooted trees also have extensive laterals and sinkers that permit them to survive on shallow soils and soils with fluctuating water tables. Although roots do not usually persist in a zone of permanent water saturation, taproots of pitch and slash pine do develop below the depth of perched water tables (McQuilkin, 1935; Van Rees and Comerford, 1990). Cypress (*Taxodium*) and some other species common to swamps have special root adaptations that permit them to grow in saturated zones. If young taproots are injured, several descending woody roots may occur at the base of the tree in place of a single taproot. Taproots of slash pine are often fan-shaped from the capillary zone to below the mean water table. These fan-shaped roots were reported by Schultz (1972) to be two-ranked, profuse, and often dichotomously branched. Boggie (1972) reported that where the water table was maintained at the surface of deep peat, root development of *Pinus conrorta* was confined to a multitude of short roots from the base of the stems, giving a brush-like appearance. Laterals developed only on trees where *Sphagnum* hummocks had raised the soil surface above the mean water level.

Tree species with inherently shallow or flat root systems and those with root systems under weak genetic control have particular advantages on shallow soils. Black spruce is an example of a species that can be found growing over a wide range of soil conditions, from peats with high or fluctuating water tables to deep sands. However, trees with shallow root systems may have a real disadvantage compared to other species on deep, easily penetrated soils, for they are poorly equipped to exploit these conditions and are more subject to windthrow than taprooted trees.

Species possessing a heart root form with a number of lateral and oblique roots arising from the root collar grow best on deep, permeable soils. However, they are also capable of exploiting fissures in fractured bedrock to a greater extent than other types of root systems.

ROOTS GROW AND GROW

Tree roots tend to grow in the direction in which they are pointed until conditions become unfavorable. While a tree's rooting habit may dictate the actual volume of soil occupied by its root system, the number and distribution of small roots determine the intensity with which the occupied soil volume is used. In this respect, root systems can be thought of as either intensive or extensive systems, regardless of rooting habit. Intensive rooting occurs when a relatively small volume of soil is penetrated by a large number of roots, while extensive rooting occurs when a large volume of soil contains but a few roots. The intensity of fine roots in the surface soil is perhaps as much a function of soil nutrient and moisture supplies, temperature, and aeration as it is of genetic control.

Although some roots may extend to great depth (10–15 m), the bulk of the root system of most trees is within less than 1 m of the soil surface, and on many sites the majority of the fine roots (<2 mm) lie in the upper 20 cm of the soil. Schultz (1972) reported that 50 percent of the total root surface area of slash pine was in the surface 30 cm of the soil, while only 6 percent was below 135 cm. Roots less than 2 mm in diameter made up 50 percent of the total root surface area. Spruce trees growing on Spodosols or Inceptisols generally have 70 to 90 percent of their fine roots within the forest floor and the upper 5 to 10 cm of mineral soil. However, in the middle Rocky Mountains, spruce, as well as aspen and Douglas-fir, often occur on Mollisols, where their fine roots are equitably distributed throughout the upper 30 to 50 cm of mineral soil.

The lateral roots of conifers (Figure 8.3) often extend to great distances. Pine root systems are among the least branched (most extensive) of the important forest species. They may possess up to 20 or more first-order lateral roots somewhat evenly spaced around a taproot and extending horizontally for a distance of 15 m or more. They apparently rely heavily on fungal mycelia and associated mycorrhizae for nutrient absorption in the forest floor and surface

Figure 8.3 Lateral roots of a white spruce extending more than 20 m (parent tree with white spot in background) at 0 to 10 cm below the forest floor. (The root is painted white for contrast.)

soil. The advantage of an extensive lateral root system on a heterogeneous site is that it permits a tree more flexibility in utilizing favorable microsite conditions. In a study of the extent of root systems of northern hardwoods, Stout (1956) noted that the mean root weight:crown weight ratio was about 4.5:1. He concluded that under each unit area of ground surface, in closed stands, at least four trees were competing for the available space, water, and nutrients.

SOIL CONDITIONS ALTER ROOT GROWTH

The physical, chemical, and biological properties of soils have profound effects on the rate of root growth and development, root habit, and intensity of root systems of established trees. Consequently, variations in root systems among individuals of a pliable species grown in different soils may be as great as those among different species grown on the same soil.

Physical Impedance

Soil bulk density strongly influences root growth. The bulk density that limits root penetration varies with tree species, soil moisture content, and soil texture (Sutton, 1991). The range of bulk densities that has been reported to limit root penetration is rather wide, 1.1 in silty clay to about 2 in clay loam. However, in general, roots grow well in soils with bulk densities of up to 1.4, and significant root penetration begins to cease at bulk densities of around 1.7. Basal tills and fragipans are examples of soil layers with high bulk density that often limit root growth.

Soil strength is now often used to determine whether or not roots might penetrate a given soil layer. Soil strength is measured with a penetrometer. The values obtained with a penetrometer are determined not only by the bulk density and moisture content of the soil but also by the pore pattern of the soil and the shape and size of the penetrometer probe. Some roots are able to penetrate soil that has soil strength in excess of 300 MPa; however, the force that roots can actually exert is in the range of 50 to 150 MPa. Obviously, roots penetrate soil using mechanisms other than that of the penetrometer; nonetheless, soil strength is a convenient guide to root growth limitation. In medium- and fine-textured soils, soil strength of 250 MPa, as measured by a penetrometer, curtails root extension. In coarse-textured soils, the pores permit the entry of the root tip into crevices in which wedging action permits the root to elongate. Thus, roots can extend into coarse-textured sand that has a soil strength of 500 MPa, as measured by a penetrometer.

Texture and structure may influence rooting through physical impedance and by their influence on soil aeration The soil does not need to become anaerobic before root growth and function are harmed. Oxygen concentration needs to be approximately 20 percent to avoid interfering with respiration at the root tip; however, oxygen concentrations as low as 10 percent may be sufficient to sustain respiration of older root tissues (Waisel et al., 1996). Trees differ in tolerance to reduced aeration. Loblolly pine apparently has a greater tolerance of poor aeration than does shortleaf pine, and red and ponderosa pines are among the most sensitive of all conifers to low-oxygen conditions in the root zone.

Soil compaction may affect both the size and distribution of root systems of planted trees, regardless of soil texture. Minore et al. (1969) grew seedlings of Douglas-fir, Sitka spruce, western hemlock, western redcedar, lodgepole pine, Pacific silver fir, and red alder in soil columns compacted to bulk densities of 1.32, 1.45, and 1.59 g/cm. In 2 years, the roots of lodgepole pine, Douglas-fir, red alder, and Pacific silver fir roots penetrated soil densities that were found to prohibit the growth of Sitka spruce, western hemlock, and western redcedar roots. Machine logging may so compact soil that the following generation of trees experiences reduced growth. Heilman (1981) found that soil compacted by logging equipment to a density of 1.74 to 1.83 g/cm^3 prevented the penetration of roots of Douglas-fir seedlings.

Wetting and drying, freezing and thawing, organic matter increase, and biological activity gradually return compacted soils to their precompacted state; however, the effects of compaction can persist for decades (Froehlich and McNabb, 1984). In most soils, moisture content is an important factor controlling the degree of compaction that occurs under a particular load. The greatest compaction seems to occur at moisture contents near but below field capacity. Compaction may be four or five times greater at 15 percent moisture content than at 5 percent moisture content. Above field capacity, little compaction may take place, but soil aggregates may be destroyed.

Soil Moisture

Moisture has a greater influence on root development and distribution than most other soil factors. Often, slow tree growth rates associated with shallow soils are not due directly to lack of space for root development but rather to the limitation of the water (and nutrient) supply associated with shallow rooting. Lyford and Wilson (1966) suggested that red maple roots grow rather well over a broad range of soil texture and fertility conditions if the soil is maintained at near-optimum moisture levels. Rooting habits change appreciably when maples are grown under very moist, poorly aerated conditions. Kozlowski (1968) reported that large root systems develop in tree seedlings grown in soils maintained close to field capacity, in contrast to the sparse root systems found in soils allowed to dry to near the wilting point before rewetting. Many investigators have noted that small roots die almost immediately in local dry areas (Sutton, 1991). Lorio et al. (1972) found that mature loblolly pine trees on wet, flat sites were nearly devoid of fine roots and mycorrhizal roots compared to neighboring trees on mounds. It is doubtful that any significant amount of root growth takes place when the soil moisture drops to near the wilting point (Waisel et al., 1996), and growth may be exceedingly slow during the dry summer months that would otherwise be favorable for rapid root growth.

There is considerable variation in the response of root systems to changes in soil moisture. Steinbrenner and Rediske (1964) found that roots of Douglas-fir seedlings were concentrated near the surface of well-watered soils but penetrated to considerable depth when surface soil moisture was below optimum. This characteristic has also been observed in red pine (Figure 8.4). On the other hand, ponderosa pine developed deep root systems regardless of the moisture regime.

Optimum soil moisture content for root development depends on soil texture and temperature, among other factors, but soil moisture near field capacity is optimum for most species. Roots of most trees grow best in moist, well-aerated soils, and they generally proliferate in layers providing the greatest moisture supply only if well aerated. Water saturation of the soil results in a deficiency of oxygen and an accumulation of carbon dioxide. Such conditions usually result in reduced root growth and, eventually, in root mortality. Some species, such as bald cypress and water tupelo and certain species of willow and alder, are capable of obtaining oxygen and growing in saturated soils; however, roots of most tree species will not survive long under such conditions. Small lateral roots of slash pine grow above the forest floor into moist but aerated clumps of grass during periods when the water table approaches the soil surface (Figure 8.5).

Frequent fluctuations in the depth of the water table, such as those found in some soils of coastal wetlands, tend to restrict the deep development of tree roots. Roots that develop when the water table is low are later killed when the

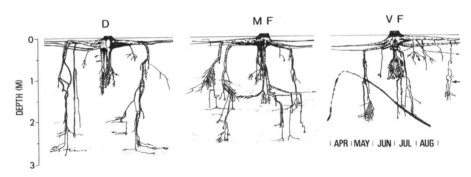

Figure 8.4   Vertical root systems of red pine on dry (D), moderately moist (MF) and very moist (VF) sites. Layers of finer-textured material are shown for D and MF soils and for the location of a moderately cemented layer in the VF soil. The typical seasonal fluctuation in water table level is indicated for the VM soil (Fayle, 1975). Reproduced by permission of the National Research Council of Canada from the *Canadian Journal of Forest Research* 5:109–121.

Figure 8.5 Slash pine roots with ectomycorrhizae and fungal mycelia growing in a clump of grass above the forest floor on a poorly drained site.

water table rises for extended periods. This periodic root pruning tends to maintain an imbalance in the shoot:root ratio. White et al. (1971) reported that stabilizing the water table at 46 and 92 cm below the surface of a coastal plain flatwoods soil (Spodosol) increased the slash pine root biomass by 69 percent and 43 percent, respectively, over the root biomass of trees on soils with fluctuating water tables. On the other hand, the spodic horizon offered little resistance to root penetration once water table levels were lowered and controlled.

Boggie (1972) found that some water control was necessary for the initial establishment of tree seedlings in peat soils, but that once vigorous tree growth was started, drying as a result of evapotranspiration was progressive; that is, improved tree growth was accompanied by further drying of the peat. Pritchett and Smith (1974) reported similar conditions in the wet savanna soils of some coastal areas.

Soil Temperature

Soil temperature affects many aspects of root growth and distribution both directly and indirectly. The minimum, optimum, and maximum soil temperatures for best tree growth vary with species and environmental conditions. The minimum soil temperatures for root growth range from slightly above 0° to 7°C; the optimum from 10° to 25°C; and the maximum from 25° to 35°C (Lyr and Hoffman, 1967). Barney (1951) reported that roots of loblolly pine seedlings grew most rapidly at 20° to 25°C, while growth at both 5° and 35°C was less than 10 percent of the maximum. The optimum temperature for red maple root growth was reported to be about 12° to 15°C (Lyford and Wilson, 1966). Root growth in cool-climate species begins and ceases at lower temperatures than in warm-climate or tropical species.

Roots may not experience true dormancy such as that of buds, but a type of quiescence may be induced by environmental factors (Zimmerman and Brown, 1971). Roots of warm-climate and fast-growing species, such as radiata pine and many eucalypts, apparently never completely stop growing during winter months, although their rate of growth may be slowed by low soil temperature. The root growth of southern pine is slowest during winter months, but roots seldom completely stop growing during most of this period (Kaufman, 1968). Loblolly pine roots continue to elongate until average weekly minimum air temperatures fall below $-2°C$ (Bilan, 1967); roots resume growth in the spring when the daily minimums no longer fall below $-1°C$ (Figure 8.6).

Lyford and Wilson (1966) found that day-to-day variations in the growth rate of red maple root tips were more closely related to soil temperature than to air temperature fluctuations. They grew parts of lateral root systems of large maple trees in sheltered trays and found that the rate of root growth in unheated trays closely paralleled the variation in daily air temperature outside the shelter. When the trays were heated, however, the red maple roots continued to grow in spite of low outside temperatures. During the winter, the temperature of the outside soil, where the trees were located, was below 1°C. While some growth probably resulted from the starch stored in the roots, it

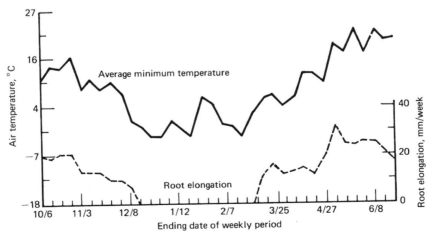

Figure 8.6 Weekly mean minimum air temperature and root elongation of a 1-year-old loblolly pine (Bilan, 1967). Used with permission.

was assumed that translocation of metabolites took place despite the low outside soil temperature.

Cool night temperatures are important in root regeneration of newly planted tree seedlings. Ponderosa pine root regeneration was significantly increased after exposure to at least 90 nights with air temperature below 6°C (Krugman and Stone, 1966). Loblolly pine seedlings also regenerate new root tips more rapidly during periods of cool night temperatures. Soil temperature may also influence the severity of attack from soil-borne organisms and affect the morphogenesis of root systems in some species (Sutton, 1991).

Soil Chemistry

Acidity and nutrient deficiencies or imbalances are the chemical conditions of the soil most likely to restrict plant root growth and development in humid regions. Acid soil may inhibit roots because of the toxicity of aluminum, the solubility of which increases with increasing soil acidity. Root tolerance to acidity differs widely among plant species, but it appears that many tree species are relatively tolerant to acid conditions (Fisher and Juo, 1995; Tilki and Fisher, 1998).

Nutrient availability in the soil affects both the growth rate and distribution of roots. Roots appear to proliferate in zones of high nutrient availability. These increases in rooting density are due to the increased biological activity that takes place in these zones rather than to the tree's activity. Binkley (1986a) has suggested that low nutrient availability might require trees to develop larger root systems than they would under conditions of high fertility. However, fertilizer applications appear to stimulate the growth and development of roots. Kohmann (1972) reported that nitrogen applications increased the number of fine roots of Scots pine (*Pinus sylvestris*) after 2 years. He also found that, while small applications of nitrogen stimulated root growth in forest floor humus, large amounts of nitrogen had a negative effect on root development. Roots in the upper mineral soil layer tolerated higher levels of nitrogen than did roots in the forest floor.

In a field study in which the water table in a Spodosol was controlled at varying heights, White et al. (1971) reported that total root biomass of 5-year-old slash pine was 58 percent greater in fertilized than in control plots. Furthermore, the greatest increase in root biomass resulting from fertilization occurred on those plots where the roots were restricted by a high water table to a small volume of soil.

Soil nutrient levels may also affect the biomass of the shoot:root ratio. The dry weights of slash and loblolly pine seedling roots were approximately one-half as great in phosphorus-deficient soil as in similar soil to which phosphorus had been added (Pritchett, 1972). In these pot tests, the addition of nitrogen alone suppressed root development, but the addition of nitrogen and phosphorus had no significant effect on the shoot:root ratio. It has generally been assumed that the application of nitrogen to deficient soils increases the shoot:root ratio as a consequence of an increase in growth hormones that promote top growth at the expense of root growth (Wilkinson and Ohlrogge, 1964). It appears, however, that the shoot:root ratios of trees may be less affected by soil nutrient levels than the shoot:root ratios of agronomic plants.

Hoyle (1971) reported that deficiencies of calcium and nitrogen prevented yellow birch primary root development in the lower substratum of New England Spodosols. Aluminum was toxic to birch root development, but the levels of other elements in the substratum influenced the degree of toxicity. The greatest reductions in root growth associated with high aluminum levels were found where magnesium and sulfur levels were low. In most cases, forest humus adequately supplied with nutrients in the upper layers did not compensate for nutrient deficiencies in the lower substratum.

ROOTS CONTRIBUTE SIGNIFICANTLY TO SOIL PROPERTIES

Tree roots grow in a variety of forms and to varying degrees of intensity, governed to a large extent by genetic forces but modified by local conditions of soil and site. The influence of these latter factors on the nature and abundance of roots is not well understood because of the difficulty of studying roots under field conditions.

The fact that woody plant roots are perennial, penetrate the soil to great depth, and have the ability to concentrate in soil layers most favorable to growth makes subsoil horizons much more important in forest soils than in agronomic soils. Other functions of roots that are often overlooked, but that are of tremendous importance, are those related to stability and development of the soil.

Tree roots provide a significant stabilizing force in mountainous areas where the soil is subject to erosion or mass movement. Fine roots and fungal mycelia serve as binding agents for surface soil, and where larger roots penetrate the surface horizons, they can anchor the soil mantle to the substrate (Swanston and Dyrness, 1973). It has been noted that the number of landslides from cut-over areas increases within 3 to 5 years of logging. This increase is attributed to a reduction in soil shear strength caused by the decay of tree roots following logging (Figures 8.7 and 8.8).

The value of living tree roots in anchoring shallow soils to the underlying subsoil is particularly important in small drainage areas, where winter storms can cause a sharp rise in groundwater level. Clear-felling further contributes to a decline in soil retention by roots through greater exposure of the soil surface and increased soil moisture levels following removal of the intercepting and transpiring tree canopy.

The form and extent of the tree root system also influence the amount of soil disturbance resulting from windthrow. Soil mixing and the development of a pit and mound microrelief are brought about by the uprooting of individual trees during storm periods, with the greatest soil modifications wrought by

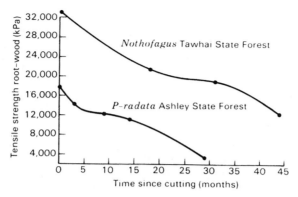

Figure 8.7 Tensile strength curves for two species showing the strength of living roots and the loss of strength as roots decay (O'Loughlin and Watson, 1981). Used with permission.

those trees with deep taproots or heart roots. A tree with a shallow root system may disturb a large area of soil when it is windthrown, but generally only the surface layer is affected.

MYCORRHIZAE—UNIQUE ROOT FORMS WITH AN IMPORTANT FUNCTION

Mycorrhizae are specialized root-like organs formed as a result of the symbiotic association of certain fungi with the roots of higher plants. Specific fungi

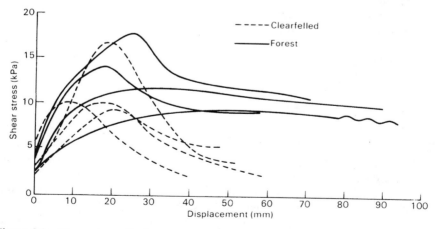

Figure 8.8 Shear stress displacement curves from field tests of the upper 5 cm of soil in a *Nothofagus* forest (O'Loughlin et al., 1982). Used with permission.

grow on and vigorously invade portions of the root in that area of the root system primarily responsible for nutrient absorption. In 1885, Frank, a German forest pathologist, coined the term mycorrhiza, meaning fungus root, to denote these particular associations of roots and fungi. A number of excellent surveys of the subject have been published, including the early work by Hatch (1937) and more recent treatments by Gerdemann (1968), Harley (1969), Haeskaylo (1971), Marks and Kozlowski (1973), Malloch et al. (1980), Mikola (1980), Smith (1980), Schenck (1982), and Allen (1990).

Without mycorrhizae, most of our important tree species could not survive long against the dynamic, fiercely competitive biological communities that inhabit forest soils. Furthermore, the mycorrhizal condition is the rule, not the exception, in nature. Roots of most plants are infected with mycorrhizal fungi. The morphology of mycorrhizae varies among plant species, and each species tends to have characteristic groups of fungi capable of producing mycorrhizae. However, Melin (1963) pointed out that many different Basidiomycetes are able to form mycorrhizae with the same tree species. For example, more than 40 species have been proven to form mycorrhizae with *Pinus sylvestris*. A single tree may be associated simultaneously with many fungal species, and more than one species of fungus has been isolated from an individual mycorrhiza.

THERE ARE TWO TYPES OF MYCORRHIZAE

Mycorrhizae have traditionally been divided into two broad classes based on the spatial interrelation of thread-like fungal hyphae and root cells. One group, the ectomycorrhizae, is characterized by fungal hyphae that penetrate the intercellular spaces of the cortical cells and usually form a compact mantle around the short roots (Figure 8.9). The other, the endomycorrhizae, is characterized by hyphae that penetrate the cells of the root cortex and do not form a fungal mantle (Figure 8.10). The latter group is also called vesicular-arbuscular mycorrhizae, V-A mycorrhizae, or VAM (although, many do not form vesicles and arbuscular mycorrhizae); AM is now the prefered acronym (Sylvia et al., 1998).

Ectomycorrhizae

Ectomycorrhizae have a much more limited distribution among plant species and a greater morphological uniformity than do endomycorrhizae. Ectomycorrhizae are restricted almost entirely to trees, but they occur naturally on many forest species. More than 2000 species of ectomycorrhizal fungi are estimated to exist on trees in North America. Most of these are *Basidiomycetes*, but certain of the *Ascomycetes* also form ectomycorrhizae. The fruiting bodies of these fungi produce spores that are readily and widely disseminated by wind and water. Ectomycorrhizae are characteristic of the families Pinaceae,

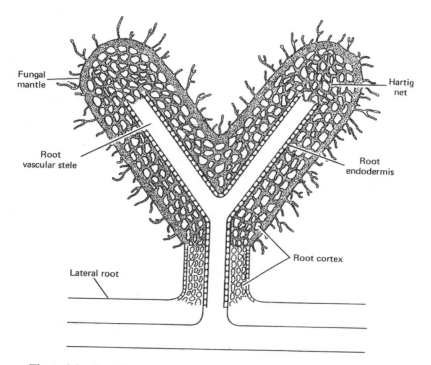

Figure 8.9 Graphic presentation (not to scale) of an ectomycorrhiza.

Fagaceae, arid Betulaceae. **Eucalyptus** and some tropical hardwood species are also ectomycorrhizal, while such angiosperm families as Salicaceae, Juglandaceae, Tiliaceae, and Myrtaceae may be either ectomycorrhizal or endomycorrhizal, depending on site conditions.

Ectomycorrhizal infection is initiated from spores or hyphae of the fungal symbiont in the rhizosphere of feeder roots. Contact between hyphae of a mycorrhizal fungus and a compatible short root may originate from spores germinated in the vicinity of the roots, by extension through the soil of hyphae from either residual mycelia or established mycorrhizae, or by progression of hyphae through adjacent internal root tissue. Fungal mycelia usually grow over the feeder root surfaces and form an external mantle or sheath (Figure 8.11). Following mantle development, hyphae grow intercellularly, forming a network of hyphae (Hartig net) around root cortical cells. The Hartig net, which may completely replace the middle lamellae between cortical cells, is the major distinguishing feature of ectomycorrhizae (Kormanik et al., 1977). Ectomycorrhizae may appear as simple, unforked feeder roots or as bifurcated, muitiforked coralloids, nodule-shaped modifications of feeder roots (Figure 8.12). Hyphae on short roots often radiate from the fungal mantle into the soil, thereby greatly increasing the absorbing potential of the roots.

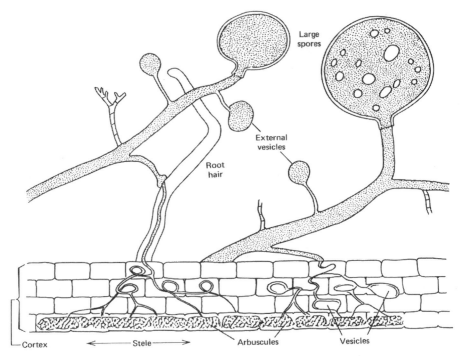

Large
spores

External
vesicles

Root
hair

Cortex   ←——— Stele ———→   Arbuscules   Vesicles

Figure 8.10   Graphic presentation (not to scale) of an endomycorrhiza. From T.H. Nicolson, Vesicular-Arbuscular Mycorrhiza — a Universal Plant Symbiosis. In *Science Progress* 55 (1967):561–581. Blackwell Scientific Publications Ltd., Oxford, UK. Used with permission.

Endomycorrhizae

The endomycorrhizal fungi are the most widespread and important root symbionts. They are found throughout the world in both agricultural and forest soils. They occur on most families of angiosperms and gymnosperms, including nearly all agronomic and horticultural crops. Among the forest tree genera with these types of mycorrhizae are *Acer, Alnus, Fraxinus, Juglans, Liquidambar, Lirindendron, Platanus, Populus. Robinia, Salix,* and *Ulmus* (Gerdemann and Trappe, 1975).

Endomycorrhizae of trees are produced most frequently by fungi of the family Endogonaceae. They generally form an extensive network of hyphae on feeder roots and extend from the root into the soil, but they do not develop the dense fungal sheath typical of ectomycorrhizae. Endomycorrhizal fungi form large, conspicuous, thick-walled spores in the rhizosphere, on the root surface, and sometimes in feeder root cortical tissue (Kormanik et al., 1977). The fungal hyphae penetrate epidermal cell walls of the root and then grow

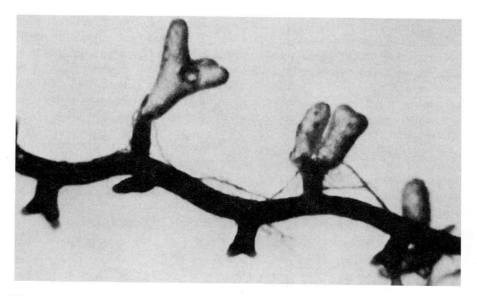

Figure 8.11 Ectomycorrhizae on slash pine. Note the fungal mantle on the short roots.

Figure 8.12 Clumps of ectomycorrhizae on a conifer root, showing a bifurcated or coralloid short root.

into cortical cells, where they develop specialized absorbing structures called arbuscules in the cytoplasmic matrix. In some instances, the fungus completely colonizes the cortical region of the root, but it does not invade the endodermis, stele, or meristem (Gray, 1971). Thin-walled, spherical vesicles may also be produced in cortical cells, leading to the use of the term vesicular-arbuscular to denote this type of endomycorrhizae. Endomycorrhizal infection does not result in major morphological changes in roots; therefore, the unaided eye cannot detect it.

The fungi that form endomycorrhizae do not produce aboveground fruiting bodies or wind-disseminated spores, as do most ectomycorrhizal fungi. Because spores are produced belowground, the spread of endomycorrhizae is not as rapid as that of ectomycorrhizae, and they may be slow to invade fumigated soil and mine spoils. Although endomycorrhizal fungi depend mainly on root contact, moving water, or soil fauna for dissemination, they are nonetheless widespread in soils.

SOIL FACTORS AFFECT MYCORRHIZAL DEVELOPMENT

Environmental factors influence mycorrhizal development by affecting either the tree roots or the fungal symbionts. After the formation of a receptive tree root, the main factors influencing susceptibility of the root to mycorrhizal infection appear to be photosynthetic potential and soil fertility (Marx, 1977). High light intensity and low to moderate soil fertility enhance mycorrhizal development, while the opposite conditions may reduce or even prevent mycorrhizal development.

The effects of soil fertility and fertilizer additions on the development of ectomycorrhizae appear to vary with the original fertility of the soil and the nutrient content of the host plant. Mikola (1973) reported that ectomycorrhizal formation on white pine seedlings is stimulated by applications of phosphate fertilizers to soils containing a low population of relatively inactive mycorrhizal fungi. It appears that growth suppression of pine following nitrogen applications to phosphorus-deficient soils results from a reduction in mycorrhizal development (and phosphorus absorption) caused by the high nitrogen:phosphorus ratio in host plant tissue (Figure 8.13).

There is no evidence that light has a direct effect on mycorrhizal development in soils. However, temperature can have a profound effect on the growth of certain mycorrhizae. Optimal temperatures for mycelial growth lie between 18° and 27°C for the majority of species. Growth of most mycorrhizae ceases above 35° and below 5°C (Figure 8.14). Some species may have wider temperature tolerance, for it has been pointed out that alpine and arctic timber lines are formed by ectomycorrhizal tree species (Moser, 1967). Such ectomycorrhizal fungi as *Pisolithus tinctorius*, which develop at soil temperatures of 34°C or higher, offer advantages in forestation of some adverse sites (Marx et al., 1970).

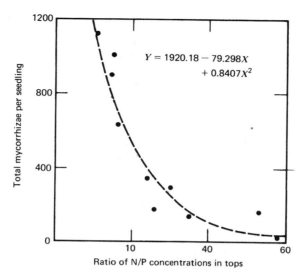

$$Y = 1920.18 - 79.298X + 0.8407X^2$$

Figure 8.13 The relationship of the number of ectomycorrhizae to the nitrogen: phosphorus ratio in loblolly pine seedling tops (Pritchett, 1972). Used with permission.

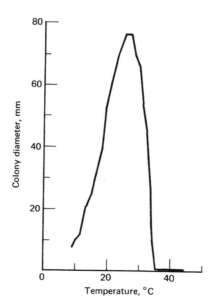

Figure 8.14 Relationship of temperature to growth of *Rhizopogon luteolus* in 11 days (average of two strains) (Theodorou and Bowen, 1971). Used with permission.

TABLE 8.1 Effect of Soil Acidity on Growth, Mycorrhizal Infection, and Type of Infection of *Pinus radiata* Seedlings

Soil Acidity (pH)	Seedling Weight (g)	Mycorrhizae per Seedling (%)	Mycorrhizae Type	
			Brown (%)	White (%)
4.5	1.16	41	43	57
6.2	1.61	53	67	33
8.0	0.33	18	100	0
LSD ($P = 0.05$)	0.15	15		

Source: Theodore and Bowen (1969). Used with permission.

Apparently, all mycorrhizal fungi are obligate aerobes, and it is presumed that mycelial growth decreases as oxygen availability decreases. It is also likely that the requirements of mycorrhizal fungi for nutrient elements are not very different from those of the host plants, although little research has been conducted on this point. It is generally conceded that formation of ectomycorrhizae on tree roots is greatest under acidic conditions; although, this may be due to low nutrient availability rather than to the soil reaction (Table 8.1). Richards (1961) concluded that poor mycorrhizal formation in alkaline soils was due to nitrate inhibition of mycorrhizal infection rather than alkalinity per se. However, Theodorou and Bowen (1969) reported that alkaline conditions in the rhizosphere severely inhibited the growth of some types of mycorrhizal fungi, apart from the possible effect of nitrate inhibition of infection. They further stated that nitrate inhibition of mycorrhiza formation under acid conditions is mainly due to a paucity of infection, not to poor fungal growth.

The presence of antagonistic soil microorganisms can influence survival of the mycorrhizal symbiont as well as root growth of the host plant. Fungicides used in plant disease control can inhibit mycorrhizal fungi under some conditions, or they may stimulate mycorrhizal development by reducing microbial competition. Eradicating ectomycorrhizal fungi from nursery soils by fumigation is generally not a problem because these fungi produce wind-disseminated spores that soon colonize the soil. The soil-borne spores of most endomycorrhizal fungi are eliminated by fumigation of nursery soils, and artificial inoculation is essential if species requiring endomycorrhizae are to be grown successfully.

MYCORRHIZAE BENEFIT THE HOST

The dependence of most species of forest trees on mycorrhizae to initiate and support healthy growth has been most strikingly illustrated by the problem encountered in introducing trees in areas devoid of symbionts. Kessell (1927) failed to establish *Pinus radiata* in western Australian nurseries that lacked mycorrhizal fungi. The seedlings grew normally only after soil from a healthy pine stand was added to the beds and ectomycorrhizae formed. Similar results have been obtained where exotic species have been used in the Philippines, Rhodesia, New Zealand, South America, and Puerto Rico (Vosso, 1971). Other areas where ectomycorrhizal trees and their symbiotic fungi do not occur naturally include former agricultural soils of Poland, oak shelterbelts on the steppes of Russia, and the formerly treeless plains of North and South America, (Marx, 1977).

The observed benefit of mycorrhizae (Figure 8.15) in the growth and development of trees has been ascribed to several factors: (1) increased nutrient and water absorption by virtue of an increased absorptive surface area resulting from the formation of bifurcated or coralloid short roots and as provided by mycelia permeating the soil in the vicinity of short roots; (2) increased nutrient mobilization through biological weathering stimulated by the fungal symbiont; and (3) increased feeder root longevity by provision of a biological deterrent to root infection by soil pathogens. There is general agreement that mycorrhizae increase the capacity of infected plants to absorb nutrients, and this can be especially important on infertile and adverse sites. Carbon compounds synthesized in the green tissue of the host not only nourish the host itself but also serve as a source of carbon for fungal mycelia. In turn,

Figure 8.15 Effect of mycorrhizal fungi on the growth of 6-month-old Monterey pine seedlings; (A) fertile prairie soil, (B) fertile prairie soil plus 0.2 percent by weight of Plainfield sand from a forest, and (C) Plainfield sand alone (courtesy of S.A. Wilde).

soil-derived nutrients absorbed by the mycelia in the soil pass into host tissue. Ectomycorrhizae are able to absorb and accumulate various elements in the fungus mantle, such as nitrogen, phosphorus, potassium, and calcium, and then translocate these elements to host plant tissue (Harley, 1969). Fungal hyphae completely permeate the F and H horizons of the forest floor, and minerals mobilized in this zone may be absorbed before they reach the mineral soil.

The fungi of ectomycorrhizae may be able to break down certain complex minerals and organic substances in the soil and make essential elements from these materials available to their host plant (Voigt, 1971). However, the importance of the solubilization of minerals by mycorrhizal fungi to the nutrition of the host plant is not well understood, and there seems to be little evidence that endomycorrhizae can exploit less soluble forms of phosphates. Nor is there any evidence that mycorrhizal fungi are themselves directly involved in nitrogen fixation, but there is evidence that the mycorrhizal system somehow stimulates nitrogen fixation in the rhizosphere (Voigt, 1971).

Mycorrhizal fungi have been reported to protect delicate root tissue from attack by pathogenic fungi (Zak, 1964). Zak suggested four means by which ectomycorrhizae could provide protection to roots: (1) by using surplus carbohydrates; (2) by providing a physical barrier; (3) by secreting antibiotics; and (4) by favoring protective rhizosphere organisms. Marx (1972) identified an antibiotic, diatretyne nitrile, that afforded some protection to mycorrhizae. Evidence also exists that endomycorrhizae can help control root disease (Schenck, 1983).

Symbiotic fungi may supply higher plants with more than inorganic nutrients from the soil. They may also provide host plants with growth hormones including auxins, cytokinins, gibberellins, and growth-regulating B vitamins (Kormanik et al., 1977).

Mycorrhizae are also important contributors to organic matter turnover and nutrient cycling in forest ecosystems (Fogel and Hunt, 1979; Vogt et al., 1982). Mycorrhizal fungal tissue may account for as much as 15 percent of the net primary production, and the majority of this material enters the SOM pool annually. Mycorrhizae accounted for about one-third of the nitrogen and phosphorus turnover in both young and old *Abies amagilis* stands (Vogt et al., 1982).

SUMMARY

Plant roots supply the connecting link between the plant and the soil. When grown under favorable soil conditions, each tree species tends to develop a distinctive root system. Root systems can be characterized on the basis of rooting habit and root intensity. The rooting habit of a tree has considerable influence on the type of habitat in which it will thrive. Tree roots tend to grow in the direction in which they are pointed until conditions become unfavorable. Lateral roots may spread more than twice the height of the tree, and vertical roots may penetrate many meters into favorable substrates.

The physical, chemical, and biological properties of soils have profound effects on the rate of root growth and development, root habit, and intensity of root systems. Soil bulk density strongly influences root growth. Texture and structure may influence rooting through physical impedance and by their influence on soil aeration. Moisture has a greater influence on root development and distribution than most other soil factors. Acidity and nutrient deficiencies or imbalances are the chemical conditions of the soil most likely to restrict plant root growth and development in humid regions.

Mycorrhizae are specialized root-like organs formed as a result of the symbiotic association of certain fungi with the roots of higher plants. Without mycorrhizae, most of our important tree species could not long survive against the dynamic, fiercely competitive biological communities that inhabit forest soils. Mycorrhizae have traditionally been divided into two broad classes, ectomycorrhizae and endomycorrhizae, based on the spatial interrelation of fungal hyphae and root cells. Mycorrhizae of both types benefit the tree in several ways. They increase nutrient and water absorption by providing an increased absorptive surface area, increase nutrient mobilization through biological weathering stimulated by the fungal symbiont, and increase feeder root longevity by providing a biological deterrent to root infection.

CHAPTER 9

Forest Biogeochemistry

Forest productivity depends on supplies of water, light, and more than a dozen vital elements (nutrients). Soils with high supplies of water and nutrients are more productive than poorer soils, as evidenced by regional relationships between water supply and annual patterns of growth in wet and dry years and between nutrient supply and growth. The dependence of growth on nutrient supply is shown clearly in Figure 9.1.

Growth depends on the nitrogen supply for two reasons: the deployment of a large canopy to intercept and assimilate light requires a large supply of nitrogen, and the accumulating biomass of the forest contains large quantities of nitrogen. Gonçalves et al. (1997) plotted the relationship between above-ground biomass increment and nitrogen uptake for a variety of *Eucalyptus* species (from age 2–10 years) across a range of sites and silvicultural treatments. The patterns between biomass increment and uptake of calcium and potassium were as strong as those for nitrogen (Figure 9.2).

The dependence of growth on water and nutrient supplies is also apparent from experimental trials (Figure 9.3). The intersection of chemistry, geology, and life is the subject of biogeochemistry. This chapter expands the focus on soil solution chemistry from Chapter 5 to consider the broader sources and sinks for nutrients within forests, and the factors that influence these flows.

ENERGY FLOWS WITH ELECTRONS

The biogeochemistry of forests is dominated by flows of energy driven by biological processes such as photosynthesis and respiration. These flows of energy entail the flows of electrons from high-energy and low-energy states, commonly called oxidation and reduction reactions, or redox reactions, respectively. Photosynthesis utilizes radiant energy to boost electrons from low-energy bonds in CO_2 into high-energy bonds in carbohydrates. The oxidation state of carbon goes from $4+$ to 0 as the electrons released in the splitting of water are transferred to the carbon:

$$2H_2O \rightarrow O_2 + 4e^- + 4H^+$$

$$4e^- + 4H^+ + CO_2 \rightarrow CH_2O + H_2O$$

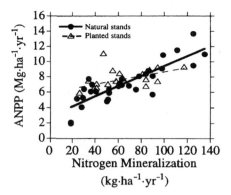

Figure 9.1   Aboveground net primary production (ANPP) of 50 forests in the north-central United States depends strongly on the nitrogen supply in the upper soil (Reich et al., 1997).

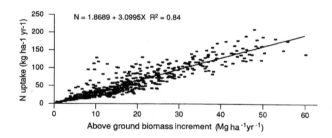

Figure 9.2   Aboveground increment related strongly to nitrogen uptake in over 200 stands of *Eucalyptus* species in Brazil (Gonçalves et al., 1997).

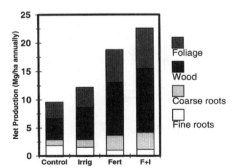

Figure 9.3   The productivity of an 8-year-old loblolly pine stand in North Carolina, responded well to addition of water and very well to addition of fertilizer; addition of both water and nitrogen gave the strongest response (data from Albaugh et al., 1998).

These two equations represent half-reactions that generate electrons (the splitting of H_2O) and that accept electrons (reduction of CO_2).

The tendency of a chemical to release or accept electrons is gauged by the electrode potential. The electrode potential (abbreviated Eh; also called reduction potential and redox potential) can be pictured as a staircase, where substances higher on the stairs (more negative potential) tend to donate electrons to chemicals farther down (more positive potential; Figure 9.4). The difference in potential between the half-reaction for oxygen to gain or release electons and the half-reaction for CO_2 to gain or release electrons is 1230 mV. Substantial energy is released when electrons cascade down the staircase (respiration), and energy is required to boost electrons up the staircase (photosynthesis).

Many compounds accept and donate electrons (Table 9.1). Where oxygen is present in soils, the reduction of oxygen to form water provides the greatest potential change for electrons from organic compounds. If all the oxygen is consumed in the soil (or at the microsite where the chemical reaction occurs), then alternative electron acceptors such as NO_3^-, Fe^{3+}, and SO_4^{2-} can be used as electron acceptors, releasing progressively less energy.

The redox potential of soils is commonly measured with a platinum electrode and may be expressed as pe:

$$pe = -\log A_e.$$

where A_{e-} is the activity of e^- (activities are slightly lower than concentrations, and they approach concentrations at infinite dilutions). At dilute concentrations, pe = (Eh in V)/0.059. At pH 7, oxygen-rich soil has a pe of about 12. As oxygen becomes depleted, pe drops. Aerobic conditions pertain until pe drops

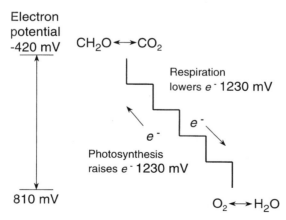

Figure 9.4 Photosynthesis uses sunlight to boost electrons from water (at 810 mV Eh at pH 7) to reduce CO_2 and form sugar (at Eh -420 mV). Respiration reverses this process, releasing energy as sugar is oxidized to CO_2 and oxygen is reduced to water.

TABLE 9.1   Reduction Potentials for Important Half-Reactions in Soils

Half-Reaction	Reduction Potential (Eh in mV at pH 7)	Change in Free Energy If Coupled to CH_2O Oxidation (kJ/mol)
$CO_2 + e^- \rightarrow CH_2O$	-420	0
$H^+ + e^- \rightarrow H_2$	-410	-4
$CO_2 + e^- \rightarrow CH_4$	-244	-23
$SO_4^{2-} + e^- \rightarrow HS^-$	-220	-25
$Fe^{3+} + e^- \rightarrow Fe^{2+}$	-5	-42
$Mn^{3+} + e^- \rightarrow Mn^{2+}$	525	-97
$NO_3^- + e^- \rightarrow NO_2^-$	750	-118
$O_2 + e^- \rightarrow H_2O$	810	-125

Source: Modified from Schlesinger (1997).

below 10 as anaerobic conditions develop and may fall as low as 0 to -4 under very reducing conditions (Lindsay, 1979). The ability of a soil to buffer changes in pe is referred to as poise; a soil with a high supply of oxygen is well poised against changes in redox potential.

The sum of pe + pH tends to be constant for a given soil, which means that if soil oxygen is depleted and pe falls, then pH must rise. For example, Chen (1987) compared adjacent conifer stands on flat topography. A stand of loblolly pine was on a shelf about 1 m above a bottomland planted to bald cypress. She found that the pH was 6.05, with a pe of 10.22 in the cypress stand. The soil appeared to be more acidic under loblolly pine, with a pH of 5.02 and a pe of 11.25. In fact, the pine soil had a lower pH, but the reason was that the redox potential of the soil was higher. The pe + pH for both sites was about 16.25, indicating that the redox and acid systems of the two soils were essentially the same.

NUTRIENT CYCLES INVOLVE POOLS AND FLUXES

Forest nutrient cycles are commonly diagrammed as pools and fluxes indicating major inputs, storage pools, and outputs. Separate diagrams may be used for each element, and the relative importance of various pools and fluxes differs substantially among elements (Figure 9.5). A 16-year-old stand of *Eucalyptus* in India took up about 44 kg/ha of nitrogen annually (not counting the unmeasured nitrogen requirement for fine-root production); this stand also had the benefit of another 12 kg/ha annually that was recycled from foliage prior to litterfall. The potassium cycle included very large losses of potassium from the canopy in throughfall and stemflow, which is typical for this ion that is not structurally bound within organic molecules. The uptake of calcium was very

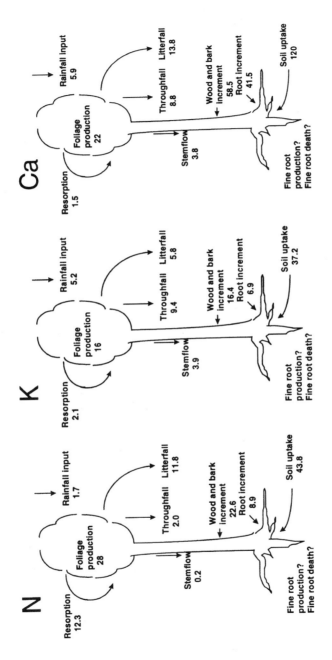

Figure 9.5 Nutrient cycles of a 16-year-old plantation of a *Eucalyptus* hybrid in India. Major features include substantial resorption of nitrogen before litterfall, major throughfall loss of potassium, and large uptake of calcium for wood and bark increment, with very little resorption from leaves before litterfall. Nutrient requirements for the growth of fine roots and nutrient return in fine root death were not estimated (data from Negi, 1984).

large, resulting primarily from the high calcium content of *Eucalyptus* bark and wood; almost no calcium was resorbed prior to litterfall because calcium is bound in insoluble organic molecules. This chapter examines the processes behind the patterns observed in nutrient cycling.

ANNUAL NUTRIENT CYCLING IS GREATER THAN ANNUAL INPUTS

Nutrient inputs from the atmosphere and rock weathering are important to the long-term development of soils and ecosystems, but on an annual basis, nutrient recycling within ecosystems provides the major source of nutrients for plant use. As noted in Chapter 5, an exception to this generalization may be sulfur, where inputs near coastlines and in industrialized regions may be greater than annual uptake by plants. Only a small portion of the total nutrient content of an ecosystem is cycled each year, so it is more valuable to know the rate at which nutrients are cycled each year than to know the total quantity of nutrients within an ecosystem.

At the decomposition stage of a nutrient cycle, nutrient atoms are released into the soil solution largely as by-products of microbial scavenging for energy. The rate of breakdown of organic matter by microbes depends on:

- The chemical quality of the material;
- The availability of energy sources (such as readily oxided carbon compounds) to fuel the microbes;
- The activities of meso- and microfauna; and
- Environmental conditions.

Microbes such as bacteria and fungi excrete enzymes that digest organic molecules into smaller units that may then be absorbed by the microbes. If the nutrient content of the decomposing litter is high, the microbes will find an abundance of nutrients relative to their energy needs and nutrient release will be rapid. If the litter is low in nutrients, the microbes will retain most of the nutrients to grow new cells, and availability to plants will be low (at least until the microbes turn over). This idea is used to explain why decomposition and nutrient release are relatively rapid for low-carbon:nitrogen material (microbes have access to a good supply of nitrogen to decompose the carbon), and slow for high-carbon:nitrogen material (insufficient nitrogen to allow decomposition of carbon, and the nitrogen that is released is immobilized by microbes to degrade more carbon). It's important to note that these carbon:nitrogen patterns may also simply index the types of carbon compounds present, and these types of compounds may simply differ in how easily they are broken down by enzymes. High-carbon:nitrogen materials typically contain polycyclic aromatic compounds that are very difficult to degrade, and low-carbon:nitrogen material contains higher-quality compounds (cellulose, proteins) that are readily used by microbes.

It may be good to keep in mind that the pool with the lowest carbon:nitrogen ratio in forest soils is the well-humified, very recalcitrant fraction of soil organic matter. Hart (1999) had compared the rates of net nitrogen mineralization in well-decayed wood and mineral soil humus in an old-growth Douglas-fir forest. The woody material had a carbon:nitrogen ratio of 117 compared to just 26 for the mineral soil, but the nitrogen in the wood was released more than twice as fast as the nitrogen in the soil.

As with many issues of scale in forest soils, a generalization that works for part of the system (such as rapid decomposition of fresh litter with a low carbon:nitrogen ratio) does not extend to other parts (humus with a low carbon:nitrogen ratio decays very slowly; most of the carbon in woody materials with a high carbon:nitrogen ratio is not relevant to current microbial activity). Although microbes do not have absolute control of nutrient availability, they are strong competitors for available nutrients. As noted in Chapter 5, microbes are strong competitors with trees for available nitrogen (reviewed by Kaye and Hart, 1997), but the short life span of microbes provides better opportunities at a seasonal time scale for trees to compete effectively.

LITTER DECAYS LIKE RADIOACTIVE MATERIAL BUT FOR DIFFERENT REASONS

A large number of litter decomposition studies have found that litter disappearance rates generally follow either a linear or an exponential decay trend. In an exponential pattern, the relative rate of weight loss is a constant proportion of the weight remaining at any given time, and the absolute weight loss is rapid in early stages but slows with time. This can be expressed by the equation

$$B_{T+1} = B_T \cdot e^{-kt}$$

where B is the biomass remaining at time T (or one unit of time later, $T + 1$), e is the base of natural logarithms, t is the time from the beginning of decomposition, and k is the rate constant. This curve can be shaped to fit the decomposition curve of most types of litter merely by changing the k value. The k value is useful for comparing the decomposability of various litter types and for estimating the time required for a given percentage of the litter to be decomposed. For example, $0.6931/k$ estimates the time needed for 50 percent disappearance, and $3/k$ estimates the 95 percent point.

Why should decomposition follow this pattern? If limiting nutrients accumulate in litter at the same time that biomass is respired, decomposition rates should *increase* with time rather than decrease. The answer to this puzzle is based on the chemistry of the organic molecules present in litter. Molecules that are readily respired, such as sugars, disappear quickly, whereas recalcitrant lignin and phenolic molecules are degraded very slowly (Figure 9.6). If litter were composed of a single compound, the classic exponential decay pattern probably would not apply.

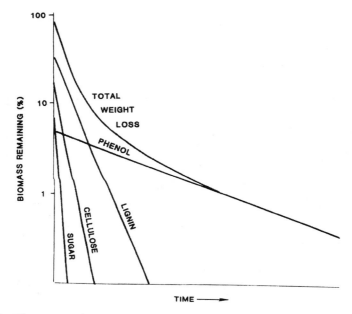

Figure 9.6   The common pattern of exponential decay of litter derives from rates of decomposition of different compounds within the litter rather than from a decrease in the rate of oxidation of particular compounds (after Minderman, 1968; used by permission of the *Journal of Ecology*).

Measurements of the remaining litter biomass are net measurements and include the residue of the original litter, material synthesized by microbes during decomposition, and dead microbial cells. As decomposition progresses, a large proportion of the remaining weight is composed of freshly synthesized materials. Paul and Clark (1996) estimated that by the time half of the mass of litter is gone, about half of the residual mass is actually composed of newly synthesized materials. Therefore, the actual rate of processing of the original litter molecules is faster than the rate of mass loss would indicate.

The overall rate of decomposition, then, is influenced by the types of organic molecules and the nutrient content of the litter (as well as by environmental factors discussed later). The parameter of litter quality that relates best to decomposition and nutrient supply is the lignin:nitrogen ratio (Figure 9.7). Lignin represents a pool of complex compounds with high concentrations of phenolic rings and variable side chains. Lignin is more difficult to decompose than simpler compounds such as cellulose, so litter high in lignin-like compounds decomposes slowly. The nitrogen in litter is present in a wide range of compounds, many of which are relatively easy to decompose. The combination of the lignin and nitrogen concentrations in a single ratio provides a surprisingly strong prediction of rates of litter decomposition and even rates of

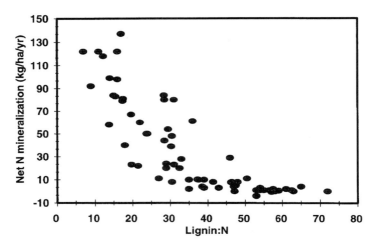

Figure 9.7 Net soil nitrogen mineralization as a function of the lignin:nitrogen ratio of aboveground litterfall from nine common-garden experiments with temperate forest species (from Scott and Binkley, 1997, used by permission of Springer-Verlag).

nitrogen supply in soils. Relatively small variations in lignin:nitrogen within a tree species across sites do not explain the variations in nutrient supply, but the differences in lignin:nitrogen across species provide moderately strong relationships across environments (Figure 9.6). These broad-scale patterns in nitrogen supply and lignin:nitrogen are typical of many patterns in forest ecosystems; the relationship across a very broad gradient in lignin:nitrogen (from 10 to 70) provides a strong correlation with the nitrogen supply ($r^2 = 0.74$). This broad insight into nutrient cycles may not be very useful in specific situations, where the range in the X variable (lignin:nitrogen) is narrow. For example, a site with a lignin:nitrogen ratio of 20 would average about 80 kg of nitrogen per hectare in annual net nitrogen mineralization compared with 40 kg of nitrogen per hectare for a lignin:nitrogen ratio of 35. However, the ranges of the observations overlap very strongly. The range for a lignin:nitrogen ratio of 20 was 25 to 85 kg of nitrogen per hectare annually, which overlaps substantially with the range of 5 to 65 kg of nitrogen per hectare annually for a lignin:nitrogen ratio of 35.

Across species, litter with high nitrogen concentrations tends to decompose faster than litter with low nitrogen content. Would an increase in the nitrogen concentration of litter following fertilization lead to faster decomposition? Similarly, would addition of nitrogen directly to litter increase decomposition rates? Both of these expectations may be appealing, but empirical evidence does not support either one. This is a classic issue of logic: species that produce high-nitrogen litter show faster decomposition, but this correlation does not prove that high nitrogen concentrations cause rapid decomposition. Prescott (1996) noted that decomposition of jack pine litter was not increased with fertilization (over 1 Mg of nitrogen per hectare added!) and that litter from fertilized lodgepole pine trees decayed at the same rate as litter from unfertilized trees despite a fivefold increase in litter nitrogen concentration. She also found that birch leaves decomposed faster in microcosms with red-alder forest floor material than in microcosms with Douglas-fir material because of greater animal activity in the alder material, not because of the nitrogen supply (which was higher in the Douglas-fir material). A lack of relationship between nitrogen supply and decomposition was also reported for red fir branches in California (McColl and Powers, 1998). After 17 years, branches (from a thinning operation) had lost about 40 percent of their original mass in unfertilized plots and in plots that received 300 kg of nitrogen per hectare.

Another error of logic and scale could develop if one expected the correlations with decomposition of fresh litter to work equally well for the decomposition of stabilized, humified organic matter. Melillo et al. (1989) found that the decomposition of red pine needles appeared to have two phases. In the first 3 years of decomposition, the litter lost mass, immobilized nitrogen, and lost acid-soluble carbohydrates. In the second phase, the residual material was high in lignin (lignin comprised a constant 70 percent of the lignin + cellulose fraction), and net nitrogen release from the litter began.

General correlations between decomposition and litter nitrogen content can lead to mistaken inferences about the influence of the nitrogen supply on decomposition. Berg and Matzner (1997) reviewed this relationship and concluded that a high nitrogen supply might speed the decomposition of labile (readily decomposed) carbon compounds but that nitrogen actually *retards* the decomposition of lignified and humified compounds.

Microclimate also regulates litter decomposition and nutrient release. For example, the decay of aspen leaves increased with temperature up to 30°C and with moisture contents up to three times the weight of the leaves (Bunnell et al., 1977). Of course, decomposition is impeded when soils are too warm or too wet. This strong response of fresh litter decomposition to temperature may not apply to the decomposition of old, well-humified soil organic matter (see Chapter 16). The quantity of clay in soils may also influence the accumulation of well-stabilized soil carbon that is resistant to decomposition. Giardina et al. (1999b) reviewed laboratory incubation studies and concluded that when clays comprise less than 30 percent of the soil mineral fraction, the decomposition of carbon relates poorly to clay content. Where soil clay contents exceed 30 percent, the rates of carbon release are quite low (when expressed on a gram of carbon loss/gram of soil carbon basis). Torn et al. (1997) examined a 4-million-year chronosequence in Hawaii and concluded that soil carbon increased for 150,000 yr as the concentration of amorphous clays increased and then declined for the next 4 million years as these clays weathered into crystalline clays.

Soil animals can have a major impact on rates of litter disappearance and decomposition. When earthworms ingest fresh litter and soil, the litter is broken into smaller pieces and blended with mineral particles into moister conditions below the soil surface. Only a small fraction of the ingested organic matter is actually decomposed during passage through the earthworm guts, so the disappearance of litter from the soil surface does not represent immediate decomposition. However, the net effect of mixing by the worms may indeed speed up decomposition. For example, the addition of earthworms to soils developed on minespoils reduced litter accumulation from 6300 kg/ha to 1175 kg/ha over 5 years (Vimmerstedt and Finney, 1973).

Studies of litter disappearance rates from leaves confined in nylon mesh bags (which exclude the larger microfauna) often show disappearance rates inside the bags to be about half those of unconfined litter.

FOREST FLOORS ACCUMULATE AND MAY OR MAY NOT REACH STEADY STATES

The organic layers (forest floor, or O horizons) at the soil surface typically accumulate in forests, sometimes to substantial depth. In general, forest floors are deepest where environments are cool and wet and where mixing by soil fauna is slight. Tropical forests (both plantations and natural forests) tend to accumulate 5 to 15 Mg/ha, equal to about 1 to 2 years of aboveground litterfall inputs (reviewed by O'Connell and Sankaran, 1997). Temperate forests commonly accumulate 20 to 100 Mg/ha of forest floor materials, sometimes including inputs from both aboveground litterfall and fine roots and mycorrhizae that grow into the organic layer. The forest floor accumulation in temperate forests roughly equals 5 to 25 years of aboveground litterfall inputs. The forest floor may include substantial quantities of partially decomposed wood, ranging from less than 20 Mg/ha in tropical forests and areas subject to frequent fires to more than 500 Mg/ha for some cool rain forests with large, long-lived trees.

Extensive research efforts have focused on forest floor morphology and decomposition rates (see Chapter 7, review by Handley, 1954), mostly in the hope that the forest floor would prove to be a key unlocking the insights needed to understand and predict nutrient availability to trees. This expectation has not been borne out; no universal pattern of forest floor morphology or decomposition has proven to be of much use in forest nutrition studies. Old expectations of mull forest floors (where mixing by soil animals comminutes the organic matter and buries it) being somehow richer than mor forest floors (where organic matter remains separate from mineral soil) have been supported only in the broadest terms: the richest forest soils have animals that mix litter into the mineral soil, and the poorest forest soils do not. Between these extremes, many examples have been found in which one forest floor type appears to be more fertile than another, with other cases showing the reverse pattern. For example, the supply of phosphorus to sugar maple trees is lower on mull soils than mor soils in Quebec (Figure 9.8), in contrast to classic expectations of higher fertility of mull soils. We suggest that after more than

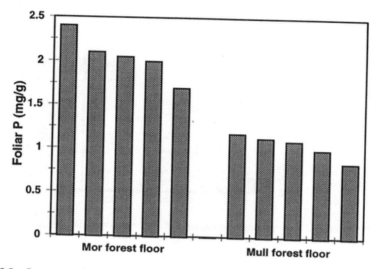

Figure 9.8 Sugar maple phosphorus deficiency develops in mull forest floor because mixing of organic phosphorus pools with mineral soil allows phosphorus to precipitate with iron and aluminum (data from Paré and Bernier, 1989).

75 years, hopeful (but unsupported) ideas about litter decomposition rates and forest floor morphology should be set aside and the processes that determine the turnover of soil organic matter should be examined in both surficial O horizons and mineral horizons.

In the long run, the accumulation of nutrients in tree biomass must come either from inputs to the ecosystem or from a net transfer from some pools in the mineral soil. The aggradation of forest biomass cannot be fueled by nutrients from a forest floor horizon that is itself aggrading (or in steady state). In the short run, the annual growth of leaves and other tissues can be supported in part by nutrients released from the forest floor through an internal ecosystem cycle. In a series of stands of lodgepole pine, the forest floor contributed about two-thirds of the nitrogen mineralized in the forest floor and 0- to 15-cm depth of mineral soil, which was about the same as in a sequence of Englemann spruce-subalpine fir stands in the same region (Figure 9.9). The forest floors of aspen stands contributed a large quantity of nitrogen annually, but the proportion of the total nitrogen mineralized in the forest floor was lower.

This comparison of net nitrogen mineralization in the forest floor and mineral soil has an important limitation that relates to scale. Fungi in forest floors may "import" nitrogen from the mineral soil, providing a nitrogen supply for use in degrading material with a high carbon:nitrogen ratio in the forest floor. Hart and Firestone used ^{15}N techniques to estimate the rate of nitrogen transfer from the forest floor of an old-growth, mixed-conifer forest

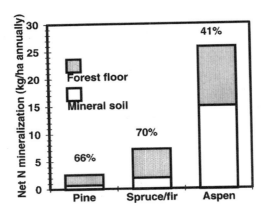

Figure 9.9 Annual net nitrogen mineralization of the forest floors and 0- to 15-cm mineral soil for 10 stands of each type, spanning a range of site productivities. Forest floor mass did not differ significantly among species, ranging from 95 to 110 Mg/ha across species (percentage values represent the proportion of net nitrogen mineralization from the forest floor; data from Stump and Binkley, 1993).

into the mineral soil—and the reverse flow. They found that leaching of organic and inorganic nitrogen from the forest floor transferred about 26 kg/ha of nitrogen annually into the mineral soil. The reverse flow, of nitrogen from the mineral soil into the forest floor via fungal hyphae, was estimated to be as large as 9 kg/ha annually. The import of nitrogen into the forest floor offset about one-third of the release of soluble nitrogen from the forest floor.

Many forests have substantial accumulations of woody material on the soil surface, commonly ranging from 50 to 100 Mg/ha and much higher in some coastal rain forests of the Pacific Northwest. Does the large amount of this material indicate a major role of woody material in annual nutrient cycling? Over a period of 200 years, the nitrogen content of Douglas-fir boles on the soil surface increases by about twofold (Sollins et al., 1987); if a site contained 100 Mg/ha of downed Douglas-fir logs with a nitrogen concentration of 1 mg/g wood, then the nitrogen content of the wood would increase from an initial 100 kg/ha to 200 kg/ha over two centuries, for an annual rate of increase of 0.5 kg/ha. Subsequent decomposition in later centuries would release the nitrogen, with measured rates of release of 1.6 to 2.6 kg/ha annually leaching from well-decayed logs (Hart, 1999). In the absence of large amounts of woody material, the nitrogen would not be immobilized during initial stages of log decomposition or released during later stages. Overall, these rates of nitrogen removal and release are small relative to other fluxes in these forests, such as atmospheric deposition or even nitrogen uptake and release by the moss layer.

HOW DOES LITTER BECOME SOIL HUMUS?

Almost all of the organic compounds deposited in the soils as leaves, roots, and stems die and are returned to the atmosphere as carbon dioxide in a matter of months, years, or decades. Only a very small fraction of this material ends up as stable, humified soil organic matter. The factors that determine the formation (and decomposition) of this major ecosystem carbon pool are not well known. Broad-scale studies show that more carbon tends to accumulate in locations with higher precipitation, but temperature differences do not produce broad differences in soil carbon (see Figure 3.2). More favorable environments lead to greater productivity and organic matter input to soils, but we don't know if the quantity of carbon that enters long-term, humified pools is simply a fraction of the carbon added to the soil or if more complex issues (such as soil nitrogen content, mineralogy, and soil community) play major variable roles.

LITTERFALL AND ROOT DEATH ARE MAJOR PATHWAYS OF NUTRIENT RETURN

Nutrients taken up by trees face several fates: incorporation into accumulating biomass, recycling to the soil via litterfall or root death, leaching from leaves or roots, or recycling from leaves or roots for use the following year. These patterns vary with each nutrient, and general trends can be illustrated using a 14-year-old slash pine forest (Table 9.2). For nitrogen, the total annual uptake was 40 kg/ha; about 40 percent was incorporated into accumulating biomass, none was leached from leaves, and about 3 percent was resorbed into the tree before the remaining 60 percent was lost in litterfall. Internal recycling is often more important in other forests; typically, about 20 to 35 percent of the

TABLE 9.2   Cycles of Nitrogen and Potassium

Component	Age 14 Years		Age 26 Years	
	Nitrogen	Potassium	Nitrogen	Potassium
Stand uptake	40	10.6	30	10.0
Biomass accumulation	16	3.7	8	3
Foliage leaching	0	3.0	0	3
Retranslocation	1	0	16	2.5
Litterfall	23	3.9	22	4.0

Note: Cycles of nitrogen and potassium differ substantially in slash pine plantations (foliage leaching is much more important for potassium), and change with age (kg/ha annually.
Source: Data from Gholz et al. (1985).

nitrogen in leaves is recycled prior to leaf fall. Potassium in the slash pine forests exhibited a very different pattern. Of a total potassium uptake of 10.6 kg/ha, about 35 percent went into the biomass increment, 35 percent was returned to the soil in litterfall, and the remaining 30 percent was leached from the leaves by rainfall. None was resorbed from needles before litterfall.

These patterns change with stand development as forests progress through stages of relative nutrient abundance and scarcity. The role of internal recycling becomes particularly important in older slash pine ecosystems. Some of the mechanisms behind these shifting patterns of nutrient use remain poorly understood.

Rates of litterfall have been examined in scores of forests around the world. A review of studies from tropical forests found that plantations tend to have lower rates of nitrogen cycling in litterfall than natural forests do (Figure 9.10).

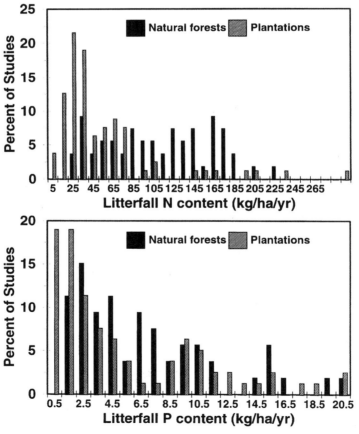

Figure 9.10 Patterns of nitrogen and phosphorus return to soils in aboveground litterfall in tropical plantations and natural forests (from a data compilation of Sankaran and O'Connell, in Binkley et al., 1997).

Tropical plantations tend to cycle 10 to 80 kg/ha of nitrogen annually in litterfall compared with 25 to 170 kg/ha in natural forests. The distinction between plantations and natural forests is less clear for phosphorus, though low rates of phosphorus cycling in litterfall are more common for the plantations than for natural forests.

Of all the carbon that reaches soil over the life of a stand, how much comes from aboveground litterfall, root death, and stem death? These inputs should equal the average net production of each type of tree tissue over the life of the stand. In an average year, foliar production (and death) might be 6 Mg/ha, root production (and death) might be 6 Mg/ha, and wood production might be 8 Mg/ha. The wood tends to accumulate each year, and in the absence of forest harvest, all the wood production eventually reaches the forest floor. In this example, the leaves and roots would comprise about 60 percent of the carbon added to the soil and the wood would comprise 40 percent. But how much would each of these sources contribute to the carbon that ends up as well-humified, long-term soil organic matter? This question remains unanswered; we simply don't know if soil organic matter tends to form more from one type of carbon input than from another. Some researchers believe that woody inputs play a large and important role to the long-term formation of humified soil organic matter, but this hypothesis needs to be tested.

TREES ADJUST TO NUTRIENT LIMITATIONS

The productivity of most forests is limited simultaneously by a variety of resources, including light, water, and one or more nutrients. For a century, the idea of the "law of the minimum" (popularized by Justus von Liebig) dominated thinking about forest production. Droughty sites were assumed to be so limited by the lack of water that addition of nutrients would not increase growth. We now know that forests (and other ecosystems) may be limited by the lack of a single major resource or by the interacting effects of low supplies of several resources (Chapin et al., 1987; Seastedt and Knapp, 1993). The effects of increased supplies of multiple resources are often not linear. For example, nitrogen alone could not increase the growth of *Eucalyptus saligna* in New Zealand, but addition of phosphorus increased growth by twofold. Addition of nitrogen and phosphorus increased growth by sevenfold (Figure 9.11), much more than the sum of the individual nutrient responses. Simple linear thinking (X and only X controls Y) is probably not very useful in forest soils or ecology.

If the supply of nutrients increases on a nutrient-limited site, how do the trees respond? The most commonly examined response is wood production; fertilization on nitrogen- or phosphorus-limited sites often increases wood production by 50 percent or more. This increased production in wood may derive from increased net photosynthesis as a result of improved biochemistry

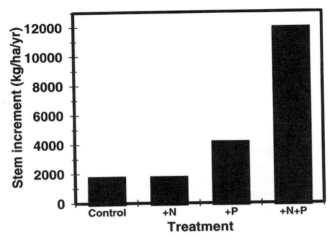

Figure 9.11   Growth of *Eucalyptus saligna* from age 1 year to age 2 years in New Zealand (extrapolated from data in Knight and Nicholas, 1996).

in the leaves; from increased light interception by an expanding canopy; or from a shift in allocation away from roots and mycorrhizae to wood production. See Chapter 13 for detailed consideration of the ecology of the forest response to fertilization.

NUTRIENT TRANSPORT AND MOBILITY WITHIN PLANTS ARE ALSO IMPORTANT

Each nutrient serves unique functions in plants, and these roles affect their mobility. For example, potassium is an enzyme activator and regulator of osmotic potential, and in both these roles it remains a free, mobile cation. Therefore, some potassium can be recycled from senescent foliage before abscission. Calcium usually binds two organic molecules together, remaining relatively immobile in plants, with little recycling before leaf fall. Nitrogen plays a wider variety of roles, and its mobility varies accordingly. If taken up as ammonium, it must be bound into amino acids (which then form proteins) to prevent ammonia toxicity. The simple organic nitrogen compounds then can be transported through the plant. If nitrogen is taken up as nitrate, it may be reduced (gain electrons) and converted to amino acids in the roots or it may simply be loaded into the xylem and transported to other parts of the plant. In either case, the nitrogen eventually is incorporated into various proteins and nucleic acids; some of these can be broken down within the plant, liberating the nitrogen for recycling, but other forms cannot be reused.

INTERNAL RECYCLING INCREASES
AS NUTRIENT AVAILABILITY INCREASES

It seems logical that nutrient conservation through internal plant recycling should be more important on low-nutrient sites, but this appealing idea appears to be wrong. Birk and Vitousek (1984) surveyed the literature from around the world for patterns in resorption of nitrogen and phosphorus as a function of leaf concentration (a measure of nutrient availability). They found no trend for nitrogen resorption when expressed as the percentage of nitrogen recovered from leaves. However, they found that leaves with higher concentrations of nitrogen showed greater retranslocation of nitrogen out of leaves when expressed as milligrams of nitrogen recovered per leaf. Fertilizer studies also show greater recovery of nitrogen from senescing needles in fertilized trees than from control trees. For example, a greater percentage of needle nitrogen was retranslocated from senescing needles of fertilized Scots pine than from needles of control trees (Näsholm, 1994). However, when the process was expressed as milligrams of nitrogen recovered per needle, the fertilized trees were found to be retranslocating more nitrogen from needles than were the control trees. Nambiar and Fife (1991) found the same pattern of greater nitrogen retranslocation from radiate pine needles and went a step further to estimate the total retranslocation from the canopies. Fertilized trees had more nitrogen in their canopies, and total nitrogen recovered per tree was more than three times greater for fertilized trees than for control trees.

Why aren't nutrients resorbed more efficiently on poor sites? As noted earlier, the mobility of nutrients within plants depends on their functions and on the structure of the molecules they comprise. Nitrogen is present in a variety of compounds, some of which are mobile (or can be broken down into mobile parts) and some which are very insoluble. In the Scots pine example above, the fertilized trees had higher concentrations of arginine (a soluble amino acid form of nitrogen), which should be readily removed before senescence (Näsholm, 1994). The recycling of nitrogen may depend more on the form of the nitrogen compounds in the tissues than on the overall nutritional status of the plant or ecosystem. Plants growing under luxuriant nitrogen regimes typically accumulate more "mobile" nitrogen compounds, whereas nitrogen-stressed plants often have a larger proportion in structural, insoluble forms.

It is also important to keep in mind that nutrition is part of the overall physiology of the plant. Nutrient uptake, transformation, and recycling represent major energy costs, and if plants are energy limited, a complete evaluation of any nutrient-use strategy should consider these energy costs. This efficiency can be examined as the grams of glucose required to produce a gram of nitrogen contained in each type of molecule. For example, a root taking up ammonium expends about 1.5 g of glucose for every gram of nitrogen incorporated into glutamine (an amino acid). If the plant takes up nitrate, it must expend an additional 4.4 g of glucose to reduce 1 g of nitrate nitrogen to ammonium nitrogen. Processing the glutamine into more complex amino acids and proteins requires an additional 10 to 15 g of glucose, for a grand total of about 15 to 20 g of glucose to process each gram of nitrogen taken up from the soil (Barnes, 1980). To complete the picture, some added cost should be included for the production and maintenance of roots to obtain the nitrogen from the soil.

What would be the energy cost of internal recycling of nitrogen? The conversion of protein containing 1 g of nitrogen into other forms (such as mobile amino acids or proteins) costs about 1 to 2 g of glucose, or roughly 10 percent of the cost (without including root costs) of taking up new nitrogen from the soil. Therefore, for any tree without an overabundance of energy, internal nitrogen recycling always would be more efficient than using soil nitrogen. Differences in internal recycling probably relate more to mobility of nitrogen compounds than to overall ecosystem fertility.

NUTRIENT INPUTS HAVE THREE MAJOR VECTORS

Nutrient inputs come from three major sources: deposition from the atmosphere, weathering of soil minerals, and, for nitrogen, biological nitrogen fixation. All three can be altered by vegetation. The "input" of nutrients from mineral weathering might be considered an internal ecosystem transfer. Weathering is conventionally called an input because the atoms are released from a pool that is unavailable to plants into pools that are readily available for uptake, leaching, and cycling within the forest.

The atmosphere contains a diverse soup of dust particles, ions, and gases, and the entrance of these into forest nutrient cycles has long-term importance for all forests. The dust particles (or aerosols) have high concentrations of nutrient cations, and sometimes nitrogen and sulfur in polluted environments. The ions dissolve in rainfall to provide substantial inputs of ammonium, nitrate, and sulfate to ecosystems. In some areas, high concentrations of ammonia, nitric acid, and sulfur dioxide may result in gaseous inputs of nitrogen and sulfur to forest nutrient cycles. A stand of loblolly pine in a moderately polluted area in Tennessee had twice the nitrogen deposition of a Douglas-fir stand in Washington, and the majority of the difference resulted from higher deposition of nitric acid vapor in the loblolly pine stand (Figure 9.12).

In the United States and Europe, networks of sampling sites are used to monitor the deposition of ions in precipitation. In the United States, the National Atmospheric Deposition Program (NADP) has documented the major patterns of deposition across the country, and has verified a very substantial decline in sulfur inputs to the eastern United States as a result of reduced emissions of sulfur dioxide from industrial sources. Rates of sulfur deposition to a forest in Denmark peaked in about 1970, at 14 kg/ha annually and declined to about 5 kg/ha annually in the mid-1990s (Hovmand, 1999). At Whiteface Mountain, New York, sulfur deposition declined by an average of

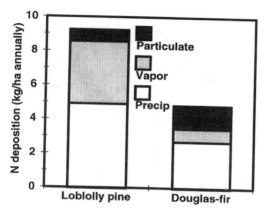

Figure 9.12 Estimated input of nitrogen to a loblolly pine stand in Tennessee and a Douglas-fir stand in Washington (data from Johnson and Lindberg, 1992).

0.27 kg/ha annually from 1985 to 1997 ($p < 0.01$; Figure 9.13). The apparent decline in nitrogen deposition was 0.06 kg/ha annually, but the trend was not significant ($p = 0.25$). Over the same period, calcium deposition decreased significantly (by 0.02 kg/ha annually) and potassium deposition increased (by 0.01 kg/ha annually).

In the eastern United States, nitrogen input to forests tend to average 5 to 10 kg/ha annually for low-elevation forests (including precipitation, vapor, and particulates) and 15 to 30 kg/ha for high-elevation forests annually (Johnson and Lindberg, 1992). Overall, less polluted areas commonly receive less than 5 kg/ha annually from the atmosphere, and more polluted areas receive more than 20 kg/ha annually.

Vegetation can affect nutrient inputs from the atmosphere by altering the capture of particles, gases, and ions. The amount of surface area exposed by a forest canopy to the atmosphere can vary substantially, between high-leaf areas in conifers and lower-leaf areas in hardwoods and between evergreen and deciduous species. A red spruce forest in the Smoky Mountains of the United States had 20 percent greater sulfur deposition and threefold greater nitrogen deposition than a nearby beech forest (Johnson and Lindberg, 1992). In the Solling area of Germany, a stand of Norway spruce experienced more than twice the sulfur input of a nearby beech forest (Ulrich, 1983).

Inputs of nitrogen can vary among species when symbiotic nitrogen fixation enters the picture. Some species of trees are capable of forming symbiotic relationships with bacteria that are capable of "fixing" atmospheric N_2 into ammonia (NH_3). The operation takes place in root nodules, where the plants supply carbohydrate energy and oxygen protection for the microbes. The fixed nitrogen is assimilated rapidly into organic nitrogen compounds and transported into the plant. Subsequent death of plant tissues makes the nitrogen available for recycling in the ecosystem (see Chapter 14).

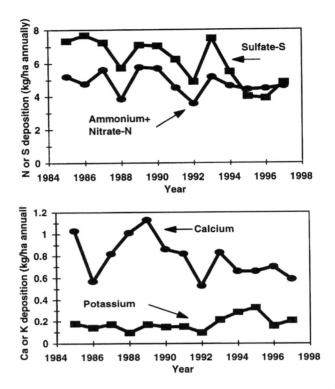

Figure 9.13 Trends in wet deposition for Whiteface Mountain, New York, showed declining inputs of sulfur and calcium, and perhaps nitrogen, whereas potassium deposition increased (data from NADP, 1999).

The weathering of minerals transfers significant quantities of nutrients from mineral lattices to more available soil pools (Table 9.3; Figure 9.14). A variety of approaches have been used to estimate the rates of mineral weathering and to examine the malleability of weathering under varying levels of acid inputs. The simplest approach is to simply estimate the rate of loss of cations from a watershed (the denudation rate), which should equal the rate of weathering minus the rate of atmospheric deposition and accumulation in vegetation or on the exchange complex. The classic watershed studies at Hubbard Brook in the 1960s used this approach (Likens et al., 1977). It was estimated that about 20 kg/ha of calcium annually must come from mineral weathering to balance the calcium lost in streamwater or accumulated in vegetation.

A second approach to estimating long-term rates of weathering involves comparing the ratios of nutrient cations (such as calcium) to zircon in the parent materials. As mineral material weathers, calcium may be released, cycled, and lost over the years, but all the zircon remains because of its unreactive insolubility. If the A horizon has a calcium:zircon ratio that is 10

TABLE 9.3 Estimated Rates of Mineral Weathering (kg/ha Annually) for Forest Soils

Site	Parent Material	P	K	Ca	Mg	Al	Method	Reference
Idaho, USA	Feldspars	—	23	26	10		Watershed balance	Clayton, (1979)
Colorado, USA	Quartz monzonite		4	20	2.4		Watershed balance	Clayton (1984)
	Granite, gneiss calcite intrusions		0.6	4.6	0.7		Watershed balance, mineralogy	Mast (1989)
New Hampshire, USA	Quartz, feldspars	—	7	21	4		Watershed balance	Likens et al. (1977)
Oregon, USA	Andesitic tuff	0.2	2	12	7		Watershed balance	Sollins et al. (1980)
California, USA	Dolomite		4	86	52		Watershed balance	Marchand (1971)
	Quartz monzonite		8	17	2			
Maryland, USA	Schist		2.3	1.3	1.7		Watershed balance	Cleaves et al. (1970)
New York, USA	Serpentine		<0.1	<0.1	34			Cleaves et al. (1974)
	Glacial outwash		11	24	8		Lysimeter leaching	Woodwell and Whitaker (1967)
Virginia, USA	Granite		1.3	8	2.6		Watershed balance	Pavich (1986)
Czech Republic	Biotite-muscovite gneiss		9.4	3.7	2.9		Watershed balance	Paces (1986)
	Biotite gneiss		23	19	14			
	Quartzitic gneis		13	8.5	6.3			
British Columbia	Quartz diorite			34			Watershed balance	Zeman and Slaymaker (1978)
Scotland	Till		2.4	17	5.3		Watershed balance	Creasey et al. (1986)
	Gabbro		1.2	20	5.9			
Luxembourg	Metashale		0.2	8.7	15.7		Watershed balance	Verstraten (1997)
Norräker, Sweden	Glacial till	0.092	0.8	5	7.92	6.03	Watershed balance	Bergholm et al. (1997)
Åseda, Sweden		0.054	3.6	2.2	3.6	8.01	Mineralogy relative to zircon	Hallbäcken and Bergholm (1997)
Farabol, Sweden		0.005	1.2	0.6	0.72	1.8		
Loten, Norway		0.03	1.2	0.6	1.2	1.53		
Vardal, Norway		0.004	0.8	0	1.32	1.89		
Marnardal, Norway		1.13	28.4	18.2	12.12	73.08		
Birkenes, Norway		0.19	4	3.6	4.2	6.57		
Sodankylä, Finland		0.011	0.62	1.17	1.35			
Kemijärvi, Finland		0.012	0.71	1.21	1.16			
Heinola, Finland		0.03	1.49	1.39	1.32			
Australia	Dacite	3	21	19	8		Watershed balance	Feller (1981)
Indonesia	Volcanic ash		12	18	25		Watershed balance	Bruijnzeel (1982)
Tropical, subtropical, 25 sites	Ultisols, Oxisols of various mineralogy		9-21	5-15	3-10		Watershed balance	Bruijnzeel (1991)

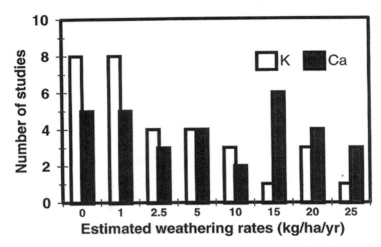

Figure 9.14 About half of the reported rates of mineral weathering in forest soils are 1 kg/ha annually or less for potassium, whereas half of the estimates for calcium weathering are over 10 kg/ha annually. The full ranges for both elements are similar (from studies in Table 9.2).

percent lower than the calcium:zircon ratio of the parent material, then 10 percent of the original calcium was lost from the parent material during weathering.

With more complex approaches, different weathering rates are estimated for the major minerals present in a soil, and the presence of minor constituents is considered. For example, Mast (1989) estimated that about 40 percent of the calcium released in mineral weathering in a Rocky Mountain alpine watershed came from minor intrusions of calcite into the granite/gneiss bedrock.

How malleable are rates of mineral weathering? The concentration of H^+ in soil solutions clearly affects the rates of mineral weathering; greater acidity leads to greater rates of mineral weathering. In soils, the weathering story may depend very heavily on localized activities of roots and microbes that involve excretion of organic compounds that can directly degrade some types of minerals (see Chapter 5).

Direct assessments of rates of mineral weathering under different tree species have not been made, but ecosystem budgets indicate substantial differences. For example, Eriksson (1996) examined the calcium contained in vegetation, forest floor, and mineral soil exchange pools under five species in a replicated, 23-year-old experimental plantation in northern Sweden. Norway spruce plots contained about 500 kg of calcium per hectare (25 $kmol_c$/ha) more than stands of Scots pine or lodgepole pine. Another species comparison by Berqvist and Folkeson (1995) included estimates of deposition and leaching as well as internal pools, and the Norway spruce in this case needed about 1.6 $kmol_c$/ha more weathering input of base cations to balance the nutrient budget. These replicated species trials do not provide conclusive evidence of differences in weathering rates, but until direct assessments are available, the null hypothesis should perhaps be that tree species do indeed affect rates of mineral weathering.

Vegetation also affects nutrient output rates, but most undisturbed forests retain nutrients so efficiently that losses are small relative to the quantities cycled. Despite this general pattern, important exceptions of high nutrient losses from vigorous forests do occur. For example, Nadelhoffer et al. (1983) examined nitrate leaching losses from several forests in Wisconsin and found that a fast-growing stand of white pine was losing about 10 kg/ha of nitrogen annually in soil leaching. Another example involves high rates of nutrient leaching from vigorous stands of nitrogen-fixing red alder. Fixation of nitrogen increases both nitrogen capital and nitrogen mineralization rates while reducing the plant's need to take up nitrogen from the soil. Nitrate leaching losses from red alder ecosystems can exceed 50 kg/ha of nitrogen annually (Bigger and Cole, 1983; Van Miegroet and Cole, 1984; Binkley et al., 1985).

Nutrient losses usually increase after harvests, fires, and other disturbances because tree uptake is a major sink for available nutrient ions. Exposure of the forest floor and soil to direct sunlight can increase temperatures, which combine with moister conditions in the absence of water uptake to accelerate decomposition and nutrient release. Thus, nutrient availability can be at a peak when plant uptake is minimal. Fortunately, microbial and geochemical immobilization are usually large enough to prevent large nutrient losses.

Is there a limit on how much nitrogen an ecosystem can retain? Intuitively, we might expect that forests will stop retaining added nitrogen when the inputs plus net mineralization exceed plant uptake. However, empirical studies have found very few systems that have reached a stage of nitrogen saturation. Fertilization with very heavy, repeated applications of nitrogen typically increases nitrogen leaching losses, but the losses remain lower than the input rates (see Chapter 13). European forests retain most of the nitrogen deposited from the atmosphere even under high deposition regimes. As noted in Chapter 16, the mechanisms that account for nitrogen retention in forests ecosystems need more investigation.

CYCLES DIFFER SUBSTANTIALLY AMONG NUTRIENTS

The majority of the mass of plants is water. The dry mass of plants is composed primarily of oxygen and carbon (about 45 percent each), hydrogen (about 6 percent), and nitrogen (about 1.5 percent for nonwoody tissues). The reliance of plants on these elements derives from their ability to form kinetically stable linear polymers (such as starch, cellulose, proteins, and nucleic acids) that can be readily controlled by catalysts (Williams and Fraústo da Silva, 1996). About a dozen other elements comprise minor proportions of plant tissues, playing a variety of roles in modifying the structure and function of these polymers.

TABLE 9.4 Summary of Features of the Major Nutrient Cycles

Element	Major Pool Used by Trees	Major Long-Term Source for Tree Uptake	Key Biochemical Roles	Solid Chemistry	Limiting Situations
Carbon	Atmosphere	Atmosphere	In all organic molecules; fundamental to structure, energy flow, genetics	Complex organic carbon flows and storage, major role in overall soil fertility; bicarbonate and carbonate ions also important	Atmospheric concentrations may limit forest growth; limitation may decline with CO_2 enrichment of atmosphere
Oxygen	Atmosphere	Atmosphere	Oxidative phosphorylation, respiration, by-product of photosynthesis	Diffusion limited when soil water is high; major control of soil redox potential	Waterlogged soils
Hydrogen	Water	Water	In all organic compounds	Free H^+ influences many biological and chemical reactions	Extremely acidic or alkaline conditions
Nitrogen	Soluble nitrate, exchangeable ammonium; N_2 for nitrogen-fixing species	Soil organic matter; atmospheric N_2 for nitrogen-fixing species	Proteins, enzymes, nucleic acids	Availability dominated by microbe/enzymatic factors; most nitrogen in soils recalcitrant; important redox reactions, gaseous phases	Most temperate forests, many boreal forests, some tropical forests, (esp. plantations also receiving phosphorus fertilizer)
Phosphorus	Soluble phosphate	Soil organic matter, adsorbed phosphate, mineral phosphorus	Nucleic acids, lipids, energy flow	Major organic and geochemical control; no gas phase, no redox reactions	Old soils high in iron and aluminum, common in semitropical and tropical situations
Potassium	Soluble K^+	Soil organic matter, exchange complex, mineral potassium	Enzyme cofactor, membrane regulation, ionic strength buffer	Remains in ionic form in plants; easily leached from leaves and soils	Miscellaneous situations, old soils, particularly if nitrogen and phosphorus fertilizers added

Calcium	Soluble Ca^{2+}	Soil organic matter, exchange complex, mineral calcium	Cell walls, also present as calcium phosphate and calcium oxalate	Dominates solution cations in nonacidic soils; precipitates with carbonate in dry alkaline soils	Rarely limiting; in some Oxisols; perhaps in some alpine spruce forests
Magnesium	Soluble Mg^{2+}	Soil organic matter, exchange complex, mineral magnesium	Enzyme cofactor in chlorophyll, other enzymes	Similar to calcium	Rarely limiting; perhaps on some pumice soils
Sulfur	Soluble sulfate, some SO_2 gas	Soil organic matter, atmospheric deposition, some mineral sulfur	Three amino acids; sulf-hydryl bonds provide three-dimensional structure to proteins	Sulfate is a major anion in solutions; specifically adsorbed on oxides; important redox reactions in anaerobic situations	Rarely limiting
Manganese	Soluble Mn^{2+}		Component of three enzymes; cofactor in several dozen enzymes including one involved in photosynthesis	Redox reactions are important; manganese solubility increases as pH decreases	Rarely limiting; perhaps in some limestone-derived soils in New Zealand; South Africa
Iron	Soluble ferrous Fe^{2+}; chelated Fe^{3+}	Soil mineral phase	Critical in many electron transport enzymes	Major constituent of mineral soils; reduced form (Fe^{2+}) 10 times more soluble than oxidized form (Fe^{3+}); organic chelates important for uptake	Limiting only in some alkaline situations for some species; addition with an organic chelate may be needed

209

Each nutrient element is characterized by a unique biogeochemical cycle; the key features of each element are highlighted in Table 9.4. Potassium is the simplest, with only one form (K^+) that remains ionic throughout its entire cycle. Phosphorus cycling is more complicated, with the phosphate anion (PO_4^{3-}) playing several roles in organic compounds and with major biotic and geochemical controls on cycling. The nitrogen cycle adds the complexity of major oxidation and reduction steps. The sulfur cycle is perhaps the most convoluted, sharing some features of both the phosphorus and nitrogen cycles. Although hydrogen technically is a required plant nutrient, it never limits ecosystem production because of its abundance in water. The cycling of hydrogen ions (H^+), however, merits attention for two reasons. First, H^+ activity, generation, and consumption interact with all the nutrient cycles, and H^+ budgets can be used to integrate the overall biogeochemical patterns of forest ecosystems. The H^+ budget also provides the key to understanding and assessing soil acidification from both natural and human-caused sources (covered in Chapter 16).

CARBON FLOWS THROUGH FORESTS

Most nutrients in forests have the opportunity to cycle through the ecosystem; the same atom of nitrogen may be incorporated in tree biomass, return to the soil, and reenter the trees. Carbon cannot be cycled in the same way (for the most part) because when organisms use the carbon fixed by photosynthesis, it tends to be lost as carbon dioxide to the atmosphere. Of course the carbon dioxide in the atmosphere can be refixed (the average residence time of carbon dioxide in the atmosphere is only a few years), perhaps within the same ecosystem if the canopy leaves "catch" the carbon dioxide released from the soil. In addition, some autotrophic organisms such as nitrifying bacteria may fix carbon dioxide that is respired by tree roots. Overall, however, the vast majority of the carbon flowing through ecosystems does not reenter the same trophic level of the system without first passing through the vast atmospheric pool.

As the carbon flows through the ecosystem, it accumulates for shorter or longer periods of time in the biomass of plants and animals and in the soil. Most of the carbon fixed by photosynthesis leaves the trees the same year it was produced for deciduous trees, with a fraction (always less than half) accumulating for longer periods in wood. Evergreen trees may retain leaves as well as stemwood for more than a year, so the flowthrough rate of carbon may be somewhat longer than for deciduous forests.

How long does carbon take to cycle through the soil and back to the atmosphere? An average turnover time for the whole soil carbon pool can be calculated by dividing the pool size by the annual input rate (and assuming that net changes in the pool size are small). This gives a surprisingly rapid turnover rate for most forests. The net primary carbon productivity of forests typically ranges from 5 to 25 Mg/ha annually, and the carbon content of forest biomass plus soils is commonly on the order of 50 to 500 Mg/ha. The flux of carbon into forest systems, then, is on the order of 1 to 50 percent of the carbon stored in the system, for average turnover rates of 2 to 100 years for all the carbon in the system. As noted above, decomposition of plant detritus typically follows an exponential decay pattern, so most of the carbon will be lost from the system rapidly, but small pools will persist for centuries.

The chemistry of the carbon compounds that comprise soil organic matter is very important for microbes, soil fauna, and trees. These compounds differ in rates of turnover, and the nutrient supply to trees depends very heavily on these rates. If a pool of 50,000 kg of carbon per hectare contained 5000 kg of nitrogen per hectare and turned over every 100 years, the annual release of nitrogen might be on the order of 50 kg/ha. If the same soil had an average carbon turnover rate of 50 years, the nitrogen availability might be more like 100 kg/ha annually.

Soil carbon has a substantial effect on soil pH. Soil organic compounds are weak acids, and soils with high carbon contents have large capacities to (1) adsorb or release H^+ and (2) sorb nutrient cations. The acid strength (tendency to release H^+) of soil organic matter in forest soils can differ substantially among sites and under the influence of different species (see Chapter 5).

The flow of carbon through ecosystems has a major inorganic component. When carbon dioxide dissolves in water, carbonic acid is formed. This weak acid tends to dissociate to bicarbonate (HCO_3^-) and H^+. The bicarbonate typically leaches from soils in association with nutrient cations such as K^+, with the H^+ either replacing the nutrient cation on the exchange complex or driving the release of the nutrient cation by weathering of minerals. The respiration of roots and decomposition of organic matter lead to concentrations of carbon dioxide in the soil atmosphere that are orders of magnitude higher than those in the open atmosphere. These high concentrations lead to large production of carbonic acid, and to a large supply of H^+ for mineral weathering and soil development.

The carbon cycle is the primary focus of forest management for tree growth; accumulation of wood and fiber represents the rate of growth of carbon storage in this pool. The supply of carbon in the atmosphere is not amenable to management (at least not at a scale smaller than the globe), but the uptake of carbon through photosynthesis is very responsive to additions of nutrients and water. Each atom of nitrogen present in foliage can yield 50 to 100 atoms of photosynthetically fixed carbon each year. Addition of nitrogen in fertilizers involves substantial energy inputs to synthesize, transport, and apply the nitrogen, but despite these inputs, the net energy recovery in wood is still on the order of 15:1 (see Chapter 13).

Trees require far more water per unit of carbon fixed than they do of nitrogen; 1 kg of fixed carbon commonly "costs" about 500 to 1000 kg (= 500 to 1000 L) of water. For this reason, forest irrigation is currently practiced only in high-value situations, such as fast-growing plantations of *Eucalyptus* or

Poplar to provide fiber to mix with conifer fibers in making high-quality paper.

Forest practices also manipulate the carbon cycle when soil carbon is increased or lost and when forest products are removed. The removal of carbon in harvested biomass is fairly straightforward to measure, but the effects of harvesting on later net losses or gains of carbon in the soil are less clear. For the variety of sites reported in the literature, Johnson (1992b) concluded that just about as many sites showed net increases in soil carbon as showed decreases (see Chapter 16).

OXYGEN FUELS ENERGY REACTIONS

Oxygen is similar to carbon in that it tends to flow through ecosystems rather than cycle. Unlike carbon, the reservoir of oxygen in the atmosphere is vast, with a much longer turnover time (about 100,000 years). Oxygen has one key role in soils with many facets: oxidation and reduction.

Oxidation and reduction reactions have to occur together, providing both a sink and a source of electrons, and molecular oxygen (O_2) has the highest potential of any electron acceptor in ecosystems ($+810\,mV$ at pH 7). Oxidation of sugar with oxygen gives five times the energy release of oxidizing sugar with carbon dioxide (forming methane).

Oxidation reactions with oxygen as the electron acceptor are also fundamentally important in soil development over time. Most mineral soils develop accumulations of iron and aluminum oxides over time. In some situations, iron oxides are present in the original mineral. In others, oxygen is necessary to oxidize reduced ferrous iron to oxidized ferric iron.

The diffusion of oxygen through soils depends heavily on soil aeration, as gases diffuse through air about 10,000 times faster than through water. The supply of oxygen in soils has two features: concentration (or partial pressure) and rate of resupply. Even small concentrations of oxygen in soils can maintain aerobic conditions with oxygen as the dominant electron acceptor in chemical reactions. However, the consumption of oxygen can deplete the supply unless the pore space remains filled with air (not water) or unless the rate of water flow is fast enough to bring in more oxygen. Oxygen diffusion into wet soils is typically too slow to prevent development of anaerobic conditions, leading to reduced microbial activity and gleying of the soil.

Soil oxygen supplies can be affected by management in two key ways. Soil compaction can reduce poor volume and restrict oxygen diffusion into the soil. The oxygen supply can also be increased intentionally in waterlogged areas through intensive systems of ditches and raised beds for planting seedlings. The improved oxygen supply then fosters microbial oxidation of organic matter (sometimes lowering the surface level of Histosols), net release of nutrients, and better root and top growth.

HYDROGEN ION BUDGETS INTEGRATE BIOGEOCHEMICAL CYCLES

Hydrogen is a critical nutrient in plants, but its ubiquitous availability in water prevents it from limiting plant growth. The more important feature of hydrogen and ecosystems is the availability of H^+ for chemical reactions. Many reactions in the soil solution depend on soil acidity, expressed as pH (negative logarithm of the H^+ activity). The pool of H^+ in soil solutions is typically small at any point in time, but this pool is strongly buffered by vast quantities of solid-phase soil acids. These solid-phase acids include the mineral and organic components of the exchange complex (see Chapter 5). Differences in pH between soils, or within a soil over time, depend in part on the equilibrium reactions between the soil solution and these large pools of solid-phase acids. Over the long term, the input and output of acids and H^+ determine the size and acid strength of the solid-phase acids.

The subject of H^+ budgets is complicated, but the basic features are clear. First, photosynthesis fixes carbon in the form of sugar. Next, some of the sugar is transformed into a variety of acids for biochemical use in cells, including acids such as malic or oxalic acid. These acids may dissociate, releasing H^+ and forming an acid anion such as malate or oxalate. The negative charge on the acid anions must be balanced by another cation if the H^+ is missing; candidates include K^+ and Ca^{2+}. When litter is produced that contains dissociated acids, these leaf acids decompose readily, consuming H^+ to return to CO_2 and H_2O, releasing the nutrient cation. The organic acids that accumulate in soil organic matter are produced through the incomplete oxidation of litter, which commonly results in proliferation of carboxyl (—COOH) groups that behave as weak acids. Soil clays also act as weak acids, releasing and adsorbing H^+, depending on the concentration in the soil solution.

This solid-phase acid complex (or exchange complex) in the soil is then "titrated" by processes that add H^+ to or remove it from the system. The net sources of H^+ are:

1. Formation of carbonic acid; high concentrations of soil CO_2 form H_2CO_3, which dissociates at pH levels higher than about 4.5. The flux of bicarbonate leaching from the soil represents the net H^+ generation.

2. Net accumulation of base cations in vegetation; the uptake of calcium, magnesium, and potassium exceeds the uptake of phosphate, leaving a net uptake of cations that must be balanced by the excretion of H^+ into the soil. (Nitrate and sulfate are not counted in uptake because they are mostly reduced inside the plants, which consumes any H^+ equivalent associated with uptake.)

3. Atmospheric deposition of H^+.

4. Net oxidation of reduced compounds; oxidation of ammonia to nitrate produces H^+. If the nitrate is assimilated or denitrified, H^+ is consumed,

with no net effect. If the nitrate leaches (and therefore ammonia oxidation exceeds nitrate reduction), a net increase in H^+ occurs.

The major processes accounting for H^+ consumption are:

1. Release of base cations from decomposing biomass; this is the reverse of the biomass accumulation in the list of H^+ sources. Over the course of stand development, biomass accumulation of cations is a major source of acidification. When the biomass is burned or decomposes, this reverse process consumes the equivalent amount of H^+. Therefore, the biomass accumulation story is important at a time scale of years to centuries, but it has no net effect on long-term soil development unless biomass is removed.
2. Specific anion adsorption; the ligand exchange of sulfate or phosphate for OH^- groups in the coordination sphere of iron and aluminum consumes H^+ and protonates the OH^- to make H_2O.
3. Mineral weathering; as noted below, the weathering of silicates involves production of silicic acid, which consumes H^+.
4. Unbalanced reduction of oxidized compounds; nitrate deposited from the atmosphere may be taken up and used by microbes or plants or denitrified. In both cases, H^+ is consumed in the reduction of nitrate, so deposition of nitric acid rain can have no acidifying effect unless nitrate leaches from the soil.

The components of H^+ budgets differ with the perspective of the budget. At the level of plant cells, H^+ flows are associated with the synthesis of adenosine triphosphate (ATP) and the maintenance of electrical balances between cations and anions. These processes can be overlooked from an ecosystem perspective because they have no net effect at this higher level of resolution.

With a focus on the root-soil interface, H^+ excretion or absorption is used to maintain electrical balance between the uptake of cation and anion nutrients. In cases where nitrate is the major form of nitrogen, anion uptake usually exceeds cation uptake. With ammonium nutrition, this pattern is reversed. A balance is achieved with H^+ flows. If too many cations are taken up, plants synthesize organic acids that dissociate and supply H^+ to be secreted into the soil in exchange for the excess cations. When anion uptake exceeds uptake of cations, H^+ can be absorbed from the soil to accompany the anions. In reality, plants may excrete OH^- (which combines with CO_2 and water to form HCO_3^-) rather than take up H^+. In either case, accounting for the flow of H^+ equivalents is sufficient for charge budgeting.

At a higher level of resolution, the source of the nutrient cations and anions must also be accounted for. Some surprising features emerge. For example, ammonium uptake requires excretion of H^+ into the soil, which increases the soil pool of H^+. However, if the ammonium came from mineralization of organic nitrogen, then it also involved the consumption of one H^+ ion when ammonia became ammonium. Plant uptake of ammonium merely replaces the H^+ consumed in the formation of ammonium. Similarly, nitrate uptake involves the uptake of one H^+ ion, but nitrification produces two H^+ ions (Figure 9.15). One of these balances the consumption of the ammonia-to-ammonium step, and the other balances the H^+ taken up by the plant with the nitrate.

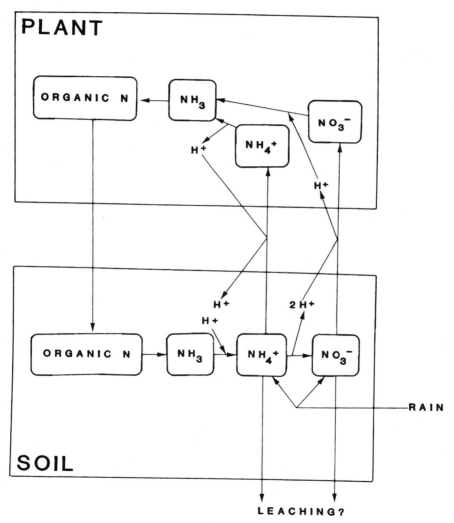

Figure 9.15 H^+ flux associated with transfers and transformation of organic and inorganic nitrogen. The nitrogen cycle has a net effect on H^+ only when it involves additions to or losses from the system, as the only form that accumulates in large quantities is organic nitrogen (for details, see Binkley and Richter, 1987).

What about the plant? With ammonium uptake it loses one H^+, but when ammonium is converted to organic forms, one H^+ ion is released inside the plant. With nitrate uptake, one H^+ ion is taken up, but subsequent nitrate reduction consumes one H^+ ion, also leaving the plant in balance.

For these reasons, the transfer of soil organic nitrogen to plant organic nitrogen involves no net generation or consumption of H^+, regardless of the intervening transformations, because there was no net change in the redox state of the nitrogen atom.

If inorganic nitrogen is added directly to the ecosystem (in rain or fertilizer) rather than mineralized within the ecosystem, the H^+ budget may not balance so nicely. Added ammonium will not have consumed one H^+ ion previously, so H^+ excretion associated with ammonium uptake would represent a net increase in soil H^+. Similarly, nitrate added to a system would consume one H^+ ion from the soil on uptake, which would not be balanced by previous H^+ production.

This H^+ pattern has important implications for acid rain. Nitric acid (HNO_3) added to an ecosystem has no acidifying effect if the nitrate is used by the vegetation; one H^+ will be consumed for each NO_3^- taken up and utilized. If nitrate comes in as a salt (such as KNO_3) rather than as an acid, then use of the nitrate will consume H^+ from the soil. Ammonium enters as a salt (such as NH_4Cl), and plant use of the nitrogen results in production of H^+. In fact, if the ammonium is nitrified, two H^+ can be generated for each ammonium ion added. If the nitrate is later utilized, the net production drops back to one, but if the nitrate leaches from the ecosystem, the net production remains at two.

What about the other nutrient cycles? In general, the H^+ budget aspects of sulfur cycling resemble those of nitrogen cycling. Internal ecosystem cycling results in production and consumption of H^+, but these processes largely balance. Phosphorus cycling is a bit more complicated, but fortunately, the magnitude of H^+ flux in the phosphorus cycle is a very small part of the overall budget.

The accumulation of nutrient cations from inorganic pools in the soil into vegetation does represent a net flow of H^+ into the soil. Any flow of cations from organic pools in the soil into vegetation resembles the nitrogen cycle, with no net change in H^+. Over the course of a rotation, this can result in substantial H^+ production. When the biomass is decomposed or burned, though, release of cation nutrients is coupled with consumption of H^+, again completing the cycle, with no net change in the overall H^+. Forest harvesting prevents completion of the cycle and leaves the soil H^+ pool increased in proportion to the cation content of harvested biomass.

Harvesting may increase decomposition rates, and this pulse of H^+ consuming activity may decrease soil acidity by two processes. First, release of cations through the oxidation of organic matter requires consumption of H^+ to form CO_2 and O. Second, some dissociated organic acids ($R\text{-}COO^-$) are oxidized, also consuming H^+ to form CO_2 and O. These processes can neutralize much of the acidity that was produced during the development of the previous stand. Indeed, if no biomass were removed, the magnitude of the neutralizing effect should be close to that of the acidifying effect. Little information is available from field studies on these dynamics, but limited work demonstrates the general tendency. For example, Nykvist and Rosen (1985) found that clearcutting several Norway spruce stands increased the pH of the humus layer at several sites by about 0.5 unit. They also compared exchangeable H^+ pools on plots with and without logging slash; decomposition of slash reduced the pool by about 30 percent. These trends should partly neutralize the acidity produced during the development of the previous stand and provide a more neutral starting point for the natural acidification that should begin with the development of the new stand. In contrast, harvesting a northern hardwoods stands in New Hampshire resulted in a decrease in pH in upper soil horizons following nitrification and nitrate leaching, along with an increase in pH in the B horizon (Johnson et al., 1991), perhaps as a result of nitrate assimilation.

Two case studies illustrate the major aspects of H^+ budgets in forests. A deciduous forest near Turkey Lakes, Ontario, experienced moderate inputs of nitrate in atmospheric deposition (about 0.35 $kmol_c$/ha annually; budget from Binkley, 1992), but the output of nitrate was large (1.75 $kmol_c$/ha annually). This net loss of nitrate indicates a net H^+ generation of 1.4 $kmol_c$/ha annually at this site as a result of the unbalanced oxidation of nitrate. Inputs and outputs of sulfate were similar, leaving little net H^+ generation or consumption. The average rate of cation accumulation in biomass was 0.65 $kmol_c$/ha/yr, or about half of the acidification effect of the nitrate budget. Accounting for these and the net H^+ effects of other processes, this ecosystem was acidifying at a rate of 2.65 $kmol_c$/ha annually. This rate represents the quantity of H^+ available to "titrate" the exchange complex, and the rate of change in soil pH depends on the acid strength of the soil and the total quantity of acid stored in the soil. The net generation of H^+ is large relative to the low quantities of base cations in the exchange complex (about 41 $kmol_c$/ha), indicating that substantial soil acidification may be occurring.

The H^+ budget for a loblolly pine plantation in Tennessee showed a strong accumulation of nitrate; virtually all of the 0.5 $kmol_c$/ha annual input was retained, for a net consumption of 0.5 $kmol_c$/ha of H^+. Mineral weathering (or other processes associated with dissociation of carbonic acid) consumed another 0.3 $kmol_c H^+$/ha annually. Sulfate output exceeded input, generating 0.4 $kmol_c$/ha of H^+ annually. The rate of cation accumulation in biomass generated 0.65 $kmol_c$/ha of H^+. The output of base cations in soil leachate exceeded inputs by about 1.5 $kmol_c$/ha, representing the largest net source of H^+ (the loss of base cations entails an increase in H^+, which is necessary to balance the lost base cations). Including other processes in this forest, the net rate of acidification was about 1.5 $kmol_c$/ha of H^+. This rate of H^+ generation is very small relative to the large pools of exchangeable base cations (about 545 $kmol_c$/ha), so any changes in soil pH at this site are likely to be slight.

In summary, for H^+ budgets, most nutrient transfers and transformations either consume or generate H^+, but fortunately, many of these changes are balanced later in the nutrient cycles and can be overlooked. For scientists interested in the intricacies of nutrient cycles, following H^+ cycles can ensure a clear understanding of the dynamics of all nutrients. For forest nutrition managers, it is important to realize that natural ecosystem processes generate and consume H^+, and evaluating the impacts of pollution or management treatments (such as fertilization) should be based on an understanding of the whole H^+ picture.

THE NITROGEN CYCLE DOMINATES FOREST NUTRITION

For a variety of reasons, nitrogen cycling has received more attention in forest research than any other nutrient. Nitrogen availability limits growth in more forests in more regions than any other nutrient, and it can be important even when it is not limiting because substantial leaching of nitrate-nitrogen can occur when nitrogen availability exceeds plant uptake. Nitrate leaching is undesirable for several reasons. The nitrogen is lost from the site. The nitrate anion is also accompanied by cation nutrients, such as K^+ and Ca^{2+}, and by potentially toxic cations such as Al^{3+}. This sequence also generates H^+ and may acidify the soil. At high concentrations, nitrate may be toxic in drinking water.

Nitrogen is particularly important in plant nutrition as a component of amino acids, enzymes, proteins, and nucleic acids. In foliage, most of the nitrogen is present in the carboxylating enzyme RUBISCO; higher nitrogen in leaves leads to higher RUBISCO concentrations and higher rates of photosynthesis (depending on light and water supplies, of course).

The soil nitrogen cycle begins when nitrogen is added to the soil in organic form (from litterfall, root death, or throughfall) or inorganic form (throughfall; Figure 9.16). The cycling of nitrogen within the soil is very murky, with rapid and slow transfers between soluble and insoluble nitrogen, soil solution ammonium, and microbial nitrogen. The actual rates of these transfers are poorly known, as the best isotope techniques still provide net transfer information.

The release of ammonium from organic nitrogen pools is called gross mineralization, such as the oxidation of glycine to form CO_2, NH_3, and water:

$$CH_2NH_2COOH + 1.5\,O_2 \rightarrow 2CO_2 + NH_3 + H_2O$$

At the pH levels common in soils, the ammonia immediately absorbs one H^+ from the soil solution to become ammonium, NH_4^+.

Much of the ammonium produced in soils is oxidized to nitrite and then to nitrate in an oxidation reaction called nitrification:

$$NH_4^+ + 2O_2 \rightarrow NO_3^- + 2H^+ + H_2O$$

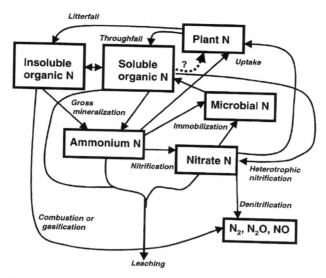

Figure 9.16 Basic features of the soil nitrogen cycle (the dotted line with "?" notes the current lack of knowledge about the importance, if any, of direct uptake of organic nitrogen by trees).

Electrons are donated from the nitrogen atom to the oxygen molecule, releasing energy for use by the microbes. Both nitrate and ammonium may be used as nitrogen sources for protein formation. This equation represents autotrophic nitrification, a process carried out by microbes that utilize ammonium as an energy source. Another type of nitrification is performed by heterotrophic microbes, using ammonium or organic nitrogen as a substrate (Paul and Clark, 1996).

Both ammonium and nitrate are subject to leaching from the soil, but most soils lose more soluble organic nitrogen in leaching water than ammonium + nitrate (Figure 9.17). Typically, only soils with very high rates of nitrogen leaching, such as those in areas with high nitrate deposition, lose more inorganic nitrogen than organic nitrogen in soil leachate.

Many compounds serve as electron acceptors in the absence of oxygen; oxygen is preferred merely because it gives the largest release of energy. An electron cascading down the redox staircase may stop at any level short of oxygen if soil aeration conditions are poor. In denitrification, an electron from some reduced carbon source is donated to nitrate, reducing it first to nitrite and then to nitrous oxide (N_2O) or nitrogen gas (N_2):

$$2HNO_3 + 10H^+ + 10e^- \rightarrow N_2 + 6H_2O$$

How far the reaction goes along this denitrification pathway varies primarily with the carbon (electron donating) status of the soil. High carbon availability

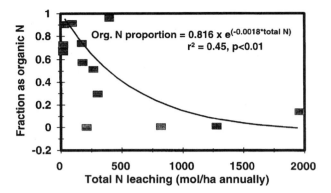

Figure 9.17   Fraction of total nitrogen leaching as organic nitrogen from 13 forests; low proportions of organic nitrogen leaching occur most often in high-leaching situations (data from Johnson and Lindberg, 1992).

favors complete reduction to nitrogen gas. Denitrification losses are usually small in nonwetland forests (Figure 9.18). Some current research suggests that some reduction of nitrate may produce ammonium rather than N gases, but more work is needed (W. Silver, personal communication).

Aside from representing a loss of valuable nitrogen from a forest, denitrification is also a concern due to the production of nitrous oxide. This gas can rise into the stratosphere and react with ozone, reducing the quantity of ozone available for absorbing potentially harmful ultraviolet radiation. On balance, forests probably produce much less nitrous oxide than do agricultural soils.

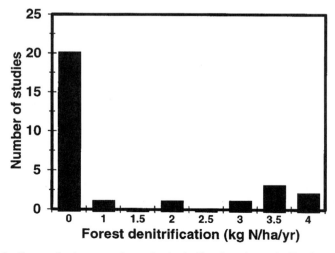

Figure 9.18   Loss of nitrogen through denitrification is typically low for forests, although a few sites may experience sizable losses (data from Davidson et al., 1990).

Nitrous oxide production by forests has also been fairly constant historically, while production from agricultural soils may have increased as a result of heavy use of fertilizers.

In some situations, plants appear capable of taking up small organic molecules that are dissolved in soil solutions (reviewed by Kaye and Hart, 1997). The extent and importance of soluble organic nitrogen in tree nutrition should become clearer over the next decade.

One source of nitrogen that is important in some situations is nitrogen fixation, whereby plants supply energy in the form of carbohydrates to nitrogen-fixing prokaryotes housed in root nodules. Nitrogen fixation adds electrons to N_2, reducing it to ammonia, which is then added to proteins and used and recycled within the ecosystem:

$$N_2 + 8H^+ + 8e^- \rightarrow 2NH_3 + H_2$$

Once plants have taken up ammonium, it is aminated onto an organic molecule, such as amination of glutamate to form glutamine:

$$(CH)_2(COOH)_2CHNH_2 + NH_3 \rightarrow (CH)_2(COOH)_2CH(NH_2)_2 + H_2$$

Various other nitrogen compounds are then produced by transamination.

Nitrate uptake is followed by nitrate reduction. Some nitrate reduction may occur in roots, using carbohydrates from photosynthesis as an energy source. Nitrate reduction also occurs in the leaves of some plants, where the reductant generated by the light reaction of photosynthesis can be used to reduce nitrate without consuming carbohydrates. This vector appears very important for *Rubus*, which in one study had four times the leaf capacity for nitrate reduction of competing pin cherry trees (Truax et al., 1994).

As noted in Chapter 10, nitrogen combustion occurs when protein nitrogen is oxidized to N_2 or nitrogen oxides in fire:

$$4CHCH_2NH_2COOHSH + 15O_2 \rightarrow 8CO_2 + 14H_2O + 2N_2 + 4SO_2 + energy$$

On an annual basis, leaching losses and denitrification are the major vectors of nitrogen loss, but on a scale of centuries, fires remove more nitrogen from many types of forests.

The various oxidation and reduction reactions of the nitrogen cycle can be confusing, especially because terms such as nitrification and denitrification do not refer to precisely opposite reactions. Some of the confusion can be removed by identifying the role of each nitrogen compound as an energy source, or as an electron acceptor or donor (Table 9.5).

Nitrogen forms a major part of all plant tissues, but leaves (and fruits) always have the highest concentrations. The canopy often contains more than half of a tree's entire nitrogen content. As leaves develop throughout a growing season, high nitrogen concentrations at bud burst are rapidly reduced as the biomass of expanding leaves dilutes the nitrogen content. A plateau usually is reached sometime after full leaf expansion; then leaf nitrogen gradually declines

TABLE 9.5 Major Oxidation and Reduction Reactions in the Nitrogen Cycle Mediated by Microbes

Process	Microbe	e^- Donor	e^- Acceptor	By-products	Purpose
Nitrogen fixation	Bacteria, cyano-bacteria, actino-mycetes	Glucose or another high-energy carbon compound	N_2	NH_3, CO_2 (and some-times H_2)	Provide nitrogen for use by microbe or plant
Nitrification, autotrophic	Bacteria, some fungi	NH_3	O_2	HNO_3	Obtain energy from NH_3 oxidation
Nitrification, heterotrophic	Primarily fungi, some bacteria	Organic matter, NH_3	O_2	HNO_3, H_2CO_3	Unclear
Denitrification	Bacteria	Glucose or another high-energy carbon compound	NO_3^-	N_2 or N_2O	Obtain energy from carbon compounds in absence of O_2

due to leaching from the leaves or resorption back into the twigs. Many trees recover substantial amounts of nitrogen from leaves before abscission. For evergreens, nitrogen concentrations in foliage gradually decrease with leaf age.

Ammonium is also sorbed onto cation exchange sites. These negatively charged sites arise from irregularities in the structure of clay minerals (such as broken edges and isomorphous substitution) and from dissociated organic acids. Forest soils commonly have 5 to 20 kg of nitrogen per hectare of ammonium sitting on exchange sites at any one time, but the flow through this very active (labile) pool is usually much greater than the average pool size would suggest. Some ammonium may also be "fixed" between the layers of clay particles, and this ammonium is only marginally available for plant uptake over long periods. Fortunately, ammonium fixation is generally not a large problem in forest soils.

The production of nitrate in soils was classically assumed to be the domain of autotrophic bacteria, but more recent work has shown that heterotrophic nitrification may be quite important in some forest soils. Heterotrophs may produce nitrate from organic nitrogen, particularly from oxidation of amines and amides (Paul and Clark, 1996). Both fungi and bacteria are capable of heterotrophic nitrification, and some of the nitrate produced in acid forest soils may be the work of heterotrophic fungi rather than autotrophic bacteria. Nitrate production by autotrophs is blocked during incubation in the presence of low concentrations of acetylene, but heterotrophic nitrification is not. Hart et al. (1997) used this acetylene-block technique to examine nitrification in conifer and alder-conifer stands in the northwestern United States and found that about two-thirds of the nitrate production in all sites appeared to be heterotrophic. More work on this basic feature of the soil nitrogen cycle is needed.

It was once assumed that low-nitrogen soils did not nitrify because trees were more effective competitors than microbes in obtaining ammonium when supplies were low. Net nitrification rates from incubations supported this idea; many low-nitrogen soils (particularly under conifers) showed no net accumulation of nitrate during 10 days or 30 days of incubation. However, this lack of nitrate accumulation (called net nitrification) commonly results from high rates of nitrate production coupled with high rates of nitrate immobilization by microbes. For example, incubations of soils from an old-growth Douglas-fir site in Oregon showed no net accumulation of nitrate during incubations for the first month. The lack of nitrate accumulation resulted from a high rate of nitrate production that corresponded with a high rate of nitrate immobilization by microbes (Figure 9.19; Hart et al., 1994). The residence time for nitrate became much longer after the microbial community had depleted the supply of available C and reduced their "demand" for nitrogen.

Recent use of $^{15}NO_3^-$ has shown that most (although not all) soils produce respectably large amounts of nitrate, and the lack of positive net nitrification results from high rates of microbial uptake (immobilization) of nitrate. For example, Stark and Hart (1997) examined gross and net rates of nitrate production in soils from 11 forests in New Mexico and Oregon spanning a wide range of soil fertility. In the summer, 5 of the 11 forests showed net positive rates of nitrification, ranging from 0.3 to 1.0 kg of nitrogen per hectare daily. The sites with net nitrate immobilization had a nitrogen reduction of 0.4 to 1.2 kg/ha daily. However, *all* of the sites showed strong rates of nitrate (gross) production, ranging from 0.3 to 2.0 kg/ha daily.

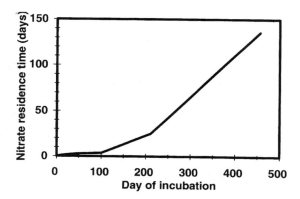

Figure 9.19   Nitrification appeared unimportant in an old-growth Douglas-fir forest in Oregon, not because nitrification did not proceed at a rapid rate but because nitrate immobilization by microbes was rapid. Nitrate residence time increased as the microbes depleted the soil supply of readily available carbon, giving the impression that nitrification began only late in the incubation period (data from Hart et al., 1994).

The first product of autotrophic nitrification, nitrite (NO_2^-), is a toxic compound. Fortunately, conversion to nitrate releases additional energy, and microbial processing continues fairly nonstop, with little accumulation of nitrite.

Nitrate not taken up by plants is likely to leach as water passes through the soil because forest soils have only slight exchange capacities for anions. Most intact forests are very tight with respect to nitrogen cycling, and soil leaching removes less than 10 percent of the nitrogen cycled annually. Exceptions occur only on sites with very high rates of nitrogen mineralization, such as some hardwood forests, some areas receiving very high atmospheric deposition of nitrogen, and some ecosystems with nitrogen-fixing trees.

Many studies have estimated in-field rates of net nitrogen mineralization on an annual basis. Within a forest type in a single region, the distribution of rates may follow a normal distribution curve (Figure 9.20). At a larger scale including many forests around the world, the pattern is more of an F-distribution. The values above 400 kg/ha of nitrogen annually all came from forests in the humid tropics; only a few such studies are available.

The final process in the nitrogen cycle is the fixing of atmospheric nitrogen into ammonia. Only a few types of microbes can perform this reaction. Some, such as the bacterium *Azotobacter*, fix N_2 without assistance from higher plants. Others, such as *Rhizobium* and *Bradyrhizobium* bacteria (found in root nodules of legumes) and *Frankia* actinomycetes (found in actinorhizal nodules on alder and *Ceanothus*), require a symbiotic relationship with plants. Symbiotic nitrogen fixation has the potential for higher rates than any free-living system for two reasons: the process is very energy expensive, and the presence of oxygen will break down the nitrogen-fixing enzyme (nitrogenase). Microbes in root nodules on legumes or other plants receive reduced carbon (energy) from the host plant, as well as protection from oxygen. Indeed, most host plants have evolved the ability to synthesize a form of hemoglobin to regulate the oxygen content of nodules.

Rates of nitrogen fixation in forest ecosystems range from near zero to several hundred kilograms per hectare annually (see Chapter 14). In the absence of symbiotic nitrogen fixers, rates are usually less than 1 kg/ha each year. Some symbiotic nitrogen fixers are much more reliable than others. For example, any member of the alder genus (*Alnus*) invariably has nodules, whereas members of the genus *Ceanothus* occasionally lack nodules.

PHOSPHORUS CYCLING IS CONTROLLED BY BOTH BIOTIC AND GEOCHEMICAL PROCESSES

The only form of phosphorus of importance in forest ecosystems is phosphate (PO_4^{3-}), and the phosphate cycle includes both biological and geochemical components (Figure 9.21). Phosphate enters into a wide variety of compounds, but the phosphorus atom remains joined with four oxygen atoms.

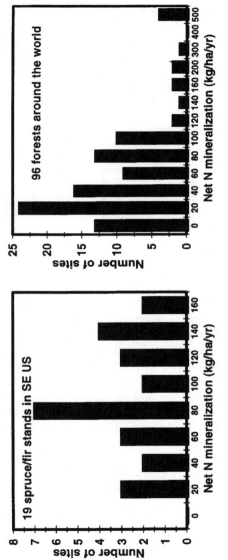

Figure 9.20 Patterns in net nitrogen mineralization from in-field incubations for a series of red spruce–Fraser fir stands in the southeastern United States (data from Strader et al., 1989) and for forests around the world (data from Binkley and Hart, 1989).

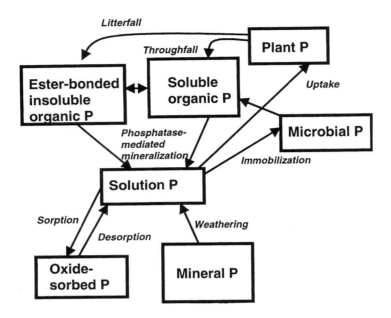

Figure 9.21   Soil phosphorus cycle.

Inside plants, phosphorus is found as inorganic phosphate, simple esters (C—O—phosphate, such as sugar phosphate), and pyrophosphate-bonded ATP (Marschner, 1995). When phosphorus supplies are limiting, leaves may allocate about 20 percent of their phosphorus to lipids, 40 percent to nucleic acids, 20 percent to simple esters, and 20 percent as free inorganic phosphate. As the phosphorus supply increases, the allocation to lipids and esters increases relative to that of nucleic acids, but the biggest increase is the pool of inorganic phosphorus, which may rise to about half of the total phosphorus in the leaves.

   At the stage of soil organic matter, phosphorus is released primarily by direct action of phosphatase enzymes. Some researchers believe that in the absence of phosphatases, it might take centuries for half of the organic phosphorus in soil to be released (Emsley, 1984). Others are skeptical of the true importance of phosphatases and maintain that these enzymes play a very limited role (Barber, 1995). Both microbes and plants can secrete various types of phosphatases into the soil. The relative importance of phosphatase types and sources in the phosphorus cycle of forests is not well known. Assays for phosphatase activity often reveal differences among soil types or vegetation types but usually show no correlation with plant-available phosphorus. This lack of correlation probably represents the difficulty of unraveling complex soil-plant-microbe interactions rather than the lack of importance of phosphatases. This complicated story of organic phosphorus mineralization is probably very important; Yanai (1992) estimated that about 60 percent of the phosphorus cycling in a northern hardwoods forest came from organic phosphorus pools.

Depending on the pH of the soil, dissolved phosphate will associate with one, two, or three H^+. At pH levels typical of forest soils, the $H_2PO_4^-$ form dominates. Phosphate in the soil solution faces four possible fates: uptake by plants or microbes; precipitation as barely soluble salts with calcium, iron, or aluminum; adsorption by anion exchange sites or sesquioxides; or leaching from the rooting zone of the forest.

Uptake of phosphorus by microbes or plants is followed by bonding one or two ends of the phosphate group to carbon chains. A single phosphate-carbon (P—O—C) bond is an ester, and two bonds are a diester. In these forms, phosphorus plays a pivotal role in plant energy transformations, protein and nucleotide synthesis, and cell replication. Much of the phosphorus in plants is bound to other phosphate groups through anhydride bonds. The energy of the phosphorus-phosphate bonds (anhydride linkage) is high, and is the source of energy when a phosphate group is removed from ATP to form ADP (adenosine diphosphate). Phosphorus transformations within plant cells are very dynamic. About 50 percent of the total phosphorus content of the plant is in the inorganic form. If cell pH is near 7, most of the free phosphate will be in the form HPO_4^{2-}. The leaves of many trees fall in the range of pH 5 to 6, so free phosphate would be found as $H_2PO_4^-$. Even in plants with severe phosphorus deficiencies, about 20 percent of the phosphorus content remains inorganic. Phosphorus compounds are fairly mobile, and a substantial portion of leaf phosphorus is resorbed prior to abscission.

In the soil, phosphate forms salts with calcium, iron, and aluminum that are only barely soluble (see Figure 5.5). Calcium phosphate salts are the most soluble but are important only in soils of near-neutral pH. Aluminum salts are the least soluble and dominate the low end of the pH spectrum. The solubility of iron phosphates depends on the redox state of the iron; reduced Fe^{2+} (ferrous iron) is 10 times more soluble than oxidized Fe^{3+} (ferric iron).

If phosphate precipitates as an almost insoluble salt, do plants have access to these pools? Three mechanisms may help. First, simply by absorbing phosphorus, the plants create a disequilibrium that causes additional small amounts of potassium salt to dissolve. Second, many plants and microbes (including mycorrhizal fungi) also secrete large amounts of simple organic compounds such as oxalate into soils. Oxalate has a strong tendency to form salts (particularly with calcium) and may liberate phosphate by grabbing the salt cation. Finally, the chemical status of the root zone (rhizosphere) may be substantially changed from that of the bulk soil; higher pH of the rhizosphere and other changes may favor the solubility of phosphorus salts.

Both the organic and the geochemical control of phosphorus cycling were important in explaining the potassium supply to sugar maple stands in a study from Quebec, Canada. Paré and Bernier (1989a,b) found higher availability of phosphorus in stands with a mor-type forest floor, where the organic matter was not mixed by worms into the mineral soil, in comparison with stands with a mull-type forest floor. The mor sites had about four times more phosphorus in the forest floor than the mull sites, and one might mistakenly expect that a larger accumulation in the forest floor pool would mean a lower supply to trees because of apparently slow decomposition. However, trees on these mor-type sites had twice the foliar phosphorus concentration of those on the mull-type sites (see Figure 9.8). The mor-type forest floors were also more acidic (pH 3.6 to 4.2) than the mull-type forest floors (pH 4.9 to 5.5), which would again lead one to expect a lower phosphorus supply from the mor-type forest floors. Mixing by earthworms at the mull-type sites substantially increased the forest floor contents of extractable iron and aluminum by three to five times compared to the mor-type site. More iron and aluminum in the mull-type forest floor bound phosphorus into forms that were less available for cycling (and uptake by trees) than the mor-type forest floors, where organic phosphorus cycled more readily. These studies again demonstrate that forest floor morphology may not provide much insight into soil fertility.

Given the high demand for phosphorus in forests and its low availability, losses of phosphorus are usually minimal. Even when the plant uptake component of the phosphorus cycle is removed, geochemical processes are often sufficient to retain all the phosphorus. For example, the phosphorus cycle of a northern hardwood forest in the northeastern United States was both dynamic and very tight. Uptake of phosphorus was about 12.5 kg/ha annually; 1.5 kg/ha accumulated in biomass, and 11 kg/ha returned to the soil in litter (Wood and Bormann, 1984). Only 0.007 kg/ha of phosphorus leached from the ecosystem because phosphorus uptake by roots and microbes dominated in the upper soil, and aluminum and iron compounds adsorbed any phosphorus that reached the B horizon. When plant uptake of phosphorus was removed by deforestation, geochemical processes retained the extra phosphorus and prevented increases in streamwater concentrations.

Not all ecosystems retain phosphorus this effectively, especially following phosphorus fertilization. Humphreys and Pritchett (1971) examined the phosphorus content of soil profiles 7 to 11 years after fertilization with superphosphate or ground rock phosphate (see Chapter 13 for fertilizer formulations). They found that almost all of the superphosphate phosphorus was retained in the top 20 cm of soils with low, moderate, or high phosphorus adsorption capacities, but that very little was retained in sandy soils with negligible phosphorus adsorption capacities. All soils retained the rock-phosphorus. Chapter 13 notes that phosphorus concentrations in streamwater can be elevated following phosphorus fertilization of forests.

Phosphate inputs are also small in forests, commonly ranging from 0.1 to 0.5 kg/ha annually. The rates of release through mineral weathering are less well known but appear to be of similar magnitude (see Table 9.3). Nutritional requirements for phosphorus are also much smaller than those for potassium or nitrogen, but even so, the ratio of the annual phosphorus requirement to the annual phosphorus input is usually much higher than for other nutrients. In an old-growth forest of Douglas-fir and other species in Oregon, the ratio of the phosphorus requirement for growth to the input of phosphorus was 18.2 (Sollins et al., 1980). The ratios were 6.5 for nitrogen, 3.0 for potassium, and 0.6 for calcium. This pattern emphasizes the importance of internal recycling and conservation in maintaining high phosphorus availability; it also illustrates the possible impacts of high phosphorus removals in harvesting or erosion.

The complex suite of phosphorus compounds in soils is often characterized by a sequence of extractions with bases and acid. The Hedley fractionations (Hedley et al., 1982; Cross and Schlesinger, 1995) first remove the phosphorus that is readily adsorbed by ion exchange resins. Next, the soil is extracted with bicarbonate and the extract is analyzed for inorganic and organic phosphate. Sodium hydroxide is used to remove less soluble phosphorus, which is analyzed in both inorganic and organic forms. The next step is addition of hydrochloric acid; the final, insoluble phosphorus fraction is termed residual phosphorus. These fractions probably represent a series of pools with varying turnover rates; bicarbonate-extractable phosphorus is probably more readily released for plant uptake than phosphorus from pools that are extracted by sodium hydroxide.

How well do these fractions represent the actual quantity of phosphorus that will be released and available for plant uptake? An elegant greenhouse experiment with *Brassica* plants examined the depletion of the phosphorus fractions within the rhizosphere (Gahoonia and Nielsen, 1992). After 2 weeks, the roots reduced the pH of the rhizosphere from 6.7 to 5.5. Bicarbonate-extractable inorganic phosphorus within the rhizosphere dropped by 34 percent (to a distance of 4 mm from the root surfaces), and the residual phosphorus (not extractable by bicarbonate or hydroxide) was also depleted by about 43 percent to a distance of 1 mm from the root surface. When a buffer solution was used to prevent acidification of the rhizosphere, plant uptake of phosphorus declined by about 17 percent, with less phosphorus obtained from the bicarbonate-extractable inorganic phosphorus and residual-phosphorus pools. Overall, the plants obtained a little more than half of their phosphorus from the pool that would often be assumed to be the major source (bicarbonate-extractable inorganic phosphorus). This study underscores the point that rates of flow (or flux) into and out of pools cannot be determined simply by pool size.

A more dynamic approach to evaluating phosphorus dynamics attempts to separate biotic controls from geochemical influences (Zou et al., 1992). The resin-extractable phosphorus in a soil sample derives from the combined influence of microbial uptake of phosphorus, enzymatic release of phosphorus from organic pools, and net sorption/desorption of phosphorus from geochemical pools. Irradiation (sterilization) of the soil sample increases the resin-extractable phosphorus by removing microbial uptake while leaving phosphatase activity and the geochemical pools unaltered. Autoclaving the soil stops phosphatase activity, leaving only the geochemical control on resin phosphorus. This technique was applied to soils from a loblolly pine plantation, and microbial uptake into organic phosphorus pools was about twice the

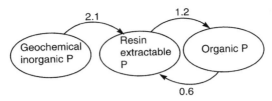

Figure 9.22 Net transfers of resin-extractable (readily available) inorganic phosphate from geochemical pools and organic pools in an Ultisol under a loblolly pine plantation (after Zou et al., 1992).

rate of mineralization from the organic pool. The most important net source in this soil appeared to be the geochemical sorption/desorption from mineral phosphorus pools (Figure 9.22). It is hoped that further work on methods of assessing phosphorus in forest soils will include more testing of current methods and development of new approaches.

POTASSIUM IS THE MOST MOBILE SOIL NUTRIENT

Entering the potassium cycle at the stage of forest floor and soil organic matter, decomposition releases K^+ to the soil solution. Potassium is present only in free, ionic form in plants, and its release from litter is usually faster than that of any other nutrient. In fact, about half of the potassium in leaves leaches prior to litterfall.

Potassium plays a wide variety of roles in plant biochemistry and ecophysiology. The pumping of K^+ into and out of the guard cells of stomata changes their osmotic potential, resulting in the opening and closing of the stomata. Free K^+ also balances the charges of major organic and inorganic anions within the cytoplasm, playing a major role in pH buffering. Potassium is also involved in enzyme activation, protein synthesis, and photosynthesis (Marschner, 1995).

Once in the soil solution, potassium has three possible fates. Plants and microbes may take up K^+, it may be retained on cation exchange sites, or it may leach from the rooting zone of the forest. Once taken up by plants, potassium remains unbound to any organic compounds and moves as a free cation through the plant to catalyze reactions and regulate osmotic potential. (Osmotic potential refers to the balance of water and dissolved compounds that help maintain the turgor, or rigidity, of plant cells.) If adsorbed on cation exchange sites, K^+ remains available for later use by vegetation. Some K^+ may become "fixed" within the mineral lattices of some types of clays and may be relatively unavailable for plant uptake. If not adsorbed, K^+ may be readily leached from the soil. Vigorous forests on relatively young soils tend to lose about 5 to 10 kg/ha of K^+ annually by leaching. Sandy soils and old soils are typically low in K^+, and leaching losses may be very small.

Potassium enters forest ecosystems from two sources: atmospheric deposition and mineral weathering. Precipitation involves a very dilute salt solution, and potassium inputs range from about 1 to 5 kg/ha annually. Dust and aerosols also contain K^+, and such "dry deposition" is especially important near marine environments. Weathering of soil minerals typically adds another 5 to 10 kg/ha each year to young soils and less than 1 kg/ha to sandy soils and old, depleted soils (see Table 9.3).

In general, K^+ inputs exceed outputs. This element limits forest growth only in some very sandy soils, some organic soils, and some very old, weathered soils where millennia of leaching have depleted the K^+ supply.

Management of potassium in forests includes fertilization, often in association with additions of nitrogen and phosphorus (see Chapter 13).

CALCIUM AND MAGNESIUM HAVE SIMILAR BIOGEOCHEMISTRIES

Sources for plant uptake are the cations released by decomposition from organic matter and cations displaced from the cation exchange complex (Figure 9.23). Rates of weathering tend to be between 1 and 30 kg/ha annually, depending primarily on mineralogy and perhaps secondarily on climate and biota. Controls on weathering are easy to list but hard to quantify, as discussed above.

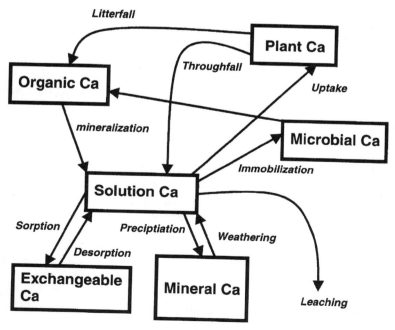

Figure 9.23 Soil calcium cycle; the cycles for other cations have a similar structure.

Within plants, magnesium forms ionic bonds with nucleophilic ligands (such as phosphoryl groups in ATP), acting as a bridge to form complexes that vary in stability (Marschner, 1995). Magnesium is also a key, covalently bonded component of chlorophyll; the chlorophyll-magnesium pool accounts for 10 to 50 percent of the magnesium contained in leaves. Calcium within plants is found as exchangeable calcium at the surface of cell walls and membranes, bound within structures, and in vacuoles. The distributions tend to be about one-quarter water-soluble calcium, half pectate calcium (primarily in cell walls), about 15 percent bound with phosphate, and the rest consisting of calcium oxalate and miscellaneous other pools. In some cases, the oxalate calcium pool can be much larger, perhaps as high at 90 percent in Norway spruce (Fink, 1992).

The biogeochemistry of calcium and magnesium includes large quantities bound in minerals in most soils (Oxisols are a major exception). This mineral pool is the long-term source for the moderate quantities adsorbed on cation exchange sites, which typically equal one to five times the calcium and magnesium contained in the biomass of a mature forest. These "base cations" cannot consume H^+ or donate electrons, so they are not bases in the chemical sense. However, exchange complexes dominated by these cations maintain higher soil pH levels than those dominated by H^+ and Al^{3+}. Calcium and magnesium do not volatilize or oxidize in fires, so the primary fire-related losses occur if nutrient-rich ash is blown away by wind before rainfall leaches the cations into the soil.

SULFUR CYCLING IS MORE COMPLICATED THAN NITROGEN CYCLING

The sulfur cycle blends features of the phosphorus cycle with some from the nitrogen cycle (Figure 9.24). Like phosphate, sulfate anions can be specifically adsorbed by soils, resulting in high retention of sulfate in largely unavailable forms. Most sulfur in plants is in a reduced form (C—S—H), but some remains as free SO_4^{2-} and some is in compounds that are difficult to identify. Like the nitrogen cycle, the sulfur cycle involves oxidation and reduction processes.

Sulfur is found in the amino acids cysteine and methionine (and therefore in proteins), as well as in enzyme cofactors, sulfolipids, polysaccharides, and secondary chemicals (Marschner, 1995). Sulfur in amino acids and proteins is in the reduced form, and sulfhydryl bonds among amino acids are responsible in part for the three-dimensional structure of proteins. Oxidized sulfate is the form in sulfolipids and polysaccharides. Reduced sulfur can be oxidized to sulfate, a "safe" form for storage.

Sulfur is present in litter in three forms: as free sulfate (SO_4^{2-}), as reduced sulfur bound to amino acids, and in unidentified compounds, possibly including ester-bonded sulfate (C—O—SO_3). Free sulfate rapidly leaches from fresh litter; the nitrogen-bound reduced sulfur can be released as microbes scavenge

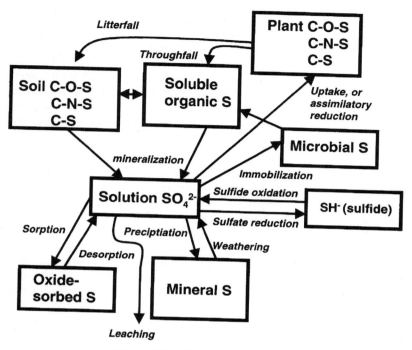

Figure 9.24   The soil sulfur cycle.

for carbon energy sources. Ester sulfur is released probably released primarily through the activity of sulfatase enzymes in soils.

Sulfate is the only free form of sulfur that is common in soils. Even when reduced sulfur is mineralized from nitrogen-bound compounds, microbes rapidly take advantage of sulfur's high-energy status to oxidize it to sulfate, much as nitrifying bacteria oxidize ammonium. Free sulfate in soil faces the same potential fates as phosphate: uptake by plants or microbes, precipitation as a salt, specific or nonspecific anion adsorption, or leaching from the rooting zone of the forest. Unlike phosphate, sulfate salts of calcium, iron, and aluminum are fairly soluble, so sulfate precipitation in forest soils is unimportant.

Sulfate is not toxic in plants, and plants with sulfur supplies in excess of current needs may accumulate sulfate in needles. Most organic sulfur compounds also contain nitrogen, and foliage generally contains about 1 sulfur atom for every 30 nitrogen atoms.

Turner et al. (1979) noted that some Douglas-fir stands that failed to respond to nitrogen fertilization may have been limited by low sulfur availability. Nonresponsive stands had less than $400\,\mu g$ of sulfate sulfur per gram of foliage, and responsive stands had up to $1000\,\mu g/g$. These authors speculated that sulfur + nitrogen fertilization might produce larger growth responses than nitrogen fertilization alone, but careful, well-designed sulfate fertilization trials have not clearly supported this hypothesis (Blake, 1985).

Forests typically require only 5 to 10 kg/ha of sulfur each year, and in unpolluted regions atmospheric inputs often range from 1 to 5 kg/ha annually. Sulfur resorption prior to litterfall usually recycles about 20 to 30 percent of the sulfur in leaves. Sulfur dioxide (SO_2) pollution results in two sources of increased sulfur inputs: direct absorption of SO_2 gas and conversion of SO_2 to sulfuric acid (SO_4) in rainfall. Forests in industrialized regions often experience sulfur inputs of 20 to 50 kg/ha annually. Where does the sulfur go? Only a minor proportion can be stored in plant biomass; the rest is either adsorbed in the soil or leached from the ecosystem. Leaching of sulfate sulfur is often in the range of 10 to 20 kg/ha annually in polluted regions.

Sulfate in unpolluted regions typically enters as a salt with sodium or calcium. In polluted areas, sulfuric acid is the dominant form. In both situations, sulfate tends to leave as a salt (in company with base cations). Sulfuric acid inputs coupled with sulfate salt outputs result in a net increase in ecosystem H^+. If sulfate leaches in company with H^+ or aluminum, acidification of streams and lakes may result. For this reason, the mobility of sulfate in forest soils is a major concern in the study of lake and stream acidification.

SMALL QUANTITIES OF MICRONUTRIENTS PLAY LARGE ROLES

The supply of micronutrients limits forest productivity in some notable cases, but overall these limitations are not common. See Marschner (1995) for more details on the features described in this chapter.

Iron is needed in plants for redox systems (cytochromes, peroxidases) and for some proteins. Key features of the iron cycle include redox reactions. The reduced form of iron (ferrous iron, Fe^{2+}) has a solubility of about 10^{-15} mol/L at neutral pH levels. Oxidized iron (ferric iron, Fe^{3+}) is only one-tenth as soluble as ferrous iron. This incredibly low solubility in water underscores the importance of organic chelates, which can increase iron solubility by several orders of magnitude (Lindsay, 1979). In acidic, waterlogged soils, some crops may experience iron toxicity owing to the increased solubility of ferrous iron at low pH; iron toxicity has not been reported in trees.

Manganese occurs in ecosystems in three oxidation states, with Mn^{2+} being most common. Manganese is a constituent of two plant enzymes: a manganese protein that splits water in photosystem II and superoxide dismutase. About three dozen enzymes are catalyzed by manganese, including RNA polymerase in chloroplasts. High levels of manganese in leaves of agricultural crops can be toxic (ranging from 200 to >5000 mg/kg, depending on the species), but manganese toxicity has not been reported in forests.

One of copper's primary functions in plants involves electron transport chains. The last step in electron transfer systems in cells occurs when copper enzymes perform the final step of transfering an electron to oxygen. About half of the copper in chloroplasts is bound in plastocyanin. Other roles for copper include a copper-zinc superoxide dismutase and phenol oxidases. Like manganese, copper at high levels can be toxic in agricultural crops, but no toxicities in forests have been reported.

The biochemical role of zinc in plants derives from its tendency to form tetrahedral complexes with nitrogen, oxygen, and sulfur. Many enzymes require zinc as a structural component, including carbonic anhydrase, alkaline phosphatase, and RNA polymerase. Some agricultural studies have shown that phosphorus fertilization may lead to zinc deficiencies, either by reducing zinc availability in the soil or by reducing zinc uptake and function in the plant.

Molybdenum is critical for two plant enzyme systems: nitrate reductase and, in nitrogen-fixing plants, nitrogenase. The oxidized state (molybdenum VI) is the most common state, with the majority of soil chemistry pertaining to molybdate (MoO_4^{2-}). Small quantities of molybdenum fertilzer (on the order of 10–20 g/ha) can substantially increase the productivity of clover pastures, but applications to nitrogen-fixing tree plantations have not been tested (because forest soils have not been molybdenum deficient). Molybdate chemistry is similar to phosphate chemistry, including substantial sorption on iron and aluminum sequioxides.

Boron is the least understood of the elements required by plants. Plants that are deficient in boron show a variety of symptoms that may be related to cell wall thickness, lignification, membrane integrity, carbohydrate metabolism, and other features. The specific role of boron in supporting normal plant functioning remains unclear, even though deficiency symptoms are easy to induce in hydroponic culture, and forest deficiencies have been reported from New Zealand to Canada to Europe.

Nickel and chlorine are also regarded as necessary for normal plant functioning, but the amounts required are never limiting. Some other elements, including sodium, silicon, selenium, cobalt, and even aluminum, can be beneficial to some species even if an absolute requirement is difficult to demonstrate.

NUTRIENT USE AND NUTRIENT SUPPLY CHANGE AS STANDS DEVELOP

A classic study of an age sequence of loblolly pine stands illustrates a classic pattern: forest production rises to a peak early in stand development, followed by a substantial decline even though the forest remains vigorous and healthy (Figure 9.25). The pattern of nutrient use and accumulation tends to follow the same trend, but is this a cause, an effect, or a simple covariation? The decline in nutrient use in older forests may derive from a lower demand by the slower-growing forest. Alternatively, the decline in nitrogen use may come from a decline in nitrogen supply, which drives the decline in production. Not enough evidence is available to sort out these possibilities, but the state of

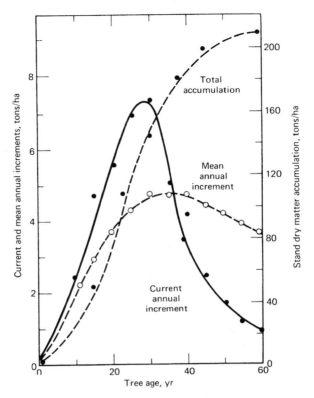

Figure 9.25 Annual and total biomass accumulation in stems of loblolly pine across 60 years of stand development (1 ton = 1 Mg; after Switzer et al., 1968).

current knowledge includes these insights in trends over stand development (from Ryan et al., 1997):

1. Both increases and decreases in soil nutrient supply are common; no universal pattern is likely to emerge.

2. The response to fertilization is very common for a given soil across a wide range of stand ages.

3. The response to fertilization in old stands does not tend to be proportionally larger than that in younger stands, so changes in nutrient supply and limitation probably do not drive the decline in growth in older stands.

This story should become clearer in the next decade or so as current research projects are completed.

THREE RULES OF BIOGEOCHEMISTRY

The construction of nutrient budgets involves a wide range of issues, including the choice of processes to measure, experimental design, and accuracy of measurements. Some major mistakes can be avoided, or insights obtained, if three simple rules are used:

1. Pools do not represent fluxes (or supplies);
2. Nutrient budgets need to balance; and
3. Attention to issues related to scales of space and time is important.

In nutrient cycling studies, it's often easier to measure the size of a pool at a given time than it is to measure the flux of nutrients into and out of the pool. For example, the size of the cation exchange pool is easy to measure, but the flux of cations into this pool from the decomposition of organic matter or the weathering of soil minerals is very difficult to measure. The quantity of organic and inorganic phosphorus that can be extracted by sodium bicarbonate is easy to measure, but not the rate at which phosphorus enters or leaves either pool. We are often very interested in the rate at which nutrients become available for uptake, but it is important to remember that estimates of pool size are often poor indicators of availability; one cannot tell what the supply of nutrient cations is by measuring the pool that sits on the exchange complex at a particular time. If the supply of a nutrient has to be estimated by the size of a pool because direct measurement of fluxes is too difficult, one needs to retain healthy skepticism about the power of any interpretations. A good example comes from an agricultural study in Hawaii, where 15 rotations of a crop removed 4 Mg/ha of potassium, and exchangeable pools of potassium declined from only 0.5 to 0.4 Mg/ha (Ayers et al., 1947).

Two examples illustrate the second rule about balancing nutrient budgets. When only part of a nutrient budget is estimated, it may be useful to consider the implications of the measured components for the unmeasured components. A research project examined the loss of nitrogen from soils following harvesting. The authors sampled soil before harvesting and 2 years later. They reported that 1300 kg/ha of nitrogen had disappeared from the soil (to a depth of 1 m). Only a few studies worldwide had ever reported rates of nitrogen loss even 10 percent as large as this one. If such a large loss had occurred, several implications for other parts of the nutrient budget would need to be considered. First, gaseous loss of this magnitude of nitrogen as N_2 or N_2O (denitrification) would be greater (by several orders of magnitude) than those found in studies that have estimated denitrification in such forests. If denitrification is unlikely to account for much of the loss, then nitrification and nitrate leaching would be the most likely causes. Leaching of dissolved organic nitrogen can be important where overall leaching rates are low, but no reports have suggested rates higher than 10 kg/ha annually for leaching of organic nitrogen from forest soils. The generation of 1300 kg/ha of nitrogen as nitrate would involve a net production of over 100 kmol of H^+. The generation of so much acidity would have reduced soil pH substantially, and leaching that much nitrate would have entailed equivalent losses of base cations and aluminum. If this amount of nitrate were dissolved in 1 m of runoff from the site (a ballpark figure for precipitation minus evapotranspiration), the nitrate nitrogen concentration would have averaged about 150 mg/L, 15 times the drinking water standard, and peak concentrations would probably have been much higher. Such large nitrogen losses are unlikely, and if the authors had high confidence in the world-record rate, readers would have benefited from the development of a balanced idea of the implications for nitrate leaching, soil acidification, and streamwater quality.

The value of looking for balanced accounting is also evident from a study that examined the cation nutrient budget for adjacent plantations of Norway spruce, beech, and birch. The authors used a model to estimate the rate of mineral weathering, and they concluded that the weathering release of base cations would be about 0.3 kmol$_c$/ha annually. How did this balance other parts of the nutrient budget? The spruce stands had greater amounts of base cations on the exchange complex, as well as greater base cation content in the larger accumulation of tree biomass. The leaching losses of base cations from the soil under spruce were more than twice the losses under the hardwoods. Summing the greater pools of base cations in soil and biomass with the greater loss rates showed that the spruce stand gained about 1.5 kmol$_c$/ha more base cations annually than did the other stands. This balancing of the nutrient cation budget indicates either that weathering was far higher under spruce (at least five times greater than estimated in the model) or that one or more of the nutrient budget components was misestimated. In many studies, the degree of balance in nutrient budgets may not be known until the late stages of a project. In these cases, consideration of the accounting balance can help gauge the confidence that is warranted in the components of the budget, highlight areas of high uncertainty, and suggest areas for further research.

The important components of nutrient budget calculations depend on the scale of interest, and extrapolation across scales needs careful attention. The first example of scale shows that important components of nutrient budgets differ across time scales. What is the rate of nitrogen loss from a lodgepole pine forest in Wyoming? At an annual scale, the average loss might be on the order of 0.5 kg/ha from leaching, or 50 kg/ha in a century. But at the time scale of a century, the nitrogen loss is likely to be dominated by a single wildfire event that would remove over 100 kg/ha of nitrogen in a few hours.

The second example of scale illustrates that patterns across stands may not match those that should be expected within stands. A pattern between soil nitrogen availability and net primary production across sites may be strong (Figures 9.1, 9.2). What would be the effect of increasing nitrogen availability within a single site? It may be tempting to expect that the pattern across sites would adequately represent the pattern within a single site, but such an expectation would need to be tested before much confidence is warranted.

The third scale issue is that the controls on a nutrient flux at one time scale may differ from the controls at another time scale. Short-term incubations show that carbon release often doubles when the incubation temperature increases by $10°C$, and this relationship could be used to estimate the increase in carbon release from soils over long terms. However, the temperature responsiveness of readily oxidized carbon may be quite different from that of stabilized, humified carbon (Giardina et al., 1999b).

Budgets of H^+ illustrate the importance of scales in both time and space. In most forests, one of the major components of the H^+ budget on an annual scale is the net accumulation of base cations in tree biomass. This acid production is commonly on the order of $0.5\,kmol_c/ha$ annually, or about $50\,kmol_c/ha$ in a century. Over the same time period, the H^+ generated from dissociation of carbonic acid (formed from the high concentration of soil carbon dioxide dissolved in soil water) might generate $25\,kmol_c/ha$, giving the impression that cation accumulation in biomass has twice the acidifying effect of bicarbonate production and leaching. However, when the forest is consumed by fire or other disturbances, the base cations are returned to the soil and H^+ is consumed in the production of carbon dioxide and water, erasing the century-long legacy of acid production by biomass accumulation. The legacy of bicarbonate formation and leaching would not be erased when the forest was consumed. So, on an annual time scale, biomass accumulation was very important in the H^+ budget and soil acidification, but on the century scale, this process became a cycle with no net effect, and only the bicarbonate production needed to be accounted for to describe long-term soil acidification.

SUMMARY

The productivity of forests depends on the supply of nutrients available within the soil for plant use; high rates of nutrient supply lead to rapid growth. The flow of chemical energy through the ecosystem involves oxidation and reduction reactions in which the flow of electrons is associated with the release and consumption of energy. Oxygen is the electron acceptor that provides the greatest amount of energy release for each electron. The nutrient cycles of forests involve pools of nutrient elements and flows of nutrients between these pools. The annual cycling of nutrients from the soil into trees is commonly greater than the annual input of nutrients into a forest, but this flow is only a tiny fraction of the total pool of nutrients in the soil. The annual decay of litter from leaves and roots provides most of the nutrients used by plants in a given year. However, the quantity of nutrients stored in the floor of most forests is either accumulating or in a steady state (until a fire or another disturbance creates drastic losses), and the increase in nutrient content of the accumulating forest biomass must come from annual inputs from the atmosphere, from mineral weathering, and from a declining pool of nutrients in slowly decomposing soil humus. The overall biogeochemical cycles differ among elements. Carbon, nitrogen, and sulfur have major oxidation and reduction reactions within ecosystems. Other elements, such as calcium and potassium, undergo no redox reactions but have strong interactions with geochemical pools (see Table 9.3 for a summary of the major elements). The accounting for pools and fluxes of elements in ecosystems includes three important rules: (1) pools do not represent fluxes (or supplies), (2) nutrient budgets need to balance, and (3) issues of scales in space and time need to be considered explicitly.

CHAPTER 10

Fire Effects

Fires are major features of the life, death, and rebirth of most forests around the world. Repeated low-intensity fires may shape the forest by killing small trees, rejuvenating fire-tolerant grasses, and accelerating the cycling of nutrients. High-intensity fires kill the majority of trees, oxidize large quantities of nutrients such as nitrogen, and disturb plant-soil interactions for decades.

Fire oxidizes organic matter to form carbon dioxide and water, releasing tremendous amounts of energy as heat. Nitrogen in organic matter is oxidized to N_2 and various nitrogen oxides are lost from the system. Calcium in organic matter is converted to calcium oxides and bicarbonates, which are lost in wind-blown ash or retained on site. Phosphorus in organic matter may be lost as a gas or released as barely soluble phosphate salts. Fires drastically alter the composition and activity of the soil biota, but little information is available on these changes. Despite the loss of nutrients in fires, nutrient availability to plants typically increases after fire as a result of heat-induced release of nutrients, reduced competition among plants, and perhaps sustained changes in soil conditions (such as temperature and water content).

MOST FORESTS BURN

Most forests of the world experience fires, with climate-driven differences in fire frequency and intensity. Seasonally dry regions such as portions of the coastal plain of the southeastern United States often experience low-intensity fires every few years, and longleaf pine forests are well adapted to frequent fires. Most temperate conifer forests experience fire regimes with moderate- to high- severity fires every 50 to 200 years. Montane and boreal conifer forests often burn at intervals of several hundred years. Deciduous temperate forests have variable fire regimes, with fire return intervals ranging from a few decades (or less) to several centuries (or longer). In tropical regions, fires are common in seasonally dry environments and rare in perennially moist areas. Even the wettest tropical rain forests may burn at a time scale of thousands of years (Sanford et al., 1985), and the annual extent of fire may be greater in tropical and subtropical regions than in the rest of the globe (Goldammer, 1995).

Figure 10.1   Controlled burning using a back fire with a light fuel load.

Frequent low-intensity fires (Figures 10.1, 10.2) tend to cover small areas, of up to a few hundred hectares. Less common, severe fires burn much larger areas (Figure 10.3).

In recent times, direct human use of fire may have had a greater effect on forests and forest soils than natural fires. Humans have historically used fire to clear land and perhaps manage wildlife (Pyne, 1995). With the development of major agriculture in forested regions, fire became the tool of choice for removing trees and wood from soils in Europe, North America, South America, and around the tropics. Fire is also a tool for forest management, where it is used to decrease the risk of wildfire to alter species composition, and to reduce the woody debris left after logging (Figure 10.4)

The effects of fire on soils have intrigued scientists since the beginning of forest soil science. In Sweden, Hesselman (1917) showed that although fire volatilized a large portion of a forest's nitrogen capital, it also increased nitrogen availability. He also found that the production of ammonium seemed highest between pH 4.5 and 4.9, whereas nitrate production was favored in the range of pH 5.5 to 6.9. In the western United States, Isaac and Hopkins (1937) reported that about 500 kg/ha of nitrogen was lost from a slash fire in a Douglas-fir clearcut. Heyward and Barnette (1934) examined the effects of repeated surface fires in mature longleaf pine forests and concluded that the total nitrogen content of the soil was not substantially reduced.

Figure 10.2 Periodic burning under controlled conditions results in park-like stands of pine.

FIRE PHYSICS LARGELY DETERMINE FIRE IMPACTS ON FOREST SOILS

The oxidation of organic matter releases large amounts of energy as heat. The actual amount of energy released in a fire depends on fuel consumption and the amount of moisture in the fuel (substantial energy can be absorbed in evaporating water from fuels). Plant biomass typically contains about 18 MJ/

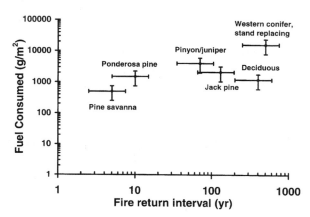

Figure 10.3 Fuel consumption (a measure of fire intensity) generally increases with the fire return interval in North American temperate forests (based on Olson, 1981, and Christensen, 1987).

Figure 10.4 Burning is commonly used to reduce fuel loadings after harvesting; such hot fires remove substantial quantities of nutrients but also decrease competition with noncrop plants and reduce the risk of wildfire.

kg of dry material (Agee, 1993). How much energy is released during a wildfire? A severe wildfire in an old-growth Douglas-fir forest may consume about 17 percent of the total biomass of 1000 Mg/ha, for a fuel consumption of about 170 Mg/ha. The energy release would be on the order of 3 million MJ/ha, or 300 MJ/m^2. Gasoline has an energy content of about 30 MJ/L, so the forest fire's energy release would equal that of burning about 10 L of gasoline across every square meter of area.

This tremendous release of energy as heat raises the temperature of the soil. The pattern of temperature increase depends on the rate of burn, the amount of fuel consumed, soil moisture, and conductivity properties of the soil. Rapidly advancing fires that consume little fuel have little effect on soil temperatures. Slower fires that consume more fuel may have temperatures exceeding 700°C at the soil surface, declining to 200°C a few centimeters into the mineral soil and to normal levels below 15 or 30 cm depth. For a variety of slash-and-burn operations in tropical forests, soil temperatures were highest in dry tropical forests and lower in monsoonal tropical forests and humid tropical forests (Figure 10.5).

The conversion of liquid water to vapor consumes large amounts of energy (2.5 MJ/L), so the release of heat in moist soils may have a limit on temperature at 100°C until all the water has evaporated. Water also has high conductivity for heat, so burning on moist soils may limit the temperature rise

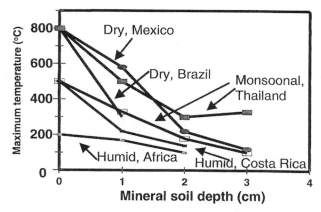

Figure 10.5 Soil temperatures at various depths for slash-and-burn treatments tend to be higher in drier climates, as drier fuels burn more intensely (data from Giardina et al., 1999a).

while at the same time increasing the depth of penetration of the heat. After 5 minutes of burning atop mineral soils the maximum soil temperature reached 80°C in a wet mineral soil, and 100°C in a dry soil (DeBano et al., 1976). After 10 minutes, the wet soil temperature reached a maximum of 120°C compared with 200°C for the dry soil.

FIRES REMOVE NUTRIENTS BY FIVE PROCESSES

Losses of nutrients in fires result from the combined effects of these processes:

1. Oxidation of compounds to a gaseous form (gasification)
2. Vaporization (volatilization) of compounds that were solid at normal temperatures
3. Convection of ash particles in fire-generated winds
4. Leaching of ions in solution out of the soil following fire
5. Accelerated erosion following fire

The relative importance of these processes varies with each nutrient and is modified by differences in fire intensity, soil characteristics, topography, and climatic patterns.

NITROGEN LOSSES ARE PRIMARILY FROM OXIDATION, NOT VOLATILIZATION

Organic compounds contain nutrients, and some of these nutrients are in a reduced state ($R-NH_2$, $R-SH_2$). These reduced forms of nitrogen and sulfur are oxidized at the temperatures reached in fires, releasing energy as gaseous oxidized compounds are created. For example, the combustion of an amino acid (cysteine) releases N_2 and SO_2 gases:

$$4CHCH_2NH_2COOHSH + 15O_2 \rightarrow 8CO_2 + 14H_2O + 2N_2 + 4SO_2 + energy$$

For some reason, these oxidation losses of nitrogen came to be called volatilization losses in the literature, as though the nitrogen simply evaporated when heated.

Three approaches have been used to examine nutrient losses from fires. The simplest method involves heating samples in a furnace and measuring the change in nutrient content. Knight (1966) used this method to examine nitrogen losses from small samples of forest floor materials heated for 20 minutes at various temperatures (Figure 10.6). No nitrogen oxidized to gas 200°C, 25 percent was lost at 300°C, and about 65 percent disappeared at 700°C. Tiedeman et al. (1979) examined sulfur losses in a furnace from the foliage of various species and found from 25 to 90 percent of the sulfur was lost as temperature increased from 375 to 575°C.

One problem with using furnaces is that a sample is heated from all directions, whereas fires generate gradients of temperatures within the forest floor. Mroz et al. (1980) placed forest floor samples in clay pots and then placed the pots in a furnace preheated to 500°C for 30 minutes. This allowed rapid heating of the surface while lower levels remained cool. They found a substantial loss of nitrogen from the upper portion of the samples, but most of this nitrogen was recovered in the lower portion. On average, the net loss of

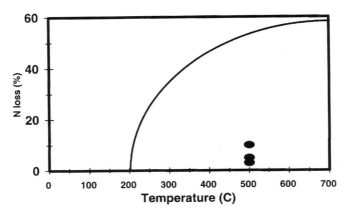

Figure 10.6. Nitrogen oxidation and loss rise rapidly as temperatures increase above 200°C (data from Knight, 1966) when small samples are heated from all sides in a furnace. Losses are much smaller when samples are heated from above for three different intervals of time at 500°C (data from Mroz et al., 1980).

nitrogen was only 3 to 10 percent of the total. In contrast, Knight's pattern in Figure 10.6 would indicate that about 45 percent of the nitrogen should have been lost. Losses of nitrogen in actual fires are probably much less than would be suggested by furnace experiments on small samples which do not allow for the importance of temperature gradients.

The second approach to examining nutrient losses is to measure the nutrient contents of fuels in the field, before and after burning. In general, nitrogen loss under field conditions appears to be related to the amount of organic matter consumed. Table 10.1 summarizes nutrient losses from fires, and Figure 10.7 diagrams this relationship for a variety of fires reported in the literature. The variation in organic matter loss accounts for about 85 percent of the variation in nitrogen loss, which equals about 1 percent of organic matter loss (for more discussion, see Raison et al., 1985). Most studies of nutrient losses from fires

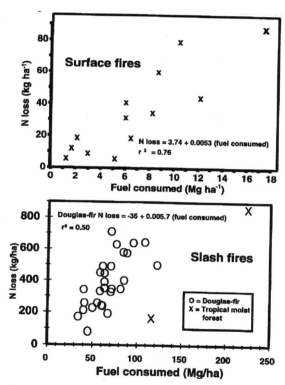

Figure 10.7 Nitrogen loss from fires is generally proportional to fuel consumption, with about 5 kg of nitrogen lost per Megagram of fuel consumed. Tropical slash fires may have somewhat lower nitrogen losses per Megagram of fuel consumed, but more information is needed (modified from Binkley, 1999; data for the lower graph from Little and Ohmann, 1988, and Kauffman et al. 1995).

TABLE 10.1 Nutrient Losses from Fires

Fire Type Vegetation	Nutrient Loss (kg/ha)					
	N	P	Ca	Mg	K	S
Prescribed/loblolly pine (Richter et al., 1982)	10–40	—	—	—	—	2–8
Slash fire/Hemlock, Douglas-fir (Feller, 1983)						
Hot	980	16	154	29	37	—
Moderate	490	9	87	7	17	—
Slash fire/*Eucalyptus* (Raison et al., 1985)	75–100	2–3	19–30	5–10	12–21	—
Slash fire/*Eucalyptus* (Harwood and Jackson, 1975)	—	10	100	37	51	—
Wildfire/Douglas-fir (Grier, 1975)	855	—	75	33	282	—
Slash fire/tropical moist forest (Kauffman et al., 1995)	800–1600	5–20	—	—	—	90–125

are from relatively low-intensity prescribed fires and high-intensity slash fires (particularly good examples include Harwood and Jackson, 1975; Covington and Sackett, 1984; Raison et al., 1985; and Little and Ohmann, 1988).

The third approach consists simply of comparing the nutrient content of burned areas with that of unburned areas. Several studies have documented the long-term effects of repeated low-intensity fires on nitrogen availability. In a 30-yr study in a loblolly pine–longleaf pine ecosystem, fires at intervals of 1, 3, or 4 years reduced the nitrogen content of the forest floor by as much as 85 percent (Figure 10.8b). But even 30 years of annual fires reduced the total nitrogen content of the forest floor and the 0- to 20-cm mineral soil by only 300 kg/ha (or 10 kg/ha for each fire). A second 30-year study (Figure 10.8a) found no net loss of nitrogen.

Annual losses of carbon from ecosystems due to fire are a function of fire intensity and fire return interval. Based on rough estimates of these parameters, annual losses of organic matter to fire vary from less than 0.01 Mg/ha in deserts and mesic deciduous forests to over 1 Mg/ha in ecosystems where intense fires with comparatively short return times are the norm. Assuming that a loss of 6 kgN/Mg fuel is a reasonable estimate for most canopy fires (this may actually be a conservative estimate of such losses given the high concen-

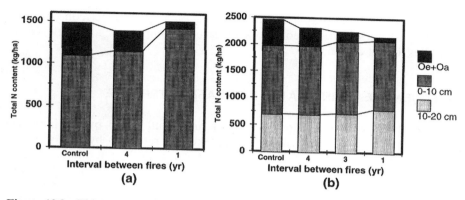

Figure 10.8 Thirty years of repeated prescribed fires in loblolly pine–longleaf pine forests reduced the nitrogen content of the forest floor and increased that of the mineral soil, for no net loss in one study (a, data from Wells, 1971), or a net loss of up to 300 kg/ha over 30 years (b, data from Binkley et al., 1992).

tration of nitrogen in leaves), annual nitrogen losses due to fire can be estimated. Thus, average nitrogen losses due to fire may vary from approximately 0.12 to 12 kg/ha/yr. Soil leaching commonly removes about 1–3 kg/ha of nitrogen annually from forests. Depending on the fire return interval and the quantity of fuel consumed, fires remove about as much nitrogen as leaching does (Johnson et al., 1998).

Gasification losses of phosphorus compounds are more complex. All phosphorus is in the phosphate (PO_4^{3-}) form in ecosystems, but high temperatures in fires generate a variety of oxide (and organic oxide) compounds. The most important oxide formed during fires is probably P_4O_{10}, which can volatilize at 360°C (Cotton and Wilkinson, 1988).

Vaporization (or volatilization) losses occur when a compound evaporates with no chemical change. For example, nitrate (NO_3^-) present in soils vaporizes at temperatures as low as 80°C (Greenwood and Earnshaw, 1984), and amino acids may vaporize at temperatures below 200°C (Weast, 1982), well below their combustion temperatures. Metal cations such as potassium, calcium, and magnesium remain stable at higher temperatures. Potassium hydroxide (KOH) vaporizes above 350°C (Cotton and Wilkinson, 1988), whereas oxides of calcium and magnesium are stable even at 2500°C (Greenwood and Earnshaw, 1984). The actual vaporization temperature may be lower for these metals when they are part of organic molecules (Raison et al., 1985).

Convection losses of nutrients in ash can be very large owing to the high concentrations of nutrients in ash and the large quantities of ash that may be lost from a site. For example, the calcium concentration in leaf litter of *Eucalyptus pauciflora* was 470 mg/kg, compared with 5740 mg/kg in black ash and 14,750 mg/kg in gray ash (Raison et al., 1985). The higher concentration in ash indicates large gaseous loss of carbon compounds and little or no gaseous loss of calcium.

NUTRIENT LEACHING RATES INCREASE AFTER FIRE

The leaching of nutrients from soil after fire is influenced by the increased quantity of ions available, changes in uptake and retention by plants, adsorptive properties of the forest floor and soil (both microbial and mineral), and patterns of precipitation and evapotranspiration. However, even in the most extreme cases, the extent of loss by this process is generally small relative to other loss pathways and the total nutrient capital. Leaching losses following wildfire in a western Washington old-growth Douglas-fir forest appeared to be small; the majority of ions released from ash were retained in the upper 20 cm of the soil (Grier, 1975). Leaching of NO_3 nitrogen from an intensely burned Douglas-fir–pinegrass ecosystem in eastern Washington accounted for a loss of approximately 0.5 percent of the total soil nitrogen capital (Tiedemann et al., 1978). These leaching losses appear to have been due to fire-caused acceleration of rates of net nitrification and the high mobility of the nitrate ion. Chorover et al. (1994) reported that a prescribed understory fire in a mixed conifer forest in the Sierra Nevada of California increased streamwater outputs of nitrate nitrogen by 1000-fold, but the postfire nitrate nitrogen loss was only about 5 kg/ha annually.

Knight et al. (1985) found large increases in soil solution concentrations of some chemical species in recently burned stands of lodgepole pine. They attributed the comparatively high fire-caused leaching losses to increased nutrient mobility, low soil ion exchange capacity, and flushing associated with snowmelt. Increased concentrations of anions and cations in soil and groundwater following fires in eastern pitch pine forests (Boerner, 1982) and longleaf pine forests (Lewis, 1974) are likely a consequence of the very coarse texture (and therefore low adsorption capacity) of the coastal plain soils.

Leaching losses of cations may be influenced by variations in soil heating. Stark (1977) observed that leaching of calcium, magnesium, and iron from soils beneath Douglas-fir–larch forests was unaffected by moderate to low soil heating during fire. However, leaching of calcium and magnesium considerably increased, while leaching of iron decreased at soil temperatures above 300°C.

EROSION MAY INCREASE NUTRIENT LOSSES AFTER FIRE

Accelerated rates of erosion following fire can cause significant nutrient losses (Wright and Bailey, 1982) because of changes in vegetation, soil properties, hydrology, and geomorphic processes (Swanson, 1981). The actual amount and duration of any increase in erosion vary widely among sites as a consequence of fire intensity, soil infiltration capacity, topography, climate, and patterns of vegetation recovery. Loss of plant cover and forest floor exposes soil to increased kinetic energy of raindrops, which may increase sediment movement. Sediment and nutrient losses may be ameliorated by surface litter and postfire needlecast. Fires may increase the erodibility of soils if mineral soil is exposed, or if the capacity for water infiltration is substantially reduced. Porosity and infiltration rates may decrease, and soil aggregates may be dispersed by beating rains. Pores may become clogged by tiny particles. The amount of erosion following a fire shows great variability across landscape positions, among soil types, and with fire intensities. An intense rainstorm could generate large amounts of erosion from a burned watershed that would show little erosion in the absence of a large storm.

Where surface organic layers are not completely consumed, changes in pore space and infiltration rates are too small to be detected. Even relatively mild fires can expose mineral soil and reduce infiltration rates. For example, Ursic (1970) found that annual burning in watersheds supporting hardwoods on hilly terrain increased streamflow and sediment yields by 50 to 100 percent. DeBano et al. (1998) tabulated results of postfire studies of erosion, and the four available studies showed large increases; burned sites yielded between 30 and 200,000 kg/ha more sediment than unburned sites. More studies are needed to flesh out the distribution of frequency of small increases versus large increases.

Sykes (1971) reported that there was little information or agreement on the hydrological effects of fire in northern forests. Burning off the heavy moss layer in these forests may alter the distribution of summer runoff, with erosion accompanying flash floods. However, there are indications of increases in water infiltration rates in burned sites compared to unburned soils, which may result in surprisingly little erosion In boreal forests (Sykes, 1971). He pointed out that these infiltration data are in contrast to those reported by workers in temperate zones, where infiltration rates on burned areas have been slower than on unburned areas.

What is the significance of soil lost to erosion after fires? Few studies have compared the productivity of soils in eroded and uneroded sites after forest fires. An intriguing study by Amaranthus and Trappe (1993) provides some clues to possible effects. They used a postlogging site in Oregon swept by a severe wildfire that completely consumed the forest floor across 85 to 95 percent of the area. Erosion estimates found that about 100 Mg/ha of mineral soil moved downslope in the year following the fire. The scientists planted seedlings of incense-cedar and Douglas-fir in soils from which the top had eroded, in the soil that accumulated in sediment traps (i.e., the former topmost mineral soil), and in the trapped soil that had been sterilized to kill a mycorrhizal inoculum. The Douglas-fir seedlings formed only a few ectomycorrhizal root tips in any of the soils, but they grew well in all three soils (height increments of about 28 mm/yr). The incense-cedar seedlings grew less well, and height growth differed strongly among soils (14 mm/yr for topsoil, whether sterilized or not, compared with 12 mm/yr for eroded soil). Colonization of incense-cedar root tips was highest in the topsoil (37 percent of root tips), intermediate in the sterilized topsoil (21 percent), and lowest in the eroded soil. Survival of incense-cedars followed the pattern of mycorrhizal infection, with twice as many surviving in the topsoil as in the eroded soil. This study may indicate that recolonization of the site by Douglas-fir may be relatively unconstrained by erosion and mycorrhizal inoculum, but that the establishment of incense-cedar may depend strongly on competition with faster growing Douglas-fir seedlings, as well as the effects of erosion on a mycorrhizal inoculum.

FIRES MAY DECREASE WATER INFILTRATION INTO SOILS

Water infiltration rates are often diminished following fire owing to the plugging of surface pores and increased fire-induced water repellency (Krammes and DeBano, 1965). Increased sediment loss as a consequence of the formation of hydrophobic soil layers has been documented in ponderosa pine forests (Campbell et al., 1977; White and Wells, 1981). Hydrophobic properties appear to develop when organic molecules volatilize (evaporate) as a result of heating; as the vapor comes in contact with cooler soil surfaces at depth, they condense and form nonwettable surfaces. The nature of such hydrophobicity has been elucidated in laboratory experiments (DeBano et al., 1998). For example, heating soils to less than 175°C has little effect on rates of water infiltration into soils. Temperatures between 175° and 200°C may substantially reduce infiltration rates, while temperatures over 290°C lead to combustion of the organic molecules that would create hydrophobic conditions at lower temperatures. Fire characteristics that promote hydrophobicity include high (but not extreme) severity, coarse soil texture, and low soil water content. Despite these insights from controlled experiments, the contribution of hydrophobicity to overall postfire erosion of forest soils remains largely unknown. The generation of hydrophobic patches in burned landscapes is relatively straightforward, and water infiltration rates in hydrophobic patches can be measured and compared with those in unburned or burned but not hydrophobic soils. Water infiltration rates may be lower in hydrophobic patches for a period of months to several years under field conditions (Dyrness, 1976; McNabb et al., 1989).

But how much erosion would result from a burned watershed if hydrophobic patches did not develop? The current state of the science does not allow us to assign various levels of importance to the factors that generate postfire soil erosion. Anecdotal insights from foresters and soil scientists working in burned areas suggest that hydrophobicity probably does play an important role in postfire erosion (R. Powers, personal communication, 1999).

Increased overland flow and loss of soil binding by root systems can result in increased rill and sheet erosion and facilitate debris flows (Swanson, 1981;

Wells, 1987). Postfire loss of ash due to wind can also be important (see Ewel et al., 1981; Kauffman et al., 1995).

The total losses resulting from fire-caused increases in leaching and erosion depend on the rates and patterns of postfire vegetation recovery. Uptake of nutrients by reestablished vegetation may significantly diminish nutrient losses, and recovery of leaf area is critical to the reestablishment of prefire hydrologic conditions (Knight et al., 1985).

ASH MAY CONTAIN LARGE QUANTITIES OF NUTRIENTS

Fires leave behind large amounts of ash, typically ranging from 2 to 15 Mg/ha (Raison et al., 1985). The concentrations of nutrients are unusually high in ash, giving nutrient contents of postfire ash layers of 20 to 100 kg/ha, 3 to 50 kg/ha, and 40 to 1600 kg/ha for nitrogen, phosphorus, and calcium, respectively. Some of the cation nutrients in ash are water soluble or readily released by microbial activity, and this should provide a large supply of available nutrients for recovery of vegetation. Other nutrients in ash, such as phosphorus may not be readily soluble and may become available only slowly (Giardina et al., 1999a).

DIRECT NUTRIENT RELEASE FROM SOIL HEATING MAY BE IMPORTANT

The obvious layers of ash following fire have led many researchers to focus on the nutrient content of ash as the major pool of nutrients released by fire. However, the direct effects of heating may release comparable quantities of some nutrients that were formerly bound in soil organic matter. Heat may release ammonium and phosphorus from organic matter even if the organic matter is not consumed in the fire.

DECOMPOSITION AND MICROBIAL ACTIVITY CHANGE AFTER FIRE

The environmental conditions of the residual forest floor and soil may be very different after the fire. Charred and darkened organic materials may absorb radiation better than do unburned materials, resulting in warmer conditions. For example, Neal et al. (1965) reported that soil temperatures (at a depth of 5 cm) in a burned portion of a Douglas-fir clearcut averaged 6°C higher than in unburned portions. Fires may also increase soil moisture by decreasing water use by vegetation.

A pulse of increased nutrient availability typically follows fires as a result of reduced competition among plants, the release of elements from organic matter, and altered activity of soil biota. Changes in microbial properties of soils after fire have been documented, but few generalizations seem supportable. Decomposition is generally expected to increase after fire because of increased temperatures at the soil surface and increased soil moisture. Such an increase was demonstrated by Van Cleve and Dyrness (1985), who placed cellulose strips in mesh bags in the upper soil horizons of unburned and adjacent burned white spruce stands. After 2 months, most of the cellulose had decomposed in the burned plots, whereas the cellulose in the forest remained practically undecomposed. In a more detailed study, Bissett and Parkinson (1980) showed that higher rates of cellulose decomposition in burned sites occurred only in the field; laboratory incubations actually had slower decomposition rates for burned soils. They concluded that the microenvironmental effects of fire were more important than the chemical effects. Decomposition studies using leaf materials generally have not shown increased rates of decomposition in burned areas. For example, Grigal and McColl (1977) found that aspen leaves decayed more slowly in burned than in unburned plots in Minnesota, and Weber (1987) found no differences in the rate of decomposition of jack pine needles on burned and unburned plots in Ontario. Covington and Sackett (1984) reported that 8 months of decomposition after a fire in a ponderosa pine stand released 108 kg/ha of nitrogen more than that found for an unburned stand. Schoch and Binkley (1986) reported a similar increase in nitrogen release of 60 kg/ha 6 months after a fire in a stand of loblolly pine.

The impact of high-intensity fires on soil microbial activity has received little attention. Bissett and Parkinson (1980) found no difference in microbial biomass between burned and unburned plots in a spruce-fir forest in Alberta, British Columbia, after 6 years, but the ratio of bacteria to fungi was higher in the burned plots. Further study is needed.

Plant nutrition after fire depends on changes both in nutrient availability and in the distribution and activity of plant roots and mycorrhizal fungi. Few studies have examined root responses to the nutritional effects of fire. Chapin and Bloom (1976), and Chapin and Van Cleve, (1981) found that excised roots from plants in young postfire ecosystems displayed higher rates of phosphate adsorption than did the same species in older ecosystems. Increased nutrient availability may result in higher concentrations of nutrients in plant tissues following fire, and these higher concentrations may (or may not!) increase the rates of decomposition and nutrient turnover (Chapin and Van Cleve, 1981).

More work is needed on the effects of fire on microbial activity and decomposition before generalizations can be supported.

SOIL ACIDITY DECLINES AND pH RISES AFTER FIRE

Soil pH typically increases immediately after fire and then declines to prefire levels over a period of months, years, or decades. Soil pH influences the availability of some nutrients, both through direct geochemical effects and through indirect effects on microbial activity. For example, Montes and Christensen (1979) raised the pH of incubated soils from beneath several different vegetation types by 0.5 to 1.7 units and found that nitrate production increased by severalfold.

Soil pH depends on the equilibrium between the exchange complex and the soil solution (see Chapter 5). An exchange complex dominated by H^+ and Al^{3+} maintains a low (strongly acidic) pH in the soil solution, whereas an exchange complex dominated by so-called base cations (K^+, Ca^{2+}, and Mg^{2+}) maintains a higher pH (less acidic) soil solution.

Fires increase soil pH by two processes. The combustion of undissociated organic acids (such as acetic acid) in litter and soil removes the organic acids from the ecosystem:

$$CH_3COOH + 2O_2 \rightarrow 2CO_2 + 2H_2O$$

In this case no H^+ ions were removed, but the removal of the organic acid component of the exchange complex may increase the pH of the soil solution.

The second fire process actually consumes H^+ from the soil, essentially titrating the soil. The extent of this process is commonly calculated as the release of base cations by the fire (Chandler et al., 1983). Although these cations are not bases in any chemical sense, their release is associated with the consumption of H^+. For example, the combustion of an organic compound (acetate) containing K^+ consumes one H^+ for every K^+ released:

$$CH_3COOK + 2O_2 + H^+ \rightarrow 2CO_2 + 2H_2O + K^+$$

The H^+ is consumed in the production of water, not by any reaction with the K^+. Through a series of reactions with water and carbon dioxide, the released cations form dissolved bicarbonate salts that may readily leach into the mineral soil, where the cations may be exchanged for other cations (such as aluminum) on the exchange complex.

The fire-induced change in pH depends on the consumption of organic acids (with H^+ associated with the anion), the consumption of organic anions (with other cations associated with the anions), the original pH of the soil, and the buffering capacity of the soil. Forest floor fuels generally contain about 1–3 moles of negative charge per kilogram. Depending on the pH, perhaps half of the charge will be balanced by H^+ (representing undissociated acids), and about half will be balanced by cations such as K^+, Ca^{2+}, and Mg^{2+} (representing dissociated acids). Combustion of 20,000 kg/ha of such fuel would consume about 20 kmol/ha of undissociated acid and would also consume 20 kmol/ha of free H^+ associated with the release of the nutrient cations. Binkley (1986b) estimated that a series of 12 surface fires during 24 years in a loblolly pine stand lowered the H^+ content of the soil by about 120 kmol/ha, resulting in an increase in pH from 3.8 to 4.1.

The increase in pH depends on the original soil pH for two reasons. Soils with pH values of about 6.5 tend to be strongly buffered by the presence of carbonate minerals, diminishing the effect of the fire. Soils with a very low pH

(<4.5) are strongly buffered by the presence of either organic acids or aluminum. The largest changes in pH after canopy fires would probably occur in the range of pH 4.5 to 6.5.

BASE CATIONS MAY INCREASE AFTER FIRE

The supply of nutrient cations (calcium, magnesium, and potassium) generally increases following fire. This increase results from direct release from burning organic matter and from any increase in subsequent rates of organic matter decomposition. Heating may also directly alter soil exchange properties, causing cation release (Khanna and Raison, 1986). In fact, the nutrient cation content of the forest floor and ash layer may be higher after fires due to the addition of material from the combustion of the vegetation and organic debris. For example, Van Cleve and Dyrness (1985) examined the nutrient content of forest floors in an unburned white spruce forest and in an area where the spruce forest was consumed by a canopy fire. Both unburned and lightly burned areas had about 1300 kg/ha of calcium in the forest floor, whereas the heavily burned areas had over 6700 kg/ha of calcium. As a side note, it is difficult to explain such a large increase in forest-floor calcium in the heavily burned areas, as the vegetation probably contained less than 1000 kg/ha of calcium. Nutrient studies need to consider the magnitude of apparent changes in light of the conservation of mass (see Chapter 9). Dyrness et al. (1989) found that cations were increased in the forest floor and mineral soil in Alaskan white spruce and black spruce stands only in intense fires. After a canopy fire in an old-growth Douglas-fir forest, Grier (1975) found no increase in total forest floor calcium.

A portion (typically 10–30 percent) of the cations contained in ash dissolve readily in water and typically leach into the mineral soil (Walker et al., 1986). Grier (1975) estimated that 670 mm of snowmelt moved 75 kg/ha of calcium out of the ash layer into the upper mineral soil and also moved 15 kg/ha deeper than 20 cm into the mineral soil. Portions of the site with a heavier ash layer leached about four times more calcium. A similar study of the effects of a severe canopy fire in a jack pine stand showed about 12 kg/ha of calcium leaching from the forest floor into the mineral soil (Smith, 1970).

Leaching losses of nutrient cations also increase after canopy fires. For example, the loss of Ca^{2+} and K^+ to streamwater rose by 26 percent and 265 percent, respectively, after the Little Sioux fire in northern Minnesota (Wright, 1976). However, even the large relative increase in K^+ loss amounted to only 1.5 kg/ha of extra loss, which is unlikely to affect site fertility.

Changes in the cation nutrition of plants after fires have received little attention, but some insights are available from more extensive research on slash-burned areas. For example, Vihnanek and Ballard (1988) found that 5 to 15 years after slash burning, the concentrations of calcium and potassium in Douglas-fir foliage on burned sites exceeded the concentrations on unburned sites in 80 percent of the sites examined. The availability of calcium, magnesium, and potassium does not limit forest growth in most temperate ecosystems (see Chapter 12), so the effects of fires on losses or cycling rates of these cations are probably not important to ecosystem recovery and productivity in most cases. However, the nutrient cation supply in many Ultisols and Oxisols is very low, and cation losses from fires on these highly weathered soils may be critical. We know of no data on the effects of fires in tropical forests on supply rates of nutrient cations, but the impacts of fires could be critical to ecosystem recovery and future production.

PHOSPHORUS AVAILABILITY INCREASES AFTER FIRE, AT LEAST TEMPORARILY

The availability of phosphorus limits production in many ecosystems around the world, yet the effects of fire on phosphorus availability have been poorly characterized. In laboratory studies, Humphreys and Lambert (1965) found that available phosphorus increased when soils were heated in a furnace to between 200°C and 600°C. Humphreys and Craig (1981) heated soils in a furnace and showed that much of the reduction of organic phosphorus was matched by increases in aluminum and iron phosphates; very little phosphorus was lost from the samples at temperatures up to 600°C. In a field study of a slash-and-burn treatment in a wet tropical rain forest, the concentration of readily available phosphorus in the upper soil declined following the first heavy rains after burning, probably resulting from conversion of phosphorus to less available forms rather than to actual losses of phosphorus (Ewel et al., 1981). In a study of harvesting and burning in *Eucalyptus*, Ellis and Graley (1983) found that increased availability of phosphorus relative to that in uncut forests lasted for only about 1 year. In addition to the direct losses of phosphorus in fire, subsequent effects of fire on phosphorus cycling include changes in pH, which may alter the solubilities of inorganic phosphates as well as microbial phosphorus transformations.

Increases in pH should generally increase phosphorus availability when phosphorus is bound with iron and aluminum and should decrease the availability of phosphorus bound with calcium (Lindsay, 1979). Increased microbial activity in burned soils may increase the release of phosphorus from organic matter, and competition between microbes and plants for the increased supply of phosphorus could determine the degree of enhancement of plant nutrition.

In soils high in iron and aluminum oxides, fires may increasing phosphorus sorption and decrease availability. For example, Sibanda and Young (1989) found that heating soils to just 200°C doubled the phosphorus sorption capacity of a soil from Zimbabwe, and heating to 400°C drastically increased phosphorus sorption. Increased sorption resulted from removal (oxidation) of organic matter that was bound with soil iron and aluminum, allowing these metals to bind phosphorus.

NITROGEN AVAILABILITY ALSO INCREASES AFTER FIRE

The concentration of soil ammonium generally increases greatly after fire, sometimes by an order of magnitude or more. Walker et al. (1986) demonstrated that soil ammonium may increase as a direct result of soil heating. Ammonium may also be added to soil in ash. Concentrations of ammonium may remain elevated as a result of both the increase in ammonium production and any decrease in ammonium consumption by plants and microbes (Knight et al., 1985; Dyrness et al., 1989). A severe canopy fire in a *Eucalyptus* forest in Australia increased soil ammonium concentrations by fourfold, but concentrations declined to prefire levels within 6 months (Adams and Attiwill, 1986).

Nitrate concentrations sometimes rise over a period of weeks or months following fire. Immediate increases are generally slight due to the low vaporization temperature of nitric acid (as low as 80°C) and the very low concentration of NO_3 nitrogen in ash. Net nitrification increases following fire in a wide array of ecosystems as a consequence of increased pH and availability of ammonium.

Pulses of ammonium and nitrate availability are ephemeral, often lasting for only a few growing seasons or less (Adams and Attiwill, 1986; Christensen, 1987). Subsequent changes in nitrogen availability following fire depend on the quantity of nitrogen lost in the fire, changes in rates of microbial mineralization after the fire, and competition between microbes and plants for mineralized nitrogen.

NITROGEN FIXATION INCREASES AFTER FIRE ONLY IF SYMBIOTIC NITROGEN-FIXING SPECIES INCREASE

Some authors have speculated (and cited hopeful papers) that nonsymbiotic nitrogen fixation increases after fire (Woodmansee and Wallach, 1981; Boerner, 1982; Christensen, 1987), but available evidence does not support such speculation. Most studies of fire effects on nonsymbiotic nitrogen fixation have found that the rates are so low as to be important only on a time scale of decades or centuries, and none has shown ecologically meaningful increases in rate following fire. For example, Wei and Kimmins (1998) estimated that free-living nitrogen fixation would average about 0.3 kg/ha annually in a lodgepole pine forest without fire, or 0.6 kg/ha annually if the forest burned in a wildfire.

On the other hand, burned areas are often colonized by plants capable of symbiotic nitrogen fixation. Longleaf pine forests in the southeastern United States may develop large populations of understory legumes following regular prescribed fires. Relatively low densities of herbaceous legumes probably contribute less than 1 kg/ha of nitrogen annually, but where densities are very high (more than two plants per square meter), rates of 5–10 kg/ha of nitrogen annually are likely (Hendricks and Boring, 1999). Even relatively low rates of nitrogen fixation can greatly accelerate the cycling of nitrogen. Nitrogen-fixing plants often require greater quantities of phosphorus than other plants require, and the potential effects of fire-related changes in phosphorus availability to nitrogen cycling should be explored.

LONG-TERM EFFECTS OF FIRE ON SOIL PRODUCTIVITY REMAIN UNCERTAIN

Fires remove nutrients from forests but typically increase nutrient turnover rates, at least in the short term. But what are the overall longer-term effects? Surprisingly little information has been collected on this fundamental question, and the available information suggests that no single generalization will be appropriate across forests and soils.

Studies with prescribed surface fires in pine stands have generally shown little if any change in foliar chemistry (Landsberg and Cochran, 1980; Binkley et al., 1992b). Few studies have documented the changes in growth in pine stands following surface fires, and no broad generalizations are supported. One study of longleaf pine stands in Alabama found that burning (five fires in 10 years) reduced volume increments by one-quarter to one-third (Boyer, 1987), whereas fire in an oak forest in New Jersey increased growth by one-third to one-half (Boerner et al., 1988).

Slash fires are typically hotter and consume more fuel than surface fires, so the impacts of slash fires might be larger than those of prescribed surface fires. W.G. Morris initiated a study of effects of slash fires in Douglas-fir forests in 1947 in which he recorded and mapped fire intensities in 62 pairs of burned and unburned plots. Resampling showed that severely burned plots had 7 to 50 percent less total nitrogen in 0- to 10-cm depth mineral soil relative to lightly burned or unburned soils (Kraemer and Hermann, 1979). Miller and Bigley (1990) examined growth on 44 of these plots, comparing tree characteristics. The mean annual increment of Douglas-fir was significantly higher on burned plots, where growth exceeded that on unburned plots by 0.85 m^3/ha annually (a 27 percent difference). Was the greater growth a result of fire-induced increases in nutrient supplies or the effect of fire in reducing competition from nonconifer vegetation? Burned plots did in fact have less nonconifer vegetation, but no experiment was done to separate the contributions of different aspects of fire's impact on conifer growth.

In another study, Vihnanek and Ballard (1988) examined Douglas-fir growth and nutrition on 20 sites (age 5–15 years after harvest) on burned and unburned plots. Nutrient contents in foliage were generally greater on burned plots and were not lower than those on unburned plots at any location. This study also could not separate any effects of fire on soils and nutrients from those of fire in reducing tree competition with nontree vegetation.

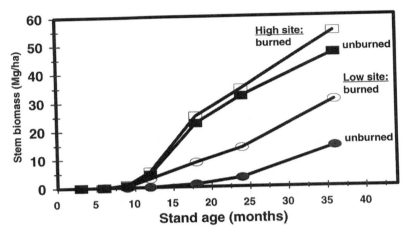

Figure 10.9   Productivity of clonal *Eucalyptus grandis* × *urophylla* was greater on sites prepared with slash fires than on sites without burning; photo is from the low-productivity site (means of 4 replicate plots/site; J. L. Stape, personal communication, 1999).

One of the best studies on the growth effects of postharvest slash fires comes from *Eucalyptus* plantations in Brazil. Postharvest fires of low to moderate intensity increased growth by about 15 percent on a high-productivity site and by more than 100 percent on a low-productivity site (Figure 10.9). The greater growth on burned plots resulted from better development of leaf area (Figure 10.10) and from less allocation of carbohydrates to belowground production (J. Stape and M. Ryan, personal communication, 1999). In this case, competing vegetation was thoroughly controlled by herbicides, so greater growth on the burned plots probably resulted from better nutrition of the trees. On the lower-quality site, the phosphorus concentrations in leaves was almost twice as high on burned plots as on unburned plots, so the major nutritional benefit of fire was probably an increased phosphorus supply in the soil (nitrogen concentrations in foliage did not differ). Many more studies of this sort (including competition control) will be needed to provide a clear picture of the situations in which fire increases or decreases soil productivity.

SUMMARY

Fires are fundamental to most forests on Earth, and the effects of fire on ecosystems and soils are often large. An inverse relationship between fire frequency and intensity is common; short intervals between fires lead to low-intensity burning, and long intervals lead to very severe fires that consume large amounts of biomass. Fires remove nutrients through several mechanisms,

Figure 10.10 *Eucalyptus grandis × urophylla* clones, with and without the use of fire for site preparation (see growth information in Figure 10.9). Better growth on burned plots resulted from improved nutrient availability, which allowed greater canopy development (top panels, looking up through canopies), and from less carbohydrate allocation belowground.

including oxidation of compounds to gaseous forms and convection of ash particles in fire-generated winds. Nitrogen losses are commonly proportional to the quantity of fuel consumed in a fire, with 5 kg of nitrogen lost for every Megagram of fuel consumed. The immediate effects of fires on soils may include changes in soil moisture conditions, as a result of reduced infiltration or reduced evapotranspiration by plants, and increased short-term availability of most nutrients. The long-term effects of fire remain poorly understood; despite unequivocal losses of nutrients in fires, forest productivity has increased after fire in some cases and decreased in others.

CHAPTER 11

Forest Soil Classification

Attempts at forest site classification can be traced back to the very beginning of silviculture as a science. Early classifications were based on characteristics of the forest stand, such as growth rate, tree height, tree species composition, or type of ground vegetation. Systems that utilize physiography, physical and chemical properties of the soil, and multiple physical and biological factors have also been developed. As forest management has become more intensive, systems of site evaluation have progressed from productivity ratings based on obvious characteristics of the forest to systems based on environmental factors including, and sometimes limited to, soil properties.

In the first quarter of the twentieth century, the question of site evaluation was subject to a great deal of research and discussion (Zon, 1913; Watson, 1917; Bates, 1918; Frothingham, 1918), with a general acceptance in North America of the site index as the primary means of evaluating site productivity. This empirical approach, based on the height of dominant or codominant trees at a given age, as proposed by Frothingham (1918), is still widely used as a measure of forest productivity and is generally used as the standard to which other methods of site quality evaluation are compared. Unfortunately, site index has some major limitations as a method of site evaluation.

Recent societal pressures have focused attention on attributes of forest sites other than their productivity measured in wood yield. This has resulted in the development of a number of complex systems of evaluation, some of which purport to approach total site assessment. All of the systems have some usefulness given certain site conditions and management objectives. They can be conveniently classified into three general categories, listed in approximate order of complexity: (1) forest productivity, including site index and vegetation types, (2) soil properties, and (3) multifactor systems.

Methods of measuring productivity of forest sites are generally grouped into direct or indirect methods, depending on whether the estimate is based on some stand measurement or on some features of the local environment. However, short of harvesting a mature stand, there is really no direct method of measuring site productivity. Nonetheless, estimating productivity on a stand volume or weight basis is a reasonable approach, in spite of the difficulty associated with these methods.

The main objection to the direct approach is that both volume and weight are affected greatly by variations in stand density. If attempts are made to hold stand density constant by measuring only well-stocked stands, one will be limited to a relatively small number of areas in developing evaluation and classification methods. On the other hand, if stocking is allowed to vary, forest productivity will have to be expressed as a function of both stand density and site quality. However, intensively managed forests usually have their density controlled within rather tight limits throughout their lives. This means that after several rotations in areas with domesticated forests, we should be able to develop site curves based on volume or weight.

SITE INDEX IS A TROUBLING BUT PREFERRED WAY TO MEASURE SITE QUALITY

Site index is the term used to express the height of dominant and codominant trees of a stand projected to some particular standard age. This index or base age may be 25, 50, 100, or any other age appropriate to the growth rate and longevity of the species being considered. Site index is extremely important in site quality analysis in North America because it forms the standard against which all other forms of site evaluation are measured. In stands younger or older than the index age, a family of height/age curves is required for projecting measured height to height at the index age (Beck, 1971). These curves are developed in a variety of ways. One common method is to measure the height and age of many stands at single points in time, fit an average curve of height-on-age to these data, and construct a series of higher or lower curves with the same shape as the guide.

There is evidence that these anamorphic curves often do not represent actual stand growth conditions accurately. For example, the guide curve is likely to be accurate only if the ranges of site indexes are equally represented at all ages. Unequal sampling may occur because of the timber harvesting and land abandonment trends in a particular region. It has been pointed out that trees reach merchantable size faster, and are often cut at a younger age, on high-quality sites. Consequently, a sample of stands selected at a particular time could result in a biased curve that would tend to underestimate the site index of stands younger than the index age and to overestimate the site index of older stands (Beck, 1971). It may also be important to apply the curves in a manner consistent with their construction. That is, if a set of curves is developed on the basis of the 10 tallest trees per acre, it should be applied on other sites by using trees selected in a similar manner.

Furthermore, the assumption that the shape of the curve does not vary from site to site is generally false (Spur, 1952; Beck and Trousdell, 1973). The degree of diversity in curve shape seems to vary with species and location, but the pattern of growth with change in site quality may be similar for many species.

Instead of being proportional at all ages for all qualities of sites, as generally depicted by conventional curves, the rate of height growth rises rapidly on the best-quality sites and then becomes relatively gradual. On the other hand, growth rates on poorer sites increase slowly during the early years but may be maintained for a longer time. One might expect to find trees on sites with the poorest site indexes about equal to those on sites with the best indexes at some age older than the index age (Beck and Trousdell, 1973) (Figure 11.1a).

The bias introduced through the use of proportional rather than polymorphic curves is probably of little importance in the relatively short rotations of most intensively managed forests. Nonetheless, variations in growth patterns among sites due to differences in certain soil conditions can be quite striking in the early years of stand development. For example, slash pine growth during the first 10 to 20 years is often quite slow on the wet savanna soils of the coastal flats of the southeastern United States. The developing stand gradually draws down the mean groundwater table, thus increasing the effective rooting volume of the soil. The increase in soil volume provides for a faster tree growth rate through improved nutrition and aeration during the later stages of stand development. In contrast, slash pine planted on the well-drained sand hills of the coastal plain grow rather well during the first 5 to 7 years after planting. However, growth tends to stagnate, as moisture becomes a limiting factor for good stand development (Figure 11.1b). That different sites can have height-age curves with several shapes, even though these sites may be of equal quality when measured

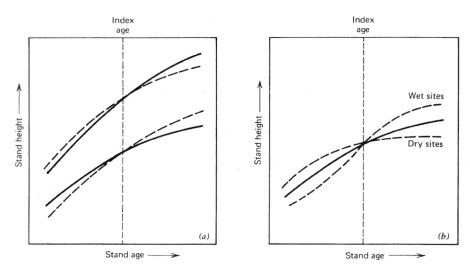

Figure 11.1   (a) Comparison of normal proportional or anamorphic curves (solid lines) with nonbiased polymorphic curves (dashed lines); (b) polymorphic curves (dashed lines) for wet and dry sites and the normal proportional curve (solid line) obtained by combining them.

at a common index age, has been reported for several species (Carmean, 1975). It is apparent that the site index system of classification is highly empirical and subject to many different kinds of error. Nonetheless, we often determine site index to the nearest 30–60 cm from site index curves that have confidence intervals of 1 m or more about each site index line. We then compare these "accurate" determinations of site quality to determinations made using other methods in order to gauge the accuracy of the method being tested. In so doing, we cannot hope to obtain a method of site quality determination that is more accurate than the site index method, which is fraught with error.

SOIL CLASSIFICATION AND SURVEY ARE ESSENTIAL TO SOUND FOREST MANAGEMENT

Although soil surveys have often been characterized as too imprecise to serve as a basis for site evaluation (Carmean, 1975; Grigal, 1984), they have proved to be quite valuable to forestry in many parts of the world (Valentine, 1986). Classification, mapping, evaluation, and interpretation of forestlands are prerequisites to sound forest management. Collectively, they comprise an inventory program that is essential to proper intensive management programs and important for the development of effective, multiple-use land management plans. Classification and mapping are related but distinct steps in conducting a soil inventory. **Classification** is analogous to taxonomic systems used for plants and animals and consists of grouping land areas based on similarities and differences in properties. Mapping is an exercise in geography wherein homogeneous land areas are delineated on a map using the classification system.

The units within the classification are pure, although the properties that define the unit are allowed to vary over some prescribed range. **Map units**, the geographical areas delimited on the ground, are seldom pure. They almost always contain areas that belong to other units of the classification system or that remain undefined. Thus, the degree of purity of map units must be defined. The homogeneity of map units depends not only on the character of the area being mapped but also on the scale of mapping. At small scales, 1:50,000 or greater, map units must be large, 10^3–10^5 km^2, and of necessity quite heterogeneous; however, at large scales, 1:10,000 to 1:20,000, map units can be small, less than 10^0 km^2, and often quite homogeneous.

Inclusions are areas within a map unit that clearly belong to a different classification unit but that are too small to be delimited at the scale being used in the inventory. Often the scale of mapping leads to map units composed rather equally of two or more soils. Such map units are commonly called **complexes** or **associations**. To be useful, such units must specify what proportion is contributed by each soil.

Evaluation is an assessment of the functional relationships within a land area. It can be carried out in the absence of formal classification and mapping.

However, for evaluative information to be retained and communicated clearly, mapping is essential. **Interpretation** is the development and presentation of interpretive management recommendations for classification of map units. It requires keen observation, statistical analysis of large volumes of data, and constant feedback to be effective. Yet, without interpretation, forestland inventory becomes a pedantic exercise of little value to land managers.

Soil and its associated environment can be thought of as the habitat or site of the forest trees and other organisms. Site includes the position in space as well as the associated environment (Barnes et al., 1998). Site quality thus is defined as the sum total of all factors influencing the capacity of the forest to produce trees or other vegetation. The soil can be seen as an integrator and recorder of the environmental factors that influence a site. As such, "reading the soil" can yield information about the history as well as the current condition of the site. It is this ability to interpret difficulty-to-observe site properties from easily observed soil properties that forms the basis of the soil survey systems used in agriculture the world over.

THE USDA NATIONAL COOPERATIVE SOIL SURVEY

This program of soil inventory and interpretation, which has parallels in Canada and many other countries, has the important advantage of employing well-established soil survey techniques based on an accepted system of soil classification. The system operates on the basic premise that soils can be identified as individual bodies and treated as integrated entities, so that knowledge of individual soil types can be interpreted in terms of their capacity to support various uses.

In the United States, detailed soil maps are produced at a scale of 1:24,000 and are published as county soil survey reports. The soil series of the classification system that are used to develop map units are correlated from one county to another so that similar soils are given a similar identifier in widely different places. Data on the management constraints associated with a particular soil series are collected widely. This allows interpretations of soil utility for a variety of uses, including forestry, to be made and published in the county survey report.

Several scientists have successfully used soil survey information and maps, by themselves or in combination with other observed properties such as drainage class, thickness of the surface horizon, and horizon depth, to predict site index (Haig, 1929; Coile, 1952; Stoeckeler, 1960; Broadfoot, 1969). Stephens (1965) reported that the soil taxonomic unit at the series level provided an accurate prediction of the Douglas-fir site index on zonal soils of the Oregon Cascades.

In spite of the considerable potential of detailed maps of the cooperative soil surveys, they have been viewed with mixed emotions by most foresters. Some forest scientists consider them a valuable aid in management planning, but others find them of little value, particularly for predicting productivity. Whether or not they are useful to an individual apparently depends to a large extent on the landform and species under consideration. Van Lear and Hosner (1967) found little, if any, usable correlation between soil mapping units and the site index of yellow-poplar in Virginia. This conclusion was prompted by the wide variation in site indexes exhibited within each mapping unit. Other workers have encountered considerable frustration in attempting to group soils by series and type. Carmean (1961) pointed out the wide range of site values that may occur on single soil types. Several researchers in other areas (Coile, 1952; Broadfoot, 1969) have noted this same problem.

Many of the shortcomings of the soil survey maps derive from the fact that the classification system and survey methods were developed primarily for agricultural use. Some soil properties important to deep-rooted trees, such as water tables and subsoil textural changes at depths of 2 to 3 m, are not considered in delineating taxonomic or mapping units. By the time soil maps are made, a considerable amount of cooperation between soil scientists and agricultural users of soils information has gone into defining the principal soil series of interest to agriculture. Similar cooperation between foresters and soil scientists has not existed historically. At any rate, it is apparent that the standard USDA National Cooperative Soil Survey reports, as presently constructed, are not as effective as they could be for forestry purposes.

FOREST SURVEYS ARE COMMON ON INDUSTRIAL FORESTLAND

In the absence of standard soil surveys, or in attempts to improve on the USDA method of mapping soils for purposes of forest management, several special forest soil survey methods have been developed. Coile (1952) developed a soil-based land inventory system that, in one form or another, is widely used in the southeastern United States, and the Weyerhaeuser Company (Steinbrenner, 1975) developed a system in the Pacific Northwest that has now been extended into the Mid-south and the Southeast.

Coile found soil-site maps desirable for both moderately intensive and intensively managed forests. He felt that soil maps should show the specific geographic extent of soil features and forest site classes useful in determining or making decisions on such things as (1) selection of species and spacing, (2) prediction of future yields, (3) drainage, (4) site preparation methods, (5) road construction, (6) definition of areas for seasonal logging, and (7) allowable costs for all phases of management based on expected returns.

Coile and his coworkers related tree growth to a limited number of easily measured or observed soil physical characteristics, such as texture of certain horizons, soil depth, consistency, and drainage characteristics, plus certain other selected features of topography, geology, and history of land use. He identified soil units with an alphanumeric code that characterized the drainage class, thickness and texture of the A horizon, depth to and texture of the B horizon, and other factors. The value of these surveys depends to a large extent on the development of working relationships through mathematical trial-and-error testing of many combinations of variables (Mader, 1964). However, relatively little attention has been given to explaining the basic biophysiological relationships involved. They are based on the premise that a few factors will satisfactorily explain site differences over a wide range of conditions.

Coile (1952) stated that "the degree of success attained in demonstrating relationships between environmental factors and growth of trees is largely determined by the investigator's judgement in selecting the independent variables that are believed to be related to tree growth in various ways and in different combinations. How well the investigator samples the entire population of soil and other site factors determines the general applicability of the result." Coile limited his paired soil-forest stand observations to pure, even-aged pine stands over 20 years old that were fairly well stocked and growing in a relatively restricted area. He and his colleagues published information on growth and yield, stand structure, and soil-site relations of southern pines that dominated southern pine management for several decades (Schumacher and Coile, 1960; Coile and Schumacher, 1964).

In Weyerhaeuser's soil survey of mountainous terrain of the Pacific Northwest (Steinbrenner, 1975), topographic features are of paramount importance and the maps are based on a strong correlation between landform and soil series within a geologic unit (Figure 11.2). The units mapped in this system are in some cases more narrowly defined than the National Cooperative Soil Survey units, but in some cases they are more broadly defined. In either event, they are given geographic place names much as are National Cooperative Soil Survey units. In the Weyerhaeuser system, topography, as evidenced by landform, is important to interpreting the survey for road construction, equipment use, and, in some cases, soil productivity. Mapping for productivity is a primary objective, and the interpretation for this purpose is developed through research. In addition, the maps are interpreted for land use, trafficability (for logging equipment), windthrow hazard, thinning potential, and engineering characteristics for road construction. Productivity interpretations are the basis for determining allowable cut and for intensive forest operations, such as regeneration methods, stocking control, and thinning.

Mapping units provide the logical basis for delineating cuts in the logging plan, according to Steinbrenner (1975). The interpretations also indicate the type of equipment required and the timing of the harvesting operation, so that the impact on site quality is minimized. The windthrow interpretation is utilized to minimize damage along harvest boundaries. Thinning potential is used to assign a priority to all lands for intensive forest practice. Engineers use the map in determining the best location for roads, drainage problems, and location and size of culverts needed. The soil survey provides information that is basic to sound forest management, and its usefulness increases as more interpretive detail is developed.

As forest management has intensified, soil maps have been constructed for a growing acreage of industrial forestland throughout the world. Systems

Figure 11.2 Topographic features are especially important in the classification of mountainous terrain (courtesy of Weyerhaeuser).

similar to Coile's and Steinbrenner's have generally been utilized. The utility of these soils maps lies in the sophisticated management interpretations associated with the map units. These are developed through research, careful record keeping, and knowledge of soil-site evaluation and the tree response to cultural practices.

MULTIFACTOR CLASSIFICATION SCHEMES HAVE GAINED POPULARITY

The multifactor approaches to forestland classification differ from soil surveys mainly in their approach and their degree or intensity of mapping. The number of land features, stand characteristics, and soil factors used in delineating mapping units is great, and as a consequence, maps based on these features provide flexibility for classifying on the basis of goals other than wood production. These schemes generally entail a combination of independent site variables, measured in the field or laboratory and superimposed as phasing elements on conventional soil series, to reflect variations within these units important to tree growth and land use. Such soil factors as soil depth, available water capacity, texture, organic matter, chemical composition, and aeration, as well as radiation, ground vegetation, and landform, have been studied individually and collectively.

Hills's (1952) **total site classification** of Ontario can be considered a multifactor approach to site evaluation. This holistic concept integrates the "complex of climate, relief, geological materials, soil profile, ground water, and communities of plants, animals, and man." Physiographic features are used as the framework for integrating and rating climate, moisture, and nutrients. The integration of the various factors of environment and vegetation at each level of this hierarchical classification scheme involves much subjective judgment and intuition (Carmean, 1975). Nevertheless, the system provides a good framework for stratifying large, inaccessible forest regions into broad subdivisions based on general features of vegetation, climate, landform, and soil associations.

An alternative multifactor approach is the **German site-type** system, which was developed in Baden-Wurttemberg in the 1940s (Schlenker, 1964). Barnes et al. (1982) have used this system successfully in Michigan. They found strong interrelationships among physiography, soils, and vegetation and used all of these factors simultaneously in the field to delineate site units. The reliance of this system on ecologists, who, in the field, simultaneously integrate ecosystem factors and are not bound to reconcile predetermined classes, leads to a high degree of subjectivity. **Ecological site classification** is common in Europe and Canada and has recently gained popularity in the United States.

SOIL FACTORS DRAMATICALLY INFLUENCE FOREST DEVELOPMENT AND TREE GROWTH

Climate, physiography, and soil comprise the abiotic factors that exert a significant influence on forest development. Where climatic and physiographic factors can be held constant by appropriate stratification procedures, soil properties become the major factor of the physical environment that has an appreciable bearing on tree growth and the one of greatest concern to the forest manager. Unfortunately, the actual soil factors that directly influence tree growth, moisture, and nutrient availability, for example, are not easily observed. There are easily observable properties, however, that are correlated to the soil factors that regulate tree growth. Since these easily observable properties are also used to delineate soil map units, soil-site relationships can be established that tell us a great deal about how well trees will grow on a given site by knowing the properties of the soil map unit. Let us consider some of the soil factors that strongly influence forest development.

Parent material is rather easily determined by consulting geologic maps or by inspection in the field. It is a major contributor to the process of soil development, and as such, it has an indirect effect on tree growth. Because of the deep rooting habit of trees, soil parent material and the condition of the geological substrata are more important to foresters than to agriculturists. The relationship between soil parent material and tree growth is most obvious in areas where the bedrock is sufficiently close to the surface to exert a continuing influence or soil properties. The effect of parent material on tree growth is also seen in soils derived from transported material such as marine sands and glacial drift. Parent material affects productivity through its effect on the chemical, physical, and microbiological properties of the soil, but the extent of this influence can be modified by climate. For example, soils derived from similar parent material, but developed under different climate conditions, may have vastly different properties and production potential because of variations in leaching of nutrients, accumulation of organic matter, and soil acidity. Nevertheless, parent material generally has a greater effect on the mineral composition of soils than other soil-forming parameters do, and there is generally a good relationship between the mineralogical composition of soil parent material and the parent rocks. For example, parent materials derived in large part from granite gives rise to soil containing a larger proportion of quartz than do parent materials derived from diorite.

Within the same climatic zone in New York State, soils derived from calcareous shale support more exacting species, such as basswood, white ash, yellow-poplar, and hickory, while soils derived from acid sedimentary rock contain a higher percentage of beech, yellow birch, red maple, and certain oaks (Lutz and Chandler, 1946). It is generally accepted that the regeneration and growth of both northern white cedar (*Thuja occidentalis*) and eastern red cedar (*Juniperus virginiana*) are better on calcareous soils than on adjacent acid soils. The presence of cabbage palm (*Sabal palmetto*) in the lower coastal plain of the southeastern United States is an indicator that limestone is sufficiently close to the surface to influence soil reaction and base saturation.

Throughout most of the southeastern United States, the effect of soil parent materials is expressed indirectly through derived properties such as soil texture or drainage characteristics. However, direct effects on soil fertility have been noted in certain areas. Parent material may have an overriding detrimental effect on the growth of certain species, such as pine growth in the base-rich chalk soils of the Black Prairies in Alabama and Mississippi and in certain highly calcareous alluvial soils in east Texas. Parent material may be beneficial to tree growth, as noted in some acid soils of Florida and South Carolina, with lower layers derived from phosphate-rich limestone. Ameliorating influences of old marine shell beds and marl deposits have also been observed on occasion around the margins and in the interiors of hardwood bays and river swamps.

If parent material origins can be distinguished in the field and have some real or suspected connections with site productivity, the effects can be identified by stratification of site measurements by parent material categories. The influence of parent materials on site productivity was noted in Sweden, where parent rocks were grouped on the basis of calcium content. Soils derived from calcium-poor rock supported a poor forest of Scots pine, intermediate groups produced good soils for pine and mixed conifers, and basic igneous and calcareous sedimentary rocks resulted in productive soils with forests of Norway spruce and hardwoods (Hills, 1961).

Soil depth is another easily observable soil property. It partially determines the volume of soil available to tree roots and affects tree growth to the extent that it affects nutrient and moisture supplies, root development, and anchorage against windthrow. Trees growing on shallow soils are generally less well supplied with water and nutrients than trees on deep soils. When soils are shallow to a restricting layer, such as a claypan, fragipan, or bedrock, depth measurements can be used with some precision to predict growth patterns in well-drained soils. Growth normally follows a trend that can be expressed as a reciprocal function of soil depth, with the greatest decline in growth found on soils with less than 25 cm of effective depth (Figure 11.3).

The **effective rooting depth** is that depth to which trees can maintain metabolically active roots during the major portion of the growing season. The absolute and effective depths of a soil are not necessarily the same because a high water table, toxic substances, or an impervious layer may completely restrict root penetration in a soil that would otherwise permit deep rooting. Some difficulty in using soil depth to estimate productivity also may be encountered where drought, erosion, or poor drainage are products of surface soil thickness or depth of soil above some restricting layer. In each of these instances, there may be soil fertility interactions important to tree growth that cannot be determined from depth measurements alone. For example, according

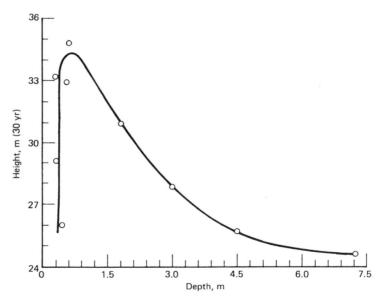

Figure 11.3 Height of *Pinus radiaita* adjusted to 30 years, in relation to effective rooting depth in soils of South Australia (Raupach, 1967). Used with permission.

to Ralston (1964), measurement of surface soil thickness as an indicator of erosion fails to discriminate between losses in fertility and reduction in effective rooting depth. Hannah (1968) found that depth of the surface soil (A horizon) was related to the growth of white and black oaks in Indiana (Figure 11.4).

Barnes and Ralston (1955) reported that site productivity for slash pine on sandy soils in Florida increased with depth to a fine-textured layer, with maximum growth obtained on soils with a fine-textured horizon at about 50 to 75 cm. They also found that depth to mottling was an index of productivity, with the best growth found where mottling occurred at 75 to 100 cm. Poor growth of *Pinus radiata* was observed on soils with less than 46 cm of rooting depth in South Australia (Jackson, 1965). However, in California, the same species grows to 20 m or more in height with as little as 15 cm of surface soil on weathered granite rubble where roots can penetrate the parent material.

Auten (1945) found that depth to compacted subsoil was the most significant factor related to yellow-poplar growth. When such a restricting horizon was less than 60 cm below the surface, the site was poorer than average. Gilmore et al. (1968) reported that the height of 18-year-old poplar varied from 3 to 17 m in an old field where effective soil depth had been reduced in some

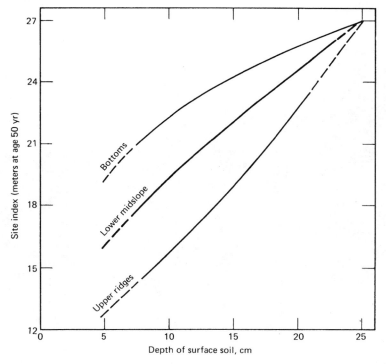

Figure 11.4 Site index of black oak as related to surface soil thickness and slope position in southern Indiana (Hannah, 1968). Used with permission.

parts by severe erosion and had been increased in other parts by deposition. Heights varied directly with depth to a fragipan, presumably due to an increase in available soil moisture with increasing effective rooting depth. The contrasting influence of depth to a fragipan on the growth of moisture-demanding sycamore and drought-tolerant loblolly pine is illustrated in Figure 11.5. Zahner (1968) found the optimum effective soil depth for loblolly and shortleaf pines growing under a rainfall regime of 117 to 132 cm/yr to be about 46 to 50 cm.

Drainage class is an inferred property, indicated by characteristic profile morphology and landscape position. It indicates the extent and duration of soil saturation in years of extreme wetness as well as in normal years (Dement and Stone, 1968). The interrelations between effective soil depth, available soil moisture, and drainage class have been reported by several researchers (Coile, 1952; Jackson, 1965; Gilmore et al., 1968). Controlled drainage for the removal of excess water may improve productivity for a particular species and ultimately may effect profound changes on stand composition and regeneration. Douglas-fir cannot tolerate shallow water tables, but red alder, western red cedar, Sitka spruce, and western hemlock all are considered adaptable to shallow water tables. Where water tables were less than 15 cm below the surface, these "species responded differently to both depth and quality of the

Figure 11.5 Effects of depth of a loess soil over a fragipan on the growth of 12-year-old sycamore (foreground) and 23-year-old loblolly pine on an Ochreptic Fragiudalf in southern Illinois. As depth of soil increases to the right, sycamore growth improves but pine growth is only slightly affected.

water table" (Minore and Smith, 1971). Alder and red cedar were little affected by water table depth and grew reasonably well where stagnant water was near the surface during winter months. Spruce, on the other hand, did not tolerate shallow stagnant water but grew well where moving (aerated) water was near the surface.

Planted cottonwood grew best when the water table was about 60 cm deep, whether the trees were planted on soil with a high water table or the water table was raised a year after planting. Yields on soil where a water table was maintained at 30 cm were about the same as those where no attempt was made to maintain a water table in tanks (Broadfoot, 1973). Both slash and loblolly pines grew better where the water table in a Haplaquod was maintained at 46 cm than where it was maintained at 92 cm during 5 years after planting (White et al., 1971).

Soil moisture influences tree growth more than any other factor on most sites, but it is not easily observable. The available water-holding capacity of a soil is influenced by a number of factors. Fortunately, it is determined primarily by texture and structure, which are easily observable. The use of soil texture as an estimator of water-holding capacity, and thus of site productivity, is complicated by the influence of texture on soil aeration, nutrient availability, and other soil fertility factors. If the growth potential of a species can be observed over a wide range of textures in a given area of well-drained soils, a curvilinear response can be expected. Growth improves with an increase in silt and clay content, as a result of more favorable moisture and nutrient supplies, to a point where further increases in the proportion of fine particles produce aeration difficulties (Ralston, 1964). Coarse fragment content also modifies soil moisture regimes. Moderate amounts of coarse fragments may favor deep penetration of light rains, thus reducing evaporation losses. However, large reductions in effective soil volume by stones decrease moisture retention storage.

Pearson (1931) wrote that "there appears to be but little difference in the ability of species to extract moisture from dry soil; that is, they all reduce it to the wilting point which in a given soil is about the same for all species. Probably the greatest difference between species lies in their ability to extend their roots and thus enlarge the sphere of their activity." The variation in drought tolerance among species is well recognized (Eyre, 1963), and soil moisture plays a prominent role in the adaptation of species to sites and their distribution among the climatic zones of the world. However, there is also considerable variation in drought tolerance among individuals of the same species (Schultz and Wilhite, 1969).

Soil aeration can have profound effects on site quality and productivity. A good supply of oxygen is necessary for root growth and the absorption of nutrients by roots. Although species vary widely in their tolerance of oxygen deficit or carbon dioxide excess in the root zone, most trees are adversely affected when deficits or excesses exceed 2 percent for a long period of time (Romell, 1922a, b). Soil aeration is impossible to observe directly, but soil color is strongly correlated to soil aeration. Gray-blue or gray-green soil colors and dull or low-chroma colors, those in the lower left-hand corner of the Munsell color charts, indicate reduction and poor aeration.

High soil density may also lead to poor aeration. Minore et al. (1969) related high soil density to poor root development of several western species. Forest site situations where such conditions may be observed in the southeastern United States are in upland soils with plastic claypans, fragipans, or cemented hardpans at shallow depths. Swamp sites may be well or poorly aerated, depending on whether the water at the site is flowing or not, since flowing water is usually better oxygenated than stagnant water. Soils with high water tables frequently exhibit a similar pattern. In both cases cultural activities, such as road building or ditch construction, may alter the flow of water and dramatically change site quality and productivity (Wilde, 1958).

Nutrient availability is of great importance to tree growth and is quite obviously difficult to observe. We can infer a good deal about nutrient availability from parent material, texture, and color. Knowing the parent material gives us a general idea about what nutrient elements might be scarce or plentiful. The color of the surface soil gives us some indication of the nitrogen status of the soil, and the texture provides some insight into the soil's cation exchange capacity and nutrient-holding power. However, additional information available only from the laboratory is needed to provide a good picture of the nutrient availability on a site.

Soil map unit descriptions generally carry some of this information. We commonly measure cation exchange capacity, base saturation, and the soil's content of nutrient cations. We may even determine extractable phosphorus and organic matter content. This allows us to make some general determination about a site's nutritional status if we know its soil type.

INTERPRETATIONS TO AID IN FOREST MANAGEMENT CAN BE DEVELOPED FOR SOIL MAP UNITS

Interpretations for land use are commonly developed for soil map units in USDA National Cooperative Soil Survey reports. These interpretations are based on data collected on a wide range of sites with similar soil types and on observations made on various soil types over a broad area. Interpretations are generally made of forestry in areas with sufficient forest cover to warrant them. Such interpretations are often based on scant data and may be quite inaccurate.

This inaccuracy arises partly because data on forests, particularly yield data, accumulate much more slowly than data on annual crops. Much of the inaccuracy, however, arises because the soil map units are not homogeneous for the properties that determine yield, response to tillage or fertilization, response to competition control, and so on. Consequently, much effort has been expended on understanding what soil properties are most closely related to yield or response to cultural practices.

Soil-site studies have received the greatest emphasis in the United States in areas where site quality, soil, and stand conditions are extremely variable. Carmean (1975) listed some 41 published reports on soil-site studies for southern pines, 34 dealing with northern conifers, 23 on western conifers, 35 on eastern oaks, and 41 on other eastern hardwoods. Features commonly correlated with site index are those site factors that significantly influence tree growth and that are easily identified and mapped. In most cases, the features of soil, topography, and climate found to be correlated with site index are indirect indices of more basic growth-controlling factors and conditions, such as available moisture and nutrients, as well as microclimate factors that affect evapotranspiration and physiological processes of trees. Possibly the significant factors determined from soil-site regression studies should be viewed merely as links in the many chains connecting tree response (site index) to causative factors such as moisture, nutrients, temperature, and light (Carmean, 1975). Soil features most important in soil-site studies are usually those concerned with depth, texture, structure, and drainage, that is, properties that determine the quality and quantity of growing space for tree roots (Coile, 1952).

Within uniform climatic zones and physiographic regions, one can usually separate site differences in productivity based on soil variables. Ralston (1964) and Carmean (1975) reviewed research in North America on soil factors affecting productivity. They found that most soil factors that correlated with site productivity were those attributes of soil profiles that reflect the status of soil moisture, nutrients, and aeration. Some of the soil properties related to productivity of several species are summarized in Table 11.1.

Recent research has placed greater emphasis on soil fertility factors, but available water still appears to be the single most important determinant of productivity of many tree species. Broadfoot (1969) reported this to be true for southern hardwoods. However, he pointed out that the use of multiple regression equations to predict the site index of new populations has generally given poor results. He attributed this fact to the impossibility of measuring the true causes of productivity, such as soil moisture and nutrient availability during the growing season, soil aeration, and physical condition, including root growing space.

In the U.S. coastal plains, subsoil properties including color, depth to fine-textured layers, depth to mottling, or some subsoil property that influences drainage, aeration, or water retention are often important to tree growth. Coile (1935) reported that the site index of shortleaf pine stands was influenced by the texture of the B horizon and its depth below the surface. He divided the percentage of silt plus clay in the B horizon by the depth of this horizon below the surface (inches) and found that there was an improvement in site quality with an increase in this texture-depth index up to about 5. Values above this index were usually associated with a decrease in site quality.

Gilmore et al. (1968) reported that the depth of incorporated organic matter and the depth to an impervious layer were the soil properties most closely related to microsite productivity for yellow-poplar in southern Illinois. Other

TABLE 11.1   Soil Properties Frequently Related to Site Productivity (from Published North American Reports)[a]

Southern pines	Subsoil depth and consistency (33); surface soil depth (23); surface and internal drainage (29); depth to least permeable horizon (14); depth to mottling (23); subsoil inhibitional water value (8); nitrogen, phosphorus, or potassium content (18); surface organic content (3)
Northern conifers	Surface nitrogen, phosphorus, and potassium content (24); surface soil texture (17); drainage class (14); depth of surface soil (8), organic content (10); thickness of B horizon (5); stone content (5)
Eastern oaks	Surface soil depth (15); depth of A + B horizon (14); subsoil texture (11); exchangeable base content (8); soil acidity (6); surface soil texture (5); organic or nitrogen content (6)
Eastern hardwoods	Depth to pan or mottling (21); surface soil texture (22); soil drainage (12); nutrient content (11); depth to water table (5); thickness of A horizon (7); subsoil texture (3); organic content (3)
Western conifers	Effective soil depth (22); available moisture (10); surface soil texture (8); soil fertility (6); subsoil texture (3); stone content (4)

[a] Numbers in parentheses indicate the number of reports up to 1998.

investigators (Auten, 1945; Smalley, 1964) found that depth to a tight subsoil or the amount of water in the rooting zone was the most important feature in evaluating sites for this species.

In a study of 124 sites in western Washington States, Steinbrenner (1975) found that total depth, gravel content, effective depth, depth of the A horizon, texture of the B horizon, and microscopic pore space in the H horizon had highly significant influences on the site index of Douglas-fir. All properties except gravel content and microscopic pore space had a positive effect on growth. Steinbrenner also found that increases in degree of slope reduced the site index at high elevations but not at low elevations. Many of these properties are used in mapping soil series in standard surveys or can be used as phases in series and type designation.

Analyses of soil acidity, cation-exchange capacity, organic matter content, and nutrient concentration have been tested as guides to classifying soils on the basis of productivity. However, the use of soil chemical analyses for this purpose has not always met with success. Gilmore et al. (1968) reported that soil acidity was the only variable in the top 20 cm of a Gray-Brown podzolic soil that was statistically correlated with yellow-poplar height. They reported that the lack of correlation between site quality and concentration of elements in the topsoil was due to the fact that trees obtain nutrients from the entire usable soil profile, which was as much as 150 cm in their area. In addition to the difficulties encountered in obtaining soil samples that adequately represent the area in which tree roots flourish, there are also problems in selecting solutions that will extract amounts of nutrients from the soil sample that can be correlated with nutrient availability to trees. Most standard extracting solutions were developed and evaluated for annual crops on agricultural soils, and they have not been calibrated for use in forest soils. Alban (1972) extracted phosphorus from surface soils collected from red pine plantations in Minnesota by 10 different methods. He reported that methods that extracted small quantities of phosphorus (water, 0.002 N H_2SO_4, and 0.01 N HCl) gave better estimates of site index than strong extractants or extractants of organic or total phosphorus. Nutrient quantification may be expressed in terms of concentration in the soil, total amounts per horizon or profile, or relative amounts in comparison to soil nutrient storage capacity or to other nutrients. Regardless of the method used in conjunction with proper sampling and analytical procedures, soil chemical analyses can be extremely useful in classifying sites on the basis of productivity as long as the variable under test is the primary growth-limiting factor (Ballard, 1971). The data-collecting phase of soil-site studies involves locating a large number of site plots, usually in older forest stands representing the range of soil, topography, and climate found within a designated forest area of region. Site index is estimated from trees on the plots, using height and age measurement or stem-analysis techniques. An effort should be made to ensure that there is a relatively wide range in site quality within the study area. The site index estimates are then correlated with associated features of soil and site using multiple regression methods. Sometimes a single soil property will have a predominant influence on site quality, and a single-variable equation can be used for field purposes. Such an example was given by Hebb and Burns (1975), in which depth to fine-textured material was used as an indicator of site quality for slash pine in the coastal plain sand hills of West Florida (Figure 11.6). Their equation showed that site index decreased approximately 1 m for every 75-cm increase in the depth to the fine-textured layer.

The use of digital computers has resulted in more accurate and complete analysis of data, and transformations expressing curvilinear trends and interactions among independent variables can now be tested with little effort. There are still shortcomings with the soil-site approach to site classification, mainly connected with the data-collecting phase. For example, the determination of site index as the dependent variable by the use of harmonized site index curves is subject to the errors previously discussed. However, the use of improved curves based on stem analysis may largely solve this problem. Carmean (1975) pointed out that quantitative values rather than qualitative rankings should be used for defining soil and topographic features as independent variables, but he acknowledged the difficulty of quantifying such features as slope shape, soil structure, and soil drainage. Regression equations have failed to predict site index accurately when applied to a very large and variable study area. There are alternative mathematical approaches to analyzing soil-site data (Verbyla and Fisher, 1989). Stratifying data or subdividing large areas into smaller, more homogeneous units should help in such instances. Soil-site results should be

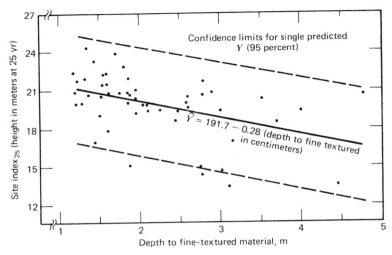

Figure 11.6 Regression of the site index for slash pine on the depth to fine-textured material in sand hills of west Florida (Hebb and Burns, 1975). Used with permission.

applied only to the particular area studied and only to the soil and topographic conditions sampled within the study area. In spite of these limitations, the soil-site equation can be an extremely useful system for evaluating soils for forest managers.

The factors that are important in mapping in one region are not necessarily important under other conditions, but to the extent that those soil factors that influence tree growth are reflected in the soil map unit, valid interpretations can be drawn from the soil map. Agronomists make very little use of soil survey information for predicting site productivity, but they use such information extensively in determining the impact of cultural practices on crop production. In the domesticated forest, foresters also use soils information as a guide in applying a variety of cultural practices. Techniques for making such predictions have been developed and, by and large, are more successful than the attempts to use soil information to predict site index (Fisher, 1984). Today in industrial forestry in the United States, soil mapping is common and soils interpretations are widely used to aid management.

SUMMARY

Attempts at forest site classification can be traced back to the beginning of silviculture as a science. In North America, the site index is the primary means of evaluating site productivity; however, this method has shortcomings, not the least of which is that trees must be present for it to be determined. Although soil surveys have often been characterized as too imprecise to serve as a basis for site evaluation, they have proved to be quite valuable to forestry in many parts of the world. Soil and its associated environment can be thought of as the habitat or site of the forest trees and other organisms. Site quality thus is defined as the total of all factors influencing the capacity of the forest to produce trees or other vegetation. The soil can be seen as an integrator and a recorder of the environmental factors that influence a site. As such, "reading the soil" can yield information about the history as well as the current condition of the site. Various soil survey systems are employed in forestry. All of them yield soil maps that show the geographic location of various kinds of soil. The value of these maps to land managers lies in interpretations—for example what species to plant, how to prepare sites for planting, anticipated tree growth, and so forth—that have been developed for various kinds of soil. Consequently, soil maps and detailed soil interpretations lie at the heart of most intensive forest management schemes.

CHAPTER 12

Nutrition Management: Nutrient Limitations

Forest growth depends on environmental conditions and on the supply of critical resources such as light, water, and nutrients. The supplies of water, nutrients, and oxygen vary substantially among forest soils and may be significantly altered by forest management activities. This chapter summarizes common nutrient limitations by region, and focuses on the assessment of the nutritional status of trees as a foundation for making decisions about management activities such as fertilization, burning, and harvesting. Chapters 13 and 14 build on this information and examine the active management of forest nutrition through fertilization and nitrogen fixation.

NITROGEN AND PHOSPHORUS ARE THE MOST COMMON LIMITING NUTRIENTS

In temperate forests, nitrogen limits forest growth more frequently than any other nutrient. Phosphorus deficiency is also common in some temperate regions and is widespread in the tropics (Gonçalves et al. 1997). Almost all tropical forest plantations are fertilized with phosphorus, and most are also fertilized with other elements including nitrogen, calcium, magnesium, and potassium. The degree of nutrient limitation in native tropical forests remains largely unknown (Grubb, 1995).

Douglas-fir forests of the Pacific Northwestern United States and western Canada are commonly limited by the soil nitrogen supply. Alleviation of this limitation by fertilization increases growth rates by 2 to $4 \, m^3/ha$ annually for 8 to 15 yr (Chappell et al., 1991). In some situations, other nutrients such as sulfur or boron may become limiting following nitrogen fertilization (Blake et al. 1990; Mika et al., 1992). In the same region, stands of western hemlock may be nitrogen limited, but nitrogen fertilization often decreases the growth of hemlock in as many stands as it increases this growth (Chappell et al., 1992). Nutrient limitations in hardwood forests in this region have not been characterized, with the exception of intensively managed plantations of hybrid poplars, which are heavily fertilized with nitrogen and zinc. Nitrogen also commonly limits forest growth in the Rocky Mountains and in the southeastern United States, often in combination with phosphorus, potassium, or sulfur (Binkley et al., 1995). Fertilization of loblolly pine plantations with nitrogen + phosphorus usually increases volume growth by $5 \, m^3/ha$ annually for 6 to 10 years.

The productivity of conifer plantations in the United Kingdom is typically limited by phosphorus or phosphorus + potassium; nitrogen limitations are uncommon, perhaps as a result of moderate levels of nitrogen deposition in rainfall (Taylor, 1991). Nitrogen limitation is the most common problem in forests of Norway and Sweden, where nitrogen fertilization may increase volume growth by 1.5 to $3 \, m^3/ha$ annually for 5 to 10 years. Very large, chronic additions of nitrogen may cause limitations of phosphorus and potassium, and some stands respond very well to low doses of boron (especially if nitrogen fertilized). Conifer forests in Denmark are less commonly limited by nitrogen, perhaps as a result of high rates of nitrogen deposition.

Many forest soils in Germany are limed each year (2.4 million ha were limed between 1984 and 1997; A. Rothe, personal communication, 1999). Liming generally does not increase tree growth (see Chapter 13), but the addition of lime is believed to benefit the soil or perhaps to reverse damage from acid rain. The distribution of lime applications across Germany depends more on differences in local concepts of acidity than on local soil chemistry; only 0.03% of Bavaria's forests have been limed, compared with 7% of the forests in Sachsen.

A great deal of interest in Germany has focused on magnesium deficiencies in conifers, especially Norway spruce (Hüttl and Schaaf, 1997). This interest derives from periodic widespread yellowing of canopies, sometimes in association with needle abscission and reduced canopy leaf area. Much of the supposed decline of spruce in Germany in the 1980s was thought to be derived from magnesium deficiencies, exacerbated either by excessive leaching from the soil or by nitrogen-induced interference with magnesium nutrition (see Oren et al. 1988; Buchmann et al. 1995). Fertilization with magnesium often reduced the yellowing of older age classes of spruce needles (Kaupenjohann, 1997). Interestingly, very few German studies appear to have considered whether alleviation of needle yellowing leads to improved tree growth (or reduced mortality); in the few case studies in which growth was assessed, no significant growth increase was found after magnesium fertilization (Makkonen-Spiecker and Spiecker 1997). Magnesium concentrations apparently have to be very low before tree growth is reduced. The actual influence of the magnesium supply on tree growth in Europe may need more direct evaluation.

Nutrient limitations are pronounced enough in conifer plantations in New Zealand and Australia that most stands are fertilized at least once in a rotation. Both nitrogen and phosphorus limitations are common, with fertilization increasing growth by 4 to $8 \, m^3/ha$ annually for 5 years or more. The low availability of micronutrients in parts of New Zealand and Australia can dramatically limit tree growth, sometimes leading to deformed leader growth and other visible symptoms. Limiting micronutrients in some locations in the region include zinc and boron. High productivity of conifer forests in Japan depends on alleviation of the nitrogen limitation; nitrogen fertilization typically raises productivity to the point where fertilization with phosphorus and potassium allows further increases in growth (Kawana and Haibara, 1995).

Intensively managed forests in tropical and subtropical areas generally require multiple-element fertilization to alleviate nutrient limitations on high growth rates. *Eucalyptus* plantations in Brazil commonly increase growth rates by 4 to 8 m^3/ha annually for 5 years or more when they are fertilized with calcium, nitrogen, phosphorus, and potassium (Barros et al., 1990; Gonçalves et al., 1997), with occasional responses to zinc and boron. Growth responses after fertilization with nitrogen, phosphorus, and potassium in South Africa are commonly 6 to 8 m^3/ha annually (Herbert and Schönau, 1989).

A great deal is known about nutrient limitations and management in tropical plantations, but surprisingly little is known about nutrient limitations in nonplantation forests in the tropics (Grubb, 1995). In Hawaii, Vitousek and colleagues (1993) documented strong nitrogen limitations in ohia forests younger than about 200 years. Forests on older substrates were limited by both phosphorus and nitrogen, and forests on very old soils were limited by phosphorus. Other studies in Jamaica, Venezuela, and Colombia suggest that nitrogen and phosphorus both limit growth in many tropical rain forests (Grubb, 1995).

BASIC APPROACHES TO DIAGNOSING NUTRIENT LIMITATION

The choice of a nutrition assessment method is not straightforward, and no single approach works in all forest types. Indeed, it is not possible to assess the nutritional status of a new, unexamined type of forest until experiments have identified a workable method.

The development of a nutrition assessment program has three basic components. The first step is the selection of criteria for defining nutritional status. The most common criterion is growth response to fertilization, and a series of fertilizer trials usually forms the foundation of a nutrition assessment program. The second step is identification of site variables that correlate well with the growth response to fertilization. As noted below, in some forest types this variable may simply be the site index, while in others it may be difficult to find any good correlates. The third step is the operational evaluation of sites through measurement of the chosen variables that relate well to fertilization response.

This sequence of steps in an assessment program considers fertilization as the nutrition treatment, but the same framework is involved in other areas of nutrition management. For example, the identification of sites susceptible to degradation through whole-tree harvesting or slash burning must be based on (1) field trials, (2) determining which characteristics (variables) denote sensitive sites, and (3) classifying operational sites into response categories.

The current state of knowledge allows useful predictions to be made about nutrient limitations for most of the major commercial forest areas of the world. For example, most (but not all) midrotation stands of loblolly pine in the southeastern United States will respond to combined additions of nitrogen and phosphorus, and intensive cropping of *Eucalyptus* plantations requires copious additions of nitrogen, phosphorus, and commonly other nutrients. Within this general framework, many case-specific exceptions are important. Some sites respond well to nitrogen alone, whereas others require phosphorus fertilization before a nitrogen response can develop (Figure 12.1).

Where local experts cannot yet provide information on likely nutrient limitations, the major approaches to relating tree nutrition to site-specific conditions are concentrations of nutrients in foliage, concentrations of labile nutrient pools in soils, and site classification. The value of each approach depends on site-specific characteristics (how well does the approach capture the variance in forest response?) and costs. Growth responses to fertilization may be correlated with site characteristics, or a simpler approach may involve defining a critical response level for classifying stands into responding and nonresponding groups. For example, Comerford and Fisher (1982) used discriminate analysis to classify slash pine stands by their response to nitrogen

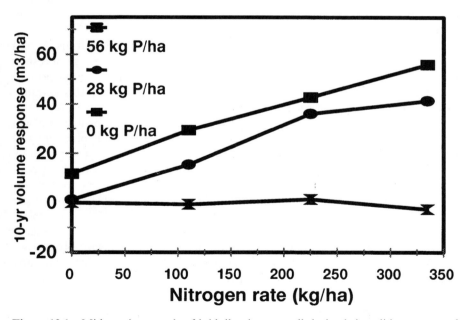

Figure 12.1 Midrotation stands of loblolly pine on well-drained sites did not respond to N alone only marginally to P alone; the response to N depended heavily on the addition of P (North Carolina State Forest Nutrition Cooperative, 1997).

and phosphorus fertilization. They found that extractable soil phosphorus or the nitrogen:phosphorus ratio of foliage could classify about 80 percent of stands correctly into responsive and nonresponsive groups. In this case, the precise magnitude of response was not critical, only whether stands responded above a defined level. The acceptability of this level of precision depends on the relative cost of fertilization and the value of response. If fertilization is cheap and response is very valuable, then the cost of fertilizing unresponsive stands will be more than offset by the value of the response that occurs in responsive stands that may not have been fertilized due to imperfect predictions (see Chapter 13).

Establishment of Field Trials Is the First Step

In establishing a set of field trials for fertilization (or other treatment), the first decision concerns the range of sites to be examined. The site factors that regulate the response of a species across its entire range may be very different from those that affect the response across the range of sites within one public forestland area or one company's lands. The range may be defined on a productivity scale (using site productivity classes) or on the basis of various soil properties such as drainage class or parent material. Choices must also be made on the range of stand conditions to be examined; nutrient limitations (and responses to treatments) commonly change during stand development.

Stratification is important at several levels in the design of field trials. Across the range of sites of interest, the experiment will produce a wide range of results that can be averaged into a grand mean and a total variance. The purpose of the second step in the nutrition assessment program is to explain this variance through association with easily measured site variables.

On a small scale, the responses of two adjacent trees to fertilization will not be identical; some of the total variability in the field trial experiment will be due to tree-to-tree differences. At this scale, two approaches are common: either fertilize many single-tree plots or fertilize a plot large enough to contain a large number of trees. The single-tree approach is used more often for examining physiological components of response, as it is not a very realistic substitute for whole-stand fertilization. What is the best size for a trial plot? Plots containing 25 to 100 trees are common, but the optimal plot size needs to be chosen for each experiment in consultation with a statistician. Most trials also use buffer strips around the actual trees to be measured, removing any boundary effects. In some cases, roots extend far beyond plot edges, underscoring the need for wide buffers.

Moving up the scale, a decision must be made about allocating effort to identify the response precisely within single study locations or across the population. If precision for an individual site is important, then multiple (replicate) plots are needed, often with three or four replicate plots for each treatment being tested. These within-site replicate plots are often "blocked" by clustering a full set of treatment plots in subsets of the site (such as lower-slope or upper-slope areas) in the hope that the block effect will remove some of the noise from the response estimate. In most situations, foresters need to know the average response across the population (or within a subset of the population, such as sites with high-clay Ultisols). A population is sampled most efficiently by installing a single replicate plot for each treatment in as many different locations as possible. Addition of extra replicates within single locations does not add "degrees of freedom" to the estimate of the response of the whole population because these replicates address only the variation within a site, not the variation among sites (they are not independent samples of the population).

At the next step up the scale, how many replications of these treatment designs should be made to ensure that the entire spectrum of sites (and the variance in response among sites) has been covered? This question requires some evaluation of statistical precision, along with decision making using uncertain information (see Chapter 13). Statisticians are crucial (and valuable) collaborators in making optimal decisions.

The next question is, which nutrients will be examined and at what levels? Before investing in a major field trial designed to answer all conceivable questions, it may be profitable to conduct some small-scale screening trials to identify the most interesting nutrients. These trials might consist of simple bioassays in which greenhouse seedlings are grown in field soils amended with various nutrients. A more intensive (and realistic) trial would involve fertilization of many single-tree plots in the field. Information from such a pilot study can be used to assess the likely value of more intensive investigations, as well as identify the most profitable lines of research (see Chapter 13).

The final decision in planning a field trial involves choosing the growth measurements to be made. Possible variables include increments in diameter, basal area, volume, and biomass. For dramatic responses, any of these parameters would demonstrate a large effect. For more subtle responses, the choice of variable is critical. Although diameter at breast height is most easily measured, it may not be a good indicator of the overall response. For example, the same diameter increment on trees of different diameters does not represent the same volume of growth. Calculation of basal area growth avoids this trap but again does not account for possible differences between plots in tree height for similar-diameter trees. Finally, some species, such as radiata pine in New Zealand, may respond to fertilization by changing stem taper rather than breast-height diameter or height.

Once growth since treatment has been measured, what is the best way to compare fertilized and control plots? Three approaches are common. The simplest method is to compare growth differences between treatments at the stand level, but if the stand conditions were not identical in each plot before treatment, some factors will be introduced to the comparison that are unrelated to fertilization. The second method compares the growth rates of similar-sized trees in fertilized and unfertilized plots, removing some of the effects of tree size from the response analysis. Finally, some studies contrast the growth of each sample tree before and after fertilization within each plot and then compare any change in within-plot growth rates between treatments. This approach uses each tree as its own control.

Relating Response to Site Variables Is the Second Step

Application of information gained in fertilizer trials is achieved by relating measured responses to variables that can be assessed on an operational basis. The first step in this second step in a forest nutrition assessment program is the selection of site variables that will be tested for correlation with the response. To a forest researcher, the best variable (or set of variables) would be the one with the highest correlation with the response. To a forest manager, the best variable would be the one that gives an acceptable correlation at a low cost.

Stand productivity is an intuitively appealing variable to correlate with nutrient limitations from fertilizer trials, as the current growth performance of a stand may well relate to nutrient limitation. A problem arises when the direction of the relationship needs to be defined. Will a vigorous, highly productive stand be more capable of responding to fertilization than a low-productivity stand? Both patterns have been reported. For example, Norway spruce in Sweden responds better on good sites than on poor ones, but in Denmark it responds better on poor sites than on good ones. At least in these cases, the contrary pattern can be explained by differing definitions of good and poor; the worst sites tested in Denmark were better than the best sites tested in Sweden (Holstener-Jorgensen, 1983). More recent data from Sweden in fact showed that high-productivity sites (site index > 32 m at 100 years) respond to 150 kg/ha of nitrogen by increasing their volume growth by about 10 m^3/ha/yr for 5 years (Pettersson, 1994).

Some studies have found that site index classes relate fairly well to fertilizer responses. For example, the average gross volume response over 8 years for Douglas-fir was twice as high on poor (Site Class IV) sites as on the best (Site Class I) sites (Peterson, 1982). The site class effect barely missed the 95 percent level of confidence, so from the standpoint of testing a scientific hypothesis, site class would not be considered a significant component of fertilization response. However, even an 80 or 90 percent correlation would be gratefully accepted by forest managers. If only a portion of potential acres will be fertilized, Site Class IV stands would have a higher probability of responding better than Site Class I stands (on average). In the same region, nutrient limitations in western hemlock showed no relationship to site class (Peterson, 1982).

Another type of site classification has been found useful in the southern Coastal Plain of the United States. Site classification based on drainage and B horizon characteristics (depth, argillic versus spodic) has proven useful (Table 12.1). All stands responded with an average increase in volume growth except for group G, and soil groups A and D were the most responsive.

TABLE 12.1   Recommendations for Fertilizing Slash Pine at Midrotation

Soil Class	Representative Soil Series	Limiting Nutrients[a]	Recommended Fertilization (kg/ha)	Average Growth Response (m^3/ha/yr)	Response Probability
A	Portsmouth, Bladen	N and P	170 N 55 P	3.1 to 5.2	100%
B	Rutledge, Plummer	N and P	170 N 55 P	1.4 to 4.2	90%
C	Mascotte	N and P	170 N 55 P	1.7 to 3.8	75%
D	Ridgeland, Leon	N and P	170 N 55 P	3.1 to 6.3	75%
E	Goldsboro	N and P	170 N 55 P	2.1 to 3.5	90%
F	Blanton, Orsino	N and P	170 N 55 P	0.7 to 2.8	50%
G	Lakeland, Eustis	None	0	0	

[a]N, nitrogen; P, phosphorus.
Source: Kushla and Fisher (1980).

FOLIAR NUTRIENTS MAY IDENTIFY NUTRIENT LIMITATIONS

On currently forested sites, direct examination of tree tissues also may be used to assess forest nutrition. Many studies have examined physiological nutrient requirements by studying development of seedlings in greenhouses or by intensive examination of a few trees in the field. Other projects have sought to identify nutrient limitations on severely deficient sites. A final category of research has used tissue analysis to predict growth responses to fertilization. These relationships are complicated by high variability and by confounding factors.

Nutrition experiments with tree seedlings usually have focused on physiological nutrition or have used seedlings as a bioassay of the fertility of soil samples. The physiological class is perhaps best represented by the work of Torsten Ingestad from the Swedish University of Agricultural Sciences. Ingestad developed a hydroponic facility where seedling roots are suspended in a chamber and misted with high volumes of very dilute nutrient solutions. By supplying nutrients directly to the roots at high rates and low concentrations, he was able to evaluate optimum supply rates and ratios of all nutrients. Three general conclusions emerged from a large number of Ingestad's studies. First, the potential growth rates of seedlings receiving optimum nutrient supplies greatly exceed those found when seedlings are grown even in very fertile soils. Ingestad concluded that the ratio of nutrients supplied to the root must be balanced precisely to match the relative levels required by the seedling. Second, these relative ratios are remarkably similar for a wide range of species (Table 12.2). Finally, plants receiving optimal nutrient supplies exhibit much greater relative growth rates than have been reported in most nutrition experiments (Ingestad, 1982).

These "Ingestad" ratios cannot be extrapolated directly to larger trees growing in soils, as the hydroponic levels appear to represent "luxury" levels. Recent work in Sweden, for example, has used ratios of 10 for phosphorus, 35 for potassium, 2.5 for calcium, and 4 for magnesium (Linder, 1995) rather than the notably higher values listed in Table 12.2 from hydroponic experiments.

Physiological studies also can be conducted in the field. H. Brix (1983) of the Canadian Forestry Service installed 85 single-tree fertilization plots to determine the relationship between foliar percent nitrogen and net photosynthesis. Photosynthesis reached a maximum at about 1.7 percent nitrogen and a minimum at 0.8 percent nitrogen, which was about 70 percent of the maximum rate.

The bioassay approach is a fertilization field trial in miniature; seedlings substitute for trees, and potted soil samples (often well sieved) substitute for on-site soils. For example, a 5-month bioassay of soils from plots with *Eucalyptus* or nitrogen-fixing *Albizia* showed a higher nitrogen supply in *Albizia* soils but a higher phosphorus supply in *Eucalyptus* soils (Binkley, 1997).

Bioassays are valuable for identifying nutrients that limit tree growth, but they often fail to provide quantitative predictions of the response to fertilization. Potential problems with bioassays include the artificial environment of the greenhouse and patterns inherent in seedling development. Mead and Pritchett (1971) found that bioassays with slash pine seedlings correlated only

TABLE 12.2   **Ratio of Various Nutrients to Nitrogen, with the Ratio of Nitrogen Normalized to 100[a]**

Species	P	K	Ca	Mg
Scots pine	14	45	6	6
Norway spruce	16	50	5	5
Sitka spruce	16	55	4	4
Japanese larch	20	60	5	9
Western hemlock	16	70	8	5
Douglas-fir	30	50	4	4

[a]A value of 10 means that the optimal concentration is 10 percent that of the concentration of nitrogen.

Source: After Ingestad (1979).

moderately with the response to fertilization in the field; a maximum of 45 percent of the variability in the field response to phosphorus fertilization could be predicted from the weight of 8-month-old seedlings. A paper birch bioassay provided somewhat better predictions, with 40 to 80 percent of the variation in field response related to bioassay results (Safford, 1982).

A further complication of bioassays is the choice of time spans. Seedling development involves a time course of growth; root:shoot ratios and other physiological patterns change. Therefore seedlings that grow more slowly in a bioassay may show effects that result in part from impaired nutrition and in part simply from retarded development.

Occasionally, plants other than trees are used to assess nutrient availability in soils; grasses and other fast-growing plants can fully exploit a pot of soil more quickly than tree seedlings can. However, comparisons of seedlings and grasses generally are advisable to make sure that rapid assessments with nontrees correlate well with seedling nitrogen uptake.

An intriguing twist on the bioassay theme uses root ingrowth bags (mesh bags filled with artificial soil) that are fertilized with various nutrients. If more roots grow into bags containing nitrogen than into control bags, then nitrogen may limit forest growth. This technique has not been used extensively, but Cuevas and Medina (1983) found that root growth responded to bags fertilized with calcium and phosphorus in a lateritic soil, and to nitrogen in a sandy soil, which matched their expectations based on foliar chemistry, soil type, and nutrient content of litterfall. Root response in a study in Hawaii also matched predictions based on the tree growth response to fertilization (Raich et al., 1994). This approach could also be tried with decomposition litter bags; if the disappearance rate from standard paper strips in mesh bags placed in the soil was greater if nitrogen (or other nutrient) was added to the bags, this might indicate a nitrogen deficiency.

VISIBLE SYMPTOMS CAN INDICATE SEVERE PROBLEMS

When nutrient limitations are severe, visible symptoms such as very low leaf area (Figure 12.2) or leaf yellowing become apparent. In the southeastern United States, forest managers may prescribe fertilization treatment based on visual estimates of leaf are a index (H.L. Allen, personal communication, 1998). Color pictures of nutrient-deficient tree leaves can be found in Benzian (1965), Bengtson (1968), Baule and Fricker (1970), Will (1985), and Dell (1996). Other symptoms may include stem deformities and loss of leaves. Although some symptoms may relate to a specific nutrient limitation, in many cases chemical analysis of foliage is needed. For example, twisted, deformed leaders in Douglas-fir have been found to relate to copper deficiency (Will, 1972), boron deficiency (Carter et al., 1983), and even arsenic toxicity (Spiers et al., 1983). A general scheme has been developed for the visual symptoms of nutrient

Figure 12.2 Phosphorus deficiency in 20-year-old slash pine limits the development of canopy leaf area and growth.

deficiencies in *Eucalyptus* (Table 12.3), depending on whether the symptoms appear primarily on old leaves or newly expanding leaves. If such visible symptoms of nutrient deficiency are present, growth is probably severely impaired.

NUTRIENT CONCENTRATIONS CAN IDENTIFY MAJOR DEFICIENCIES

Simple measures of the nutrient concentration in foliage may be sufficient for identification of nutrient limitation in some situations. "Critical" concentrations have been proposed for most major commercial forest species (Table 12.4). For most conifers, nitrogen concentrations less than 10 to 12 mg per gram of leaf mass indicate nitrogen deficiency, and critical phosphorus concentrations equal about 10 percent of the nitrogen concentrations.

Foliar analysis has shown remarkably little ability to predict the magnitude of growth response to fertilization. Part of the problem lies in the sources of variation associated with the analysis of foliage. The nutrient concentration in leaves varies with leaf age and development, location within the canopy, age and competitive status of the tree, and even year-to-year variations in precipitation (Turner et al., 1978).

TABLE 12.3   Visible Symptoms of Nutrient Deficiencies in Some *Eucalyptus* Species

Leaf Age	Symptoms	Likely Deficient Nutrient
Old leaves	Leaf coloration even, green to yellow; small reddish spots may develop secondarily	Nitrogen
	Leaf coloration even, with reddish blotches, or leaves uniformly purple to red	Phosphorus
	Leaf coloration patterned, interveinal chlorosis	Magnesium
	Leaf coloration patterned, scorched margins or interveinal necrosis	Potassium
Newly expanding leaves	Dieback at shoot apex, nodes enlarged, leaves with corky abaxial veins, apical chlorosis, malformed with incomplete margins	Boron
	Dieback at shoot apex, nodes enlarged, leaves with irregular or undulate margins, some interveinal chlorosis	Copper
	Dieback at shoot apex, nodes normal, leaves buckled due to impaired marginal growth	Calcium
	No dieback, leaves normal in size	
	Leaves pale green to yellow	Sulfur
	Leaves yellow with green veins	Iron
	Leaves with marginal or mottled chlorosis, small necrotic spots	Manganese
	No dieback, leaves small and crowded	Zinc

Source: After Dell (1997).

A decrease in the percentage of phosphorus throughout a growing season may result from removal of phosphorus or an increase in carbohydrate content (a dilution effect). This effect can be removed by accounting for changes in leaf weight. Some methods include this factor indirectly by expressing nutrient content per leaf, per needle fascicle, or per 100 leaves. Direct measurements of leaf area have allowed nutrient content to be expressed per unit of leaf area. This approach allows direct assessment of the change in leaf nitrogen content by removing the effects of varying leaf weight (density).

Careful attention to sources of variability can allow fairly precise characterization of foliar nutrient levels, but studies have not verified the ability of even precise characterization to predict the fertilization response. Part of the problem may derive from the importance of two factors: varying leaf nutrient concentrations and varying biomass of the total canopy.

No clear empirical relationship has emerged between foliage sampling schemes and response to fertilization. In the absence of clear empirical patterns, the most common choice is to sample upper crown foliage of the current or 1-year-old age classes (for conifers retaining more than 1 year of foliage) during a period of relatively stable concentrations. This period might be late summer

TABLE 12.4   Recommended Critical Concentrations of Foliar Nutrients[a]

Species	N	P	K	Ca	Mg
Loblolly pine	11	1.0			
Slash pine	8–12	0.9	2.5–3.0	0.8–1.2	0.4–0.6
True firs	11.5	1.5			
Scots pine	12–14	1.4–1.8	3.5–4.5		0.8
Douglas–fir	10–12	0.8–1.0	3.5–4.5	1.5–2.0	0.6–0.9
Lodgepole pine	10–12	0.9–1.2	3.5–4.0	0.6–0.8	0.6–0.8
Western hemlock	10–12	1.1–1.5	4.0–4.5	0.6–0.8	0.6–0.8
White spruce	10.5–12.5	1.0–1.4	2.5–3.0	1.0–1.5	0.5–0.8
Western redcedar	11–13	1.0–1.3	3.5–4.0	1.0–2.0	0.5–0.9
Norway spruce	12–15	1.2–1.5	3.5–5.0	0.4–0.6	0.4–0.6
Eucalyptus grandis, South Africa	12.5	1.0	3.6	5.6	3.5
Eucalyptus maculata, Australia	12–12	0.4–0.5	4	1.5–2.0	0.3
Radiata pine	12–15	1.2–1.4	3.0–3.5	1.0	0.6–0.8

[a]Trees with foliage below these levels are probably deficient; trees near these levels may or may not be deficient; and trees substantially above these levels probably have adequate nutrition for rapid growth.

Source: Based on Carter (1992), Powers (1992), Herbert (1996), Will (1985), and Allen (1987).

for deciduous species or winter for evergreen species. Sampling during stable periods allows greater repeatability among sites and years but does not necessarily give the best indication of nutrient limitations.

Some success has been achieved in identifying the overall nitrogen status of a tree by measuring its amino acid contents. For example, analysis of the arginine content of leaves was superior in distinguishing between fertilized and unfertilized stands than was percent nitrogen (van den Driessche, 1979b).

Critical concentrations for micronutrients are not well established. Possible deficiency in conifer needles might be indicated by levels less than 25 mg/kg of manganese, 50 mg/kg of iron, 15 mg/kg of zinc, 3 mg/kg of copper, 12 mg/kg of boron, and 0.1 mg/kg of molybdenum (Carter, 1992).

VECTOR ANALYSIS MAY HELP IDENTIFY LIMITING NUTRIENTS

V. Timmer of the University of Toronto developed a simple, rapid approach for identifying nutrient limitations under field conditions (Timmer and Stone, 1978; Timmer and Morrow, 1984). The method takes advantage of the determinate growth habit of many conifers. In these species, the number of needles that can be produced in the current year is predetermined in the bud set in the previous year. Trees that are fertilized at the beginning of the current season can have larger needles but not more needles, than unfertilized trees. The increase in needle weight correlates well with later increases in growth (Timmer and Morrow, 1984); trees that respond to fertilization with large increases in first-season needle size also respond with large stem growth increases over the next several years. Timmer expanded on this idea by including the nutrient content of foliage.

To use this approach to identify nutrient limitations, treatments plots are fertilized in the spring with a single combination of potentially limiting nutrients, and control plots are left untreated. In the late summer, needles are collected, dried, weighed, and analyzed for nutrient concentrations. Nutrient concentration and needle mass are used as axes on a graph, and the nutrient content (concentration times mass) isolines are plotted (Figure 12.3). The trajectory from the control to the fertilized value is used to diagnose the limiting status of each nutrient.

How successful is this approach? Early tests with conifers in Canada found good correspondence between changes in foliar nutrients and wood growth responses (Weetman and Fournier, 1982; Timmer and Morrow, 1984). Valentine and Allen (1990) found that an application of nitrogen + phosphorus led to different interpretations of limitations than the foliar response to single-element fertilization. They also compared predictions of nutrient limitation with actual stem growth responses, contrasting the graphical approach with the simpler critical concentration predictions (see Table 12.4). Five of nine stands responded best to nitrogen + phosphorus; three of these responses were

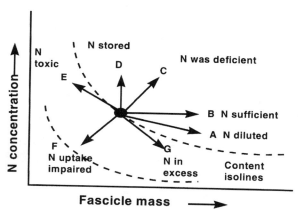

Figure 12.3 The graphical approach to gauging nutrient limitation based on foliar response. An increase in needle mass and nutrient concentration indicates that nutrients may limit growth (vector C). An increase in concentration without an increase in needle mass indicates storage of extra nitrogen (vector D). Other vectors are uncommon and indicate negative effects (after Valentine and Allen's [1990] version of Timmer's graphical analysis method).

predicted correctly by the foliar response to fertilization, and four were correctly predicted by simple critical nutrient concentrations in foliage. Four of the stands responded best to nitrogen alone, and these were all correctly identified by the foliar response to fertilization, while the critical concentration level predicted that two of these stands were also deficient in phosphorus. Binkley et al. (1995) tried the foliar response approach in a replicated age sequence of lodgepole pine in Wyoming. They found that needle mass increased with complete fertilization in older stands and that concentrations of nitrogen, phosphorus, and potassium all increased. However, factorial fertilization with nitrogen, phosphorus, and potassium showed responses only to nitrogen, which would have been consistent with simpler predictions based on critical concentrations in foliage from unfertilized trees. Overall, it's not clear that foliar responses to fertilization identify nutrient limitations much better than critical concentrations.

Timmer and Teng (1999) used the vector analysis approach to examine decline in sugar maple stands. This retrospective study used the concentration of nutrients in foliage and the degree of canopy retention, both expressed as 100 percent for the vigorous maple trees. The less vigorous trees had 30 to 50 percent of the foliage development found for vigorous trees. The smaller canopies had lower concentrations of calcium and magnesium, and in combination with the smaller canopy size produced much lower masses of calcium and magnesium in the canopies (Figure 12.4; magnesium similar to calcium, but not shown). In contrast, the smaller canopies had much greater concentrations of manganese, giving nearly the same manganese mass in the total canopies. The vector for manganese is "E" in Figure 12.3, indicating a toxic trend. Timmer and Teng (1999) concluded that these patterns were consistent with calcium and magnesium deficiencies that were exacerbated by high manganese levels and recommended field testing of their hypothesis.

DRIS COMBINES FOLIAR CONCENTRATIONS, NUTRIENT RATIOS, AND GROWTH

The Diagnosis and Recommendation Integrated System (DRIS) is a multifactorial approach to identifying optimal tree nutrition (Beaufils, 1973). The "norms" are identified by the foliar nutrient concentrations and ratios of foliar nutrients for high-productivity stands. Stands with lower than normal concentrations or ratios are expected to respond well to fertilization. Hockman and Allen (1990) tested the DRIS approach using 62 fertilization trials spread across 52 soil series. The foliar characteristics from the upper 25th percentile in growth were used to define normal values. These values were then used to make predictions about the growth response to nitrogen and phosphorus fertilization. About 80 percent of the stands predicted to respond to nitrogen fertilization actually did, and no site that was evaluated as unresponsive

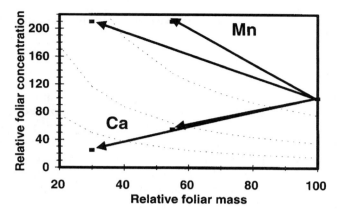

Figure 12.4 In this vector analysis of sugar maple decline, the nutrient concentrations of healthy trees are set at 100, and the mass of a full canopy is also set at 100. The relative decrease in estimated canopy mass of unhealthy maple trees was accompanied by a decrease in calcium concentration (and magnesium concentration, not shown), giving a major reduction in the relative calcium mass (dashed lines are isolines of mass contained in canopies) in the canopies of unhealthy trees. Concentrations of manganese increased by twofold, combining with the decline in canopy mass to provide about the same manganese mass in the canopies of healthy and unhealthy trees (data from Timmer and Teng, 1999).

actually responded. The DRIS norms correctly identified 90 percent of the phosphorus-responsive stands, but the analysis mistakenly predicted that 25 percent of the nonresponding stands would respond to phosphorus fertilization.

SOIL ASSAYS

Which Soil Pool Is the Best Indicator of Site Fertility and Nutrient Limitation?

In many cases, site index or soil groups can be supplemented with direct soil assays of nutrient availability, since current soil fertility may relate to fertilization response. Relatively simple chemical analyses can measure the total quantity of nutrients present in a soil sample. Unfortunately, only a small fraction of the total pools are involved in nutrient transfers each year, and measurements of total nutrient capital may not relate well to the quantities available to plants. The task is further complicated by the variation in soils with depth. Nutrients are typically more abundant in the upper profile, which coincides with the maximum occurrence of fine roots. In some cases, nutrient supplies from locations deeper than 50 cm appear important, and assessing the nutrient supply only in the uppermost part of the profile may underestimate actual availability.

How Available Is Soil Nitrogen?

As discussed in Chapter 9, the availability of inorganic ammonium and nitrate is determined by a suite of factors, including the size of enzymatically labile pool of nitrogen and the activity of the microbial community. The process of nitrogen mineralization is complicated by the fact that microbes both release and reimmobilize nitrogen. The mineralization of nitrogen occurs as a by-product of microbial scavenging for high-energy carbon compounds. Microbial replication and growth require nitrogen to synthesize new cells, so soils with abundant carbon supplies show a strong tendency to immobilize nitrogen. For these reasons, the nitrogen available for plant uptake is generally represented by a "net mineralization" rate, which is the difference between actual (gross) mineralization and reimmobilization by microbes. A difference in net mineralization rate in a forest soil could derive from a change in either gross mineralization or immobilization. These factors are probably too dynamic to be directly useful as gauges of nitrogen availability to trees, so a variety of empirical approaches have been developed (Table 12.5).

In general, about 1 to 3 percent of total soil nitrogen will be available for tree uptake each year (see Binkley and Hart, 1989, for a review of methods of assessing nitrogen availability in soils). If this percentage were constant in all soils, the total nitrogen content would provide a good index of plant-available nitrogen. This percentage is constant enough to allow relative comparisons among sites across a very wide range of nitrogen contents. The nitrogen content of forest ecosystems typically ranges from about 1500 to 15,000 kg/ha, and this 10-fold range exceeds the 2-fold or 3-fold range in annual mineralization of total soil nitrogen. Nevertheless, comparisons of sites with similar nitrogen contents, or of one site under various treatments, must use a nitrogen availability index that is more sensitive to nitrogen-cycle dynamics than is total nitrogen content.

At any given sampling date, several kilograms per hectare of ammonium nitrogen and nitrate nitrogen are present in forest soils (some soils may not have measurable nitrate). This pool of "mineral" or inorganic nitrogen is incredibly dynamic; soil nitrate typically turns over on a scale ranging from hours to 1 or 2 days (Stark and Hart, 1997). Although some researchers have inferred nitrogen availability from these ephemeral pools, this approach is simplistic, as these pools are only a small fraction of the nitrogen that passes through them on an annual basis. As noted in Chapter 9, the average residence time of nitrate in soils is on the order of hours to days. This reasoning is similar to estimating a person's income based on the amount in a checking account at one point in time. Even repeated samplings throughout an annual cycle may not reflect adequately the total flow of ammonium and nitrate through the "clearinghouse" of available pools.

TABLE 12.5 Comparison of Variability and Recovery of Total Nitrogen by Several Methods of Assaying the Soil Nitrogen Supply

| Method | Coefficient of Variation (Mean/Standard Deviation) | | Percent of Total Soil Nitrogen Recovered |
	Replicate Analyses of a Single Soil Sample	Replicate Soil Samples from the Same Stand	
Total nitrogen	5%	11%	100%
Extractable ammonium + nitrate	22%	71%	0.7%
Anaerobic 40°C, 7-day incubation	18%	52%	2.8%
Aerobic 20°C, 30-day incubation	17%	69%	2.5%
Boiling water release of ammonium nitrogen	15%	38%	4.3%
Autoclave release of soluble organic and inorganic nitrogen	10%	30%	16.6%
Seedling nitrogen uptake in 6 months	n.d.	26%	1.1%

Source: Binkley (1986a).

Despite this high variability, the concentration of nutrients in soil solution may be useful for identifying nutrient limitations even if it cannot represent the quantities of nutrients available throughout the year. For example, Smethurst (1999) followed the concentrations of soil ammonium and nitrate in soil solutions (derived from "pastes" of soils saturated with water). The concentrations were extremely variable in the first 2 years of the development of a blue gum (*Eucalyptus nitens*) plantation, ranging from 0.5 to 9 mmol/L of nitrate and 0.05 to 0.2 mmol/L of ammonium (Figure 12.5). After year 2, the concentrations of both forms of nitrogen dropped below 0.03 mmol/L, and stayed there. This two-order-of-magnitude drop was matched by the onset of a strong growth response of the trees to fertilization. This case showed that although nutrient concentrations in soil solution varied by more than an order of magnitude throughout the course of a year, the long-term trends could still be dramatic enough to determine when the nitrogen supply became limiting. This approach has also shown some promise in diagnosing the phosphorus limitation and warrants more development.

Two basic approaches are used to obtain a less dynamic evaluation of available nitrogen. One version uses chemical treatments to break down the

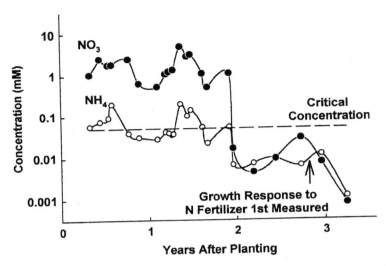

Figure 12.5   Changes in soil solution nitrate and ammonium in plantations of *Eucalyptus nitens*; responsiveness to nitrogen fertilization began after soil solution concentrations dropped below the critical level of about 0.8 mM (from Smethurst, 1999).

simplest organic nitrogen compounds, as this pool may be the source of easily mineralized nitrogen. These chemical treatments include autoclaving, boiling in water or a salt solution, and extraction with weak sodium hydroxide. Typically, these treatments release 4 to 20 percent of total nitrogen, substantially exceeding annual mineralization under field conditions.

Biological approaches incubate samples under controlled conditions in the lab or under ambient field conditions. Three types of lab incubations are common. The simplest consists of a 1-week incubation at 40°C of samples in water-filled bottles (so-called anaerobic incubation). After incubation, samples are extracted with a salt solution (such as 2 M KCl) to remove ammonium from cation exchange sites. The ammonium level is then determined by distillation, by colorimetry, or with a selective ion electrode. These anaerobic incubations differ greatly from field conditions, but the results relate moderately well to plant-available nitrogen. This method simply may release the nitrogen bound in the microbial biomass, which is a major source of available nitrogen. It's important to realize that these indexes may be dynamic, showing large variations among sampling periods. For example, Wright and Hart (1997) measured net nitrogen mineralization for 14 sampling dates in ponderosa pine stands and found very large variations among sampling times (Figure 12.6).

The second approach to biological assays is more realistic than anaerobic incubation. It employs incubation for several weeks at room temperature with just enough water to keep the sample moist. Extraction and analysis are similar

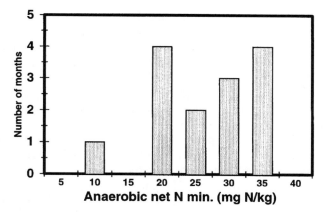

Figure 12.6 A major challenge in using soil nitrogen incubations to identify nitrogen limitation is the large variability associated with the time of sampling. Fourteen monthly samples showed a threefold range in a nitrogen availability index in ponderosa pine stands in Arizona ($n = 3$ stands; data from Wright and Hart, 1997).

to the anaerobic procedure. In addition to being more realistic, these aerobic incubations allow nitrification to occur and can provide an index of nitrification potential. Any nitrate present or produced in the anaerobic incubations may be denitrified since microbes use nitrate as a terminal electron acceptor in the absence of oxygen.

The third variation is probably the most realistic because it involves incubations that last for several months; accumulated ammonium and nitrate are periodically leached from the sample with weak salt solutions. The course of nitrogen release is followed over time, and a "nitrogen mineralization potential" (N_0) is calculated (Stanford and Smith, 1972). Although useful to researchers, this approach is too laborious for use in most forest nutrition applications.

The lab incubations are easy to standardize and usually show lower variability among replicates, but the in-field approaches are more sensitive to any environmental differences among sites. These environmental conditions are especially important in assessing the effects of stand treatments, such as harvest or fire. For example, Burger and Pritchett (1984) examined the effects of harvesting and site preparation on nitrogen availability in a slash pine site in northern Florida. They used an intensive series of laboratory incubations over a total of 125 days, where soil samples were leached with a dilute salt solution every 25 days to remove accumulated ammonium and nitrate. The control (unharvested) stand showed the highest nitrogen mineralization rates, and the intensively prepared sites had the lowest rates. Such laboratory assessments do not account for the effects of harvest and site preparation on microclimate, so the investigators adjusted the laboratory data according to the response of nitrogen mineralization to temperature and moisture conditions on each treatment plot. The pattern of mineralization among the sites was reversed after this adjustment. Two major conclusions can be drawn from these comparisons: harvesting and site preparation may reduce the mineralizability of soil nitrogen, but improved microclimate may actually increase nitrogen mineralization for a few years. Although long-term experiments are needed to evaluate effects on stand growth, the decline in mineralizability suggests that once regeneration has removed any differences in microclimate, nitrogen availability may be lower in the intensively prepared sites.

Incubations can also be performed under field conditions. Several approaches are commonly used for on-site (or in situ) incubations. The simplest one involves pounding a plastic tube into the soil, placing a loose cap on top, and collecting it 1 month later. The difference in ammonium and nitrate concentrations at the beginning and end of the incubation represents net mineralization, and the net change in nitrate is net nitrification. A more complicated version leaves the tubes open to precipitation, with an ion exchange resin bag at the bottom of the tube to catch any ions that leach from the soil column. Hart and Firestone (1989) compared closed incubations (in plastic bags) with open tube incubations (with resin bags on the top and bottom), and found that they provided the same estimate of net nitrogen mineralization in a young mixed-conifer forest, but that the estimates differed by a factor of 2 for an old-growth mixed conifer forest.

Another on-site incubation approach involves simply placing undisturbed soil cores in plastic bags and burying them for 30 days. These incubations are repeated over an annual cycle, and in combination with estimates of bulk density, annual estimates of net nitrogen mineralization can be calculated. These methods are subject to a variety of artifacts, including artificial soil moisture concentrations, the presence of severed roots (and mycorrhizae), and the removal of normal carbon inputs. The annual estimates of net nitrogen mineralization from these methods appear to be good for moderate and fertile sites, but they are too low to account for nitrogen uptake on poor sites. For example, the estimated rates of net nitrogen mineralization in two rich soils in New Zealand amounted to about 80 percent of the measured nitrogen uptake by radiata pine (Dyck et al., 1987). On two less fertile sites, the estimate of net nitrogen mineralization was only 8 to 35 percent of the measured uptake by the trees, showing a very poor relationship for low-nitrogen conditions.

The choice of incubation period is important for on-site assessments. If the periods are too short, the net nitrogen mineralization may be strongly influenced by the disturbance involved in setting up the incubation. Longer periods run the risk of running out of labile carbon for microbes. Incubations of soils from a Douglas-fir stand in Oregon revealed that a single 4-month incubation in the field showed only about half of the net nitrogen mineralization of four 1-month incubations (Figure 12.7). The different environmental conditions in laboratory incubations produced almost twice the rate of net nitrogen mineralization as the in-field series.

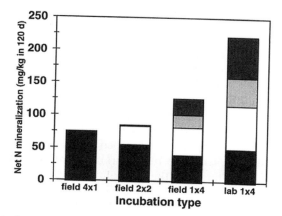

Figure 12.7 Net nitrogen mineralization in intact soil cores over a 4-month period as one long incubation; two 2-month incubations; and four 1-month incubations in the field and laboratory (modified from Binkley and Hart, 1989).

Ion Exchange Resin Bags Are Sensitive to a Range of Factors

Another on-site method of assessing the nutrient supply uses ion exchange resins in porous nylon bags (or mesh balls) to trap ions contained in water passing by. The quantity of ions adsorbed will be affected by:

1. Mineralization. In a greenhouse study, incubated soil that had been fertilized the previous year yielded three times as much nitrogen in resin bags as unfertilized soils (Binkley, 1984).
2. Ion mobility. The concentrations of ammonium in the soils in this greenhouse study were twice those of nitrate, but the higher mobility of nitrate in soils led to 10 times as much nitrate as ammonium in the resin bags.
3. Water regime of the soil. Resin bag nitrate in this study showed no relation to the amount of water leaching through; modest amounts of leaching moved as much nitrate to resin bags as heavy leaching did. In contrast, the delivery of ammonium to resins depended very strongly on the quantity of water passing through the soils and into the resin bags.
4. Competition for nitrogen. Resin bags are very poor competitors with microbes and plants. Additions of cellulose (as an energy source for microbes) reduced resin nitrogen 90 percent, and the presence of grass with the soil and resin bags essentially prevented any nitrogen from accumulating in resin bags.

How well do resin bags index tree-available nitrogen? Under controlled greenhouse conditions, one study found that they related to nitrogen uptake by Douglas-fir seedlings from well-mixed soils as well as any other method (Binkley, 1986). Resins also appear to be very sensitive in detecting treatment effects in experiments, such as the difference in nitrogen supply among tree species and sites (Binkley et al., 1986; Garcia-Montiel and Binkley, 1997) and increases in nitrogen availability following harvesting (reviewed in Binkley and Hart, 1989). A test of resin bags for determining availability of cation nutrients in a Japanese forest of *Cryptomeria japonica* found that the ratios of cations (potassium:calcium, potassium:magnesium, calcium:magnesium) were remarkably similar between cations adsorbed by resins and cations in soil solution (collected by porous-cup lysimeters), indicating that resin bags may provide a good representation of the soil solution (Wu et al., 1994).

Given the wide array of problems inherent in estimating net mineralization rates, it is unlikely that any method can be relied on to provide a precise estimate of actual mineralization rates (nutrient supply rates) in the field. However, these methods can provide useful indexes of nitrogen availability and, in some cases, can increase the precision of fertilizer prescriptions in forest nutrition management programs. As noted in Chapter 13, an assessment method may have to achieve very high levels of precision if the overall value of the response to fertilization is much more valuable than the cost of fertilization; most commonly, it is more profitable to risk fertilizing stands that may not respond well than to risk failing to fertilize very responsive stands.

Various Indexes Have Proven Useful for Different Regions

In the Pacific Northwest United States, Peterson et al. (1984) found that the growth response of Douglas-fir to nitrogen fertilization declined somewhat as the nitrogen content of the forest floor increased ($r^2 = 0.24$) and the carbon:nitrogen ratio declined ($r^2 = 0.21$). Shumway and Atkinson (1978) went a step further and found that mineralizable nitrogen (with anaerobic incubation) correlated somewhat better with growth response ($r^2 = 0.67$). The response of Western hemlock has proven more difficult to unravel, but extractable phosphorus (see the next section) in the forest floor gives a good prediction of the response to nitrogen fertilization ($r^2 = 0.59$; Radwan and Shumway, 1983). Moving south into California, Powers (1980) also found that anaerobic incubations related well to the ponderosa pine response to nitrogen fertilization. In this case, mineralization was highly correlated with total soil nitrogen ($r^2 = 0.83$), so either measure could be used to predict the response. Fir stands with net mineralization values below 10 mg/kg of soil nitrogen responded strongly to nitrogen fertilization, while stands with higher nitrogen supplies did not (Figure 12.8).

Soil assays for nutrient limitations in the southeastern United States have not been impressive. Lea and Ballard (1982) examined both soil and foliage analysis methods, and concluded that such approaches "may never be sufficiently precise to be useful." Hart et al. (1986) found that the growth of

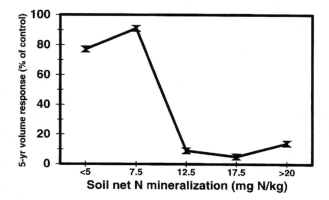

Figure 12.8 Nitrogen limitation on growth of *Abies* species in California was strong where anaerobic incubations of soils (from 18 to 22 cm into mineral soil) showed a low nitrogen supply (<10 mg/kg; >80 percent of stands responded) and a marginal growth response where the nitrogen supply was higher (20 to 50 percent of stands responded) (data from Powers, 1992).

unfertilized stands predicted the response to fertilization of 12 stands of loblolly pine better than did any measure of soil nutrient supply (slow-growing stands responded better). Fortunately, most loblolly pine sites are very responsive to nitrogen, phosphorus, or nitrogen + phosphorus fertilizers, and a profitable fertilization program can be developed without perfect information on the responsiveness of each site (see Chapter 13).

Soil Phosphorus Availability Is Usually Estimated by Extractions

The total phosphorus content of a soil sample may be determined following digestion in strong perchloric acid, but this number provides little information for site assessment because most of the total phosphorus pool is unavailable to plants. Phosphorus availability in soils is not measured easily, but a number of extraction procedures have been developed that correspond somewhat with plant-available phosphorus (Olsen and Sommers, 1982). Two dilute acid extractants are common for assaying acid forest soils. A mixture of hydro-chloric acid and ammonium fluoride (Bray and Kurtz method) dissolves most of the calcium phosphate salts. The fluoride also can replace some of the specifically adsorbed phosphorus, releasing phosphorus bound to iron and aluminum. The second dilute acid extractant, used in the so-called double acid or Mehlich method, is a mixture of hydrochloric and sulfuric acids. This method has been popular for assaying phosphorus availability in pine planta-tions in the southeastern United States. In a comparison of these methods with the loblolly pine height growth response to phosphorus fertilization, Wells et al. (1973) found a fairly close correlation between the methods ($r^2 = 0.56$ to

0.67). Nevertheless, the methods accounted for only 25 to 40 percent of the observed variation in fertilizer response. These authors recommended phosphorus fertilization for sites with extractable phosphorus levels below 3 mg/kg.

A sodium bicarbonate extraction method (Olsen method) is common for phosphorus assays of neutral to alkaline pH soils, as well as for soils high in calcium carbonate (limestone). The bicarbonate anion causes precipitation of calcium, freeing phosphorus from calcium salts. Most forest soils are acid, so this technique has not proven generally useful in forestry. For example, Kadeba and Boyle (1978) found that a sodium bicarbonate extractant dissolved about the same amount of phosphorus from acid soils as acid extractants did, but the phosphorus availability estimates were unrelated to the phosphorus taken up by pine seedlings. In this experiment, the acid extractants did not perform impressively either, but an anion exchange resin method worked fairly well.

Ion exchange resins were first used in soil analysis to assess phosphorus availability in laboratory assays (Amer et al., 1955). This method merely mixes anion exchange resins with water and soil and allows the resin to adsorb phosphorus from the soil for several hours. The resins are then removed from the soil and extracted to determine phosphorus adsorption. Of all the methods tested by Kadeba and Boyle (1978), resin phosphorus accounted for the most variation (66 percent) in pine seedling phosphorus uptake.

Ion exchange resin bags also have been used in the field for assessing on-site phosphorus availability. Hart et al. (1986) found that IER bag phosphorus correlated weakly with current loblolly pine growth in the Coastal Plain of North Carolina ($r^2 = 0.31$), as did double-acid extracts ($r^2 = 0.26$). Neither method alone predicted the response to nitrogen + phosphorus fertilization, but combining either method with prefertilization growth accounted for 80 to 85 percent of the variation in growth response.

In Chapter 9, it was noted that mineralization of phosphate from organic compounds may be mediated by phosphatase enzymes. Can assays of phosphatase activity accurately estimate phosphorus availability to trees? Little work has been done to relate phosphatase activity to tree nutrition, but agricultural experience generally has not been encouraging. One interesting twist involves addition of phosphatase enzyme to assay for the phosphatase-labile pool of phosphorus. Fox and Comerford (1992a) found that about 30 percent of the water-soluble phosphorus from one Spodosol was mineralized in the presence of excess enzyme compared with about 20 percent for another Spodosol.

Three other approaches are commonly used in phosphorus availability research, but they are probably too complex to be used for operational assessments of soil fertility. Sorption isotherms are constructed by equilibrating soil samples with solutions containing a range of dissolved phosphorus concentrations. Soils with a high sorption capacity will remove most of the dissolved phosphorus from solution, whereas soils with a low capacity will allow most of the phosphorus to remain in solution (for good examples, see Humphreys and Pritchett, 1971, and Torbert and Burger, 1984). A comparison of sorption curves among soil types may give an indication of current relative phosphorus availability, as well as the ability to retain phosphorus fertilizers in readily available pools.

The use of radioactive ^{32}P offers opportunities for very detailed analysis of phosphorus availability and dynamics. However, the difficulties associated with radioactive isotopes prevent the direct use of these methods in operational assessments.

Many researchers have used variations of the Hedley fractionation scheme to determine the proportion of soil phosphorus that resides in pools that might differ in turnover rates and availability, as described in Chapter 9.

Soil Sulfur Assays Are Similar to Phosphorus Assays

Sulfur limitations are not common in forests, so soil tests for sulfur availability have not been examined extensively for nutritional assessment of forests. In agricultural soils, soil sulfur assays have included extractions by water, acid, acetate, and bicarbonate, as well as laboratory mineralization incubations and assays of sulfatase activity. Not surprisingly, no method has proven superior for all soils and situations. Soil sulfur cycling simply involves too many pools of differing activities for one method to work on a range of soils in which the importance of each pool differs. In addition, annual sulfur inputs from the atmosphere (as sulfate or as direct gaseous absorption of sulfur dioxide) supply much of an ecosystem's annual requirement.

Cation Availability Is Also Difficult to Assess

Although the cation nutrients held on exchange sites form a readily available pool, they do not represent the cation-supplying ability of the soil. Cations removed from exchange sites often are replenished rapidly from other sources, such as organic matter decomposition, mineral weathering, or release of ions "fixed" within the layers of clay minerals. For example, 15 years of a grass crop removed a total of 4000 kg/ha of potassium without substantially depleting the exchangeable potassium pool in a Hawaiian soil (Ayers et al., 1947). Exchangeable cations may well represent the total available pool in old, highly weathered soils, but the exchangeable pool is only part of the available pool in most forest soils.

Cation supplies limit forest growth less commonly in commercial forests than the supplies of nitrogen and phosphorus, and less work has been invested in developing predictive soil assays. In agriculture, various extractants (such as weak nitric acid) have proven useful in evaluating the actual availability of cation nutrients. Also, in an interesting version of the ion exchange resin method, successive exposures of a soil sample to cation exchange resin assess a soil sample's ability to supply potassium over time (Talibudeen et al., 1978).

In Brazil, operational decisions about fertilizer application look at the size of exchangeable pools of calcium and magnesium in the soil (Gonçalves et al.,

1996). The beneficial amount of dolomitic lime (containing calcium and magnesium) is estimated as

$$\text{Dolomite application (Mg/ha)} = \frac{[20 - (Ca + Mg)]}{10}$$

where $Ca + Mg = mmol_c/kg$ (or $mmol_c/L$) of soil.

Although cation availability does not currently limit the growth of most forests, intensification of forestry (and acid deposition) may lead to cation depletion and limitations in future rotations (Chapter 16). Coping with these declines may require development of assay methods appropriate for forest soils.

Soil Micronutrient Assays Are Performed Rarely

Micronutrients are required in such small amounts that soil tests usually are not precise enough to assess the difference between sufficient and deficient quantities. Micronutrient limitations on tree growth are uncommon (but critical in some cases), and little work has been done to improve the sensitivity of soil assays. In most cases, bioassays or foliar analysis are preferred methods for assessing micronutrient status.

THE FINAL STEP IN NUTRITION ASSESSMENT IS OPERATIONAL ASSESSMENT OF SITES

Following the fertilization trials and experiments on assessment methods, a forest nutrition assessment program classifies management units into predicted response classes. If the chosen assessment method is based on stand growth or site index, response classification could be performed without collection of new data on each stand. If the method requires soil or foliage analyses, a substantial investment may be needed in statistical design, sampling, and laboratory facilities. Although such an investment may be large, it would represent only a small addition to the total cost of a fertilization program. If fertilization cost $120/ha, an operational assessment program that identified 10,000 unresponsive hectares would save $1,200,000! However, if the value of responding stands were much greater than the cost of fertilization, then the assessment method would have to identify unresponsive stands very accurately (i.e., not misclassify many stands) to justify the risk of failing to fertilize responsive stands. The value of predictive information needs to be evaluated in the economic context of the overall program (see Chapter 13).

INTERACTION BETWEEN NUTRITION AND PESTS AND PATHOGENS

Fertilization has varying effects on tree susceptibility to pests and pathogens. Overall, these interactions do not tend to be important relative to stand volume growth, but in some situations the effects can be important. For example, nitrogen fertilization of ponderosa pine forests on low-potassium soils increased the gross volume increment by $2\,m^3/ha$ in 4 years, but it also increased the incidence of bark beetles and mortality (about 10 percent of the trees died), giving a net increment loss relative to control plots of $-1\,m^3/ha$ over 4 years (Moore et al., 1994). Inclusion of potassium with nitrogen in the fertilizer gave a gross increment boost of about $3\,m^3/ha$ in 4 years, and no trees died from beetle infestation.

A somewhat different interaction between nitrogen fertilization and bark beetle mortality was found in a lodgepole pine stand by Waring and Pitman (1985). Additions of nitrogen or carbohydrates (to lower the nitrogen supply by increasing microbial immobilization) were combined with the use of pheromones to attract bark beetles. Fertilization increased stem growth per unit of leaf area, and the carbohydrate addition reduced nutrient availability and tree growth/leaf area. Pines with stem growth per unit of leaf area (indicating vigorous trees) were generally more resistant to attacks (Figure 12.9); fewer beetle attacks were required to kill low-vigor trees.

Matson and Waring (1984) showed that increasing the nutrient availability reduced susceptibility of mountain hemlock seedlings to laminated root rot. Some reports have also indicated that fertilization may increase susceptibility to pathogens, such as fusiform rust on loblolly pine (Smith et al., 1977). Other

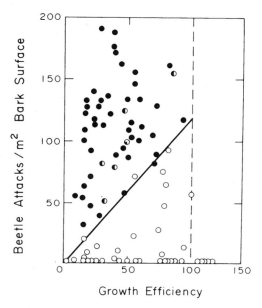

Figure 12.9 Lodgepole pines with high rates of growth/leaf area (vigorous trees) were more resistant to attacks by bark beetles than trees with lower vigor. Closed circles represent trees killed by beetles; open circles represent survivors (from Waring and Pitman, *Journal of Applied Entomology*, 1983, 96:S. 269, used by permission).

studies have found little evidence of such a problem (Kane, 1981), and no strong general conclusions are warranted. On balance, any undesirable effects of fertilization on tree susceptibility to pests are probably small, but undesirable consequences do occur.

THINNING INCREASES THE NUTRIENT SUPPLY PER TREE

Few studies have examined the effects of thinning (reducing stand density by cutting a portion of the trees) on soil nutrient supply. Most have found little effect on net nitrogen mineralization, for example, which indicates that the nutrient supply per residual tree increases in proportion to the reduction in stocking (Beets and Pollock, 1987; Martinez and Perry, 1997). In some cases, net nitrogen mineralization may increase following the removal of a portion of the stand, providing a double boost to the nitrogen supply of residual trees (Prescott, 1997).

Fertilization trials have occasionally found substantial growth increases following thinning (Chappell et al., 1992). When this occurs (the pattern is not universal), the increase in nutrient supply for each residual tree may not be enough to allow maximum rates of canopy expansion into the space vacated by harvested trees.

SUMMARY

The productivity of most forests is limited by the supply of one or more nutrients, with nitrogen and phosphorus limitations spread across the largest area. Severe nutrient limitations can be diagnosed by visual symptoms of leaves, by very low stand leaf area, or by very low nutrient concentrations in chemical analyses of leaves. Most forests that are nutrient limited have no overt visual symptoms, and fertilization experiments are needed to identify limiting nutrients. Responses to fertilizer applications may correlate with foliar chemistry, soil classes, productivity classes, or soil analyses. No method has proven useful in all situations, but in most cases, one or more approaches have been shown to correlate usefully with stand nutrient limitations.

CHAPTER 13

Nutrition Management: Fertilization

The supply rates of nutrients are variable, changing naturally across landscapes, with changes in vegetation (composition or condition), and under the influence of management. These patterns largely determine the growth and sometimes the vigor of forests. Where the nutrient supply limits production of forests (Chapter 12), fertilization is often very profitable. Single applications of fertilizer commonly increase wood production by 3 to 10 m³/ha over periods of 5 years or longer (Figure 13.1). Fertilization also increases the size of individual trees (as well as that of stands), with the largest increases on the largest, most dominant trees. The increased value of large trees may be greater than the simple increase in wood volume.

Forest fertilization is a routine silvicultural operation in practically all commercial forests that are managed intensively for high production. The value of wood increased rapidly in the United States and around the world in the latter part of the 1980s and throughout the 1990s, and the cost of fertilizers held relatively steady. These market changes dramatically increased the profitability and use of fertilizers (Figure 13.2). Over this same period, the formerly high rates of fertilization of Swedish forests dropped drastically as concerns about atmospheric deposition extended to fertilization (see Chapter 16).

Fertilization practices for established stands vary even when the same nutrient is deficient. For example, 200 kg/ha of nitrogen may add 7 m³/ha in a Douglas-fir stand in Oregon, while a similar growth response in loblolly pine in North Carolina may be obtained with only 100 kg/ha. In a survey of the literature, Weetman and Fournier (1984) found that sites that respond to nitrogen fertilization typically require 15 to 25 kg nitrogen to produce an extra cubic meter of wood. What accounts for the response differences? Important factors include stand condition (age, size, stocking, and vigor), patterns of nitrogen immobilization/mineralization, and the availability of other nutrients and water.

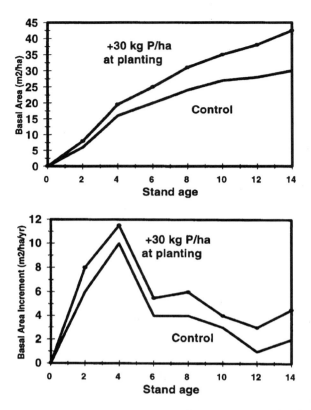

Figure 13.1 A single application of 30 kg/ha of phosphorus at the time of planting provided a sustained increase in basal area and basal area growth of *Eucalyptus grandis* through the rotation in South Africa (data from Herbert, 1996).

FERTILIZATION COMMONLY INCREASES NET PRIMARY PRODUCTION AND REDUCES ALLOCATION TO ROOTS, INCREASING WOOD GROWTH

Fertilization increases wood production by increasing total tree production (gross or net primary production) and by increasing the allocation of carbohydrates to wood production. Increases in overall production come from increases in nutrient concentrations in foliage (indicating increased concentrations of major enzymes such as RUBISCO) and increases in leaf area, which combine to increase net photosynthesis.

An intensive fertilization regime in a stand of loblolly pine increased wood production by 5.5 Mg/ha annually, in part through increased net primary production (by 9.2 Mg/ha annually) and in part from less allocation to fine root growth (40 percent lower in fertilized plots). Intensive fertilization of a

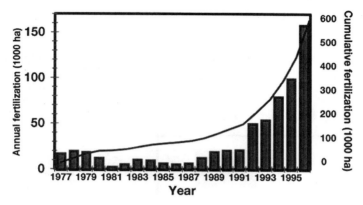

Figure 13.2   Forest fertilization has increased dramatically in the past 20 years in loblolly pine stands in the southeastern United States (data from the North Carolina State Forest Nutrition Cooperative, 1997). Fertilization has become a standard operating procedure for most regions in the world with intensive plantation forestry.

young stand of *Eucalyptus saligna* also increased wood production by increasing net primary production and reducing allocation to fine root production. Fertilization of a snowgum (*Eucalyptus pauciflora*) stand with 500 kg P/ha in a single dose led to an increase in wood production of 1.5 Mg/ha annually, even though net primary production increased by only about 0.9 Mg/ha annually (Figure 13.3). Fine root production dropped in the fertilized stand, providing a large part of the carbohydrate supply for wood growth.

The nutrient (and enzyme) concentrations in foliage typically increase with fertilization, contributing to the increase in net primary production. Heavily fertilized *Eucalyptus* trees averaged about 20 percent higher nitrogen concentration than control trees and had, on average a 15 percent greater capacity for photosynthesis (Figure 13.4). Whether the trees were fertilized or not, their photosynthetic capacity related moderately well to leaf nitrogen content.

The increase in foliar nutrient concentrations and leaf area may or may not be associated with substantial increases in total production. Not all of the nitrogen in leaves is present in photosynthetically active enzymes, not all of the canopy is well illuminated, and overall canopy transpiration may be restricted by the water supply. A simulation model is needed to account for all the interactions that follow forest fertilization, and the models currently available do not represent belowground production well.

FERTILIZATION CHANGES STAND CHARACTERISTICS

In addition to increasing stem growth, fertilization commonly alters mortality, stocking, diameter distributions, and in some cases even the maximum size

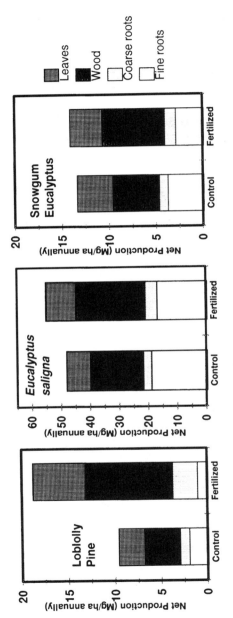

Figure 13.3 Fertilization of loblolly pine in North Carolina (470 kg/ha of nitrogen, 135 kg/ha of phosphorus, 275 kg/ha of potassium) doubled net primary production, and wood production was 2.4 times the control value (Albaugh et al., 1998). Fertilization of *Eucalyptus saligna* in Hawaii (390 kg/ha of nitrogen, 170 kg/ha of phosphorus, 455 kg/ha of calcium, 60 kg/ha of potassium) increased wood growth by increasing total production and decreasing root production (M. Ryan, D. Binkley, and J. Fownes, unpublished data). Fertilization of snowgum (500 kg/ha of phosphorus, no nitrogen added) increased net primary production by just 6 percent and wood production by 30 percent (Keith et al., 1997). In all three cases, the relative and absolute allocation to fine root production declined, contributing to the increased production of wood.

314

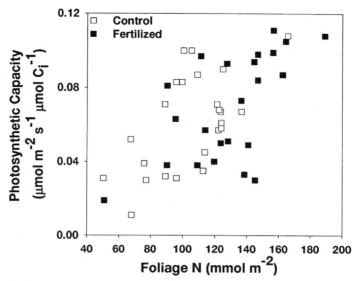

Figure 13.4 The photosynthetic capacity of 29-month-old *Eucalyptus saligna* trees in Hawaii related strongly to foliar nitrogen concentrations (80 mmol/m^2 of leaves ~ 1.2 percent nitrogen, 160 mmol/m^2 ~ 2.5 percent nitrogen); fertilized plots had about 20 percent more nitrogen in leaves and 15 percent greater photosynthetic capacity (data from M. Ryan, personal communication, 1999).

attainable by the trees. As trees become larger, fewer stems can be carried per hectare, so fertilization commonly increases the mortality of suppressed trees by enhancing the performance of dominant trees. In rare cases, fertilization has led to nutrient imbalances that increased tree stress and susceptibility to insects and pathogens.

What patterns would be expected when competition is less important? All trees should exhibit faster growth, and mortality should be low. This pattern obviously applies for comparisons of low-density and high-density stands or of thinned and unthinned stands. The pattern may also apply when fertilization drastically alters the factor that limits tree growth. For example, fertilization of loblolly pine with moderate levels of nitrogen may be considered to accelerate stand development, decreasing the time it takes to reach some ceiling of maximum tree size and density. On the other hand, phosphorus fertilization often is described as "increasing site index," which translates into not only accelerating growth but also "raising the ceiling" of the growth trajectory. In this case, site limitations would be reduced so that all trees perform better after fertilization, and competition-driven mortality need not increase. (see Figure 5.8). These two responses have also been discussed by Snowden and Waring (1984) for radiata pine.

Are fertilized stands more susceptible to drought than unfertilized stands? One can imagine that higher leaf areas might be associated with greater water use and greater sensitivity of large canopied-trees to cavitation of stemwood during drought periods. If this risk were of major importance, a variety of studies would have reported drought-related mortality from fertilized stands. Few studies have reported any interaction between drought tolerance and fertilization. Some evidence indicates that water-use efficiency increases after fertilization and that fertilized trees generally are not at greater risk of dying during droughts than trees with poorer nutritional status. Some studies have reported evidence of higher mortality in fertilized plots. During the most severe drought on record, a fertilized plot of radiata pine in Australia showed 10 percent mortality (concentrated in smaller-diameter classes of trees, not dominants) compared with no mortality in a control plot (Snowdon and Benson, 1992). This study was not replicated, so confidence in the cause of the difference between plots is hard to establish. In Brazil, a moderately severe drought killed 24 percent of *Eucalyptus* trees in control plots, 31 percent of trees in low-fertilization plots, and 42 percent of trees in high-fertilization plots; with six replicate plots for each treatment, the overall effect of fertilization on mortality was significant at a level of $p = 0.07$ (J.L. Stape, personal communication, 1999). The greater mortality of fertilized trees did not appear to be related directly to increases in canopy size (the number of dying trees was not related to the leaf area of plots), so a mechanistic view of the influence of fertilization on drought tolerance remains unclear.

NUTRIENT LIMITATION IS COMMON ACROSS ALL STAND AGES

Some early investigators expected that nutrient limitation would decline after conifer stands achieved full leaf area, as the demand for nutrients would be lower (Miller, 1981). Empirical evidence generally shows that sites that are nutrient limited at early stages of stand development are also limited later in stand development (reviewed by Ryan et al., 1996). For example, fertilization of adjacent stands of 45-year-old and 250-year-old stands of lodgepole pine showed equivalent nutrient limitation in both stands (Figure 13.5).

APPLICATION METHODS INCLUDE AIRPLANES, HELICOPTERS, AND TRACTORS

Methods of application usually are dictated by economics and terrain. Level sites can be fertilized by tractor, but steep slopes require aerial application. The method used affects the evenness of the application. Hand application ensures precise placement around individual trees (Anderson and Hyatt, 1979), reducing the fertilizer cost but increasing the application cost. Fertilizer can also be applied near seedlings as part of mechanized regeneration operations. Applica-

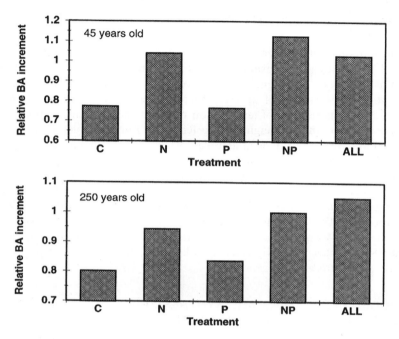

Figure 13.5 Relative basal area increment for 5 years after fertilization in proportion to the 5-year increment prior to fertilization in two adjacent stands of lodgepole pine in the Fraser Experimental Forest, Colorado. Both stands increased basal area growth by about 30 to 40 percent in response to any fertilization treatment that included nitrogen. (D. Binkley, unpublished data). C = control, N = +200 kg N/ha, P = +50 kg P/ha, ALL = +N + P + other elements.

tion from fixed-wing aircraft can be less expensive than hand application, but it tends to give uneven distribution of the fertilizer. For example, the coefficient of variation (the standard deviation as a percentage of the mean) for airplane application may be about 110 percent (Strand and Promnitz, 1979). This means that at an application rate of 200 kg/ha of nitrogen, about 10 to 20 percent of an area would receive no fertilizer, and about the same proportion would get more than double the intended rate. Helicopter applications are more precise, with a coefficient of variation of about 25 percent.

In the, 1980s, radar-controlled methods of aerial application greatly improve the precision of aerial fertilization (Hedderwick and Will, 1982). The radar system guided the helicopter along parallel, evenly spaced lines and allows the pilot to reload and return to the exact location of the previous run. Precise application allows the rate of fertilization to be reduced from 100 kg/ha to 80 kg/ha of phosphorus or less. The savings in fertilizer cost more than cover the cost of the radar system. In the 1990s, global-positioning satellites (GPS) made fertilization application even more precise.

FORMULATION IS IMPORTANT

The formulation of fertilizer is important for calculating the amount to be applied to an area. For example, 100 kg of ammonium nitrate (NH_4NO_3) contains 35 kg of nitrogen, 60 kg of oxygen, and 5 kg of hydrogen. An application of 100 kg/ha of nitrogen therefore requires 285 kg of ammonium nitrate. Most discussions of fertilization mention the quantity of the nutrient applied rather than the weight of the fertilizer. For example, a rate may be reported as "100 kg of nitrogen as urea" or simply as "100 kg urea nitrogen."

Agricultural fertilizers in the United States are sold with the percentages of "nitrogen-phosphorus-potassium" listed on the bag. A 10-kg bag of "10-10-10" would appear to have 1 kg of each nutrient. However, the phosphorus and potassium values are expressed as the antiquated "oxide" forms: P_2O_5 and K_2O (potash). To obtain the true percentage of phosphorus, the listed value is multiplied by 0.44; for potassium the multiplier is 0.83. A bag containing 10 kg of 10-10-10 fertilizer actually contains 1 kg nitrogen, 0.44 kg phosphorus, and 0.83 kg potassium.

NITROGEN FERTILIZERS ARE SYNTHESIZED FROM NITROGEN GAS AND NATURAL GAS

Nitrogen fertilizers are commonly synthesized by the Haber-Bosch process, in which N_2 and CH_4 are passed over a metal catalyst at high temperature and pressure to produce NH_3. The ammonia may then be transformed into a variety of fertilizers, such as urea (($NH_2)_2CO$), nitrate, or ammonium. The form of fertilizers affects chemical reactions in the ecosystem, and availability of the fertilizer to trees. For example, ammonium nitrate dissolves quickly in water; ammonium may be retained on cation exchange sites in the forest floor or mineral soil, and nitrate may be taken up quickly or leached into the mineral soil. The ammonium may gradually become available to plants and nitrifying bacteria, and the nitrate may quickly be absorbed by plants or leached from the soil profile. Urea must first be hydrolyzed to form ammonium:

$$(NH_2)_2CO + 2H_2O + 2H^+ \rightarrow 2NH_4^+ + H_2CO_3$$

Note that one H^+ is consumed from the soil solution for each ammonium formed. This consumption is offset in part by dissociation of carbonic acid (H_2CO_3) to form H^+ and bicarbonate (HCO_3^-). Plant uptake of ammonium also will release one H^+ back to the soil solution. Urea fertilization usually raises soil pH for a few months but has little effect more than 1 year later. The H^+ effects of ammonium and nitrate fertilizers are not balanced, like those resulting from urea transformation and use. Uptake of ammonium generates H^+, which is not balanced by previous consumption from the soil solution. Nitrate uptake consumes H^+ from the soil solution without balancing previous production (as is the case for on-site nitrification). Annual fertilization of agricultural soils can alter H^+ budgets substantially, but fertilization of forest soils is typically infrequent enough that any change is negligible. Seedling nurseries present an exception to this generalization. Annual fertilization with certain forms of fertilizer can have a significant acidifying effect.

How important are these differences in chemistry? Urea contains 45 percent nitrogen and 55 percent hydrogen, carbon, and oxygen, and ammonium nitrate is about 34 percent nitrogen. Fertilizer weight is important in aerial applications, and in this case, urea would appear to be the better choice. Any savings in application costs, however, could be offset by differences in growth responses. For example, the Scots pine growth response on some soils may be about 45 percent greater with ammonium nitrate than with an equivalent amount of urea nitrogen (Malm and Moller, 1975). Operational fertilization in Sweden in the 1970s shifted from urea to ammonium nitrate (and then to nitrogen-free fertilizers because of concerns about atmospheric deposition). In North America, up to a 20 percent greater response to ammonium nitrate was reported in some Douglas-fir studies from British Columbia (Dangerfield and Brix, 1979; Barclay and Brix, 1985), but other experiments showed no difference (Brix, 1981). Jack pine stands in eastern Canada have shown no significant difference between nitrogen forms in growth response (Weetman and Fournier, 1984). Loblolly and slash pines in the southeastern United States have shown little difference in response between the two forms (Allen and Ballard, 1982; Fisher and Pritchett, 1982).

What factors account for the varying response patterns? For the Swedish sites, soil studies indicated that ammonium, formed from urea hydrolysis, is tied up by microbes in the forest floor, whereas nitrate (from ammonium nitrate) passes rapidly to fine roots in the mineral soil. Because loblolly and slash pine stands in the United States lack the well-developed humus layers found in Sweden, ammonium immobilization may be less critical. More research using ^{15}N tracer techniques is needed to explore fertilizer immobilization dynamics.

Fertilizer formulation may affect the loss of fertilizer from the ecosystem. Nitrate may be subject to rapid leaching if heavy rainfall precedes uptake by plants. Gaseous losses from urea fertilization may occur if urea is added during a fairly dry period. Hydrolysis to ammonium can produce a rapid, very localized increase in soil pH around the fertilizer granule that will cause some ammonium to deprotonate, forming ammonia gas. Rainfall after fertilization helps prevent volatilization losses by dissolving the urea granules and by preventing localized increases in pH. The season of fertilization may be an important component of the tree response if rainfall patterns affect retention of the fertilizer. In any case, volatilization losses from urea fertilization are usually less than 10 percent of the applied nitrogen.

Some special formulations of nitrogen fertilizers (such as urea-formaldehyde or iso-butylidene di-urea) have been developed to promote slow release of nitrogen, maximizing the time course of availability to trees and minimizing the losses from high application rates. These formulations are expensive, and their use in forestry has been limited to high-value nursery and Christmas tree soils.

The size of fertilizer granules is another important consideration in forest fertilization. The precision of aerial fertilization is increased with the use of relatively large (0.5 to 1.0 cm) granules. Large granules are affected less by wind patterns and may penetrate canopies better (reducing foliage burning).

PHOSPHATE FERTILIZERS ARE MINED

Phosphorus fertilizers come in three general forms: ground rock phosphate, acid-treated phosphates, and mixtures with other nutrients. The original source of most phosphorus fertilizer materials is mined rock phosphate. In general, phosphorus fertilizers have little effect on soil acidity, but weathering of ground rock phosphate may slightly increase soil pH.

Ground rock phosphates are mined from deposits of various types of calcium apatite minerals:

$$Ca_{10}(PO_4CO_3)_6(F, Cl, or\ OH)_2$$

The solubility of apatite varies with the anion (F^-, Cl^-, or OH^-). Fluoride gives the lowest solubility, so a high content of fluoride in rock phosphate fertilizer reduces the rate of release of phosphorus for plant uptake. The rate at which various types of rock phosphate dissolve and become available for plant uptake can be determined from solubility with citric acid (see Bengtson, 1973). The rate at which ground rock phosphorus becomes available to plants is also affected by soil pH and the particle size of the rock; low pH and small particle size favor rapid dissolution. Large grains dissolve very slowly, especially in neutral and high-pH soils.

Phosphate minerals may be treated with acids to produce fertilizers with a range of phosphorus contents. Superphosphate is generated by mixing ground rock phosphate with sulfuric acid, and contains about 8 percent phosphorus, 20 percent calcium, and 12 percent sulfur:

$$3Ca(H_2PO_4)_2 \cdot H_2O + 7CaSO_4 \cdot 2H_2O$$

Rock phosphate can also be treated with phosphoric acid (H_3PO_4) rather than sulfuric acid to produce concentrated (or triple) superphosphate with 20 percent phosphorus, 13 percent calcium, and no sulfur:

$$Ca(H_2PO_4)_2 \cdot H_2O$$

These forms dissolve readily and are quickly available to plants.

Under agricultural conditions, rapid fertilizer availability is an asset. In forestry, prolonged availability at low levels may be preferred to rapid availability. Therefore, ground rock phosphate (with high citrate solubility) may have a bit of an edge over superphosphate, especially on soils with high phosphorus sorption capacity (see Figure 5.8). Differences in response to various forms of phosphorus are usually slight, and relative fertilizer costs usually should guide decisions in forest nutrition management (Allen and Ballard, 1982; Hunter and Graham, 1983; but see also Bengtson, 1976; Torbert and Burger, 1984).

Some forms of fertilizer can be bulk blended for single applications of multiple nutrients. One of the blends used most commonly in forestry, diammonium phosphate (DAP), contains about 24 percent phosphorus and 21 percent nitrogen. In most cases, more nitrogen than phosphorus is needed, so DAP is often mixed with urea or ammonium nitrate to achieve the desired nitrogen:phosphorus ratio in the fertilizer.

SULFUR FERTILIZERS ARE USED MAINLY TO LOWER SOIL pH

Three forms of sulfur fertilizers are commonly used in agriculture: elemental sulfur (S_2), calcium sulfate ($CaSO_4$, gypsum), and diammonium sulfate ($(NH_4)_2SO_4$). Sulfur limitations on forest growth are rare, so the major purpose of sulfur fertilizers in forestry has been to lower the pH of nursery soils. Elemental sulfur oxidizes rapidly to sulfuric acid.

Micronutrient amendments are needed so rarely that their formulations may be tailored for site-specific requirements. Boron deficiencies are usually alleviated with applications of borate salt or borax, with care taken not to increase boron availability to toxic levels. Deficiencies of zinc in Australia are commonly corrected with an aerial spray of a zinc sulfate ($ZnSO_4$) solution or with a mixture of granular superphosphate and zinc. Application of copper salts to soils that are high in organic matter may be much less efficient than additions of copper chelated by organic molecules or copper sulfate ($CuSO_4$) solutions applied directly to foliage (for a discussion, see Stone, 1968, and Bengtson, 1976).

FERTILIZATION WITH "LIME" MATERIAL CAN ALLEVIATE CALCIUM AND MAGNESIUM DEFICIENCIES

Various forms of fertilizer contain calcium and magnesium. The most commonly used forms are types of "lime," which include a basic anion such as carbonate or sometimes hydroxide. Other forms of calcium and magnesium fertilzers work well, but most applications come from mined deposits (or industrial wastes) that are basic. For example, application of 200 kg of gypsum (calcium sulfate) increased the volume of *Eucalyptus grandis* by 50 percent on an Oxisol in Brazil (Barros and Novais, 1996).

A wide variety of industrial wastes can be applied to forest soils to provide cation nutrients and increase growth (Gonçalves et al., 1996). These include "slags" that result from blast-furnace production of iron, "dregs" and "grits" from Kraft processing of wood for pulp, "extinct lime" (sodium hydroxide plus

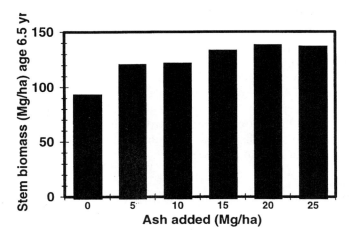

Figure 13.6 Stem biomass response to addition of forest biomass ash in Brazil (data from Moro and Gonçalves, 1995, and Guerrini et al., 1999).

calcium carbonate) from pulp processing, and ash from burning of forest biomass to generate electricity (used either directly as ash or as "burnt ash" following greater oxidation).

For example, Moro and Gonçalves (1996) added a range of wood ash to a plantation of *E. grandis* in Brazil (Figure 13.6). Stem growth increased up to application rates of about 15 Mg/ha of ash (which should contain about 270 kg/ha of calcium and 25 kg/ha of magnesium). Higher rates of application did not increase growth.

RAISING SOIL pH REQUIRES LARGE AMOUNTS OF LIME

In some cases, forest managers may seek to raise soil pH in the expectation of improving tree growth. The growth response to gypsum by Barros and Novais (1996) did not result from an increase in soil pH, as the application rate was far too low to change the pH of an Oxisol. The response resulted from improved calcium and sulfur supplies.

Massive amounts of lime are required to substantially alter soil pH (Figure 13.7). Addition of 1 to 2 mg/ha of limestone is commonly required to change soil pH by 1 unit, and this effect is restricted to the forest floor and the uppermost mineral soil.

Liming does not generally increase tree growth. In Finland, experimental liming of forest soils showed that about 65 to 75 percent of the added calcium remained in the forest floor and upper mineral soil even 25 years after treatment and reduced extractable aluminum by about 20 percent (Derome et al., 1986). The liming treatments did not benefit the trees; in fact, growth

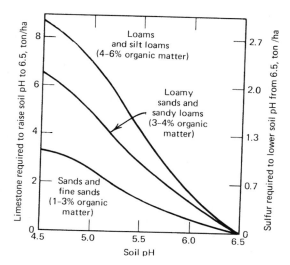

Figure 13.7 Tons (milligrams) of agricultural limestone needed to raise soil pH and tons of sulfur required to lower soil pH as a function of current pH and soil texture. The quantity required is the difference between the curved lines for the current pH and the desired pH.

declined by an average of 3 percent for Scots pine and 10 percent for Norway spruce. The effects of liming in Finland were matched by those in Sweden, leading to the conclusion that liming experiments indicate an overall reduction in growth of about 10 percent for 15 to 20 years (Popovic and Andersson, 1984; Andréason, 1988). In a review of experience from Europe and Sweden, Nihlgård and Popovic (1984) concluded that soil acidification could be prevented with applications of 5,000 kg of lime per hectare but that stand growth would probably decline. Some German studies have reported better growth responses to liming. For example, Spiecker (1991) found that growth of Norway spruce was increased about 6 years after a liming treatment, but the treatment also included addition of nitrogen, so the response may have resulted from either the liming or the nitrogen.

Liming has strong influences on forest biogeochemistry, especially within the forest floor. Liming often (but not always) increases decomposition rates in the forest floor and may double rates of nitrogen mineralization (Persson et al., 1989). In Bavaria, Germany, Kreutzer (1995) found little effect of 4000 kg of dolomitic lime per hectare on the pH of the mineral soil, but the rate of decomposition on the forest floor increased dramatically. The nitrogen content of the forest floor dropped by 170 kg/ha, and soil leachate contained extremely high concentrations of nitrate nitrogen (> 10 mg/L). Liming increased earthworm biomass by eightfold, and also increased root biomass on the forest floor (and decreased root biomass in the mineral soil).

If liming does not increase tree growth, why would foresters invest in the expense of liming? Beese and Meiwes (1995) summarized German perspectives on liming, emphasizing ideas about improving soils through liming, perhaps reversing expected negative impacts of acid deposition even if tree growth does not improve. As noted in Chapter 12, widespread liming of forest soils in Germany is related only partially to soil types and strongly to local beliefs about acidity.

ONLY ABOUT 10 TO 20 PERCENT OF THE FERTILIZER ENTERS THE TREES

Most of the nutrients applied in fertilization never enter the trees. For a variety of pine species, recovery of nitrogen and phosphorus fertilizers in trees has been reported to range between 5 percent (at very high application rates) and 25 percent (Ballard, 1984). The average across 30 studies of fertilizer retention in a range of conifer ecosystems was about 50 percent, with about half of the retained nitrogen entering the trees (Johnson, 1992). The most recent summary of acid-rain related studies with labelled nitrogen showed that about 25 percent of the added nitrogen was taken up by trees (Nadelhoffer et al., 1999). As with many features in forest soils, the variability around these averages is large enough that individual situations may not relate at all to the average. The highest rates of recovery in trees appear in applications with either low application rates or repeated applications.

The rate of nitrogen capture by trees depends on competition between trees, microbes, and chemical processes that stabilize nitrogen in soil organic matter. For example, Chang (1996) added ^{15}N to plots with seedlings of western redcedar, western hemlock, or Sitka spruce and either allowed the understory plants to compete with the seedlings or removed the understory. Removal of the understory increased ^{15}N uptake by the seedlings by two- to eightfold, but even the highest recovery in the seedlings was still just 15 percent of the added nitrogen, compared with 50 percent or greater recovery of the ^{15}N in the soil.

The retention of nitrogen fertilizer with increasing doses was examined in an experiment in Billingsjön, Sweden. Nitrogen was applied at rates of 0 to 600 kg/ha on three occasions to an old (80- to 100-year-old) Scots pine stand over a period of 20 years (Nohrstedt, 1990). The soil (forest floor down to the B2 horizon) retained about 50 percent of the added nitrogen up to a cumulative dose of 1500 kg/ha; addition of 1800 kg/ha decreased nitrogen recovery to about 35 percent (Figure 13.8). The absolute quantity of nitrogen recovered in the soils appeared to reach a maximum of about 700 kg/ha over 20 years. Nitrogen leaching (in soil water at 50-cm depth) was not increased by nitrogen rates of up to 1500 kg/ha, but it increased from background levels of about 1 kg/ha annually to about 4 kg/ha annually in the plots that received 1800 kg/ha of nitrogen. Clearfelling these stands generated no increase in nitrogen leaching from plots that had received less than 800 kg/ha, and plots

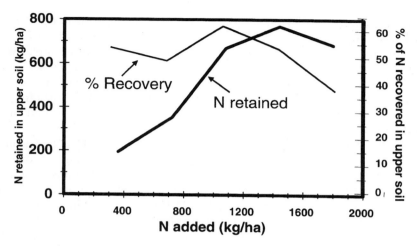

Figure 13.8 Cumulative nitrogen fertilization (over 20 years) of an 80-year-old Scots pine stand in Billingsjön, Sweden, showed high relative retention of fertilizer from the forest floor to the B2 horizon (data from Nohrstedt, 1990).

that had received higher nitrogen additions lost up to 17 kg/ha annually for 3 years after the harvest. Losses of nitrogen through denitrification were negligible (<0.1 kg/ha/yr) for all treatments before and after harvesting (Nohrstedt et al., 1994).

WHERE DID THE MISSING FERTILIZER GO?

Two approaches are commonly used to follow the fate of applied nutrients: biomass and nutrient pool measurements and the use of isotopes. Using the biomass and nutrient pool method, a researcher attempts to measure changes in pool sizes and identify any increases resulting from fertilization. The "pool change" method requires accurate sampling and low within-pool variability. For example, fertilization of loblolly pine trees at age 3 or 4 years showed that about 10 percent of the added nitrogen was found in the pines 2 years later, with about 5 to 15 percent in the herbaceous plants. This approach works for small vegetation with fairly low nitrogen content, but the nitrogen content of large trees cannot be estimated with enough precision to allow the presence of any extra nitrogen to be determined. Some very intensive studies have met with success in accounting for fertilizer recovery. For example, addition of 400 kg/ha of nitrogen to a radiata pine forest increased plant nitrogen uptake by 166 kg/ha (this 40 percent recovery in trees is unusually high), another 145 kg/ha remained in the soil, and 60 kg/ha was lost in soil leachate (Raison et al., 1990). The remaining 35 kg/ha (9 percent of the applied nitrogen) remained unaccounted for.

A few nitrogen fertilization studies have used the stable isotope ^{15}N as a label. The ratio of the isotopes ^{15}N and ^{14}N can be determined later for each pool and the contribution of fertilizer nitrogen calculated. For example, Melin et al. (1983) applied 100 kg/ha of ammonium nitrate to 130-year-old Scots pine, with either the ammonium or the nitrate labeled with [15]N. The total recovery in their single-tree plots was about 80 percent of the applied nitrogen for both forms, but about 10 percent more of the nitrate fertilizer was found in the vegetation after 2 years. The missing 20 percent of the fertilizer was either volatilized or leached below the 30-cm depth of the soils. Nadelhoffer et al. (1999) synthesized results from nine forests in North America and Europe that received additions of labelled ^{15}N to investigate ideas about excessive levels of nitrogen inputs. Less than 10 percent of the added nitrogen was lost in soil leachates at seven of the nine sites, and the other two sites (which experienced large rates of nitrogen deposition from the atmosphere) leached about 15 to 50 percent of the added nitrogen. Across all sites, the authors estimated that about 20 to 25 percent of the added nitrogen entered trees, 70 percent was retained in the soil, and 10 percent was lost to leaching or as gases.

The general pattern of fertilizer distribution is as follows:

1. Less than a quarter of the fertilizer is taken up by trees in the first few years.

2. Another quarter or so is immobilized in microbial biomass and soil organic matter.

3. A variable and difficult-to-measure amount is lost from the forest ecosystem through leaching and volatilization.

WHY DOESN'T MORE FERTILIZER GET INTO THE TREES?

This is a difficult question, but sample calculations (after Miller, 1981) may illustrate some of the factors. Consider a forest soil with 5000 kg/ha of nitrogen, 1 percent of which (50 kg/ha) mineralizes annually and is taken up by trees. About one-quarter of the 50 kg/ha goes into biomass increment, and the rest returns to the soil in litter and dead roots. In the first growing season following fertilization with 200 kg/ha of nitrogen, availability could be 250 kg/ha. This rate would exceed the uptake capacity of the trees. If uptake increased by 50 percent, then less than 15 percent of the added nitrogen would move into the trees. Of the remaining fertilizer, most is immobilized by the microbes; only a small fraction is lost. The soil nitrogen capital is now about 5175 kg/ha, and if mineralization patterns have not been altered, the nitrogen available to trees in the second growing season after fertilization would be 1 percent of 5150, or 51.5 kg/ha. In general, tree recovery of added nitrogen can be explained by increased uptake in the first season after treatment (Miller, 1981).

Fertilization may have more of a residual effect on nitrogen mineralization than this sample calculation would indicate, but the general trend probably holds unless additions are large relative to total soil nitrogen.

Given such low recovery of nitrogen fertilizers by trees, why are such large amounts of fertilizer added in operational fertilization programs? Smaller quantities may be largely immobilized by soil microbes, leaving little for tree uptake. Johnson and Todd (1985) applied urea nitrogen to recently planted stands of loblolly pine and yellow-poplar at a rate of 100 kg/ha once a year for 3 years, or 25 kg/ha four times a year for 3 years. Both species responded better to the once-a-year applications than to the quarterly applications, probably because microbial immobilization removed most of the 25 kg/ha dose but a lower proportion of the 100 kg/ha dose.

FERTILIZATION INCREASES GROWTH FOR PERIODS RANGING FROM 5 YEARS TO ENTIRE ROTATIONS

As noted in Chapter 12, operational fertilization commonly increases wood production by 3 to 10 m^3/ha annually for a period of 5 to 10 years. Responses to phosphorus fertilization sometimes last for more than a decade, whereas nitrogen responses usually taper off after a few years. What accounts for the difference? Part of the answer may simply relate to the application rate. Fifty kg/ha of phosphorus represents about 25 years of accumulation in pine biomass, whereas 100 kg/ha of nitrogen might represent only 5 years' worth. The rates of phosphorus additions may be large enough relative to the capital of "available" phosphorus in the soil to sustain a prolonged increase in phosphorus availability. However, the mechanisms by which stands respond to fertilization have been examined in only a few studies, and response patterns are open to a variety of interpretations.

FERTILIZER RESPONSES MAY DIFFER AMONG SPECIES AND GENOTYPES

Genetic differences among individuals and strains within species plays a major role in tree growth and nutrition. Most genetic selection programs focus on overall growth and stem form, without direct attention to interactions between genotype and nutrients. The importance of nutrition to genetic selection is illustrated by comparing two clones of *Eucalyptus grandis* from Brazil (Barros and Novais, 1996). The better-performing clone obtained twice as much phosphorus from the soil (and fertilizer) than the poorer clone and produced twice as much stem biomass. The poorer clone obtained only 30 percent of the potassium used by the better clone.

Superior genotypes may not be superior in all environments, so tree breeders are often concerned with Genotype × Environment interactions (Zobel and Talbert, 1984). Nutrient availability is part of this environmental component. A number of experiments have examined the Genotype × Nutrient interaction by measuring the growth response to fertilization of genetically selected trees. Some species, such as loblolly and slash pines, have shown no significant interactions (Matziris and Zobel, 1976; Rockwood et al., 1985). For these species, superior trees respond to fertilization with the same volume increment exhibited by other genotypes. Other species, such as radiata pine in Australia, show marked differences in fertilization response among genotypes (Waring and Snowdon, 1977; Nambiar, 1984). For example, the poorest-responding family of radiata pine in one study increased its growth by only 9 percent in response to nitrogen fertilization, whereas the best-responding family doubled its growth (Fife and Nambiar, 1995). Fertilization improved water use efficiencies of all families, with no interacton between water use and nitrogen response among families.

What mechanisms account for Genotype × Nutrient interactions? It is possible that some genotypes are more efficient at utilizing nutrients. In agricultural plants, this commonly involves both greater growth per kilogram of nutrient obtained from the soil and greater uptake from the soil (Saric and Loughman, 1983). However, variations in nutrition among tree genotypes have so far been related simply to differences in uptake rather than to different nutrient-use efficiencies (Nambiar, 1984; Barros and Novais, 1996). The primary mechanism appears to be genetic variations in root systems that result in variations in nutrient uptake. Increased rooting density (centimeters of roots and mycorrhizae per volume of soil) should enhance the uptake of nutrients, especially those of intermediate or low mobility, such as phosphate or ammonium (Barley, 1970; Bowen, 1984).

SITE PREPARATION MAY INCREASE THE FERTILIZATION RESPONSE

The interactions of site preparation and tree nutrition vary with site-specific combinations of factors, such as soil types, treatment intensities, and regeneration methods. Site preparation can affect the response to early-rotation fertilization in two ways: by altering the need for supplemental nutrient sources and by allowing seedlings to use more nutrients.

Site preparation may increase the ability of seedlings to utilize nutrient amendments. Soil bedding is a common practice in wet flatland areas. Bedding plows raise ridges of soil (15 to 30 cm high) for planting seedlings. This operation improves soil aeration, decreases soil resistance to root penetration, and generally accelerates seedling growth. In the Coastal Plain of North Carolina, seedlings in bedded plantations often respond better to phosphorus fertilization than do those in unbedded sites. For example, Gent et al. (1984) reported that the average growth response to bedding + phosphorus roughly equaled the sum of the response to each treatment individually. In an economic analysis, however, the combined response was worth 25 percent more than the sum of the separate treatment values. The benefits of bedding may decline with time, as the development of a full canopy increases stand evapotranspiration and may remove excess soil water as effectively as ditches.

FERTILIZATION MAY AFFECT NONTARGET VEGETATION

Understory production usually increases after fertilization unless the overstory is too dense to allow a response. In the "Garden of Eden" study in California, Powers and Ferrell (1996) provided an abundant nutrient supply to young ponderosa pine trees, with and without herbicide control of competing vegetation. At their Whitmore site, fertilization (with 1000 kg/ha of nitrogen plus a full suite of other elements) had little effect on pine volume, but the biomass of understory shrubs increased from 6.2 Mg/ha in control plots to 16.4 Mg/ha in the fertilized plots. Removal of competing vegetation increased pine volume by 3.5-fold by increasing the water and nutrient supply to the pines. Competition control combined with fertilization further increased the pine volume to 4.5 times the volume of the control plots.

Browse for deer and other animals is usually increased by fertilization, especially when applied in combination with thinning (Rochelle, 1979). Forage production for cattle may also increase; studies in the southeastern United States have found that fertilization at the time of plantation establishment increases forage yield by 350 to 5500 kg/ha annually for 5 years. Fertilization at the time of thinning may increase forage yields by 650 to 2000 kg/ha annually for a couple of years (Shoulders and Tiarks, 1984).

A very interesting effect of fertilization in Douglas-fir stands was reported by Prescott et al. (1993); they found that repeated fertilization greatly reduced understory salal shrubs, a major competitor with the trees (Figure 13.9). The mechanisms behind this surprising pattern are being explored.

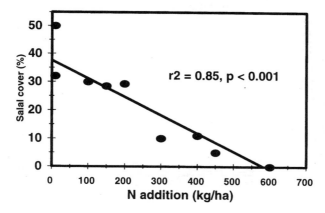

Figure 13.9 Additions of nitrogen decreased the cover of the understory salal shrubs, in Douglas-fir stands (after Prescott et al., 1993).

FERTILIZATION USUALLY HAS MINOR EFFECTS ON WATER QUALITY

Streamwater concentrations of nutrients may increase following fertilization as a result of direct input into streams or leaching through the soil. Most increases remain well below drinking water standards (10 mg nitrogen per liter as nitrate in the United States and Canada, 11.3 mg nitrogen per liter as nitrate in Europe), but some studies have found peak concentrations of nitrate nitrogen as high as 10 to 25 mg/L for short periods (Figures 13.10 and 13.11) and for short distances downstream.

Average concentrations increase much less than peak concentrations, and forest fertilization studies have found that average concentrations have not risen to levels that would threaten drinking water quality (Binkley et al., 1999). Nutrient input to streams varies with the formulation of the fertilizer; ammonium nitrate typically leads to higher nitrate nitrogen concentrations than does urea (Figure 13.12). Repeated fertilization may lead to sustained, high concentrations in streamwater, but this possibility has not been explored. Phosphate fertilization poses no drinking water threat, but increased concentrations may allow for transient increases in the productivity of aquatic ecosystems. All of the studies summarized in Figure 13.11 examined temperate forests; given massive rates of fertilization in tropical regions such as Brazil, it will be important to determine if these patterns hold for warmer, wetter situations.

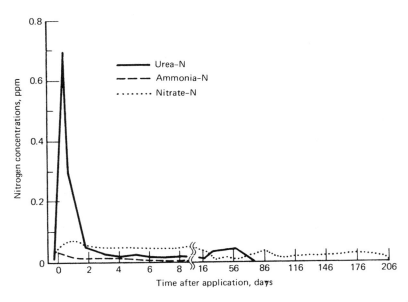

Figure 13.10 Average concentration of nitrogen forms following fertilization of two watersheds with 225 kg/ha in Washington (ppm = mg/L; from Moore, 1975).

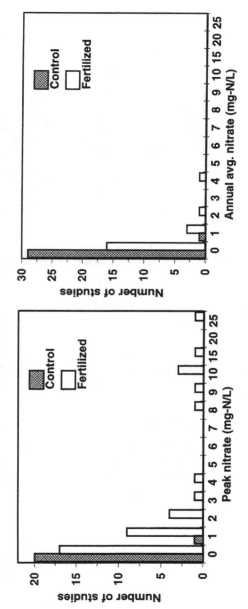

Figure 13.11 Nitrogen fertilization usually increases the peak concentration of nitrate with maximum values <2 mg/L. Several studies have documented peak increases of more than 10 mg/L. Fertilization has much less effect on the annual average concentrations of nitrate (from Binkley et al., 1999).

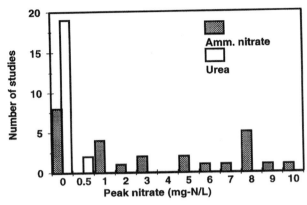

Figure 13.12 Springwater nitrate nitrogen concentrations in Sweden following fertilization with 115–175 kg/ha nitrogen (data from Tamm et al., 1974).

WASTE DISPOSAL ON FORESTLANDS MAY INCREASE STAND PRODUCTIVITY

Two types of waste materials are commonly disposed of on forest lands. Wastewater results from secondary treatment of sewage; solid materials have been settled out, and the water has been aerated to kill microbes and reduce the biological oxygen demand (metabolizable carbon compounds) in the water. Sewage sludge is the solid residual material that usually has been composted aerobically for a short period and partially digested under heated, anaerobic conditions. Both wastewater and sludge are rich in nutrients and can greatly increase forest production.

The costs of transport and application exceed the value of increased wood production, but disposal on forest lands may be a low-cost alternative to other disposal methods. For example, setting up a 25-ha irrigation system on forest land to dispose of wastewater from a town of 5000 people might cost $5000/ha, with an annual operation cost of $1200/ha (estimated from Young, 1979, and Myers, 1979). This cost far exceeds the value of increased tree growth.

In a case study on the use of sewage sludge, Schreuder et al. (1981) estimated the net present value (discounted at 7 percent; see Chapter 9 for an overview of economics) of sludge disposal on forest lands was about −$7000/ha. Disposal in a sanitary landfill might have a value of −$10,000, or $3000 more costly than disposal on forest land. The growth response to sludge treatments in this case might increase value by $325/ha to $675/ha over 2 years. This value increase is large in relation to forest production but negligible relative to the cost differences between disposal methods.

WASTEWATER IS A RICH NUTRIENT SOLUTION

A good example of the impacts of wastewater irrigation on forest lands comes from experiments on the Pack Forest at the University of Washington. In one study (Schiess and Cole, 1981), municipal wastewater with nutrient concentrations listed in Table 8.2 was applied to plots with (1) no vegetation, (2) poplar cuttings, (3) Douglas-fir seedlings, or (4) grass. The water was applied at the rate of about 5 cm/week. Retention varied greatly among vegetation types in the second year of irrigation. The barren plot retained only 35 percent of the 400 kg/ha of nitrogen added; most of it was oxidized to nitrate and leached past the 180-cm soil depth. Denitrification probably removed an additional unmeasured amount. The poplar plot retained 95 percent of the added nitrogen, and the Douglas-fir plot retained about 85 percent of the added nitrogen. The grass plot accumulated much less nitrogen in biomass, retaining only about 75 percent of the added nitrogen.

A major concern with wastewater irrigation is the leaching of nutrients through the soil and into aquatic ecosystems. In the Washington studies, no phosphate was lost from any of the plots due to the high specific adsorption capacity of the soil. Sulfate retention equaled about 60 percent of the added sulfur and also differed little between plots. The vegetation had the largest effect on nitrate concentrations of soil leachate (at 180-cm depth): 13 mg/L of nitrate nitrogen for barren plots, 0.5 mg/L for poplar, 2.3 mg/L for Douglas-fir, and 4.3 mg/L for grass. In this experiment, only the barren plot averaged higher nitrate concentrations than the maximum standard for human consumption. Other studies found similar results; nitrate leaching losses are generally low, but in cases of high application rates or poor vegetation cover, the rates may be unacceptably high (see various chapters in Sopper and Kerr, 1979).

The University of Washington researchers also found that wastewater stimulated growth much better than riverwater. After 4 years, poplars irrigated with wastewater averaged 70 Mg/ha compared to 7 Mg/ha with riverwater irrigation. Douglas-fir biomass in the same treatments was 34 8 Mg/ha, and biomass in grass plots was 40 and 8 Mg/ha. Similar increases in aboveground biomass production have been found in all studies of wastewater irrigation on forests.

SEWAGE SLUDGE RESEMBLES SOIL HUMUS

The solid residue from municipal wastes consists mostly of undecomposed organic materials and microbially synthesized compounds — the same features that characterize soil humus. A major difference between sludge and soil organic matter is that sludge has higher concentrations of salt and heavy metals. Sludge from industrial sources (rather than domestic sources) is particularly high in heavy metals.

Extensive research on the University of Washington Pack Forest has provided some of the most complete information available on the effects of sludge on forest nutrition. The municipal sludge from Seattle was composed of 26 percent carbon, 2.3 percent nitrogen, 1.5 percent nitrogen, 1.5 percent phosphorus, 0.16 percent potassium, 0.40 percent calcium, 0.29 percent magnesium, 2000 mg/kg zinc, 700 mg/kg copper, 1200 mg/kg lead, 50 mg/kg cadmium, and 150 mg/kg nickel (Zasoski, 1981). The heavy metal content was low to average for sludge from mixed domestic/industrial sources. The fresh sludge was also very high in water-soluble salts, including 530 mg/L of ammonium nitrogen and 150 mg/L of sodium. These salts leached rapidly from the sludge into the soil; after 15 months, the water-extractable ammonium had declined to 7 mg/L and sodium to 9 mg/L.

The decomposition rate of the sludge layered on top of the forest floor (in depths ranging from 5 to 20 cm) was surprisingly slow. Only 12 percent of the sludge weight was respired in the first year and even less in the second year (Edmonds and Mayer, 1981). After 2 years of decomposition, about 65 percent of the original nitrogen and 80 to 90 percent of the original phosphorus content remained in the sludge. Only a fraction of the mineralized nitrogen (about 2 to 3 percent) reached the soil; most (>95 percent) was either volatilized as ammonium (due to the high pH of the sludge) or denitrified (Vogt et al., 1981). At first glance, it might seem, therefore, that nitrate leaching through the soil should not represent a threat to water quality. A 10-cm layer of sludge, however, contained about 7500 kg/ha of nitrogen, so nitrate leaching from the sludge into the soil totaled about 35 to 70 kg/ha of nitrogen annually. This rate of nitrate input to the mineral soil probably would be retained by vegetation uptake and microbial immobilization, posing little threat to water quality. Heavier applications might exceed the retention capacity of the ecosystem.

Tree growth in a 55-year-old Douglas-fir stand on a poor site responded very well to the application of a 5- to 10-cm layer of sludge. Both thinned and unthinned control stands produced about 3900 kg/ha of stem wood over 2 years (calculated from data given by Archie and Smith, 1981). Thinned and unthinned stands treated with sludge produced about 5300 kg/ha of stemwood over the same period.

These studies demonstrate that forests effectively filter nutrients from wastewater and sludge if the rate of nutrient input to the soil does not exceed the retention ability of the ecosystem. Schiess and Cole (1981) estimated that accumulation of nitrogen in biomass by Douglas-fir trees could reach a maximum of 150 to 250 kg/ha annually and that belowground accumulation could add another 200 kg/ha annually. Wastewater application rates exceeding this maximum would probably lead to increased concentrations of nitrate in drainage water, as would excessive leaching of nitrate from sewage sludge. The ability of an ecosystem to retain nitrogen might also decline after many years of application due to increased rates of nitrogen cycling within the ecosystem.

HEAVY METAL CONCENTRATIONS IN SLUDGE
ARE A SERIOUS CONCERN

Sludge from domestic wastes is high in nutrients and generally low in heavy metals. Most municipal sewage, however, combines domestic and industrial wastes, so most sludge contains significant amounts of metals such as zinc, lead, and cadmium. Some sludge materials are used in agricultural soils, with careful attention paid to the heavy metal content and the total application rate. Forestlands are often considered a better choice than agricultural soils for disposal of sludge, as accumulation of heavy metals in trees is less worrisome than accumulation in vegetables for human consumption.

The heavy metal content of sludge does not prevent large increases in tree growth. For example, Bledsoe and Zasoski (1981) grew conifer seedlings of five different species in sludge and in mixtures of sludge and soil. In all cases, seedlings in soils amended with sludge grew much better than seedlings without sludge. The seedlings with sludge did accumulate higher concentrations of heavy metals (particularly zinc and cadmium), but these levels did not impair growth.

The addition of heavy metals to forest ecosystems also results in accumulations in ecosystem components other than trees. Zasoski (1981) examined the metal content of understory plants in a Douglas-fir plantation amended with sludge. Blackberry shrubs in sludge plots had twice as much lead and six times as much cadmium as shrubs in control plots. Thistles accumulated even more metals; lead concentrations increased 11-fold and cadmium concentrations 15-fold.

Sludge applications to forests also affect wildlife. Animal populations may benefit from increased production of browse plants or be harmed by a diet high in heavy metals. West et al. (1981), working in the Washington Pack Forest, found that the use of sludge-treated forests by deer averaged 18 deer per square kilometer, compared with 6 deer per square kilometer for a control area. Sludge areas also showed higher rates of fawn production: 1.7 fawns per doe versus 0.9 fawn per doe in a control site. Heavy metal contents of livers and kidneys were higher in deer from sludge plots (especially cadmium), but the levels were still quite low.

In general, the heavy metal content of sludge should pose little threat to forest ecosystems, but an awareness of the potential problems (and of the metal content of the sludge used in each situation) is important.

NUTRITION ASSESSMENT PROVIDES INFORMATION
FOR DECISION MAKING

All of the nutrition assessment methods involve some degree of uncertainty in their relationships with stand nutrition, and in many cases a method may fail to meet scientific criteria of significance. Scientists typically want to be certain that hypotheses have at least a 90 or 95 percent probability of being correct before rejecting the null hypothesis that an assessment method is unrelated to stand nutrition. Researchers will reject a method that gives only an 80 percent guarantee of being correct, but many forest management decisions involve greater uncertainty; any likely trends may have value.

Forest Nutrition Management Involves Unavoidable Uncertainty

Forest nutrition management entails manipulations of ecological systems to meet social objectives. The objectives are usually measured in dollars or other monetary units.

Quantification of the growth response to fertilization is much simpler than calculation of the economic value of the treatment. A case study with loblolly pine illustrates the major components of an economic analysis. The discussion starts with some basic definitions and simple calculations and then becomes progressively more realistic and complex.

The case study involves a 1-ha stand of 30-year-old loblolly pine with a site index of about 20 m at 25 years. Density was 750 stems per hectare, with an average diameter and height of 25 cm and 28 m. The stand was fertilized with 100 kg/ha nitrogen as urea and 50 kg/ha phosphorus as triple superphosphate at a total cost of \$120/ha. Fertilization increased the yield by 17 m³/ha over 5 years, and average diameter increased to 30 cm. The following analysis assumes that the management of the unfertilized stand is profitable by itself and that the added value of fertilization can be judged on an incremental basis.

If all biomass were harvested at age 35 years and used as pulpwood, the extra wood would be worth \$106 (at \$6.25/m³, or about \$15 per cord) at the time of harvest. This increased value would not cover the cost of fertilization, so complex calculations are not needed to show that the investment would lose money.

Fertilization becomes more attractive if some of the biomass can be used as chip 'n' saw (cutting a few small boards and chipping the rest) or sawtimber. About 25 percent of the biomass would go for pulpwood, 45 percent to chip 'n' saw, and 30 percent to sawtimber. The added value of pulpwood (25 percent of 17 m³) would be \$27. The chip 'n' saw value would be an extra \$111, and sawtimber would add another \$122. The total increased stumpage value from fertilization would be \$260/ha, or \$140 more than the cost of fertilization.

Since money in the present is considered more valuable than money at a future date, some form of interest or discount rate needs to be included. Two common approaches have been used to include a factor for time preference in the evaluation of forest investments. Net present value (NPV) calculations take the value of a resource at some future date and calculate the current value based on a chosen interest (or discount) rate:

$$NPV = \text{Future value} \times (1 + \text{interest rate})^{-year} - \text{current costs}$$

In the loblolly pine example, \$260/ha would be discounted over 5 years at the chosen rate of interest. Fertilization would be profitable only if the present (discounted) value of the harvest exceeded the present cost of fertilization. If an annual return of 7 percent were desired, $260/ha after 5 years would have a current value of $185/ha, for an NPV of $65/ha after subtraction of the fertilization cost. The NPV at a 5 percent discount rate would be $83/ha, or $49/ha at a 9 percent rate.

The second approach, called the internal rate of return (IRR), does not specify a desired interest rate. The IRR is simply the interest rate at which the NPV (after subraction of costs) equals zero. Ignoring the time factor, the investment of $120/ha yields $260/ha, or a 117 percent return on the investment. With simple interest calculations, 117 percent after 5 years would equal a 23 percent return for each year of the investment. Annual compounding of interest, however, would reduce the rate of return to 17 percent annually. If an interest rate of 17 percent were chosen, the current value of the extra biomass would be $120/ha, or $0/ha after subtraction of fertilization costs:

$$-120 + 260/(1 + 0.17)^5 = 0$$

Economists generally prefer to evaluate alternative investments by ranking the NPV of each option. The investment with the highest NPV (calculated with a discount factor representing the firm's cost of capital) will provide the highest profit. The interpretation of IRRs can be more complex, and IRRs are generally considered by economists to be less informative than NPVs.

INFLATION AND CHANGES IN STUMPAGE VALUES MAY ALTER PROFITABILITY

An annual rate of return of 10 percent may be attractive in periods of low inflation, but it could be uninspiring in times of high inflation. Fortunately for forestry investments, the price of timber stumpage typically keeps pace with inflation. Between, 1950 and 1980, the price of pine pulpwood in the southeastern United States matched the rise in the consumer price index at a rate of 3.5 percent annually. The price of pine sawtimber increased at a rate of 6.2 percent annually (Johnson and Smith, 1983), or 2.7 percentage points more each year than the rate of inflation. For simplicity, investment analyses may assume that increased stumpage value matches inflation and calculate values based on "real" dollars. Alternatively, both stumpage and inflation may be explicitly included and comparisons examined in "nominal" dollars. The choice of accounting procedure should match the form used to evaluate competing uses of investment capital.

Returning to the loblolly pine case study, what would be the effects of 5 years of inflation at 5 percent? If no increase in stumpage occurred, the real rate of return would drop from 17 to 11 percent, and the NPV (in current dollars, discounted at 7 percent + 5 percent inflation) from $65/ha to $28/ha. If pulpwood stumpage increased at the rate of inflation, and if chip 'n' saw exceeded inflation by 2 percent and sawtimber by 4 percent, the real rate of return would be 20 percent (NPV of $93/ha).

COMPOUND INTEREST FAVORS LATE-ROTATION FERTILIZATION

Nutrition management must combine ecologic responsiveness at varying ages with the economic context of compounding interest to determine the optimal time of fertilization. In the loblolly pine case study, what if the same gain in value occurred in the loblolly pine stand over 10 years rather than 5 years? The rate of return (without inflation or taxation) would drop from 16 to 8 percent, and the NPV (at 7 percent) would be $12/ha. A 30-year investment period would drop the rate of return to under 3 percent. These calculations illustrate the attractiveness of fertilizing a few years before harvesting.

Fertilization early in a rotation generally can be justified for short rotations (such as in tropical plantations) or if the growth response is very strong and sustained. Phosphorus fertilization of loblolly and slash pine plantations typically provides growth responses that last for several decades. Fertilization is justified economically because accelerated growth rates exceed the growth of compound interest. For example, fertilization with phosphorus at the time of planting of loblolly pine may increase the site index from 30 m to 50 m. This rise represents a yield increase of about $42 \, m^3/ha$ over 30 years, or $1.4 \, m^3/ha$ annually. The mean annual increment of unfertilized stands was about $2.4 \, m^3/ha$, so fertilization gave an average annual increase in growth of almost 60 percent. Responses of this magnitude certainly keep pace with the compounding of interest on the fertilization investment.

GROWTH RESPONSES ARE UNCERTAIN

In the loblolly pine case study, the gain from fertilization was given as $17 \, m^3/ha$. This value is the average of a series of 104 installations of the North Carolina State Forest Nutrition Cooperative. Some stands increased their growth by more than $17 \, m^3/ha$, and a few even showed reductions in growth after fertilization (due largely to increased mortality). Across the wide range of sites examined, the standard deviation of growth responses was about $14 \, m^3/ha$. Therefore, about two-thirds of all installations responded within $\pm 14 \, m^3/ha$ of the mean of all sites, and about 95 percent fell within $\pm 28 \, m^3/ha$ (Figure 13.13).

About 11 percent of the stands responded with a decrease in growth. These trials covered a wide range of stand conditions, so stratification of sites into categories of responsiveness might reduce response variability. Some stands are not limited by nitrogen or phosphorus supplies, and others are so densely stocked that fertilization accelerates mortality. For these reasons, the fertilization of each stand involves uncertainty about the magnitude of the response.

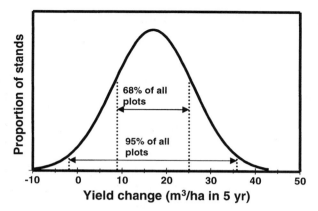

Figure 13.13 Distribution of the 5-year-growth response to fertilization, with an average response of 17 m³/ha across stands and a standard deviation of 14 m³/ha.

Formal methods are available to aid in making decisions under uncertainty and are well suited to decision making in forest nutrition management. The discussion here is limited to uncertainty in level of response, but these decision-making approaches can also be used to evaluate the importance of uncertainty about costs and values.

BREAKEVEN ANALYSIS IS THE SIMPLEST APPROACH

If the magnitude of the growth response to fertilization is uncertain, a manager may simply ask how large the response must be to cover the cost of fertilization (including the discount rate for the investment). In the loblolly pine case study, fertilization cost $120/ha, and the value of extra growth was about $15.30/m³ (without considering taxes). To account for the 5-year time period of the investment, a 7 percent discount rate would mean that the NPV of 1 m³ 5 years in the future would be $10.90. Therefore, the breakeven point would occur where the NPV of the extra growth equaled $120, or about 11 m³/ha. Since this value is about 35 percent less than the regionwide average response to nitrogen + phosphorus fertilization, a decision to fertilize would probably be wise. In fact, if a very large number of stands is to be fertilized, enough stands should respond to cover the cost of fertilizing those that did not. If only a small area is to be fertilized, the odds of hitting very responsive or very unresponsive stands are high. As more areas are fertilized, the overall response should approach an average level. Further, even if a large area would ensure overall profitability, it might be desirable to identify unresponsive stands and increase profits by efficient allocation of fertilizer. These issues illustrate the economic incentive behind developing a framework for making the best decisions under uncertainty.

A DECISION TREE IDENTIFIES CHOICES, PROBABILITIES, AND OUTCOMES

If a manager faced a decision on fertilizing all stands or none, the breakeven analysis presented above would indicate that she should fertilize if she expected an average response of 11 m³/ha or more. In reality, she would be happiest to fertilize only stands that respond with more than 11 m³/ha, saving the investment on stands that do not reach the breakeven response level. If the true average of 17 m³/ha were distributed normally across all stands with a standard deviation of 14 m³/ha, then about 67 percent of the stands would exceed the breakeven point (Figure 13.14). Integrating the area under the curve above 11 m³/ha shows that the NPV of fertilizing only responding stands is about $145/ha. The average response would now be about 24 m³. What would be the cost savings if the unresponsive stands were not fertilized? At first glance, it might appear that the savings would match the cost of fertilization ($120/ha). However, the true savings would be somewhat less because many stands that did not achieve the breakeven point would still show some increased growth. Integrating the curve below 11 m³/ha shows an average savings of $98 for each unresponsive hectare not fertilized.

These numbers can be put into a decision-tree framework to illustrate the choices faced by the manager (Figure 13.15). If she decides to fertilize all stands (the upper fork of the tree), 67 percent of the stands would respond above the breakeven point and 33 percent would not. Multiplying these proportions by the value of each type of stand gives an overall value of this decision:

$$0.67(\$145/ha) + 0.33(-\$98/ha) = \$65/ha$$

Note that the combined value of $65/ha is the same value that was calculated

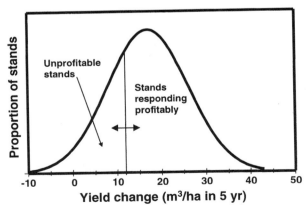

Figure 13.14 If the breakeven response were 11 m³/ha after 5 years, then about 67 percent of the stands would respond profitably.

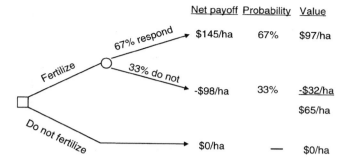

Figure 13.15 Decision tree representing the choice between fertilizing all stands or none if the average response is 17 m³/ha.

originally for this case study. If the manager had decided not to fertilize (the lower fork of the tree), no cost would be incurred and no response obtained. As the upper fork of the decision tree yielded a value $65/ha greater than the lower fork, deciding not to fertilize would have an opportunity cost of $65/ha.

FERTILIZATION ENTAILS TWO KINDS OF RISK

This example shows that there are two kinds of wrong decisions that managers might make. The first type (I) would be to fertilize a stand that would not respond with at least a breakeven increase in yield. Equally real is the second type (II), where a potentially responsive stand is not fertilized. Fertilizing a nonresponsive stand in this case costs $98/ha on average; accounting for the fact that only 33 percent of the area falls into this category gives an average cost (Type I error) of $32/ha across the entire area. Failure to fertilize a responsive stand (with a value of $145/ha on 67 percent of the area) is a Type II error with an average cost of $97/ha. In this case, the Type II error is three times as costly as the Type I error, and an optimal manager would rather err on the side of fertilizing unresponsive areas to obtain the increased value from fertilizing every responsive area.

In reality, managers often prefer to incur Type II errors rather than Type I errors, as only Type I errors lose money out of pocket. Managers are typically less accountable for Type II errors, in which potential profits are missed but no current funds are lost. Nonetheless, this basic decision framework provides the best available estimates of costs of both types of wrong decisions, allowing managers to apply their own criteria for the acceptability of each type of error.

What if the true average response were 0 m³/ha rather than 17 m³/ha? Assuming that the standard deviation remained at 14 m³/ha, the distribution in Figure 13.16 shows that about 22 percent of fertilized stands would still

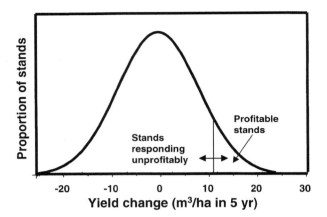

Figure 13.16 If the average response were $0 \, m^3/ha$ with a standard deviation of $14 \, m^3/ha$, then 22 percent of the all stands would respond above the breakeven level of $11 \, m^3/ha$.

exceed the breakeven point. The average value of response would be $78/ha for each responding area. The cost of fertilizing unresponsive areas would now be $156/ha, which exceeds the actual cost of applying the fertilizer. This is because half of the stands would show less growth after fertilization.

The decision tree in Figure 13.17 demonstrated that a decision to fertilize would show a value of $78/ha on 22 percent of the areas and $-$156/ha on 78 percent of the areas. Combining these figures would give $-$120/ha for fertilization of the entire area. The Type I error in this instance would be $-$122/ha ($0.78 \times -$156/ha). Failure to fertilize a responsive stand would give a Type II error of only $-$17/ha ($0.22 \times $78/ha). A wise manager in this

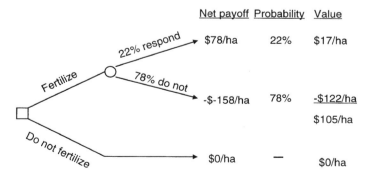

Figure 13.17 Decision tree representing the choice between fertilizing all stands or none if the average response were $0 \, m^3/ha$ (see text).

case would clearly prefer to minimize the Type I risk and would rather accept a Type II risk.

WHAT IS THE VALUE OF PERFECT INFORMATION?

In the case where a manager knows that the average response to fertilization would be $17\,m^3/ha$, the value of fertilizing all areas is $65/ha. If the manager knows precisely which stands would respond and which would not, she would obtain a value of $145/ha on 67 percent of the areas without incurring the cost of $98/ha on 33 percent of the areas. The net value would be $97/ha across all areas, so the value of perfect information on stand response would be $97/ha minus the $65/ha that would be obtained if all areas were fertilized, or $32/ha.

If the manager knows that average response would be worth $0/ha across all areas, she would probably choose not to fertilize. If perfect information were available, only the responding stands would be fertilized, with a net value of $17/ha.

In the first situation, the manager would find it profitable to spend up to $32/ha in an assessment program to determine the likelihood of response for each area. If the land base were 10,000 ha, then a research investment of up to $320,000 might be justified. In the second situation, the value of perfect predictions would be $17/ha, or $170,000 for 10,000 ha. Any assessment program is unlikely to predict responsiveness with 100 percent accuracy, so a smaller budget would actually be justified.

In both situations, the manager knew in advance what the average response and proportion of responding stands would be and needed information only on which stands happened to fall into each category. If the average response were unknown, then the decision framework would become more complicated. The expected value of perfect information is the difference between the value of perfect decisions and the value of decisions made with existing information or assumptions. As an example, if a manager thought the average response to fertilization would be $0\,m^3/ha$ when the true average was $17\,m^3/ha$, she would decide not to fertilize (with a value of $0/ha). Perfect information in this case would allow a return of $145/ha on 67 percent of the area, or $97/ha. The value of perfect information rises from $32/ha, where the manager already knew the average response, to $97/ha where the average response was unknown and was incorrectly assumed to be zero. For a 10,000-ha area, a research effort might be supported with up to $970,000 rather than $320,000. Conversely, a manager expecting an average response of $17\,m^3/ha$ when the real average was $0\,m^3/ha$ would lose $120/ha rather than make a profit of $17/ha with perfect information.

In general, the value of perfect information is high if prior assumptions about responses differ greatly from actual responses. If perfect information merely shows that prior assumptions were correct, then the research that provided the information would have added little value.

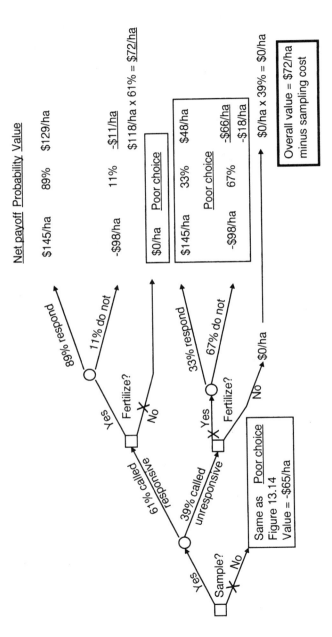

Figure 13.18 Decision tree representing the choice to fertilize with 80 percent accuracy in a response prediction, assuming that an average response of 17 m³/ha and 67 percent of stands are truly responsive. An "X" marks the poorer choice of each fork.

WHAT IS THE VALUE OF IMPERFECT INFORMATION?

No site factors correlate perfectly with fertilization response, so the expected value of perfect information will always be greater than the realized value of research. In fact, a little knowledge can be dangerous in some cases.

Consider two assessment methods — one that distinguishes between responsive and unresponsive stands (average response of $17 \, m^3/ha$) with 80 percent accuracy (correctly identifies 80 percent of the responding stands and 80 percent of the unresponding stands) and another with 60 percent accuracy. Two decisions are involved at this point: whether to sample for response assessment and whether to fertilize. If a decision is made not to sample, the lowest fork of the tree in Figure 13.18 is the same as the tree diagrammed in Figure 13.15. If a sampling for response assessment is chosen, then 80 percent of the 67 percent that are actually responsive will be identified as responsive. Therefore, 61 percent of all stands would be fertilized. Of these, 89 percent should actually respond (the probability of stand responding when it is called responding) and 11 percent would not. The payoff from responding stands would be $145/ha and that from nonresponding stands would be $-$98/ha, for a net payoff of $118/ha. As only 61 percent of the area would have been fertilized (identified as responsive), the overall value of this branch of the decision tree would be $72/ha. The lower fork of this decision would result in not fertilizing the 61 percent of the area identified as responsive and would, of course, have a value of $0/ha. The better decision, then, is the upper fork, which exceeds the lower one by $72/ha.

The next step in the analysis in Figure 13.18 is to make decisions about the 39 percent of all stands identified as unresponding. If they are fertilized anyway, 33 percent will actually respond despite the prediction and 67 percent will not. This yields a net payoff of $-$18/ha, which is $18/ha less than the option of not fertilizing. Combining the $72/ha for the fork identified as responsive with the $0/ha value for the unresponsive fork gives an overall value of $72/ha. The value of fertilization with the assessment program is therefore $72/ha, less the cost of performing the assessment. The value without the assessment (Figure 13.15) was $65/ha, so sampling would be warranted if 80 percent accuracy in predictions could be obtained for less than $7/ha.

What if the sampling method provided only 60 percent accuracy in distinguishing between responsive and nonresponsive stands? The net value from fertilizing the stands that were predicted to be responsive would be $45/ha. Many of the stands that were predicted to be unresponsive would actually respond well, giving a positive value of $19/ha for fertilizing the "unresponsive" stands. The total value of fertilization would be $65/ha (the same as in the base case, as the decision would lead to fertilization of all hectares) minus the cost of sampling.

This comparison of the accuracy of nutrition assessment methods illustrates the importance of clear decision-making frameworks in the evaluation of the profitability of nutrition treatments. The accuracy of the assessment is most important when the value of a responding stand differs substantially from the cost of fertilization. If the value greatly exceeds the cost, it may be better to fertilize all areas rather than risk missing some responsive areas due to an inaccurate prediction. If the cost is large relative to the value of a responding stand, high accuracy may be needed to minimize the risk of fertilizing unresponsive stands. If the cost and value are similar, then any predictive edge may be helpful.

THE VALUE OF RESEARCH DEPENDS ON EXISTING KNOWLEDGE AND FOREST SIZE

These examples also illustrate the importance of prior knowledge or assumptions in estimating the value of new information. If additional information merely indicates that good decisions would have been made based on existing knowledge, then little value is gained. If existing knowledge or assumptions are erroneous, then additional research may prove very profitable. The value of research in forest fertilization would vary according to the results of a pilot study (with fertilization of various nutrients on a few representative sites):

1. If the mean response is large and the variance (or standard deviation) is small, all areas should be fertilized, with little concern about predictions of response.
2. If both the mean and the variance are small, fertilization would probably be unprofitable.
3. If the mean is small but the variance is large, an assessment program might increase the profitability of fertilization.
4. If both the mean and the variance are large, a large investment in an assessment program is probably justified.

The value of research also depends on the size of the area (and the number of years) to which it will apply. Some forest nutrition research might be too costly if apportioned over a small area, whereas a great deal of research can be justified for land holdings of several hundred thousand hectares. This fact has been exploited by forest nutrition cooperatives, in which each member company gains benefits from the economy of scale of combined land holdings.

FERTILIZATION IS AN INVESTMENT OF ENERGY AS WELL AS MONEY

Decisions about fertilization are typically based on costs and expectations of the return on investment. The response to fertilization can also be examined in terms of energy invested and energy returned in the form of wood biomass. From the simplest standpoint, the energy costs can be tabulated and compared with the energy content of the extra wood obtained from fertilization. The internationally accepted unit for energy is the Joule (J); 1 million J (MJ) equals

950 BTU, or 240 kcal. The synthesis of nitrogen fertilizers involves the use of natural gas energy to reduce N_2 to NH_3, which is then processed into urea or ammonium nitrate. The total energy cost of synthesis is about 50 MJ per kilogram of nitrogen in urea. Mining phosphate minerals has an energy cost of about 1.6 MJ/kg of phosphorus, and transformation into superphosphate (8 percent phosphorus) or triple superphosphate (20 percent phosphorus) adds 2.8 MJ/kg and 8.0 MJ/kg of phosphorus, respectively (Pimentel and Pimentel, 1979).

In the loblolly pine case study presented earlier, 100 kg/ha of nitrogen and 50 kg/ha of phosphorus (as triple superphosphate) were used. The energy content of this application would be 5000 MJ/ha for nitrogen and 480 MJ/ha for phosphorus. The next step is transport to the forest. Switzer (1979) estimated the cost at about 0.0064 MJ per kilogram of nitrogen in urea for each kilometer; assuming that the factory is 800 miles away, the transportation energy for 100 kg of nitrogen sums to 50 MJ. Including the phosphorus fertilizer would raise transportation energy costs to 75 MJ. Application from a helicopter would cost another 180 MJ (Switzer, 1979). Loblolly pine plantations are typically fertilized from tractors, so the application cost is probably close to 50 MJ/ha. The growth response averaged 17 m^3/ha, and harvesting this amount (assuming that harvest energy is proportional to biomass removed) would require about 9000 MJ. The energy content of softwood is about 15,000 MJ/m^3, or 250,000 MJ for 17 m^3/ha. Summing the costs of synthesis, transportation, application, and harvest, the energy cost of fertilization + harvest is about 14,600 MJ/ha. The ratio of wood energy to the energy cost of the operation is 250,000:14,600, or about 17:1. In contrast, the return in energy in wheat or corn in response to fertilization typically falls between 5:1 and 10:1 (calculated from Pimentel and Pimentel, 1979).

SUMMARY

Most forests increase in growth rate following fertilization, and many cases provide substantial profits. The increased stem growth following fertilization derives from increases in leaf area and net primary production; increases in foliage photosynthesis; and shifts in allocation of photosynthate from root production to stem production. Fertilization also changes stand characteristics, accelerating growth, dominance, and self-thinning of stands. Nitrogen fertilizers are synthesized from air, using natural gas as an energy source. Other types of fertilizers come from mining, with varying degrees of purification and treatment before use. Most of the fertilizer applied to forests remains in the soil, with about 20 to 25 percent entering trees. Experiments in forest nutrition and fertilization include uncertainties about the responsiveness of stands and of groups of stands. Formal methods of approaching management decisions are available to guide the application of imperfect information from experiments to real-world decisions about stand management.

CHAPTER 14

Nutrition Management: Biological Nitrogen Fixation

Biological nitrogen fixation is reduction of atmospheric nitrogen (N_2) to ammonia (NH_3), which is then processed into proteins and other biochemicals (see Chapter 9). Only a few prokaryotic organisms are capable of performing this reaction, either as free-living organisms or in symbiosis with higher plants.

Some biological nitrogen fixation occurs in all forests, but usually the rates are low relative to precipitation inputs and tree requirements. Forests with species capable of symbiotic nitrogen fixation, however, may have nitrogen fixation rates that rival the annual uptake requirement for nitrogen. Some nitrogen fixers, such as red alder and black locust, can be used directly for commercial products. In other cases, nitrogen fixers may be used to increase the growth of interplanted crop trees. Crop trees mixed with nitrogen-fixing species experience increased nitrogen availability but may suffer from competition for other site resources. The value of silvicultural systems with nitrogen-fixing species depends on the balance between enhanced nitrogen nutrition of the crop trees and increased competition for other resources. Biological nitrogen fixation can be a useful silvicultural tool and is a potential alternative to nitrogen fertilization. Like all tools, however, it is not appropriate for all situations. The choice between nitrogen fixation and nitrogen fertilization requires an understanding of the ecologic and economic effects of both sources.

NITROGEN FIXATION WAS HARNESSED IN FORESTRY SHORTLY AFTER ITS DISCOVERY

The first attempts to increase forest growth through nitrogen fixation used lupines in a Scots pine plantation in Lithuania in 1894 (discussed by Mikola et al., 1983), just a few years after the discovery of nitrogen fixation in legumes. By the turn of the century, lupines were being used operationally in the restoration of forests degraded by litter removals. One of the few long-term quantitative experiments on lupines in forestry examined applications of lime, potassium, and phosphate fertilizers at the time of planting of Norway spruce, Japanese larch, and Scots pine (Siebt, 1959, summarized in Baule and Fricker,

1970). Some plots were also sown with lupine seeds, and other plots were fertilized repeatedly with nitrogen and potassium. After 25 years, the volume of spruce in lime + potassium plots was only 40 m³/ha compared with 75 m³/ha for plots treated with lime + phosphorus + nitrogen + potassium. Spruce volume on lime + phosphorus + potassium + lupine plots was 150 m³/ha. Japanese larch and Scots pine also grew best with lupines, but the increase over fertilized plots was only 50 percent (pine) to 75 percent (larch). Siebt attributed the lower response of these species to greater competition between the tree seedlings and vigorous lupines.

This early study did not compare the effects of nitrogen and lupines on tree growth without concurrent additions of lime, phosphorus, and potassium. A substantial growth response to nitrogen alone may have been possible, but lupine establishment probably required both lime and phosphorus additions. Further, subsequent studies found responses to vary greatly between sites (see Rehfeuss et al., 1984). For these reasons, the use of lupines in forestry has declined in Germany over the past two decades (Rehfeuss, 1979), although lupine use is not uncommon in some other parts of Europe (Mikola, et al., 1983).

In New Zealand, sand dunes have been stabilized with a mixture of lupines and marram grass since the 1940s. This system uses the grass to stabilize the dunes, while lupines fix about 160 kg/ha of nitrogen annually. Radiata pine seedlings are planted later, and the trees rapidly shade out the lupines. Buried lupine seeds sprout when the stand is thinned. This silvicultural system has been used to create about 60,000 ha of productive forest land in New Zealand (Gadgil, 1983; see also Figure 15.5). In one study from New Zealand, nitrogen fertilizer increased the stem basal area more than lupines did, but the basal area was greatest where both treatments were combined (Figure 14.1).

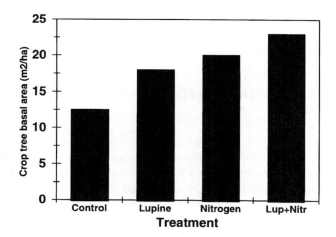

Figure 14.1 Basal area accumulation of radiata pine in New Zealand in relation to addition of lupines, nitrogen fertilizer, or both (data from Gadgil, 1983).

Biological nitrogen fixation clearly has the potential to increase forest growth, but this nitrogen source is used very little in current forest management. This is due in part to the problems associated with increasing the complexity of stand management, but also to lack of awareness of the potential benefits of nitrogen fixation. Successful harnessing of nitrogen fixation requires an understanding of the ecosystem processes that regulate the growth response of crop trees.

NITROGENASE ENZYME REDUCES NITROGEN GAS TO AMMONIA

The nitrogen fixing enzyme nitrogenase reduces atmospheric nitrogen to ammonia through addition of electrons and hydrogen ions:

$$N_2 + 8e^- + 8H^+ \rightarrow 2NH_3 + H_2$$

Because electrons are added to N_2, this is a reduction process requiring energy input. The production of hydrogen gas represents an unproductive energy drain, and some nitrogen-fixing organisms have a hydrogenase system that oxidizes hydrogen to form water and release energy. The theoretical energy requirement is about 35 kJ per mole of nitrogen fixed, but it is difficult to assess the actual cost under field conditions. A good approximation of the actual cost might be about 15 to 30 g of carbohydrates (6 to 12 g carbon) for every gram of ammonia nitrogen produced (Schubert, 1982; Marshner, 1995). About half of the carbohydrate cost is for the nitrogen fixation process, and the balance is for the growth and maintenance of nodules and for assimilating and exporting the fixed ammonium. The nitrogenase enzyme is also very sensitive to oxygen, and the high-energy requirements coupled with the need for oxygen protection set limits on nitrogen-fixing systems. Interestingly, a substantial amount (25–30 percent) of the CO_2 released during the nitrogen fixation process can be "refixed" inside the root nodules by PEP carboxylase.

THREE TYPES OF NITROGEN FIXERS MAY BE IMPORTANT IN FOREST SOILS

Nitrogen fixation can be performed only by certain strains of prokaryotic (lacking a nucleus) microbes. Cyanobacteria are photosynthetic, filamentous bacteria that use solar energy to fix N_2. They protect nitrogenase from oxygen by forming special nitrogen-fixing cells (heterocysts) with thickened walls. Other single-cell bacteria, such as *Clostridium* and *Azotobacter*, use energy obtained from decomposing organic matter to fix nitrogen. Unfortunately, the heterotrophs achieve oxygen protection through rapid respiration of carbohydrates to convert oxygen to carbon, consuming large amounts of energy. In some cases, these bacteria may be concentrated around roots (in the rhizosphere), where they obtain carbon energy sources from plants. This loose arrangement is called associative nitrogen fixation. A more highly developed system for fixing nitrogen involves plants that form nodules to house bacteria (*Rhizobium* and *Bradyrhizobium* for legumes) or actinomycetes (*Frankia* for alder, *Casuarina*, and other non-legumes). This symbiotic system supplies the endophyte with energy from the host plant, and the plant, in turn, receives an internal supply of nitrogen. The nodules even contain a form of hemoglobin to regulate oxygen concentrations.

SYMBIOTIC NITROGEN FIXATION CAN BE USED IN NUTRITION MANAGEMENT PROGRAMS

Free-living nitrogen fixation systems typically supply only small quantities of nitrogen to forests. Free-living bacteria have too much trouble coping with low-energy supplies and crippling concentrations of oxygen. In periodically flooded soils such as rice paddies, the associative nitrogen-fixing bacteria are protected from oxygen by anaerobic soil conditions and are supplied with energy from leaky plant roots. Rice paddies show nitrogen fixation rates on the order of 30 kg/ha annually. Unfortunately, associative nitrogen fixation has not been shown to be very important in well-aerated forest soils.

Cyanobacteria may contribute small amounts of nitrogen to forest soils, but measurements of their contributions usually are not separated from those of heterotrophic bacteria. In some forests, epiphytic lichens such as *Lobaria* species combine cyanobacteria with green algae and fungi to fix a few kilograms per hectare of nitrogen annually. Free-living systems face severe environmental constraints, and generally rates cannot be improved by active management.

Symbiotic nitrogen fixing systems may add more than 100 kg/ha of nitrogen annually to forests (Table 14.1). The two major groups of nitrogen-fixing plants are legumes and actinorhizal plants. Nodules on legumes such as *Robinia*, *Leucaena*, *Albizia*, and *Acacia* house bacteria of the genera *Rhizobium* or *Bradyrhizobium*. Actinorhizal species such as *Alnus* (alder), *Casuarina*, and *Ceanothus* contain actinomycetes of the genus *Frankia* in their nodules.

Rates of nitrogen fixation vary substantially across sites for the same nitrogen-fixing species. For example, 30 estimates of nitrogen fixation for red alder show that the most commonly reported rates are between 40 and 80 kg/ha of nitrogen annually (Figure 14.2). Twenty percent of the studies estimated rates higher than 100 kg/ha of nitrogen annually. No substantial differences are evident for rates in pure stands and stands mixed with conifers or poplars.

The rates of symbiotic nitrogen fixation depend on the number and vigor of the host plants, the degree of nodulation, and the environmental factors that influence overall plant carbohydrate balance. Some species, such as alders, always have large quantities of nodules and relatively high rates of nitrogen

TABLE 14.1 Representative Rates of Nitrogen Fixation (kg/ha Annually) in Forests

Type of Nitrogen Fixation	Location	Vegetation	Rate (kg/ha of Nitrogen annually)	Reference
Free-living	Southeastern USA Forest floor	Loblolly pine hardwoods	<0.01	Grant and Binkley (1987)
	Pacific Northwest USA	Douglas-fir		Heath (1985), Silvester et al. (1982), Denison (1979)
	Forest floor		0.1–1	
	Mineral soil		<0.1	
	Decaying logs		1.4	
	Canopy lichens		3–4	
	Montana/Idaho, USA Wood residue	Fir, hemlock, cedar	0.1–1.4	Jurgensen et al. (1992)
	Forest floor		0.01–0.07	
	Soil wood		0.05–0.22	
	Mineral soil		0.5–0.7	
	British Columbia	Lodgepole pine	0.3–0.6	Wei and Kimmins (1998)
Legumes	New Zealand	Lupines in radiata pine plantations	90–160	Gadgil (1976)
	Eastern USA	Black locust	35–300	Jencks et al. (1982), Boring and Swank (1985)
		Understory herbaceous legumes	<1 low density; 5–10 high density	Hendricks and Boring (1999)

Location	Species	Rate	Reference
Hawaii, USA	*Albizia*	75–150	Garcia-Montiel and Binkley (1998), Kaye et al. (1999)
British Columbia, Canada	Lupine in lodgepole pine	2	Henrickson and Burgess (1989)
Tropics	*Albizia*	65–260	Reviewed in Binkley et al. (1997)
	Leucaena	100–180	
Pacific Northwest USA	Red Alder	50–150	Reviewed in Binkley et al. (1994)
Actinorhizal	Sitka alder	20–150	Binkley (1981), Heilman and Ekuan (1983)
	Ceanothus spp.	0–110	Binkley and Husted (1983 summary)
Nepal, India	Himalayan alder	30–120	Sharma (1993)
Sweden	Gray alder	43	Johnsrud (1979)
Europe, Canada	Black alder	16–60	Akkermans and van Dijk (1979), Cote and Camire (1984)
British Columbia, Canada	*Shepherdia* in lodgepole pine	0.8	Hendrickson and Burgess (1989)
Tropics	*Casuarina*	35–95	Reviewed in Binkley et al. (1997)

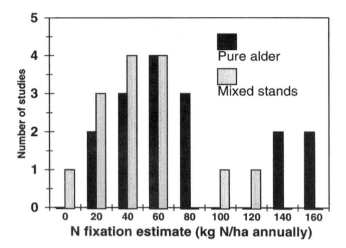

Figure 14.2   Estimates of nitrogen fixation rates for red alder in the northwestern United States and British Columbia using a variety of methods on a wide range of sites (data from Binkley et al., 1994).

fixation/plant. The *Frankia* endophyte appears to be ubiquitous. One gram of soil from Oregon may contain about 100,000 *Frankia* (Myrold and Huss-Danell, 1994). Other species, such as actinorhizal *Shepherdia* and *Purshia*, are much more variable in nodulation.

NITROGEN FIXATION RATES ARE DIFFICULT TO MEASURE

Nitrogen fixation rates can be estimated by a variety of methods, with four approaches used commonly in forest research: ^{15}N methods, nitrogen accretion, chronosequences, and acetylene reduction assays. The addition of highly enriched ^{15}N to soils may provide the best estimates of nitrogen fixation, but the cost of sufficient ^{15}N to label soils in the field has limited the use of this method for trees. Some processes in nitrogen cycling alter the ratio between naturally occurring ^{15}N and ^{14}N, and if certain key assumptions are met, this "natural abundance" approach can provide inexpensive estimates of nitrogen fixation. Unfortunately, the key assumptions are probably not met in many (if any) forests. Nitrogen accretion is also a simple approach that involves measuring the total nitrogen content of an ecosystem at two points in time. The difference between the samplings represents the addition of nitrogen from fixation, plus any precipitation inputs. This accretion method assumes that outputs are negligible and underestimates true nitrogen fixation rates on sites that experience high nitrate leaching or denitrification. The chronosequence approach is really a variation of the accretion approach, but with more assumptions. Several sites of different ages are used to represent trends to be expected within one site over time. If all site factors are constant except for stand age, then spatial patterns can be interpreted to represent temporal patterns. Finally, the acetylene reduction techniques take advantage of the nitrogenase (nitrogen-fixing) enzyme's preference for reducing acetylene to ethylene (both of which are easily measured) rather than the reduction of dinitrogen to ammonia (not measurable without ^{15}N methods).

^{15}N LABELING SHOWS HIGH RATES OF NITROGEN FIXATION FOR *LEUCAENA* AND *CASUARINA*

The best use of the ^{15}N addition method for estimating nitrogen fixation comes from the work of J. Parrota, D. Baker, and M. Fried on plots in Puerto Rico (Parrotta et al., 1996). These investigators planted pure and mixed replicated plots of *Eucalyptus*, *Leucaena*, and *Casuarina*. Small subplots were labeled with ^{15}N so that nitrogen derived from the soil would be substantially enriched in ^{15}N relative to nitrogen derived from the atmosphere. The ^{15}N:^{14}N ratio in *Eucalyptus* trees represented the isotope ratio of nitrogen derived from the soil, and the lower enrichment in the nitrogen-fixing trees allowed separation of the nitrogen fixer's reliance on the atmosphere and soil. This approach showed that *Leucaena* fixed about 90 percent of its nitrogen supply through age 2 years and about 60 percent of its nitrogen supply by age 4 years; *Casuarina* derived about 50 to 60 percent of its nitrogen from the atmosphere for both periods. This decline in nitrogen derived from the atmosphere for *Leucaena* was relative; the quantity of nitrogen fixed remained steady at about 70 to 75 kg/ha annually, but the uptake of nitrogen from the soil increased as the tree demand for nitrogen increased.

NATURAL ABUNDANCE OF ^{15}N IS AFFECTED BY NITROGEN FIXATION IN FORESTS

In the atmosphere, ^{15}N comprises about 0.3664 percent of the N_2, and ^{14}N comprises 99.6336 percent. The isotope ratio is soils is typically enriched in ^{15}N because the minor difference in the mass of the nitrogen atoms favors the reaction, mobility, and loss of lighter ^{14}N. If a soil is sufficiently enriched in ^{15}N relative to the atmosphere, it might be possible to use the natural abundance of nitrogen isotopes to estimate the input of nitrogen from fixation.

Natural abundance levels of ^{15}N are often gauged as delta ^{15}N (or σ^{15}N), which equals the parts per thousand deviation of a sample from the percentage of ^{15}N in the atmosphere. A soil sample that contained 0.3710 percent ^{15}N would have a σ^{15}N of 4.6 $((0.3710-0.3664) \cdot 1000)$. As a first approach, the contribution of the atmosphere to the nitrogen economy of a nitrogen-fixing tree could be estimated by comparing the σ^{15}N of foliage with the σ^{15}N of total soil nitrogen. The reality is much messier. The $\sigma^{15}N$ within trees may differ by several per mil from the roots to the leaves, and the $\sigma^{15}N$ of soil total nitrogen often increases by several per mil with depth. Even worse, the $\sigma^{15}N$ of soil ammonium and nitrate commonly do not match that of the total soil nitrogen, or each other. One approach to circumventing these problems is to use a non-nitrogen-fixing reference tree to gauge the $\sigma^{15}N$ of available soil nitrogen. For example, total soil nitrogen in a pure Douglas-fir plantation had a $\sigma^{15}N$ of 5.8, significantly higher than the 2.3 $\sigma^{15}N$ for a mixed stand of red alder and Douglas-fir (Binkley et al., 1985). Accounting for the higher percentage of nitrogen in the mixed stand soil, these data suggest that the alder had contributed about 60 percent of the nitrogen present in the 0- to 15-cm soil depth. The confidence in this pattern had to be tempered, however, with data from two other sites where the $\sigma^{15}N$ of total soil nitrogen was too close to the atmospheric level to be useful and by a fourth site where the $\sigma^{15}N$ in the mixed alder–Douglas-fir stand was *more* enriched in ^{15}N.

In a recent review, Högberg (1997) suggested that differences in $\sigma^{15}N$ might need to be greater than 5 per mil before interpretations about nitrogen fixation could be made with confidence.

NITROGEN ACCRETION AND CHRONOSEQUENCES CAN CATCH LARGE RATES OF NITROGEN FIXATION

These two approaches can be illustrated by several research projects on snowbrush (*Ceanothus velutinus*) in the Cascade Mountains of Oregon. Youngberg and associates (1979) intensively characterized the nitrogen pools of a recently logged site and resampled for nitrogen accretion several times over the next 20 years. Zavitkovski and Newton (1968) chose a chronosequence approach in the hope that differences among sites of varying ages would track the nitrogen accretion occurring within each site, providing a rapid estimate of nitrogen fixation. Youngberg et al. (1976, 1979) found that nitrogen accumulated in vegetation + soil (to 23 cm) at a rate of about 110 kg/ha annually for the first 10 years, and at 40 kg/ha annually for the next 5 years. Zavitkovski and Newton (1968) found a very similar pattern (in nitrogen content of vegetation + soil to a 60-cm depth), despite high variability among sites (Figure 14.3). A regression equation of ecosystem nitrogen content with age shows a significant slope ($p < 0.02$) in the chronosequence series of sites of about 80 kg/ha annually. A 95 percent confidence interval around this relationship ranges from 20 to 140 kg/ha annually. However, Zavitkovski and Newton found few nodules on any plants younger than 5 years, and concluded that the three youngest stands were probably unrepresentative of the rest of the sequence. Deleting these stands, they concluded that nitrogen fixation was negligible in the chronosequence.

The chronosequence method can be converted into the accretion method if sites are resampled at a later date. It is hoped that each ecosystem will progress

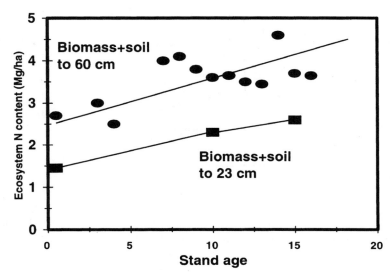

Figure 14.3 Nitrogen fixation estimates for snowbrush based on a chronosequence (ovals, based on Zavitkovski and Newton, 1968) and at a nearby site where accretion was followed within a single site over time (squares, based on Youngberg et al., 1979).

along the trajectory established for the chronosequence. Using this approach, 6 of the original 13 sites examined by Zavitkovski and Newton were resampled 15 years later. Five sites showed no significant change in soil (0- to 15-cm depth) nitrogen content, but one showed a very significant increase of 280 kg/ha (17 kg/ha annually; Figure 14.4). Some nitrogen accretion may have occurred in other ecosystem pools, but it is unlikely that high rates of nitrogen fixation took place without evidence of increased nitrogen in the upper soil. This resampling showed that the apparent trend in the original chronosequence was not in fact a reliable estimate of nitrogen fixation for these sites, and that the sites in the chronosequence probably differed fundamentally from the single site sampled by Youngberg. As with most features of forest soils, site specific details were important in addressing the rates of nitrogen fixation by snowbrush.

These snowbrush examples illustrate the difficulties experienced in obtaining estimates of nitrogen fixation rates and the importance of methodological assumptions and limitations. Researchers at the Hubbard Brook Experimental Forest in New Hampshire designed a nitrogen accretion study in which they hoped to control the major sources of variation tightly enough to allow them to estimate rates of free-living nitrogen fixation by the accretion method. Bormann and others (1993) created "sandbox" ecosystems by filling boxes (7.5 × 7.5 m) lined with Hypalon polymer to a depth of 1.5 m with gravel (for the bottom 0.15 m), glacial outwash sand (1.3 m deep), and topsoil (0.05 m deep). High rates of nitrogen accretion under black locust (100 kg/ha/yr of

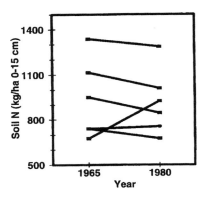

Figure 14.4   Fifteen-year changes in total soil nitrogen (0- to 15-cm depth) in six stands sampled by Zavitkovski and Newton (1968) and resampled in, 1980 (Binkley and Newton, unpublished data). Only one of the sites showed a significant accretion of nitrogen; overall, the average annual rate of nitrogen accretion across all sites was -0.7 kg/ha annually.

nitrogen) and black alder (270 kg ha^{-1} yr^{-1} of nitrogen) after 3 to 6 years were not surprising. However, the researchers found an unaccounted-for increase in total sandbox nitrogen content under red pine and pitch pine of about 55 kg ha^{-1} yr^{-1}. Intensively sampled measurements of atmopsheric deposition suggest that only about 10 kg ha^{-1} yr^{-1} of the extra nitrogen could be attributed to rainfall inputs. Where did the other nitrogen come from? Acetylene reduction assays (a method described below) of roots showed a maximum rate of 1.8 nmol of ethylene production per gram of root in an hour, and the authors concluded that this very high rate of rhizosphere nitrogen fixation probably accounted for the large nitrogen accretion under pines.

How much confidence is warranted in this conclusion about free-living nitrogen fixation? Two approaches can be used: considering the implications of these findings for other situations and double checking whether the observed rate of acetylene reduction supports the conclusion. If pine forests could obtain 40 to 50 kg/ha/yr of nitrogen) when growing on nitrogen-deficient soils (such as the glacial outwash sand in this study), it would be surprising how commonly nitrogen-deficient pines are found on poor soils and how strongly they respond to nitrogen fertilization (R. Powers, personal communication, 1998).

Is an acetylene reduction rate of 1.8 nmol per gram of root per hour a respectably rapid rate? These sorts of assays are fraught with methodological problems, including disturbance effects and oxygen partial pressures. However, a back-of-the-envelope calculation can be used, with generous assumptions, to see if this rate could be extrapolated to the same ballpark as the accretion estimate. First, let's use the highest rate observed (rather than the average), 1.8 nmol/g^{-1}/hr, and assume that this rate pertains for 24 hr/day, 180 days/year, and that the fine root biomass is 3000 kg/ha. Converting from acetylene reduction to N_2 reduction requires division by 3 to account for the number of electrons needed per mole of gas. This generous extrapolation gives 7.8 moles of N_2 fixed per hectare annually, or 0.2 kg ha^{-1} yr^{-1} of nitrogen. This maximum estimate is still two orders of magnitude lower than the estimated accretion.

What should be concluded from the Hubbard Brook sandbox experiment? The estimated rate of accretion cannot be explained by the observed rate of rhizosphere nitrogen fixation, which leads to two questions: what would we find if we examined rhizosphere nitrogen fixation in similar systems, and what would happen if the sandbox accretion experiment were repeated? The first question was answered by Barkman and Schwintzer (1998), who looked for free-living nitrogen fixation in 18 pine stands in the same region; they found an overall average rate of 0.06 kg/ha of nitrogen annually. The second question is harder to address; forest soil scientists have not rushed to repeat the sandbox experiment.

We note that a variety of studies (some weak, some strong) have reported substantial rates of nitrogen accumulation that cannot be accounted for in the measured nitrogen budgets. A recent resampling of a forest site in Tennessee found that over a period of 15 years, about 675 kg/ha of nitrogen had accumulated beyond the estimated rate of atmospheric deposition (Johnson and Todd, 1998). The annual rate of nitrogen accumulating from unaccounted-for sources was about 45 kg/ha of nitrogen, very similar to the result of the sandbox experiment.

ACETYLENE REDUCTION ASSAYS ESTIMATE CURRENT RATES OF NITROGEN FIXATION

The fourth approach to estimating rates of nitrogen fixation also involves large sources of variation. If the average rate of nitrogen fixation per gram of nodule (activity rate) were known, an estimate of the annual nitrogen fixation rate could be obtained by multiplying the rate by the nodule biomass per hectare. Nodule activity usually is assessed with an acetylene reduction assay, in which the nitrogenase enzyme reduces acetylene to ethylene rather than reducing nitrogen gas to ammonia:

$$C_2H_2 + 2H^+ + 2e^- \rightarrow C_2H_4$$

Acetylene and ethylene are easier to measure than the nitrogen compounds. Recalling that nitrogen fixation requires the addition of $6e^-$ for every mole of N_2 consumed, it would appear that the reduction of 3 moles of acetylene would be equivalent to 1 mole of nitrogen. However, H_2 gas is an unavoidable drain on the nitrogen-fixing system when N_2 is reduced, but no generation occurs when acetylene is reduced. Another potential artifact in the acetylene reduction method is called nitrogenase derepression. Normally, the product of nitrogen fixation (ammonia) has a negative feedback on further synthesis of the nitrogenase enzyme. This feedback balances the rate of nitrogen fixation with the plant's ability to process ammonia into proteins. With acetylene reduction, no ammonia is produced, no feedback develops, and the synthesis of extra nitrogenase is uninhibited. For these reasons, the conversion factor for acetylene to nitrogen is not always constant, but usually averages 3 to 10 moles of C_2H_2 per mole of N_2 under carefully controlled incubation conditions.

Nodule activity also varies on a diurnal cycle; rates usually decline at night (Figure 14.5). Seasonal cycles are also important (Figure 14.6), as are patterns in water availability. It is very difficult to account for all these sources of variability; most studies simply try to perform assays on many nodules at many points in the growing season in the hope of hitting a reasonable approximation of average rates. In most cases, it would be fortunate to achieve a confidence interval of ± 50 percent on a seasonal estimate of average nodule activity.

Nodule biomass per plant is easy to determine on small plants by excavating entire root systems, but large plants present mammoth problems. Ecosystems with large plants usually are sampled for nodule biomass by digging pits or taking soil cores at random across the site. The soil then is sieved to collect all nodules. The variability is again quite high, and a standard error of the mean is likely to be 25 to 50 percent of the mean.

Because high variability is unavoidable with the acetylene reduction method, these estimates should be viewed as very rough approximations. Acetylene reduction assays can be useful in gauging the general magnitude of nitrogen fixation (such as less than 20 kg/ha or more than 100 kg/ha). They are probably most valuable for assessing the physiological aspects of nitrogen fixation, such as response to light and water regimes.

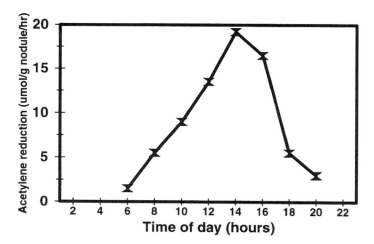

Figure 14.5 The rate of nitrogen fixation by nodules of Himalayan alder begins to rise after canopy photosynthesis begins in the morning, and tapers off as evening approaches (after Sharma, 1988).

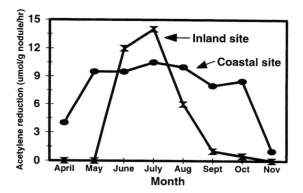

Figure 14.6 Nitrogen fixation by nodules on red alder trees occurred for a longer period at a coastal site in Oregon with a mild, wet climate compared to nitrogen fixation red alder trees at a higher-elevation inland site in Washington (data from Binkley et al., 1992a).

How well do rates of nitrogen fixation from acetylene reduction approaches compare with nitrogen accretion estimates? An adjacent 55-year-old pair of stands of Douglas-fir and Douglas-fir mixed with red alder were compared by the accretion method, and the current rates of nitrogen fixation were estimated by the acetylene reduction method (Wind River site from Binkley et al., 1992). At the less fertile site, the biomass and soil (to 0.9 m) in the mixed stand contained about 2800 kg/ha more nitrogen than the pure Douglas-fir stand, for an annual average rate of accretion of about 54 kg/ha of nitrogen. Leaching losses were about 20 kg/ha of nitrogen annually higher from the mixed stand, raising the estimated rate of nitrogen fixation to 74 kg/ha annually. The seasonal average rate of acetylene reduction was 8.3 μmol/g/hr, and the nodule biomass was 325 kg/ha. The rate of acetylene reduction at a hectare scale would be 8.3 μmol/g/hr · 325,000 g/ha of nodules · 2880 hours per season, divided by 3 moles of C_2H_2/mol N_2, times 28 g/mol of nitrogen, which totals to 73 kg/ha of nitrogen for the season. This close match between methods is coincidental in part. A similar comparison at another site yielded an accretion-based estimate of 102 kg of nitrogen fixed per hectare annually and a current acetylene reduction-based rate of 85 kg of nitrogen fixed per hectare annually (Binkley et al., 1992).

FAVORABLE ENVIRONMENTS ALLOW HIGH RATES OF NITROGEN FIXATION

Rates of symbiotic nitrogen fixation are highly variable; in general, conditions that favor plant growth promote nitrogen fixation. Nitrogen fixation is usually favored by high light intensities, adequate moisture supplies, warm temperatures, and an adequate nutrient supply (especially phosphorus). High rates of fixation occur only when nitrogen fixers are maintained in unshaded conditions.

DOES NITROGEN FIXATION DECLINE AS SOIL NITROGEN INCREASES?

Although nitrogen fixation by agricultural legumes does tend to decrease as availability of soil nitrogen increases, no feedback inhibition has been shown for actinorhizal species under field conditions. Several greenhouse studies showed that high concentrations of ammonium or nitrate can decrease nitrogen fixation, but the concentrations used in these studies far exceeded soil solution concentrations under field conditions. Abnormally high concentrations of ammonium or nitrate around the roots may have misleading effects on nitrogen fixation. For example, Ingestad (1981) found that a high supply rate of nitrogen to alder roots at a low concentration actually stimulated nitrogen fixation rates.

Under field conditions, alders have been found to fix substantial quantities of nitrogen on forest soils with some of the highest nitrogen contents in the world (Franklin et al., 1968; Binkley et al., 1985; Sharma et al., 1985). The high rate of nitrate leaching from many alder forests (sometimes in excess of 50 kg/ha of nitrogen annually) provides a final piece of evidence for lack of strong feedback regulation by nitrogen availability on nitrogen fixation. As described above, the ^{15}N experiment with *Leucaena* and *Casuarina* in Puerto Rico showed that the proportion of the plant nitrogen derived from nitrogen fixation declined as the old trees made more use of soil nitrogen, but the rate of nitrogen fixation per hectare remained high (Parrotta et al., 1996).

Why would plants expend energy to fix nitrogen on sites where soil nitrogen was abundantly available? No conclusive evidence is available, but some speculations can be advanced. Nitrogen fixation requires more energy than does assimilation of ammonium, so ammonium uptake might appear to be more efficient than nitrogen fixation. Nitrate assimilation costs may be similar to those of nitrogen fixation. The picture is actually more complex, as metabolic costs are not the only energy costs involved. From a whole-plant perspective, the cost of developing and maintaining root systems to obtain ammonium and nitrate must also be included. Ammonium is not very mobile in soils, so a larger root system may be needed to obtain soil ammonium than is needed for nitrate. Nitrogen gas is far more mobile than either ammonium or nitrate. Indeed, nodulated (nitrogen-fixing) alders have been shown to develop smaller root systems than nonnodulated (non-nitrogen-fixing) plants do. From a whole-plant energy perspective, nitrogen fixation may not be more costly than assimilation of soil nitrogen, but this speculation needs further testing.

NITROGEN FIXATION ACCELERATES NITROGEN CYCLING

What happens to the nitrogen once fixed? The ammonia is rapidly aminated to become part of organic amino acids and proteins. In free-living systems, the bacteria and cyanobacteria simply use the nitrogen as any other microbe would—to grow and synthesize new cells. In symbiotic systems, the fixed nitrogen is shipped from the nodules into the plant roots, where it performs the usual functions. Soil enrichment from nitrogen fixation comes only after microbe and plant tissues have died and decomposed to release the extra nitrogen.

Increases in nitrogen availability may have positive feedback that stimulates nitrogen mineralization. This is one of the major benefits of mixing nitrogen-fixing species with crop trees, and can it be illustrated in mixed stands of alders and conifers. Portions of a Douglas-fir plantation on Mt. Benson in British Columbia, Canada, contained shrubby Sitka alder, red alder, or no alder (Binkley, 1983; Binkley et al., 1984). The site index in the absence of alder was low, about 24 m at 50 years. Sitka alder fixed only about 20 kg/ha annually, but after 23 years the nitrogen content of litterfall was 110 kg/ha annually with Sitka alder compared to only 16 kg/ha without alder. The nitrogen fixation rate of red alder was about twice that of Sitka alder, and the nitrogen content of litterfall was about 130 kg/ha annually. What accounted for this increase in nitrogen cycling? The nitrogen content of alder leaves (and alder litter) was very high, about 3 percent of the dry weight, as compared with less than 1 percent for Douglas-fir needles. Most litter immobilizes nitrogen from the surrounding soil in the initial stages of decomposition, but nitrogen-rich litter begins mineralizing nitrogen immediately. Ecosystems with nitrogen-fixing species enjoy both greater inputs of nitrogen and produce litter rich in nitrogen; these two factors combine to greatly accelerate nitrogen cycling and availability.

NITROGEN-FIXING PLANTS ALSO AFFECT CYCLES OF OTHER NUTRIENTS

A mixture of nitrogen-fixing plants and crop trees may show changes in the availability of nutrients other than nitrogen. Increased nitrogen availability may directly affect the cycling of other nutrients, and increased ecosystem production may rapidly tie up available nutrients in tree biomass. Returning to the Mt. Benson study, the phosphorus concentration of Douglas-fir foliage was 0.22 percent without alder and only 0.12 percent with Sitka alder. Sitka alder leaves contained a healthy level of 0.30 percent phosphorus. Red alder trees in the same plantation further reduced Douglas-fir foliage phosphorus concentrations to 0.09 percent, and red alder leaves contained only 0.14 percent phosphorus. Why did the alder impair Douglas-fir phosphorus nutrition? The rate of biomass accumulation was increased about 75 percent with Sitka alder and 260 percent with red alder. The phosphorus contained in the ecosystem's biomass totaled 45 kg/ha without alder, 105 with Sitka alder, and 150 kg/ha with red alder. Thus, the decrease in soil phosphorus availability may simply have been due to accumulation of phosphorus in rapidly growing biomass. As the stand without alder developed, the accumulation of phosphorus in biomass may also have reached a point where phosphorus availability in the soil declined. Acceleration of stand production increases demands for all nutrients, and the benefits of nitrogen-fixing species may be limited by the availability of other nutrients on very poor soils. This pattern would also be expected if nitrogen fertilizer were used to greatly accelerate stand growth.

The decrease in phosphorus availability under red alder on Mt. Benson also appeared to limit red alder performance. A bioassay of soil samples from the red alder–Douglas-fir stand showed that red alder seedlings produced double the biomass and five times the nitrogen fixation activity per plant when phosphorus and sulfur were added (Binkley, 1986a).

Limitations of nutrients other than nitrogen will affect nitrogen fixation as well as the performance of crop species. Other studies with red alder have found both substantial declines in phosphorus availability (Compton and Cole, 1998, assuming that the large site difference in total phosphorus did not confound the species effect on available phosphorus) and substantial increases (Giardina et al., 1995). A strong pattern of reduced phosphorus supply was also apparent in 16-year-old plantations of *Eucalyptus saligna* and nitrogen-fixing *Albizia facaltaria*. Nitrogen availability in pure *Albizia* plots was four times higher than in pure *Eucalyptus* plots, but pure *Eucalyptus* plots had five times more available phosphorus than pure *Albizia* plots (Figure 14.7). Phosphorus availability under nitrogen-fixing species appears to be increased in some cases and depressed in others; this is another situation where simple generalizations are not helpful.

NITROGEN-FIXING TREES ACIDIFY SOIL IN SOME ECOSYSTEMS BUT NOT IN OTHERS

Conifers generally are thought to produce more acidic soils than do hardwoods on similar sites. This is probably a bad generalization anyway, and it may be more productive to characterize the factors that influence soil acidity and discuss how these differ among species and sites. About half of the studies of nitrogen-fixing species have shown significant soil pH declines, and about half have not.

The processes that influence soil pH under nitrogen-fixing trees can be illustrated with some details from a comparison of 55-year-old stands of red alder and conifers in Washington State. The pH of soil 0–15 cm in depth (measured in distilled water suspension) under the alder-conifer stand at the less fertile site was 5.1, compared with 5.4 in the pure conifer stand (Binkley and Sollins, 1990). The same comparison in dilute calcium chloride suspension

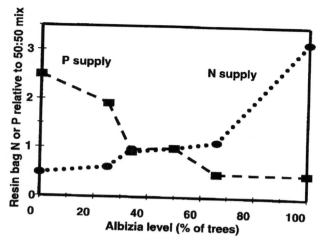

Figure 14.7 Relative changes in availability of nitrogen and phosphorus for a range of *Albizia* proportions in a mixture with *Eucalyptus*. The values for the 50:50 stand were set a 1, and values for other *Albizia* levels are shown as proportional increases in or decreases from this baseline. The supply of nitrogen increased with the amount of *Albizia* in the stand, but the supply of phosphorus decreased (data from Kaye et al., 1999).

showed no differences between species. The cause of the difference in water, but not in salt, was the higher concentrations of soluble ions in the mixed stand. The levels of soluble anions were almost twice as high in the soil from the alder-conifer stand, and of this extra anion charge, about 3 percent of the balancing cation charge came from H^+; this modest increase lowered soil pH (measured in water) by 0.3 unit. Measurement in dilute salt suspensions removed this effect of ionic strength and showed no difference between stands.

This analysis showed no effect of species on soil pH, but that doesn't mean that the species did not differ in their effects on the factors that influence soil pH (see Chapter 5). The base saturation was higher in the 0- to 15-cm soil depth in the alder-conifer stand than in the conifer stand, and higher base saturation indicates that the exchange complex is "more dissociated" as an acid and should maintain a higher pH in the soil solution. To have a higher base saturation without a higher pH, some other factor must have worked to lower the pH. In this case, the acid of the organic matter in the mixed stands was stronger than that of the conifer stand.

A related study on a more fertile site (Binkley and Sollins, 1990) found that alder acidified the soil by strongly reducing base saturation (= increasing acid saturation). This decline in base saturation probably resulted from high rates of nitrate production and leaching (about 50 kg/ha of nitrogen annually as nitrate); the H^+ generated in nitrification replaced base cations on the exchange complex, and these cations leached away with the nitrate. This fertile site also showed a strong increase in soil organic matter in the alder-conifer stand, and the acidity of this organic matter was partially responsible for the decline in pH. The acid of the alder-conifer organic matter was also stronger, contributing somewhat to the lower pH in comparison with the conifer stand.

A similar study with nitrogen-fixing *Albizia* and *Eucalyptus* in Hawaii found that soil pH was reduced somewhat under *Albizia*. The effect resulted primarily from a reduction in base saturation rather than a change in soil organic matter (acid quantity) or acid strength (Rhoades and Binkley, 1995). In a study of *Ceanothus* mixed in Jeffrey pine stands in California, soil pH was not altered by *Ceanothus*, despite a doubling of the of the organic matter and acid quantity (exchange capacity) in the A horizon (Johnson, 1997). The lack of change in pH despite a doubling of the acid quantity in the soil resulted from a 50 percent increase in base saturation under *Ceanothus*; the increase in acid quantity should have lowered the pH but the increase in base saturation should have increased it, giving no net change.

SOIL CARBON INCREASES UNDER NITROGEN-FIXING SPECIES

Soil organic matter often increases under nitrogen-fixing species, typically by 10 to 30 mg/ha. This increase is probably driven both by increases in pools of new carbon and by lower rates of decomposition of old soil carbon. A comparison of soil carbon under nitrogen fixers (*Casuarina, Leucaena, Albizia*) and eucalypts in Puerto Rico and Hawaii used differences in carbon isotopes between carbon-3 (such as the trees) and carbon-4 plants (grasses from previous land use) to trace the loss of old carbon and gain of new carbon under the influence of each tree species (Resh et al., 1999). Across all sites, the soils under the non-nitrogen-fixing *Eucalyptus* lost about 600 kg/ha more of carbon each year than the nitrogen-fixing systems. The greater retention of old carbon may have resulted from the increase in nitrogen; as noted in Chapter 9, nitrogen may help stabilize old, humified carbon at the same time that it may stimulate the decomposition of labile carbon (by improving the nitrogen supply to microbes).

The functional importance of the extra organic matter under nitrogen-fixing trees is hard to assess. Organic matter favors soil aggregation and aeration, supplies cation exchange capacity, and increases the soil's ability to retain nutrients. Long-term experiments are needed before major conclusions can be drawn about the importance of increases in soil organic matter.

MANY SILVICULTURAL STRATEGIES CAN EMPLOY NITROGEN-FIXING SPECIES

The ideal nitrogen-fixing plant and silvicultural system will vary with management objectives. In some cases, nitrogen-fixing species such as red alder, black locust, *Albizia*, and *Casuarina* can be used as the crop species. Other silvicul-

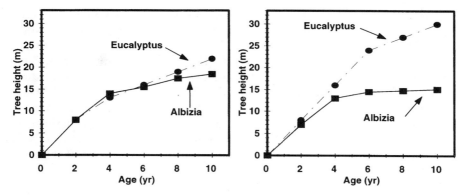

Figure 14.8   Height growth of *Eucalyptus saligna* and *Albizia facaltaria* was similar for the first 6 years in pure plots and then diverged (left graph). In mixtures, *Albizia* height growth was suppressed by *Eucalyptus*, whereas Albizia increased the height growth of *Eucalyptus* (right graph; data from DeBell et al., 1997).

tural systems may use nitrogen-fixing species to provide nitrogen to crop trees. Nitrogen may be provided by alternating rotations of nitrogen fixers and crop trees or by mixing both species in one plantation. Alternating rotations have the advantage of removing competition between nitrogen fixers and crop trees but suffer from unfavorable extensions of investment periods with the addition of a nonpaying nitrogen fixation period. Mixed plantations offer attractive investment periods, but competition between nitrogen fixers and crop trees needs to be controlled carefully.

A first approximation of relative competitive ability among species can be made by comparing height/age curves and the implications for canopy dominance. Desirable patterns in height and dominance are illustrated by pure and mixed plots of *Eucalyptus* and nitrogen-fixing *Albizia* (Figure 14.8). Height growth was similar in pure plots of each species early in stand development, with *Eucalyptus* showing greater increases in later years. In the mixed plots, *Eucalyptus* growth was higher in the presence of *Albizia*, and *Albizia* height growth was lower. This clear dominance by the valuable crop tree allowed 40 percent greater accumulation of *Eucalyptus* biomass in mixtures with *Albizia* than in pure plots (DeBell et al., 1997).

NITROGEN FIXERS CAN TAKE ADVANTAGE OF UNUSED SITE RESOURCES

The only growth factor for which nitrogen-fixers and crop trees do not compete fully is nitrogen, so any other site constraints may limit the usefulness of nitrogen-fixing species. Fortunately, the early development of plantations involves a period of several years when site utilization is below maximum.

Nitrogen fixers can be used during this period with little interference with crop trees. Similarly, thinning a stand may free resources for nitrogen fixers for a few years.

Nitrogen-fixing plants can also be used to control competition from other noncrop vegetation. For example, the no-alder stand in the Mt. Benson study site had an understory of salal (*Gaultheria shallon*) with a biomass of 2500 kg/ha. The addition of shrubby Sitka alder reduced other understory vegetation to 1000 kg/ha. In the Coastal Plain of North Carolina, an herbaceous legume (*Lespedeza cuneata*) reduced the hardwood component of a loblolly pine plantation and allowed greater leaf area in the lower crown of the pines (Figure 14.9).

WHEN DOES FIXED NITROGEN BEGIN TO BENEFIT ASSOCIATED CROP TREES?

Only a few studies have examined how long it takes for a nitrogen-fixing plant to provide significant amounts of nitrogen for associated species.

The major way that fixed nitrogen becomes available to associated trees is through decomposition; the annual litter produced by the nitrogen-fixer decomposes, and the released nitrogen is taken up by trees. Minor amounts of nitrogen may pass directly from the root system of the nitrogen fixer to associated trees through mycorrhizal connections (Ekblad and Huss-Danell,

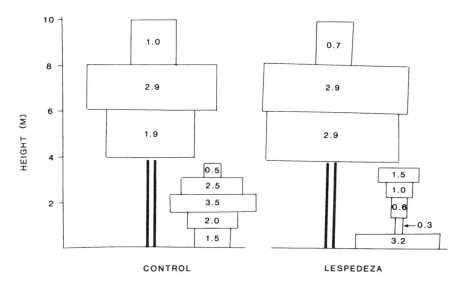

Figure 14.9 Eight-year-old loblolly pines showed greater leaf area (7.5 m²/m²) with nitrogen-fixing *Lespedeza* than without it (5.8 m²/m²). The response is probably due to both nitrogen fixation and control of competing hardwoods (data from Moser, 1985).

1995), but the rates of exchange among roots are probably too low to be important. About one to two percent of the fixed N may leak from roots into the soil (Uselman et al. 1999).

The Puerto Rico ^{15}N labeling study described above found that the *Eucalyptus* trees began taking up nitrogen fixed by *Casuarina* or *Leucaena* sometime between 2 and 3.5 years of age (Parrotta et al., 1996). A study using natural abundance levels of ^{15}N concluded that a *Leucaena* plantation derived about 70 percent of its nitrogen from the atmosphere and that by age 6 years, the understory plants were as reliant on nitrogen fixed recently by *Leucaena* as was the *Leucaena* itself (van Kessel et al., 1994).

In temperate forests, three studies have examined mixtures of poplars and alders for short-rotation biomass production (DeBell and Radwan, 1979; Hanson and Dawson, 1982; Cote and Camire, 1984). All found that poplars neighboring on alders grew better in the first 2 to 4 yr than poplars neighboring on poplars. This increased growth could have resulted from soil nitrogen enrichment by alder (DeBell and Radwan, 1979) or from the size differences between the species. Because the poplars were larger than the alders, poplars with alders probably experienced less competition than poplars with poplars (Cote and Camire, 1984).

These tropical and temperate studies indicate that nitrogen fixers may supply nitrogen to associated plants within a few years of establishment.

THE GROWTH RESPONSE TO NITROGEN FIXATION MAY HAVE THREE COMPONENTS

Nitrogen-fixing plants can produce a range of effects on the growth of interplanted crop trees. Like fertilization, nitrogen fixation may increase the photosynthetic capacity and leaf area of crop trees. In some cases, increased soil nitrogen availability might allow a reduction in the production of fine roots and mycorrhizae. The similarities between fertilization and nitrogen fixation are limited, however, if competition between the species counters the benefits of nitrogen. For example, shading of crop tree leaves may prevent any increase in photosynthesis that otherwise might result from better nitrogen nutrition. Similarly, competition for light or water could prevent canopy expansion by the crop trees. Finally, competition for water and nutrients other than nitrogen may require large investments in root systems. Very little is known about these important interactions.

The potential importance of interactions is illustrated by the leaf area patterns at the Mt. Benson site. Without alder, the Douglas-fir stand had a projected leaf area index of 5.4 m^2/m^2 and a stemwood production of 4.1 Mg/ha (Binkley, 1986). With shrubby Sitka alder, the Douglas-fir leaf biomass was not significantly increased, but stemwood production increased to 5.8 Mg/ha annually. The codominant red alder reduced the Douglas-fir leaf area index to 1.9 m^2/m^2, but stemwood growth was 5.1 Mg/ha annually. Sitka alder increased Douglas-fir stem growth per unit of leaf biomass by 40 percent, probably from a shift in biomass allocation away from roots and into stems plus a contribution from increased net photosynthesis. The dramatic 3.5-fold increase in stem growth per unit of leaf area in the red alder stand indicated a major shift in allocation.

THE CROP TREE RESPONSE TO NITROGEN FIXATION VARIES WITH SITE FERTILITY

The mixed plantation experiments at Wind River and Cascade Head allow comparisons of conifer production with and without red alder throughout 55 years of stand development. Conifers suffered in the presence of alder at Cascade Head but were transformed into superdominants at Wind River (Miller and Murray, 1978; Binkley and Greene, 1983). What accounted for the different responses? The alders fixed nitrogen at both sites, but even so, the no-alder Cascade Head plots had more than twice as much soil nitrogen as the plots with alder at Wind River. Nitrogen availability probably did not limit forest production at the fertile Cascade Head site, whereas the infertile Wind River stand responded dramatically to nitrogen fixation by alders. A similar comparison of a fertile site at Skykomish, Washington, with the infertile Mt. Benson site showed the same pattern (Figure 14.10).

HOW MUCH NITROGEN FIXATION IS NEEDED TO SUPPLY CROP TREE?

Nitrogen fixation has received much less attention than fertilization in forest nutrition research. At present, stand prescriptions for the use of nitrogen-fixing plants are only best guesses. Ideal prescriptions would balance nitrogen fixation rate (and subsequent effects on nitrogen cycling) against competition with crop trees for other resources. Lacking such ideal information, Miller and Murray (1978) noted that Douglas-fir trees 10 m from an alder stand showed greater growth than trees more than 10 m away. They expanded this observation into a recommendation of 50 to 100 red alder per hectare. They suggested that 20 to 50 kg/ha of nitrogen fixation annually might meet crop tree requirements (Miller and Murray, 1979).

The most thorough approach for evaluating the ratios of nitrogen-fixing trees to associated crop trees of another species involves replicated plantations of pure and mixed-species stands. This type of study was established in Hawaii using *Eucalyptus saligna* and nitrogen-fixing *Albizia facaltaria* (DeBell et al., 1997). The greatest accumulation of total stem biomass, as well as *Eucalyptus* stem biomass, occurred at a ratio of 34 percent *Eucalyptus* to 66 percent *Albizia*. The average tree size at this mixture ratio was far larger for the *Eucalyptus* than in the pure stand, which would accentuate the value of the *Eucalyptus* in the mixture by more than the simple increase in biomass (Figure

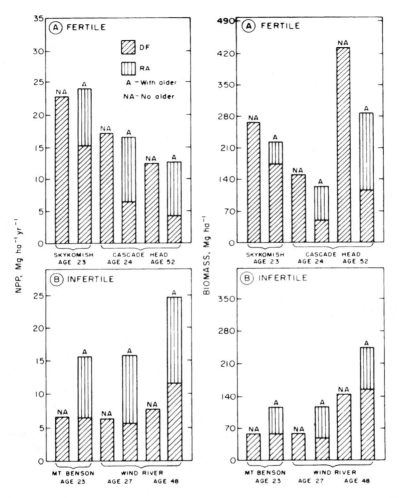

Figure 14.10 On fertile sites, Douglas-fir growth was better without alder (A). On infertile sites, Douglas-fir growth was not greatly affected by red alder for three decades and then increased with alder (B) (Binkley and Greene, 1983).

14.11). At this location, a mixture of 25 percent *Albizia* and 75 percent *Eucalyptus* showed little increase in available soil nitrogen; more *Albizia* was needed to have much effect (see Figure 14.7). At a nearby location, Ewers et al. (1996) examined the pattern of soil nitrogen enrichment across the boundary between stands of *Eucalyptus* and *Albizia*. They also concluded that *Albizia* would have little effect on soils unless it comprised more than 20 percent of the stand.

Many more field trials are needed before silvicultural prescriptions can be tailored for various species and sites.

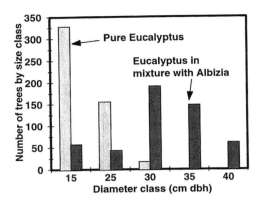

Figure 14.11 In addition to having more *Eucalyptus* biomass, the mixed *Eucalyptus-Albizia* plot had far larger trees, with commensurately greater value per cubic meter (data from DeBell et al., 1997).

IS A KILOGRAM OF FIXED NITROGEN EQUAL TO A KILOGRAM OF FERTILIZER NITROGEN?

Silvicultural use of biological nitrogen fixation differs substantially from fertilization. Nitrogen fixation supplies a moderate amount of nitrogen annually, and fertilization adds a large dose in a single year. Fixed nitrogen comes as part of litter organic matter; fertilization comes in an inorganic form and may even decrease soil organic matter by stimulating decomposition. Short-term effects of fertilization must be contrasted with longer-term effects of increased nitrogen cycling rates with nitrogen-fixing systems, but little information is available for direct comparison of nitrogen fixation and fertilization in forests.

Agricultural studies suggest similar relative values for both sources. For example, Voss and Shrader (1984) measured production of crops from various species rotations and fertilization regimes over a 25-year period. They found that two seasons of legume meadows supplied enough nitrogen for 2 years of corn and oats, and that no yield increase could be obtained with further fertilization. In the absence of legume rotations, 120 to 180 kg/ha of nitrogen was needed to maintain corn production. Indeed, after 25 years of continuous corn cropping, yields could not be maintained at maximum levels by fertilization alone; legume rotations were needed for maximum sustainable yields. The mechanism behind the corn yield decline was not examined but could have resulted from altered soil physical properties or effects of pathogens in the uninterrupted corn rotations.

The relative efficiencies of fixed nitrogen and fertilizer nitrogen should provide profitable arenas for future research. From an applied forestry perspective, it is not yet possible to draw strong conclusions about the relative ecological merits of nitrogen fixation and fertilization as sources of nitrogen. Both are useful tools that can provide attractive economic returns, and the choice between them probably will be based more on expected response (and risks) than on relative ecological value of a kilogram of nitrogen.

NITROGEN-FIXING PLANTS CAN ALTER THE IMPACT OF ANIMALS AND DISEASES

Very little work has been done to examine the impacts of destructive agents in pure plantations relative to those containing a nitrogen-fixing species. Some evidence supports the hypothesis that red alder reduces the incidence of root rot (*Phellinus weirii*) in Douglas-fir (see Nelson et al., 1978), but this hypothesis has not been well tested. Possible mechanisms include increased soil nitrate concentrations, lower soil pH, production of phenolic compounds, and simple dilution of the susceptible conifer roots with immune red alder roots.

One of the best examples of mixed stands suffering greater damage than pure stands comes from the mixed-plantations Wind River Experimental Forest, where a narrow strip of a Douglas-fir plantation was planted in the 1920s with a mixture of red alder. This plantation configuration produced an oasis for black bears searching for food in the early spring. When the stands reached age 40, bears began to strip bark (to eat the cambium beneath) from Douglas-fir in the alder zone. By age 55, most of the Douglas-fir in the alder strip showed damage from bears, and many of the best trees had been girdled and died. No trees were damaged in the pure conifer portion of the plantation.

Mixed stands may also experience less damage from animals. Snowbrush species are preferred to browse plants for deer and elk, and browsing on Douglas-fir regeneration is often reduced in association with snowbrush. For example, Scott (1970) found greater Douglas-fir stocking and growth in association with snowbrush cover. Over 70 percent of the Douglas-fir seedlings without snowbrush cover were severely browsed compared with less than 30 percent of seedlings on the edge or beneath snowbrush canopies. It would be naive to assume that interactions between crop trees and nitrogen fixers are always this beneficial. As with all silvicultural systems, it is important to consider the possible benefits and costs of nitrogen fixation on a site-specific basis.

NITROGEN FIXATION MAY HAVE MULTIPLE RESOURCE BENEFITS

Mixing nitrogen-fixing plants into forestry plantations increases plant diversity, with a wide array of consequences. Wildlife habitat can be improved by mixing nitrogen fixers with crop trees. For example, redstem ceanothus supplied about a third of the winter browse eaten by elk in one study in Idaho (Trout and Leege, 1971). Nitrogen-fixing plants in riparian (streamside) zones have major effects on stream ecosystems. In many cases, buffer strips are left undisturbed when stands are harvested to ensure shading and minimal impacts on stream temperature and siltation. Riparian vegetation is also a major energy source for stream flora and fauna in the form of litterfall, and litter from nitrogen-fixing species such as red alder is especially rich in nutrients. These nonnutrition aspects of mixed plantations have received little research attention, despite their importance to the overall costs and benefits of nitrogen fixation.

WHY ISN'T NITROGEN FIXATION USED MORE OFTEN IN FORESTRY?

Nitrogen fixation is not appropriate for every nitrogen-limited site. The risks of competition between crop trees and nitrogen fixers are real, so the value of biologically fixed nitrogen must be high to warrant the added risk. A large portion of the most productive soils in the Pacific Northwest is currently dominated by red alders that have excluded conifers since logging in the early part of the century. Fertilization can provide many of the benefits that nitrogen fixation provides at a lower risk. Management inertia is also a contributing factor; all vegetation other than crop trees is often considered a weed problem that is best kept to a minimum. Finally, forest managers are typically less familiar with the potential profitability of nitrogen fixation than they are with the profitability of fertilization.

Nitrogen-fixing trees are currently used in forestry in a few situations in temperate areas and commonly in tropical areas. About 0.5 million ha of black locust plantations are found in the former Austro-Hungarian Empire, and black locust plantations exceed all other species in Hungary (Keresztesi, 1988). The use of nitrogen fixation in silvicultural systems is more limited in other temperate areas, but it is increasing. Thirty years ago, nitrogen-fixing red alder had negative value in the eyes of foresters; now mills pay about half as much for alder logs as they do for Douglas-fir logs. Given the faster early growth of red alder, about 10,000 ha of red alder stands have been planted in the northwestern United States to be harvested on 30- to 35-year rotations (D. Hibbs, personal communication, 1999).

In the tropics, commercially planted species of nitrogen-fixing trees come primarily from four genera: *Casuarina*, *Leucaena*, *Albizia*, and *Acacia*. For each of these genera, some of the species produce wood that is valued highly enough in local markets to support silvicultural investments. Over 3 million ha of *Acacia* plantations are being managed in Asia and Africa (Brown et al., 1997).

IS NITROGEN FIXATION PROFITABLE?

If the nitrogen fixer's value rests solely on stimulating the growth of crop trees, value gains will be greatest when three conditions are met: nitrogen availability limits tree growth, nitrogen fixation rates are substantial, and nitrogen fixers offer little competition to crop trees. For example, nitrogen-fixing *Albizia* has no market value on the island of Hawaii, but mixed stands of *Albizia* and *Eucalyptus* produced more *Eucalyptus* biomass, on bigger stems, than in pure *Eucalyptus* plots (see Figure 14.10).

Creative silviculturalists could probably develop applications for specific situations in which costs, risks, and responses would favor nitrogen fixation or fertilization. Neither approach should be considered ecologically or economically superior for all situations.

Finally, it is important to keep in mind that economic analyses provide information for use in decision making; however, they are not the only factor to consider in choosing the best management options. The ecological effects of nitrogen-fixing plants extend beyond increasing tree growth, and multiple resource values may also be important.

SUMMARY

Nitrogen fixation is the reduction of atmospheric nitrogen to ammonia for use by plants. Only a few types of prokaryotic organisms can perform this reaction; some of these organisms are free-living, and others live in association with higher plants. Rates of nitrogen fixation range from less than 1 kg/ha annually for forests without symbiotic nitrogen-fixing plants to more than 100 kg/ha annnually where vigorous nitrogen-fixing trees dominate the forest. Several approaches are available for estimating rates of nitrogen fixation in forests, including long-term accretion of nitrogen in the ecosystem and short-term assays of nitrogen fixation activity using acetylene. The value of nitrogen fixation in forestry depends on whether there is a market for the nitrogen fixer biomass, and on the effects of the nitrogen-fixing species on associated crop trees of other species. Nitrogen-fixing trees compete with other trees for all resources except nitrogen. Several case studies have shown that nitrogen-fixing trees can increase the growth of associated crop trees on nitrogen-poor sites or reduce growth of neighboring trees on nitrogen-rich sites.

CHAPTER 15

Forest Soil Management

Forest soils have been distinguished from agricultural soils by the fact that they receive little if any cultivation and few amendments. However, the rapid rise of plantation forestry and the increasing human management of forest soil are beginning to blur this distinction.

SOIL MANAGEMENT PLAYS A MAJOR ROLE IN INTENSIVE PLANTATION FORESTRY

The millions of hectares of the world's forests managed as plantations include a wide range of soil and climatic conditions and a variety of tree species. In general, tree planting is used to affect (1) afforestation of open sites that have not supported forests for some period of time; (2) reforestation of sites, often with genetically improved stock, that have recently been clearcut, and (3) conversion of one type of forest to another. Examples of open lands that have been planted to forests are former agricultural lands in England, southwestern France, central Europe, and the southern United States; sand dunes in New Zealand and Libya; arid regions in Israel and other Middle Eastern countries and parts of sub-Saharan Africa; steep slopes in Europe and Japan; peatlands in the British Isles, Sweden, and Finland; grasslands in Venezuela, Colombia, and Argentina; brushlands in Brazil; former sea bottom in Holland; and spoil banks resulting from mining operations in many countries around the world.

Plantation establishment following clearcutting of the same species is standard practice in many areas where intensive silviculture is practiced. Examples of such reforestation efforts are found in the Douglas-fir forests and southern pine forests of the United States, Australian *Eucalyptus* forests, and the *Chamaecyparis* forests of Japan and Taiwan.

Plantation establishment on sites still supporting considerable forest cover often involves conversion from native to exotic species. Examples include the pine forests of Australia, replacement of marquis vegetation with pines or eucalypts in the Mediterranean, conversion of scrub hardwood and coppice in central Europe, and improvement of degraded tropical forests in several equatorial countries (Evans, 1992). Often natural forests are converted to plantations of widely different species, and from highly complex associations to exotic monocultures. These include plantations of *Pinus caribaea* and *Gmelina arborea* in Brazil; *Pinus radiata* in Australia, Chile, and New Zealand; *Pinus caribaea* in Fiji; *Tectona grandis*, *Eucalyptus tereticornus*, and *Cryptonieria japanica* in India; and *Triplochiton scleroxylon*, *Eucalyptus saligna*, and *Pinus* species in Africa. Whether regeneration is accomplished with native or exotic species, plantation forestry offers a unique opportunity to locate species and selections within species according to their soil-site requirements. It is not sufficient to plant hardwoods on "hardwood" sites and conifers on "conifer" sites because of the tremendous differences in growth potential on different soils for different genera, species, and selections.

Some reasons for the trend toward plantation forestry include (1) greater demands for forest products; (2) introduction of genetically improved stock; (3) easy manipulation of the stand for the most desired assortments; (4) higher yields; (5) full use of land by complete stocking; (6) improved conditions for the operation of machinery; (7) reduced cost of logging; (8) prompt restocking after harvesting; (9) more uniformity of tree size; and (10) improved log quality. **Plantation forestry** generally involves an integrated system of cultural practices, including slash disposal, site preparation, and careful planting of cultivars. It may include such other silvicultural practices as soil drainage, weed and pest control, pruning, thinning, fertilization, and the use of genetically superior planting stock. It usually involves even-age management and clearcut harvesting of monocultures.

Plantations are literally man-made forests in the sense that they are established and maintained as the result of site manipulation. Such efforts to improve the site and increase tree survival and growth may have profound influences on certain soil physical, chemical, and biological properties, particularly properties of the forest floor. Whenever heavy machinery is used in harvesting or site preparation, there is an opportunity for undesirable soil disturbance. Soil disturbance is less for track than for wheel vehicles, less on flat than on steep terrain, less on sandy than on clay soils, and less on dry than on wet sites. However, long-term effects on soil properties and long-term site productivity are mostly in the realm of conjecture because few well-controlled studies have been conducted over sufficient time periods to give a definitive answer.

LAND CLEARING MAY ADVERSELY IMPACT SOIL PRODUCTIVITY

Considerable quantities of vegetation or slash must often be removed from a site during stand establishment. Development of detailed plans prior to logging can be invaluable in solving disposal problems. Such plans should take into consideration the soils, terrain, climate, and cover type. The type, amount, and distribution of logging waste or unwanted vegetation and terrain conditions will influence the choice of disposal method and machinery. Proper location of roads and selection of techniques and tools suited to each vegetation-terrain combination may avoid costly damage to the soil and site.

Machines are widely used for disposal of slash and excess vegetation in preparation for plantation establishment. Machines are extensively used, along with fire, in the conversion of eucalypt stands to pine in Australia, in the disposal of low-quality hardwoods in preparation for pine plantations in the southeastern United States, and in site clearing for Douglas-fir plantations in the Pacific Northwest. Some modification of a bulldozer blade is often used for the initial operation in site preparation (Figure 15.1).

Bulldozers with **V-shaped** or **serrated blades** are often used to sever stems of residual hardwood. Root rakes and Young's teeth attached to bulldozer blades may be used to uproot trees and windrow debris. The windrows are often burned after a few months, but they still occupy a considerable portion of the land area that would otherwise be available for planting. Unfortunately, these windrows often contain considerable topsoil as well. Haines et al. (1975) reported that approximately 5 cm of the surface soil was deposited in windrows during shearing and piling operations in the Carolina Piedmont. After 19 years, loblolly pine in an unbladed area contained 85 percent more volume than trees in the bladed area.

Considerable concern has arisen over the precipitous decline in soil organic matter that often accompanies stand conversion efforts, particularly in the

Figure 15.1 Clearing woody vegetation with a shearing blade as a first step in site preparation for plantation establishment in Oklahoma (courtesy of Weyerhaeuser).

tropics (Lundgren, 1978; Chijicke, 1980). Morris et al. (1983) found that an extremely large amount of topsoil ended up in windrows of a sheared and piled slash pine flatwoods site even when the clearing had been done with care. The effects of land-clearing methods on soil erosion rates and subsequent site productivity can be quite significant (Dissmeyer and Greis, 1983). A major problem with the use of heavy equipment for uprooting and windrowing brush and logging debris is that the quality of the operation is heavily dependent on the skill and dedication of the heavy equipment operator. No matter how well planned land clearing operations may be, the potential for site degradation can be controlled only through careful supervision of skilled and dedicated equipment operators.

Successful conversion of wiregrass (*Aristida stricta*) scrub oak (*Quercus laevis*) sandhills in Florida and the Carolinas to southern pines has been accomplished by chopping (Figure 15.2). This operation generally consists of passing a heavy (up to 20 mg) **drum roller** over the soil surface one or more times. Sharp parallel blades attached along the length of the drum cut debris and vegetation, including small trees. Because only a small part of the chopped vegetation and debris is incorporated into the mineral soil, this organic material acts as mulch. It retards erosion, moderates soil temperature fluctu-

Figure 15.2 Site preparation with a rolling drum chopper on a dry site (courtesy of Weyerhaeuser).

ations, and reduces soil moisture losses. Chopping appears to be a practical means of conserving soil nutrients and is particularly useful on dry, sandy soils (Burns and Hebb, 1972).

SOIL PREPARATION FOR STAND ESTABLISHMENT

Harvesting and the elimination of logging slash and unwanted vegetation may be considered the first steps in preparing a site for regeneration. However, the term site preparation is commonly used for those soil manipulation techniques designed to improve conditions for seeding or planting that result in increased germination or seedling survival and tree growth. While this may be accomplished by manual operations in some remote localities and on relatively clean sites, soil preparation is increasingly achieved by a variety of machines such as scarifiers, discs, trenchers, bedding or tine plows, subsoil rippers, and so on.

Scarification is a Means of Exposing Mineral Soil

In cool to cold regions, exposure of mineral soil, elimination of competing vegetation, and improvement of water relations are primary goals of soil preparation operations. Because of the slow rate of litter decomposition, reactivation of the soil and release of nutrients for the young seedling are also important. **Scalping** is one method of scarification. It is accomplished with a trailing plow or V-blade that produces a furrow about 5 cm deep and 0.5 to 1 m wide, in which the seedlings are planted. Scalping displaces much of the organic-enriched topsoil from the immediate reach of the young plants and may result in prolonged slow growth on sandy soils.

Spot scarification has been accomplished with a number of devices ranging in complexity from anchor chains to the Bracke cultivator. The objective of such treatments is to expose or mound mineral soil. Mounds of mixed mineral soil and organic matter are generally superior for seedling survival and growth to mounds of mineral soil, which are superior to mineral soil exposed by scraping, which is superior to planting through the undisturbed forest floor (Sutton and Weldon, 1993; 1995). Unless scarification is carried out parallel with the slope, in which case it may initiate erosion, this type of soil manipulation does not lead to site degradation.

Cultivation Often Increases Survival and Early Growth of Planted Seedlings

Harrowing or **disking** differs from plowing mainly in the depth and degree of disturbance of the soil. Harrowing consists of drawing multiple discs through the surface soil and displacing but not necessarily inverting the disc slice. Plowing, which inverts a slice of soil, can be done by either a disc or a moldboard plow. The latter is most often used on difficult sites. Both methods are effective in mixing organic and mineral materials and reducing compaction. Harrowing and plowing may result in a temporary immobilization of soil nutrients due to the incorporation of large amounts of organic material. Soil preparation for planting is sometimes accomplished by double moldboard plows. Trees are then planted in the furrow, except in wet areas, where the furrow may serve as a water trap. **Deep-tine plowing** is often done on peats or soils with indurated layers. The plow tines may penetrate to a depth of as much as 90 cm, and the plow has a moldboard that turns out a furrow up to 60 cm deep (Figure 15.3).

Deep loosening of subsoil layers, often called **ripping**, is a relatively recent development in forest soil management. It is accomplished with chisel-like bars drawn up to 0.5 to 1 m into the soil by powerful tractors and by three-in-one plows that both rip and bed the site. Subsoil loosening is aided by a winged foot at the base of the chisel arm that vibrates as it is drawn through the soil. Soil fracturing is maximized if ripping is done when soils are relatively dry. Tree roots commonly find old root channels and proliferate in them. Ripping provides artificial channels, and trees have been shown to proliferate in such channels, particularly in compact soils (Nambiar and Sands, 1992). The fractures or artificial channels gradually disappear (Carter et al., 1996), but roots often penetrate along the fracture lines long before closure occurs.

There have been attempts to improve the productivity of degraded forest Spodosols with ortstein layers. Ripping to rupture the hardpan, plus the addition of lime, organic materials, and mineral fertilizers, improves the growth of pines and mixed stands. Attempts to improve the growth of pine by deep

Figure 15.3 Deep-tine plowed peaty-gley ironpan soils in a pastureland afforestation project in southeast Scotland. Sitka spruce has been planted on the resulting beds.

placement of lime and phosphorus fertilizers during site preparation of an Aquod soil met with only limited success (Robertson et al., 1975).

Bedding, the use of opposing discs to create a mound, is a standard soil preparation practice in many areas. Bedding concentrates surface soil, litter, and logging debris into a ridge 20 to 30 cm high and 1 to 2 m wide at the base. Planting on beds usually improves the survival and early growth of both conifers and hardwoods on a variety of sites. The beneficial effects of bedding have been attributed to improved drainage; improved microsite environment for tree roots in regard to nutrients, aeration, temperature, and moisture; and reduced competition from weedy species.

On wet sites, survival and growth responses from bedding derive primarily from removal of excess surface water. Surface drainage results in accelerated decomposition of organic matter and mineralization of nutrients. Where the A horizon is thin and the B horizon is either impervious to roots or nutrient-poor, the concentration of topsoil, humus, and litter provides more favorable conditions for root development and, subsequently, top growth. Haines et al. (1975) reported that the incorporation of 73 Mg/ha into 15-cm-high mounds of screened surface soil significantly increased slash pine growth over that obtained where no debris was added. However, incorporation of debris in excess of 73 Mg/ha, the approximate amount left after logging, resulted in no additional improvement after 4 years.

Pritchett and Smith (1974) combined three types of site preparation with varying amounts of mineral fertilizers (Table 15.1). In this test, harrowing reduced ground vegetation, but it resulted in only slightly better tree growth than at the control site. Competition for moisture was not a critical factor on this wet site. Bedding, on the other hand, resulted in a 98 percent increase in tree heights and a 74 percent increase in total volume after 9 years of growth

TABLE 15.1 Effects of Site Preparation and Fertilization on Height and Volume of Slash Pine after 9 Years on a Typic Albaquult

Nutrients Added		Site Preparation Technique					
		Nonprepared		Harrowed		Bedded	
N (kg/ha)	P	Height (m)	Volume (m³/ha)	Height (m)	Volume (m³/ha)	Height (m)	Volume (m³/ha)
0	0	2.17	1.94	2.87	3.57	4.30	8.85
90	0	2.07	1.60	3.26	4.63	4.88	12.82
0	90	5.95	14.39	5.89	20.33	7.66	49.80
90	90	6.74	21.43	7.02	32.68	7.29	37.28

Source: Pritchett and Smith (1974). Used with permission.

on unfertilized plots. The response to the fertilizer applications indicated that, while part of the growth improvement from bedding derived from improved nutrition, equally important benefits came from reduced competition and improved surface drainage.

Wilde and Voigt (1967) reported that at age 14, white pine planted on beds were significantly taller (5.2 m) and had better survival than trees planted in furrows (1.4 m). They found that planting on scalped soil resulted in improved survival, but average tree heights were only 3.2 m. There have been a number of other reports of improved tree growth from harrowing or bedding in the southern pine region (Haines and Pritchett, 1964; Worst, 1964; Derr and Mann, 1970; Klawitter, 1970a; Mann and McGilvray, 1974; Harrington and Edwards, 1996), for maritime pine in southwest France (Sallenave, 1969), and for radiata pine in Australia (Woods, 1976; Hoopmans et al., 1993; Costantini et al. 1995) (Figure 15.4).

Bedding often increases survival and early growth, but growth gains dissipate with time. This phenomenon is often associated with density-dependant mortality within the stand. Because of high survival and rapid growth, the canopy in the bedded areas closes faster than in unbedded control areas. Soon after canopy closure, destiny-dependant mortality, or in some southern pine species crown thinning, begins. Either of these processes soon leads to growth losses that allow the control plots to catch up to the bedded plots. Obviously, density management is important if one is to take advantage of growth gains from cultivation.

Improvements in tree growth in response to different types of soil management practices depend greatly on soil properties, site conditions, and the tree species involved. Soils information can be a valuable aid in deciding which intensive management practices to employ.

Fisher (1980) presented a guide for the use of soils information in silviculture in the southeastern coastal plain. He found that a system of soil groups based on soil drainage class, depth to and nature of the B horizon, and the character of the A horizon was a valuable aid in determining soil management practices for intensive silviculture. This system allows the use of existing soils information in deciding which species to plant, what harvest system to use, what water management techniques to employ, how to prepare the site, and what fertilization regime to follow.

SOIL MANAGEMENT ON EXTREME SITES

The vast majority of forest soils require no special management techniques for continued high productivity. But when disturbed or adverse sites are planted with trees in an effort to ameliorate soil conditions and improve aesthetic values, special soil management techniques are usually required. Afforestation of spoil banks resulting from mining operations, seasonally flooded areas and

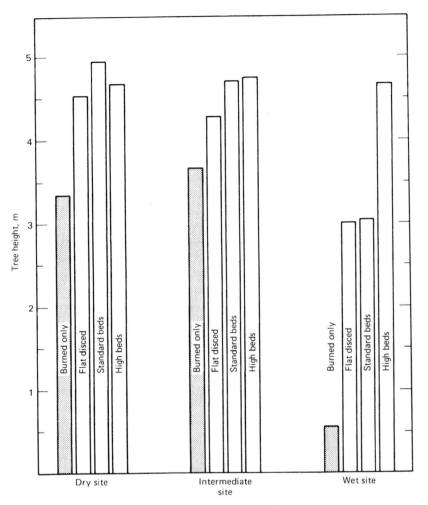

Figure 15.4   Height of 8-year-old slash pine as influenced by seedbed treatment and soil moisture conditions (Mann and McGilvary, 1974). Used with permission.

peatlands, moving sand dunes, and old cultivated fields degraded by erosion cannot be accomplished without special effort. Some of these soils are nutrient-deficient but can be made reasonably productive through the addition of fertilizers. Soils of some adverse sites contain toxic substances or lack mycorrhizal fungi, while others are too dry or too wet for good tree growth. The one thing that all problem sites have in common is the need for special soil management efforts in order to obtain acceptable plant establishment and growth.

Dry, Sandy Soils Pose Special Problems to Forest Managers

Rainfall exerts a major influence on the worldwide distribution of forests. Even in regions with ample rainfall, there are areas where available soil moisture is too low for good tree growth. These areas are mostly deep sands with minimal water-holding capacity or shallow soils with low moisture-storage capacity. Dry sands that are also exposed to strong winds offer a special problem in terms of **dune stabilization**.

The stabilization of moving sands is a special problem in certain coastal and inland areas when marine or glacio-fluvial deposits of sand are disturbed. Blow sands that have required special afforestation efforts are found in France, the Netherlands, New Zealand, North Africa, North America, and the Middle East (Mailly et al., 1994), as well as in some coastal and dry, sandy areas on most continents. Dune fixation by tree planting was underway in Europe by the middle of the eighteenth century and in America by the early nineteenth century (Lehotsky, 1972). One early attempt at dune stabilization in America was on Cape Cod (Baker, 1906), and other projects have been undertaken in the Great Lakes region of North America and on Cape Hatteras. The sandy areas around the Great Lakes became unstable and began to move following timber removal and farming in the late 1800s, and extensive dune stabilization programs began in the early 1900s.

Sand dune fixation requires the establishment of a permanent plant cover (Kellman and Kading, 1992). The cover directs or lifts the prevailing wind currents away from the easily eroded sandy surface. This usually cannot be accomplished directly by planting trees. The young trees simply are not sufficient to alter the wind currents, and the abrasive sands often damage unprotected seedlings. Consequently, blow sands are first stabilized temporarily by such devices as parallel rows of fencing, closely spaced stakes driven into the sand, surface trash, plantings of drought-resistant grasses, or a combination of these techniques. The distance between the rows of stakes or grass depends on the intensity of erosion but may be no more than 1 to 2 m.

Xerophytic clump grasses have been used successfully in stabilization efforts, but normally they cannot be established in infertile sands without the application of a mixed fertilizer. In New Zealand (Gadgil, 1971), nitrogen fertilizer was used to establish bunch grass as a preliminary step in stabilizing blow sands. After about 2 years, lupine was broadcast seeded without additional fertilizer. After 2 or 3 years, the lupine was mechanically crushed or disked in order to reduce competition for planted pines. It was estimated that the reseeding legume supplied the soil with 200 to 300 kg/ha of nitrogen during the 5 to 6 years before the pine overtopped it.

As soon as the sand movement has been halted, the dunes are normally planted with a hardy tree species (Figure 15.5). Conifers have generally been superior to hardwoods on adverse sites, although both types have been used. Radiata pine is planted on dune sands in New Zealand, and maritime pine is often used in southwest France. Jack pine, red pine, Scots pine, pitch pine,

Figure 15.5 Planting radiata pine on a stabilized sand dune in New Zealand.

Virginia pine, and sand pine have been used on less favorable sandy sites in other areas, while white pine, Norway spruce, locust, and sometimes alder have been planted on the moister sands between dunes. Planting distances for pine seem to vary from 1.5 × 1.5 m on eroding sites to 2.0 × 2.0 m on other sites. Hand application of a fertilizer equivalent to about 200 kg/ha of diammonium phosphate, or spot application of the same material applied in an opening 20 to 30 cm on either side of the seedling and 15 cm deep, is often done soon after planting. The amount of fertilizer at each spot should not normally exceed about 50 g in order to reduce the danger of salt damage to the seedling root system. The nutrient ratio should be based on soil fertility conditions as determined by experience or by a soil test.

There are vast areas of deep, dry sands in many countries that require special management techniques for the establishment of desirable tree species. These marginal sites often support woody scrub or grass vegetation and are therefore well stabilized. Droughty soils of the southeastern United States sandhills are examples of these adverse sites. These sites, mostly Psamments, support remnants of long-leaf pine stands harvested around the turn of the twentieth century, plus scrub hardwoods and wiregrass (*Aristida stricta*). Early harvesting methods made no provision for regeneration of the pine, and subsequent efforts at reforestation, especially with slash pine, have not been highly successful. The understory scrub oaks and wiregrass prevent natural establishment of most pines (Burns and Hebb, 1972; Outcalt, 1994).

Irrigation or Fertigation on Dry Sites Is Becoming Popular

Irrigation of forests has long been practiced but on a limited scale. There are nearly 250,000 ha of irrigated plantations in Pakistan (Sheikh, 1986) and probably and a somewhat larger area of irrigated plantations in India. Some of these plantations were established as early as the 1860s. Production on these sites has been excellent, but it often declines due to increasing salinity (Ahmed et al., 1985). Irrigated fuelwood plantations are common in the arid regions of Africa and the Middle East (Wood et al., 1975; Jackson, 1977). Some investigations on the use of irrigation or fertigation, irrigation with a nutrient solution, have been carried out in the United States (Leaf et al., 1970; Baker, 1973; Harrington and DeBell, 1995; Dougherty et al., 1998). In general, these practices result in improved growth, but they are costly and run the risk of nutrient loss into surface water or groundwater. Horticulturists seem to have perfected fertigation regimes for orchard crops, and forestry could probably apply these regimes to forest plantations if growth increases warranted the expense incurred.

Seasonally Flooded Soils Present Unique Problems

These periodically wet soils occupy a rather large percentage of the tropical rain and monsoon rain forests, but because of the high evapotranspiration rate of these ecosystems, excess water is seldom a limiting factor for tree growth. Other extensive areas of seasonally wet forests are found in coastal flats, basins, and soils with underlying impervious layers. Examples of wetland forests are the nearly 8 million ha along the coastal plain of the southeastern United States (Klawitter, 1970b). These complex wetlands are very broad and nearly level interstream areas of coastal flats. They are interspersed with **bays** and **pocosins**, which are characterized by high centers with natural drainage impeded by elevated rims, sluggish outlets, and impermeable subsoils, as well as swamps and ponds. These wetlands support unique vegetation, act as filters for water moving from uplands to streams, and have gained protection through environmental legislation in many countries. Such legislation often allows "normal silvicultural operations" to be carried out, but controversy as to what constitutes normal silvicultural operations continues (Messina and Conner, 1998).

The bays and swamps are generally wet year round and are often covered with one to several feet of peat. However, an excessively high water table during only a few months of the year appears to be the primary limiting factor to intensive woodland management of most wet plains. During seasonally dry periods, the plains can be exceedingly dry because their primary source of water is precipitation. Tree cover may consist of scattered suppressed stems of one of the southern pines or locally dense stands of pines and hardwoods, with an understory of grasses and shrub clumps. Occasional rises of a few centimeters in the land surface may retard surface water flow over large areas, except in the immediate vicinity of shallow streams. Water moves slowly though the soils because of slowly permeable subsoil layers and small hydraulic heads between the soil surface and the stream.

Some forms of **water control**, mostly through ditch drainage, have been applied to thousands of hectares of forests on wet flats and in the Southeast and Gulf Coastal Plains of the United States since 1960. The objectives have been to (1) improve the trafficability of woods roads so as to provide ready access for harvesting, fire protection, and tending; (2) reduce soil compaction and puddling; (3) facilitate regeneration of pine on prepared sites and ensure early survival of planted seedlings; and (4) increase growth or reduce rotation age (Hewlett, 1972). Whether or not such practices qualify as normal silvicultural operations is still hotly debated.

Surface **drainage ditches** may not remove water from soils of the wet flats very effectively because the drawdown of the groundwater table along the ditch may extend only a few meters inland from the ditch. The effective distance from the ditch is a function of the ease with which water moves laterally through the soil. The higher the percentage of large pores in the soil, the more effective ditch drainage will be.

Young and Brendemuehl (1973) reported that a drainage system consisting of main drains 1.2 m deep and 800 m apart with lateral collecting ditches 0.5 m deep spaced about 200 m apart improved the site productivity of a Rains loamy sand (Typic Paleaquult), but the system was not intensive enough to provide the pole-sized slash pine with continuous protection from excess water. Furthermore, natural aging of the system tended to return the area to its original state following several wet years. In drainage tests in northeastern France on pseudogley soils with impermeable layers at 25 to 50 cm, drainage ditches at 20-m intervals improved the growth of Norway spruce, Douglas-fir, and larch almost as much as spacings of 10 m and significantly more than spacings of 40 m (Levy, 1972).

In some cases, the use of water by a semimature stand of trees will eventually make the ditch system unnecessary or even detrimental to stand water relations. Since tree growth is best when water table depth is constant (White and Pritchett, 1970), it is wise to have some control on the level to which a drainage system lowers the water table (Fisher, 1980).

The benefits from draining wetlands generally derive from increased rooting depth, which in turn reduces windthrow, improves soil aeration, and increases the nutrient supply. The last result may come about through accelerated oxidation of soil organic matter, as well as through increases in the volume of root-exploitable soil. However, excessive drainage of permeable soils may actually reduce growth and inhibit regeneration of certain types of wet-site trees, such as tupelo and cypress. In a drainage experiment on Leon fine sand (Aerie Haplaquod), 7-year-slash pine growth was not improved by 0.5- or 1.5-m-deep ditches. Poor growth near ditches apparently resulted from excessive removal of groundwater, particularly during dry periods (Kaufman et al., 1977).

Bedding, or row mounding, may be an alternative to lateral ditching of wet flats and is certainly a normal silvicultural operation. Klawitter (1970b) found that 4-year-old slash pines planted on beds were 13 percent taller than trees on unbedded plots in a Typic Paleaquult, even though the experimental area was adjacent to a main drainage ditch. In another part of the coastal plain on an undrained soil with similar properties, slash pine averaged 67 percent taller on bedded plots than on unbedded plots after 3 years (Pritchett and Smith, 1974). Early advantages of bedding may not persist into the later stages of stand development on some sites; however, on wet flats, where trees stagnate without some relief from excess water, the response to bedding persists throughout the rotation.

An improvement in accessibility and an increase in the growth rate of stagnant stands of pine often found on the seasonally flooded flatlands appear to be desirable objectives. However, it is not at all clear whether these objectives can or should be obtained through major drainage projects. A combination of water control, through perimeter ditching for the removal of excess surface water, and bedding and fertilization, where needed for improved tree survival and growth, appears to be a more appropriate management scheme. The forest manager must strive for multiple-use land productivity while maintaining a relatively stable ecosystem on these delicate sites.

Peatlands Are Often Managed for Forest Production

Peatlands occupy vast areas of permanently wet lands, especially in cool climate regions such as those found in Scandinavia, Siberia, Canada, and parts of the British Isles. These **organic soils** occupy as much as 15 to 30 percent of the total land area of Sweden and Finland, where they often support poor stands of spruce and pine. In Finland, swamp drainage for forestry purposes has been carried out on over 2.5 million ha. Another 3 to 4 million ha of peatlands are worthy of drainage in regions in which forest production is favorable, while the remaining 3 million ha will be left to nature conservancy, according to Huikari (1973). Holmen (1971) reported that about 7 million ha, 17 percent of the land area of Sweden, may be regarded as peatlands with little or no forest production because of adverse water and nutrient conditions. Another 1 million ha have been drained. Water is generally controlled at 20 to 40 cm below the soil surface with a series of interconnecting ditch drains and check dams (Figure 15.6).

Canada has 130 million ha of peatlands, much of which is forested (MaeFarlane, 1969; Kuhry et al., 1993), and large areas of forested peatlands occur in Minnesota, Michigan, and parts of New England. These forested "muskegs" illustrate nicely the two major interests of foresters and ecologist in these unique soils. The first interest is in their stability in the face of perturbation (Sayn-Wittgeostein, 1969); the second is in the productive capacity of the muskeg forests (Vincent, 1965). Many of these organic soils are shallow accumulations over bedrock. These shallow peatlands are often quite

Figure 15.6 Main lateral ditch in a water control system in Finnish peatland planted to Scots pine.

productive of timber, wildlife habitat, and water, but when they are disturbed, they break down quickly and are reformed at an astonishingly slow rate. On deep peat, there is often such a high water table rise after logging that reforestation of the site is very difficult. The management of these complex lands is difficult, and our understanding of the proper techniques for the protection and utilization of peatlands remains incomplete.

Holmen (1969) pointed out that afforestation of drained peatlands is not generally successful without the use of fertilizers. Phosphorus and potassium are the elements most often deficient, and applications of 45 kg phosphorus and 120 kg potassium per hectare often suffice for 15 to 20 years. On poorly decomposed peat, it is also advisable to apply about 50 kg nitrogen per hectare. Micronutrients are not generally deficient.

Water control in peatland can do more than increase forest production. It can improve the habitat and supply of food for wildlife and enhance recreation possibilities and other types of multiple use. However, drainage and site preparation may increase fire danger, subsidence, and nutrient loss and can lead to irreversible drying (Charman et al., 1994; Trettin et al., 1997). There has been concern that drainage and site preparation might lead to increases in CO_2 and CH_4 emissions from peatlands (Roulet and Moore, 1995; Minkkinen et al., 1997). Although the emission of these gases is increased by forest practices, the additional CO_2 fixation by rapidly growing forest plantations more than offsets the increased emissions and makes drained and site-prepared peatland sites net carbon sinks (Laine et al., 1997).

Gelisols Present Special Problems

In regions with a long period of winter cold and a short period of summer warmth, a layer of frozen ground develops that does not completely thaw during summer. This perennially frozen ground is termed **permafrost**. Soils with permafrost are placed in a separate order, Gelisols, in the USDA classification system (Bockheim et al., 1997). Gelisols, largely a Northern Hemisphere phenomenon, occur in a circumpolar zone in North America, Greenland, Europe, and Asia. Permafrost may also occur in the periglacial environments of high mountains in either hemisphere. Nearly half of the area of Canada and the former Soviet Union is affected by permafrost (Baranov, 1959; Brown, 1967). All of Greenland and 80 percent of the land surface of Alaska are underlain by some type of permafrost (Ferrians, 1965).

Permafrost can occur in either mineral or organic soils. It is sometimes divided into three types: (1) seasonally frozen ground that is not actually permafrost since these areas may thaw during a small portion of the year; (2) discontinuous permafrost, where the frozen layer is discontinuous in horizontal extent; and (3) continuous permafrost, where the frozen layer is continuous in horizontal extent. In Alaska, the tree line and the division between continuous and discontinuous permafrost generally coincide; however, in Asia, the zone of continuous permafrost extends far south of the tree line (Figure 15.7).

The management problems associated with permafrost generally arise from the important role that vegetation plays in insulating the soil from temperature change and thus perpetuating the permafrost layer. When forests are harvested or burned or when tundra vegetation is destroyed by vehicular traffic during the summer season, the permafrost layer may thaw. This leads to soil subsidence and either flooding or cryoturbation of the soil (Nicholas and Hinkel, 1996; Swanson, 1996: Arseneault and Payette, 1997). Either of these changes in soil physical properties usually precludes new plant growth for decades while the soil structure and permafrost layer redevelop.

The placement of structures such as roads, buildings, and pipelines on permafrost soils also presents great difficulties if the soil is to be prevented from thawing (Brown, 1970). It is imperative that management practices that preserve the permafrost layer be developed. This means only partial removal of the tree canopy, the construction of insulating pads beneath buildings, and the restriction of vehicular travel over the ground.

Turning Mine Spoil Back Into Soil Is a Challenge

Revegetation of land that has been drastically disturbed by open pit mining for coal, iron ore, sand, phosphates, and other resources is a slow process at best. These wastelands must often be graded, partially resurfaced with fertile soil, fertilized and limed, and stabilized before afforestation is attempted. The kind and degree of amelioration depend on the physical, chemical, and biological properties of the spoils. These properties vary, depending on lithology, spoil

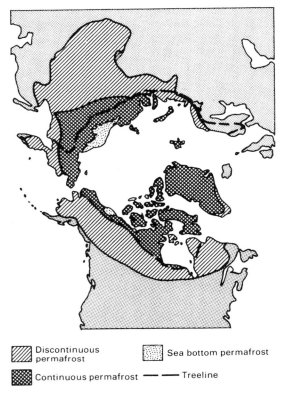

Figure 15.7 Distribution of permafrost and location of the tree line in the Northern Hemisphere (French, 1976). Used with permission.

age, degree of weathering, and erosion processes. Bare spoils cause serious environmental problems in many places by drastically affecting adjacent land, water, forests, or aesthetic values. Because of the widespread concern about their degrading effect on the environment, reclamation of spoil banks has received considerable research attention in recent years (Burger and Torbert, 1992). There are grass, shrub, and herbaceous species adapted to most classes of spoils, but there are few tree species adapted to the most severe spoil sites.

It is common practice to establish grasses or mixtures of grasses and legumes on severe spoil sites. After these species are seeded, slopes are generally mulched with straw (4 to 5 mg/ha) and the mulch is anchored with asphalt emulsion. After a vegetative cover becomes well established, trees are often planted. Care must be taken to avoid allelopathic cover crops, or the establishment of trees on stabilized spoils may be very difficult (Fisher and Adrian, 1980; Larsen et al., 1995).

Some spoils support excellent tree growth without any site preparation or soil amendments, while others require grading, terracing, harrowing, or ripping

(Dunker et al., 1995). Grading of spoils high in clay content may cause compaction, thereby reducing air and water infiltration, root penetration, and tree growth. By contrast, Czapowskyj (1973) reported on studies in Pennsylvania indicating that coarse-textured anthracite spoils were improved by compaction resulting from grading operations. He also found evidence that calcium, magnesium, and potassium concentrations were higher in certain graded spoil types than in their ungraded counterparts.

When the soil drainage class is altered by the removal of impeding soil layers, regrading often does little to re-create sites that will support the tree species that occupied the area before mining. Darfus and Fisher (1984) found that when Spodosol soils that supported slash pine were dredge mined, the spodic horizon was disturbed and postmining soil moisture regimes did not support slash pine.

All essential elements are found in most overburden, but concentrations vary from very low to toxic. Most mine spoils are deficient in nitrogen, and vegetative growth on these spoils results from nitrogen added by rainfall, biological fixation, or fertilizer addition. However, Cornwell and Stone (1968) found that certain anthracite and black shale spoils supplied sufficient nitrogen for good tree growth as a result of weathering of these materials. Spoils of other mining operations are generally devoid of organic material and nitrogen. May et al. (1969) reported that while kaolin mining spoils were extremely variable, most of them were deficient in nitrogen, as well as in phosphorus, potassium, and calcium. Marx (1977) obtained improved survival and growth of pine seedlings planted on spoil banks when the seedlings were inoculated in the nursery with select strains of ectomycorrhizal fungi (Figure 15.8).

Levels of available iron, manganese, zinc, and other metals may be sufficiently high in very acid soil materials to pose a serious problem in the establishment of vegetation. Fortunately, toxic strip mine spoils are not widespread, and most toxic conditions can be eliminated with applications of limestone. Although lime is a relatively inexpensive material, the extremely high application rates sometimes needed on the more acid mine waste can be quite costly. From 10 to 90 ton/ha may be required to neutralize the acids contained in spoil materials before leguminous plants and grasses can be grown successfully (Czapowskyj, 1973).

Vegetation usually grows better on reclaimed spoils if a variety of amendments are used to ameliorate the site (Schoenholtz et al., 1992). A wide variety of organic amendments have been tested including papermill sludge, municipal sewage sludge, mushroom compost, pulverized fuel ash, wood chips and sawdust, and shredded municipal garbage (Sabey et al., 1990; Feagley et al., 1994; Pichtel et al., 1994; Stark et al., 1994; Sort and Alcaniz, 1996; Perkins and Vann, 1997). All of these help promote vegetation establishment to some degree.

Inoculation of the sites with mycorrhizal fungi appears to be as important as the addition of organic matter (Shetty et al., 1994). Ectomycorrhizae reinvade the site rapidly if there are undisturbed areas nearby to serve as

Figure 15.8 Reclaimed coal mine spoils in northern Bavaria. After leveling, a mixture of alder, birch, oak, and larch was planted.

sources of spores (Gould et al., 1996). The soil-borne spores of arbuscular or endomycorrhizae do not reinvade a site readily. In addition, the spores of these organisms have usually been eaten or killed in stockpiled topsoil, so it is usually necessary to plant mycorrhizal seedlings or to inoculate the site artifically (Allen, 1991; Mehrotra, 1998).

MANAGEMENT OF NURSERY SOILS

The successful operation of a forest nursery involves large investments in labor, buildings, and equipment, and only an operation sufficiently large to make effective use of the specialized machinery can normally be justified. Except in special situations where part of the equipment may be rented or made to serve double duty, a nursery of 5 to 10 ha might be considered a minimum size for an economic operation. A nursery of this size can produce 8 to 16 million 1-0 pine seedlings per year, providing seedlings for 4000 to 8000 ha of plantations. A substantially larger area is needed if a rotation with a green manure or cover crop is followed, or in cool climates where two years or more are needed to produce seedlings of satisfactory size for field planting. Armson and Sadreika (1974) have presented excellent instructions for nursery soil management for northern coniferous species.

Figure 15.9 View of a southern pine nursery with a capacity of about 25 million seedlings per year.

Although seedling production in the United States has been steadily increasing, especially in the private sector, the production of both hardwood and coniferous seedlings must increase to meet the need for expanded areas of plantation forestry. Figure 15.9 shows a southern pine nursery. The production of containerized nursery stock has also been steadily increasing, but in many parts of the temperate zone, "bare root" production of nursery stock still predominates.

Nursery Site Selection Is Important to Successful Production

Because nurseries are not easily moved from one site to another, it is especially important that much thought be given to soil and site factors that make for efficient operations in the production of quality seedlings (Morby, 1984). **Soil depth** and **texture** are of prime importance for nurseries. A soil depth of 1.0 to 1.5 m is normally desired, without a radical textural change between horizons, to ensure adequate drainage and aeration and space for root development.

Areas exposed to extreme temperatures and strong winds should be avoided. In cold climates, location near large bodies of water will moderate temperature extremes. On the other hand, terrain should be selected that permits free drainage of cold air, thus reducing the risk of frost injury. The nursery site should be located near a reliable transportation network and as close as possible to major areas of reforestation, both for convenience and to minimize differences in growing season length and other climatic conditions between the two areas.

Soil Texture Is Often the Key to Success or Failure

Soil texture influences water regime, cation exchange capacity and nutrient retention, susceptibility to sand splash, surface washing, compaction, and the ease of lifting. Loamy sands to sandy loams are generally preferred textures for nursery soils. Coarse-textured sands are acceptable for nurseries, but coarse textures place special demands on management of water, fertility, and organic matter. Soils at the other end of the textural range, those containing more than 30 to 40 percent silt plus clay, are definitely to be avoided. Soils containing a high percentage of silt often become very hard when dry, erode easily, and demand a meticulous watering regime. Soils high in clay are often difficult to cultivate, make lifting difficult, and require considerable effort to adjust moisture, acidity, and nutrient levels. They are also slow to warm in the spring, and in cold climates, seedlings may suffer from frost heaving. Furthermore, seedling root systems are likely to be injured during lifting from fine-textured soils.

Site Leveling May Be Necessary but Is Problematic

Level terrain is preferred for nurseries on sandy soils, but a slight slope may be necessary to provide surface drainage for finer-textured soils. Sufficient grading and leveling should be done to minimize the erosion hazard to beds, translocation of fertilizer salts, and accumulation of water in depressions. However, exposing large areas of subsoil should be avoided. The movement of large amounts of topsoil may result in nutrient-poor or even calcareous subsoil being near the surface. Where extensive areas of topsoil must be removed during leveling operations, care should be taken to stockpile the topsoil and then redistribute it over areas that were exposed during grading. If such redistribution is not possible, a special effort must be made to adjust soil acidity and to increase the organic matter content of the disturbed soil before any attempt at seedling production is made.

Ideally, a **soil map** will have been used in siting the nursery. A more detailed soil map of the nursery area, including any reserve area for later expansion, should be made as soon as possible after site selection. The map should indicate the soil texture in both surface and subsurface horizons, depth to restricted drainage, changes in soil acidity, and nutrient status and other properties critical to seedling growth. For example, hardwood seedlings generally require more fertile and less acid soils than conifer seedlings, so that if both hardwoods and conifers are grown in the same nursery, it may be possible to reserve the more fertile areas for hardwoods. Furthermore, a detailed map makes a convenient base for planning and record keeping—two essential steps in any nursery operation.

Soil Fertility and Acidity Management Are Crucial to Nursery Success

A reasonable concentration of nutrients in the original soil is desired, but since they can be added as fertilizers, the soil reserve of nutrients is not of overriding importance. It is more important that the soil not contain high concentrations of soluble salts, free carbonates, or toxic materials and that it contain a reasonable level of organic matter.

Optimum soil acidity for most tree species lies between pH 5.2 and 6.2. The lower half of this range appears to be best for conifers, while many deciduous trees grow best in the pH range from 5.8 to 6.2 (Figure 15.10). Although many species thrive when soil acidity is high, the efficiency of fertilizer use is decreased in strongly acid soils. The amounts of native calcium, magnesium, and potassium are usually low in very acid soils; furthermore, added potassium is more easily leached and the availability of added phosphorus is reduced under these conditions. Aluminum, iron, and manganese tend to be much more soluble in acid soil. Moderately high concentrations of these elements probably

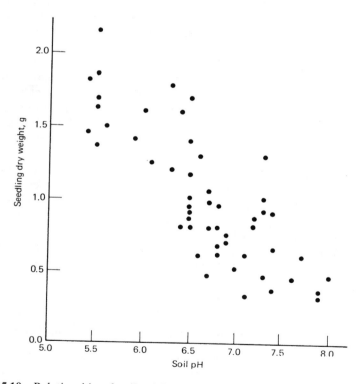

Figure 15.10 Relationship of soil acidity to mean total dry weight of 2-0 red pine seedlings (Armson and Sadreika, 1974). Courtesy of the Ontario Ministry of Natural Resources.

have little direct effect on conifer growth, but they may well affect some deciduous seedlings and the cover or green manure crops grown in rotation with the seedlings.

At the other end of the range, near neutrality, some species have difficulty obtaining sufficient iron, and sometimes manganese, for normal growth. Pine seedlings are particularly subject to iron deficiencies in soils above approximately pH 6.5. Most tree species, including pines, can be grown in soils above pH 7.0 if their nutritional requirements for iron, manganese, and phosphorus can be met with chelates or foliar sprays. In fact, they can be successfully grown in some finer-textured soils near neutrality without micronutrient additions if the soils are well supplied with micronutrients and organic matter. Some root diseases are more prevalent in near-neutral and alkaline soils. Obviously, soils selected for nursery sites must permit control of acidity. If irrigation is anticipated, the water should be checked for soluble salts. For example, 25 cm water containing 500 ppm soluble salts adds 1250 kg/ha of these materials to the soil. This is more than enough salt to significantly affect the pH value of weakly buffered sandy soils.

Simple indicator test kits and pH meters are useful for checking soil acidity, but it is well to keep in mind that changes in soil acidity of as much as 0.5 to 1.0 pH units can take place seasonally. Furthermore, microsite differences caused by leveling or organic matter applications often result in wide variations in pH value among soil samples. Where adjustments in soil acidity are needed, agricultural limestone or sulfur is usually added. However, small changes in soil acidity can be made by using fertilizer sources that have acidic or basic residues. The application of 200 kg/ha of nitrogen as ammonium sulfate, for example, can increase the acidity of sandy soil as much as 1.5 pH units during a season.

Care should be exercised in using limestone or sulfur to alter the acidity of sandy soils to prevent excessive change in reaction. The kinds and amounts of clay and organic matter control the soil's cation-exchange capacity, which in turn determines the amount of lime or sulfur required to change soil acidity by a given amount. Generalized curves useful in determining the amount of agricultural limestone required to lower the acidity of some soil textural groups are presented in Figure 15.11. The same curves may be used to predict the amount of sulfur required to raise soil acidity from the present level to a desired level. About one-third as much sulfur as agricultural limestone is required to lower the soil reaction by a given amount.

Soil Organic Matter Is Generally the Key to a Productive Nursery

The maintenance of a reasonable level of organic matter is particularly important in sandy nursery soils to retain fertilizer elements against leaching and to buffer the soil against rapid changes in acidity. Because these soils contain little clay, their exchange capacity resides almost entirely with the organic fraction. Organic matter additions also improve soil structure, friabil-

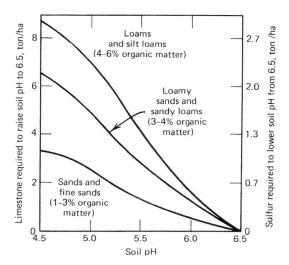

Figure 15.11 Tons of agricultural limestone required to lower soil acidity or tons of sulfur required to increase soil acidity. Differences in requirements correspond to differences between present and desired pH values.

ity, and water intake and retention and reduce soil crusting and erosion. The organic matter of a soil is the seat of microbial activity, and it contains the reserve of most nutrients in sandy soils.

The practical level of organic matter that can be maintained in sandy soils is largely dictated by climatic conditions. For example, maintaining 4 to 5 percent organic matter in Oregon nursery soils may be less difficult than maintaining 1 to 2 percent in southeastern coastal plain soils. Soil organic matter can be increased briefly by the use of green manure crops. Significantly longer increases can be achieved by the addition of compost, sawdust, peat, or other organic material. However, such increases are temporary because of the decomposition of these materials by soil organisms, and they must be replaced on a regular schedule (Davey, 1984; Rose et al, 1995).

Cover Crops Serve Several Purposes

Nursery seedlings are often grown in rotation with green manure and cover crops in attempts to replace organic matter lost by the high biological activity associated with well-aerated and fertilized nurseries. Each 1 percent of organic matter in the top 20 cm of a sandy soil is equivalent to 24 to 28 mg of dry matter per hectare. Therefore, a nursery soil may normally contain 50 or 60 mg of organic matter. While a good green manure crop may add 20 to 30 mg of fresh material to the soil, only 5 to 6 mg of this is dry matter. Moreover, when mixed into the soil, 50 to 75 percent of this material will decompose within a few months, leaving only 1 or 2 mg of relatively stable humus to add to the soil's organic matter reserve. This is little more than that required to replace normal decomposition losses during a cropping season. While the benefits from green manure organic materials are rather short-lived, the use of green manure is, nonetheless, the most practical means of maintaining the productivity of many nursery soils.

Rotations with green manure crops can also be an effective means of reducing infestations of nematodes and soil-borne diseases that often infest nursery soils, provided that the crops selected are resistant to the growth and development of such organisms. A cover crop may also be selected that will increase the population of certain endomycorrhizal fungi useful to the hardwood seedlings that follow. The selection of a suitable green manure crop should also be related to soil and climatic conditions. For example, nursery soils are often too acid for good growth of many leguminous crops. Some species of vetch, lupine, soybeans, and field peas may be used, but to obtain rapid plant growth and abundant production of organic material, one must usually resort to crops such as oats, rye, millet, corn, sudangrass, sorghum, or a sorghum-sudan hybrid.

Cover crops also protect the soil from wind and water erosion and reduce nutrient leaching during any extended period when an area is not used for the production of nursery stock. A cover crop should be plowed under before it reaches maturity to avoid the nuisance of a volunteer crop in later nursery beds. However, the crop should be allowed to grow as long as feasible because of the increased dry matter production and because mature and lignified materials do not decompose as rapidly after being turned into the soil.

Other Organic Additions Are Generally Necessary

Composted and raw organic materials such as sawdust, straw, peat, and ground bark have long been important sources of materials for improving the physical conditions of soil and increasing organic matter in nurseries. Since these are often waste materials, they are relatively inexpensive to obtain but costly to handle. More recently, the supply of these materials has diminished to the point where they are used principally on nursery soils with critically low levels of organic matter or to correct problem areas, such as those resulting from leveling operations. At the same time, digested sewage sludge has become a common amendment for increasing soil organic matter content.

Composted materials consisting of sawdust, straw, and other materials with a wide carbon:nitrogen ratio are mixed with topsoil and/or mineral fertilizers or animal manure and allowed to partially decompose or ferment. These materials range in carbon:nitrogen ratio from over 1000:1 for raw sawdust, to 500:1 for fresh bark, 60:1 for peat, and 10:1 for digested sewage sludge. During the fermentation process, carbon dioxide develops, the carbon:nitrogen ratio is decreased, and the mineral constituents are made available. Composted materials can be applied at fairly heavy rates of 40 to 60 mg/ha and incorporated into the seedbed without danger of significant nitrogen immobilization. There is a danger of introducing root pathogens with some composted materials. Fumigation of the materials prior to application, or of the nursery soils after application, will minimize this problem. Moreover, composted materials are quite expensive to use, so raw organic materials are the more common amendments for nursery soils. In using raw organic materials, care must be exercised to prevent excessive nutrient immobilization during critical periods.

Relatively inert materials such as acid peat, sawdust, and digested sludge are slow to decompose and thus persist in the soil for longer periods. While they are less expensive to use than composted materials, they can result in nitrogen shortages at critical periods in seedling development, and more caution is required in their use. Allison (1965) reported that the nitrogen requirements of microorganisms that decomposed sawdust ranged from 0.3 percent of the dry weight of Douglas-fir wood to 1.4 percent for red oak wood. Digested sludge often has a high content of heavy metals that may damage tree seedlings or cause other environmental problems (Rose et al., 1995). Great care must be taken in choosing the source of digested sludge for use in nursery soils.

When raw sawdust or other highly carbonaceous materials are to be used directly in nursery soils, they should be applied sufficiently in advance of the seeding operation so that partial decomposition takes place prior to seed germination. These applications may be made prior to planting of a cover crop or green manure crop or more than a month prior to seeding the nursery beds. In any event, 10 to 20 kg of extra nitrogen is needed per milligram of dry organic materials added in order to hasten decomposition and minimize the danger of a nitrogen shortage to the nursery crop. Since this extra nitrogen will become available to nursery stock after the added organic material has largely decomposed, care must be taken to prevent excessive levels of nitrogen in the planting bed at later stages of seedling development. Monitoring the concentration of nitrogen in seedling tissue, or the available nitrogen in the soil, and applying supplemental nitrogen only when the results indicate a need can accomplish this.

A Fertilization Program Is Essential to Excellent Nursery Stock Production

While added composted materials, manure, and organic residues are acceptable means of maintaining soil organic matter levels that contribute to good physical properties, they generally do not contain sufficient nutrients to replenish those lost during cropping. Consequently, the maintenance of nursery soil fertility depends largely on the use of commercial fertilizers, applied alone or in combination with organic materials and/or green manure crops (van den Driessehe, 1984).

Relatively large amounts of fertilizers are often used in nurseries to replace nutrients removed in the harvested crop, as well as those lost by leaching or

TABLE 15.2   Nutrient Concentrations and Contents, per Hectare, in 8-Month-Old Slash Pine Seedlings

Component	N	P	K	Ca	Mg	Al	Cu	Fe	Mn	Zn
				Nutrient Concentration (mg/kg)						
Tops	1.20	0.15	0.60	0.43	0.09	408	4.8	118	161	48
Roots	0.86	0.23	0.59	0.36	0.08	1947	10.8	617	66	40
				Nutrient Content (kg/ha)						
Tops	44.6	5.6	22.3	16.0	3.3	1.5	0.18	0.44	0.60	0.18
Roots	8.7	2.3	6.0	3.6	0.8	2.0	0.01	0.62	0.06	0.04
Total	53.3	7.9	28.3	19.6	4.1	3.5	0.19	1.06	0.66	0.22

fixed in the soil. Leaching losses can be rather high in irrigated sandy soils even where the organic matter content is maintained at a reasonable level. Crop removal may also be substantial because both seedling roots and tops are removed. Slash pine seedlings from a Florida nursery, grown at the rate of 2.2 million plants per hectare, weighed 16,500 kg (green weight) when harvested at 8 months of age. Oven-dry weights averaged 3715 and 1012 kg/ha for seedling tops and roots, respectively. This results in the removal of large amounts of nutrients (Table 15.2). Armson and Sadreika (1974) reported even larger removals of nutrients in 2-year-old red pine and spruce (Table 15.3).

Soil Testing Is a Must

Addition of relatively large amounts of organic materials and mineral fertilizers, along with removal of nutrients in the crop and by leaching, result in considerable fluctuation in nutrient levels in nursery soils during a cropping period. A soil test provides information on the fertility status of soils that is particularly useful prior to seedbed preparation. Soil acidity and nutrient status are normally adjusted prior to seeding the tree crop. Since plants are not available for tissue tests at this time, soil tests are the principal diagnostic tool. Soil samples should be collected from seedbeds 4 to 6 weeks prior to seeding to allow adequate time for sample analyses and the application of any necessary fertilizers.

A composite soil sample should be taken from each nursery management unit or from areas within a management unit that differ in soil type, past management, or productivity. The composite sample should consist of 12 to 15 subsamples (cores) taken at random over a uniform area. Cores are normally taken to a 15- to 20-cm depth with a 2.5-cm-diameter soil tube and collected in a clean container. After air-drying and mixing, about 250 ml is packaged,

TABLE 15.3 Amounts of Nitrogen, Phosphorus, and Potassium, in Kilograms per Hectare, Contained in 2-Year-Old Red Pine and White Spruce Seedlings Planted at Four Seedbed Densities

Nutrient	Seedbed Density (Plants/m^2)			
	108	215	430	861
Red Pine				
N	113	119	164	220
P	16	22	27	35
K	56	80	109	133
White Spruce				
N	88	107	150	167
P	19	24	31	41
K	44	60	71	114

Source: Armson and Sadreika (1974). Courtesy of the Ontario Ministry of Natural Resources.

labeled, and submitted to a laboratory for testing. The location from which the sample was collected should be identified on a map of the nursery on which each management unit is designated. A detailed soils map can serve as a basis for record keeping as well as management planning. Chronological records of soil test results, fertilizer treatments, fumigation, seeding date and rate, irrigation, and other cropping practices are useful to the manager in diagnosing problem areas and planning for future crops. Soil fertility tests should be carried out annually.

Fertilization Is a Two-Step Process

Limestone or sulfur, when used to adjust soil acidity, and base fertilizers are normally applied broadcast before fumigation and seeding or transplanting. In this way, the amendments can be well mixed throughout the rooting zone. If limestone or sulfur is needed, it should be mixed with the soil 2 or 3 months in advance of seeding, to allow time for them to react with soil components. By contrast, most fertilizer materials are applied near the time of seeding or transplanting to minimize the danger of leaching losses.

Phosphorus deficiencies are generally corrected by the application of 40 to 100 kg/ha of phosphorus as ordinary or concentrated superphosphate. The actual amount needed depends on soil-fixing or retention properties and the degree of deficiency, as indicated by a soil test. In acid sandy soils, it is often more economical to apply larger amounts of less soluble forms of phosphorus such as ground rock phosphate. Phosphorus in this form is less soluble and less likely to leach beyond the seedlings' shallow root system than phosphorus in soluble phosphates. Rates of 500 to 2000 kg/ha are normally applied at 1- to 3-year intervals, or as indicated by soil test results.

Nitrogen and potassium needs may be met by mixing part of the material into the seedbed prior to planting and applying the remainder in two or more "top-dress" applications during the growing season. Split applications are particularly effective for nitrogen because of the ease with which it can be lost from the system and the high salt index of most nitrogen sources. Soluble fertilizer salts may result in damage to transplants when present in high concentrations. Usually no more than 30 to 50 kg of mineral nitrogen is applied to sandy soils prior to seeding. Higher rates may be used on finer-textured soils or when less soluble sources, such as urea-formaldehyde or sulfur-coated urea, are used.

Most sources of nitrogen and potassium are completely water-soluble, and their solutions can be injected into the irrigation system for direct application. Solid materials can be applied as a top-dress to the seedlings. The number of top-dress applications needed depends on crop performance, soil conditions, rainfall, and irrigation patterns. However, three or four applications per year are not uncommon for sandy soils. Potassium is sometimes top-dressed on seedlings a few months prior to lifting from the nursery bed in an effort to harden and condition them to withstand cold temperatures and transplanting shock. Because the application of nitrogen can induce other nutrient deficiencies (Teng and Timmer, 1995; van den Dressche and Ponsford, 1995), asseys of seedling foliar nutrient content during the growing season are useful for maintaining balance.

Tree Conditioning for Transplanting Improves Survival and Growth

Undercutting, wrenching, and root pruning are widely used methods of conditioning seedlings of certain species. When properly scheduled and conducted, these cultivation treatments can be used to control top-root ratio, carbohydrate reserve accumulation, and the onset of dormancy. A reduction in top-root ratio may improve the success of plantings on droughty soils. Increased food reserves and early dormancy generally reduce cold damage and ensure better survival on adverse sites but, most important, an efficient root system is the key to success where transplants are stressed by low soil moisture.

Undercutting consists of passing a thin, flat, sharp blade beneath seedlings at depths of 15 to 20 cm. Undercutting is usually performed only once in a bed of seedlings several months before lifting. The purpose of the operation is to sever seedling taproots, reduce height growth, and promote development of fibrous root systems. It has been successfully used in nurseries for pines, oaks, and other species that tend to form taproots.

Wrenching has been described in detail by Dorsser and Rook (1972) as a method of conditioning and improving forest seedlings for outplanting. Their research on wrenching was largely confined to radiata pine, but the technique has also been used on other species, including hardwood seedlings. Wrenching can be performed with a sharp spade inserted beneath the seedlings at an angle that severs their taproots and then slightly lifting to aerate the root zone. However, it is usually accomplished by passing a sharp, thin, tilted blade beneath seedbeds at certain intervals following undercutting to prevent renewal of deep rooting or height growth.

Timing of the conditioning sequence is important, and it varies with tree species and soil and climate factors. The sequence generally begins when seedlings are near the height desired for outplanting, but with at least 2 months of growing season remaining. With radiata pine in New Zealand, the operation begins when seedlings are about 20 cm in height. A thin, flat blade is passed beneath seedlings at a depth of about 8 to 10 cm with minimum disturbance to lateral roots. After undercutting, wrenching is performed at 1- to 4-week intervals for the remainder of the seedlings' growing season, using a thicker, broader blade, tilted 20° from horizontal, with the front edge lower than the rear. The type of seedling desired dictates the interval between wrenchings.

The term root pruning is sometimes used to describe the undercutting operation outlined above. It has also been used to describe the trimming operation performed after seedlings have been removed from the ground. However, in New Zealand and other countries where wrenching is commonly practiced, root pruning is accomplished by vertical cutting with a coulter, or rolling circular blade, between rows of seedlings at 4- to 6-week intervals. The vigorous lateral root growth initiated after the taproot is severed is partially controlled by cutting the roots 8 to 9 cm from the trees. Wrenching and vertical root pruning are often used with slow-growing conifer species, such as spruce and northern pines, to produce 2 + 0 seedlings with a root area index equal to that of 1 + 1 or 2 + 1 transplants (Armson and Sadreika, 1974).

Biocides Protect Seedlings but Alter Nursery Soil Processes

Nursery crops are often grown as a single species of densely populated, rapidly growing succulent plants. Consequently, they are particularly vulnerable to the ravages of diseases, insects, and competing vegetation. Many nursery pests can be controlled, or their damage minimized, by manipulating the soil environment to ensure sturdy, vigorous seedlings. For example, the soil reaction should be kept below pH 6.0 to discourage the development of pathogenic fungi. Damping-off injury of very young seedlings, commonly associated with *Pythium, Rhizoctonia, Phylophthora, Fusarium,* and *Sclerotium* fungi, is rarely serious if the nursery soil is maintained at pH 5.5 or lower. High soil nitrogen concentrations at the time of germination should also be avoided. Low to moderate plant density and adequate moisture combined with good drainage of the surface soil generally result in optimum seedling growth without favoring disease and pests. Rotating tree seedlings with green manure and cover crops that are not hosts of disease organisms, insects, and nematodes is an accepted method of reducing the pest problem. However, in older nurseries, pests often cannot be controlled by soil management techniques alone, and biocides must be employed.

Soil fumigation to control weeds, soil-borne diseases, nematodes, and soil-inhabiting insects is a common and economically feasible practice in tree nurseries in many parts of the world. In addition to controlling damping-off organisms, fumigation with such compounds as methyl bromide, ethylene dibromide bichloropropene, and chloropicrin controls most root rots, probably the most destructive of all nursery soil pathogens. Fumigants are applied as gases under plastic covers or injected into the soil as a solution. In either event, the best results are obtained if the chemicals are applied when surface soil temperatures are between 10° and 30°C, when soil moisture is adequate for good seed germination, and when the soil has been recently cultivated to ensure good penetration. Penetration is not as good in fine-textured or compacted soils as in sands and sandy loams, and organic matter reduces the effectiveness of fumigants by sorption and degradation. Therefore, higher than normal amounts of fumigants may be needed on fine-textured soils with high levels of organic matter.

Methyl bromide use is being phased out because of its potential to destroy the ozone layer; however, other broad-spectrum biocides will undoubtedly take its place. All biocides alter the soil flora and fauna to a greater or lesser extent. Many microorganisms reestablish themselves in the soil within a few weeks after fumigation. The biological diversity of the soil microbial population is reduced, but the functional diversity is apparently not altered and decomposition processes are not greatly effected (Degens, 1998). However, mycorrhizal fungi may not recolonize nursery soils in some regions for several months or even years. This is particularly true of endomycorrhizal fungi that produce soil-borne, rather than air-borne, spores.

Long time lags for the reestablishment of ectotrophic mycorrhizae are more likely to be found in cold climates or where exotic species have been recently introduced. The time-honored methods of inoculating seedbeds by mixing a small amount of forest soil into the beds or using forest litter as protective mulch can hasten reestablishment. Because these methods may introduce pests into the nursery, commercial inoculum, which is now available in some areas, may a better source of mycorrhizal fungi (Bowen, 1965; Marx, 1977). Unfortunately, no source of endotrophic mycorrhizal inoculum other than soil or litter exists. Nurseries that produce endomycorrhizal tree species—most hardwoods—must either avoid the use of broad-spectrum biocides or systematically inoculate their nursery beds.

SUMMARY

Forest soils have been distinguished from agricultural soils by the fact that they receive little if any cultivation and few amendments. However, because of the rapid rise of plantation forestry and the increasing human management of forest soil, this distinction is beginning to blur. Plantations are literally human-made forests in the sense that they are established and maintained as the result of site manipulation. Such efforts to improve the site and increase tree survival and growth may have a profound influence on certain soil physical, chemical, and biological properties, especially properties of the forest floor. Some soils offer particular management problems. These include dry, sandy soils that are prone to wind erosion, seasonally flooded soils, organic soils or peatlands, gelisols or soils that contain permafrost, and mining spoils or wastes. Many unique techniques for managing forest growth on such soils have been developed. Forest tree nursery soils are not forest soils, but their management is essential to plantation forestry. The management of these soils employs many agronomic techniques and has become a highly developed subdiscipline within forest soils.

CHAPTER 16

Long-Term Soil Productivity

Perhaps the most important feature of soil ecology and management is long-term sustainability of productivity. This productivity may be defined in terms of soil biodiversity, ability to support the rapid growth of trees, or ability to support the same vegetation community that existed at some point in the past. The length of time over which productivity is to be sustained has variable components. Soil management for profitable growth of trees will have a primary focus on the current generation of trees, but productivity over several generations has greater importance to societies. On a sufficiently long time scale, soils change substantially through pedogenesis, including periods of increasing and perhaps decreasing fertility.

These issues of sustaining soils go beyond our current ability to measure and predict the effects of common management practices across landscapes and around the world. We know that fires remove nutrients from sites, but we are far from a comprehensive understanding of how this loss combines with all the other effects of fire to determine overall changes in soil fertility. We have an increasing number of very well studied soils and forests, but the number of these intensively characterized sites is dwarfed by the range and diversity of forest soils around the world.

The practice of forest soil management is invariably wrapped in great uncertainty, and the success or failure of operations depends on how well we apply what we know and how critical the unknown aspects are. A forester may do a conscientious job of minimizing nutrient losses during a harvesting operation, but soil productivity may be harmed by unexpected soil compaction.

This chapter examines some of the major long-term questions on the ecology and management of forest soils:

- How rapidly do forest soils change in the absence of management?
- How do forest soils change under the influence of different species?
- How does reforestation improve former agricultural soils?
- How does harvesting affect soil fertility?
- Will acid deposition degrade forest soils?
- If the climate warms, how will forest soils change?

Given the unavoidable uncertainty about forests soils, we begin with consideration of uncertainty and error.

THERE ARE THREE MAJOR WAYS TO BE WRONG

From statistics we are familiar with Type I and Type II errors. A Type I error involves accepting a hypothesis (and rejecting the null hypothesis) that is actually false or simply believing in things that aren't true. As noted in Chapter 13, a decision to fertilize a stand with a nutrient that is not limiting would be a Type I error. A Type II error involves rejecting a hypothesis (and accepting the null hypothesis) when in fact it is true or simply failing to believe in things that are true. With fertilization, a decision not to invest $100 to fertilize a hectare that would respond with $250 worth of extra growth would be a Type II error. We would also like to add a Type III error that may be common in forest soils: the null hypothesis is rejected for the wrong reason or the hypothesis is accepted for the wrong reason (this Type III idea came from Powers et al., 1994). A classic example in forest soils would be the conclusion that soil removal into windrows has no effect on subsequent tree growth because the soil nutrient supply is not critical. The prediction may be correct, but the inferred mechanism may not have been true; the lack of reduction in tree growth on windrowed plots could result from lowered nutrient supply affected by improved competition control.

HOW RAPIDLY DO FOREST SOILS CHANGE IN THE ABSENCE OF MANAGEMENT?

Over pedogenic time frames, we have reasonably clear information about changes in soil structure, chemistry, and fertility (see Chapter 2). These changes include increasing organic matter and nitrogen content, improving soil structure, and an overall increase in soil fertility. Beyond a certain point, a variety of process may lower soil fertility in old soils, including impeded drainage, development of anaerobic conditions, and long-term loss of unreplenished nutrients. What changes are common on shorter time periods—decades to centuries?

Given the brief history of forest soil science (and of statistics in general), very few sites have been followed for more than three decades; we simply don't have a direct opportunity to test ideas about soil changes over periods of 50+ years. The longest record for a single site comes from the Rothamsted Broadbalk "wilderness" plots, where in 1880 an agricultural field was allowed to revegetate with trees and shrubs. Soil carbon accumulated at a rate of about 3.5 Mg/ha annually (Jenkinson, 1994), and this classic experiment has influenced the way many soil scientists expect soil carbon to increase with afforestation. Unfortunately, the size of the wilderness is too small to separate the normal changes in forest soils from edge effects of neighboring fields (Figure 16.1), and the observed rates of change in the soils are probably much larger than would be found in a stand large enough to avoid edge effects.

The next-longest record is probably the work of O. Tamm and C.O. Tamm from Sweden. O. Tamm sampled soil pits in southwest Sweden in 1927 and had the foresight to record locations precisely enough that a resampling in 1982 could be made within a few meters of the original pits (Tamm and Hallbäcken, 1985). Over the 45-year period, soil pH declined by almost 1 unit in the O horizons and by about 0.5 unit even in the designated C horizons. These rates may not represent patterns that would be expected for unmanaged forest soils; some of the Norway spruce stands were plantations beyond the natural extent of the species, and all of the sites had been managed to some extent in the past. Further, atmospheric deposition may have contributed to soil acidification. Independent of the change over time was a decline in pH with stand age; at both sampling periods, older stands had lower soil pH. This pattern points to a possible cyclic effect of forest age on soil pH: acidification as the trees age but alkalinization during the disturbance that replaces an old forest with a young one. Soil alkalinization has been documented in some cases in Sweden, where the decomposition of organic matter and release of base cations were associated with consumption of H^+ (Nykvist and Rosén, 1985). Forest harvesting may have an opposite, acidifying effect when net nitrate production leads to increases in nitrate (and aluminum and H^+) in soil solutions (Brouwer, 1996; see later in this chapter). The lower pH in the C

Figure 16.1 The Broadbalk wilderness is too small to separate actual changes over time under trees from the edge effects of adjacent fields (S. Hamburg).

horizons suggests that the pH decline may have resulted in part from increases in ionic strength (see Chapter 5); current precipitation has higher ionic strength than in the past, and pH measured in water is typically sensitive to ionic strength.

An extensive resampling program looked for changes in the base cation content of forest floors across New England included the Hubbard Brook Experimental Forest, a repeatedly sampled chronosequence, and a regional sampling (Yanai et al., 1999). The precision of sampling and resampling varied among the sites and studies; the calcium content of the forest floors would have had to decline by about 1 percent annually to be detected with the sample sizes in these studies. The authors concluded that there was no evidence to support claims of a declining base cation content of forest floors in this region. They also noted that if they could find no difference with 2000 samples over a period of 21 years, it would be uneconomical to look for changes of less than about 1.5 percent per year. A simple fact of forest soils is that large variability means that only large changes can be detected with confidence, leaving a great deal of room to debate Type I and Type II errors associated with our ideas.

In the absence of long-term study sites, our best information may come from chronosequence studies, in which different locations of different ages are assumed to represent the changes that would happen within single sites over time (see Chapter 14). Most of these studies have examined soil development over time spans of primary succession or following agricultural abandonment (see later).

In rare cases, creative scientists have been able to find opportunities to ask long-term questions from already-available settings. For example, Johnson et al. (1981) wanted to know how much of the sulfate adsorbed in soils in the southeastern United States came from recent atmospheric deposition. They sampled beneath, adjacent to, and below an old house and compared the pools of adsorbed sulfur. In this case, losses of sulfur from beneath the house could be abnormally low owing to a lack of soil leaching, which would tend to minimize the difference between the soil under the house and beyond the house. Despite the possible confounding effect of altered leaching, they found far more adsorbed sulfate beyond the house, which matched well the rates expected from input/output budgets in a nearby watershed.

HOW DO FOREST SOILS CHANGE UNDER THE INFLUENCE OF DIFFERENT SPECIES?

Since the time of the pioneering work of Ebermayer in Germany, forest researchers have been interested in the reciprocal effects of soils and vegetation. As noted in Chapter 2, earlier ideas about soils and forests included notions of the trees shaping the chemistry and fertility of soils (Handley, 1954; Remezov and Progrebnyak, 1965). Soils and vegetation have mutual influences, and it has proven difficult to separate the effect of various species on soil development. Early workers acknowledged that certain tree species generally occur on specific types of soils and that such soils may indeed be developed by the dominant species. For many decades, European foresters and soil scientists commonly thought that conversion of beech forests to spruce would degrade the soil. Most early studies, however, examined stands where the observed differences in the development of soil profiles occurred over several centuries rather than within the tenure of the current stand. In addition, declining productivity in monoculture plantations was often due to litter raking, not to the species planted (Baule and Fricker, 1970). Stone (1975) summarized the situation and concluded that the effects of different species on soils had not been well demonstrated; more recent work has supported some ideas and refuted others.

Our best insights about reasonable rates of change in soil properties come from common garden experiments in which different species are planted on the same soil. These plantations usually represent artificial conditions, sometimes incorporating the effects of former agricultural practices or the introduction of exotic species. The rates of divergence seen in common garden experiments are probably near the upper bound of the rates of change that would occur under normal forest development, where changes in rates of growth and nutrient cycling would take longer to develop.

These common-garden experiments have supported the idea that tree species differ substantially in their effects on soils. For example, adjacent plantations of 30-year-old white pine and red oak in Ontario, Canada, showed a much greater accumulation of organic matter in the forest floor (O horizon) under pine than under red oak but much more accumulation of humus in the A horizon under oak (Figure 16.2). Common-garden experiments have supported some of the classic generalizations (such as soil acidification under spruce relative to hardwoods) but not others (such as greater nitrogen availability under hardwoods than under conifers; see Table 16.1).

Trees may differ in their effects on soils by a variety of mechanisms (Table 16.2). How substantial would the combined effects of these processes be? The effects of species on nitrogen mineralization are quite large when one species in the comparison fixes nitrogen; nitrogen-fixing species commonly increase soil nitrogen mineralization by severalfold. In the absence of nitrogen fixers in comparisons among species, the differences are still large (Figure 16.3); half of the available studies found differences of twofold or more. Soil base saturation is very malleable under the influence of different species, with one species often having twice the base saturation of another. More studies have compared differences in soil pH among species than other factors; common differences are on the order of 0.2 unit, with many larger differences.

Somehow soil acidity has been linked with soil fertility in much of the literature. This expectation probably came from agricultural soils, where reduction of acidity often increases crop production. However, most tree species that occur on acidic soils are well adapted to these soils, and the relationship from agriculture may not apply to forestry (except where acid-

Figure 16.2 After 30 years in a common garden at the University of Toronto Observatory, white pine plots accumulated much larger forest floors than red oak plots, and earthworm mixing of soils led to greater humus accumulation in the A horizon under oak.

intolerant species are planted on acidic soils). Do changes in forest soil pH affect changes in net nitrogen mineralization? The common garden data do not support this hypothesis. Prescott et al. (1995) examined differences in soil nitrogen availability and pH across a species trial with 11 tree species. The trend between these soil properties was far from significant, and the data support *declining* nitrogen availability with increasing pH more than they do any increase in nitrogen availability (Figure 16.4).

Relatively few common garden trials have included the same species, so generalizations about the effects of a particular species across a range of soils are limited. Norway spruce has been included in four common garden experiments with a variety of other species; in all cases, spruce lowered soil pH relative to other species (Binkley and Giardina, 1998). In these cases, the productivity and cycling of nutrients in the spruce plots matched or exceeded those of the other species, so acidification was coupled with increased productivity.

Tree species also differ substantially in their effects on soil fungi and bacteria. A common garden experiment with five species in Wisconsin showed a threefold range in fungal biomass and a 50 percent range in bacterial biomass (Scott, 1996). The biomass of bacteria declined as the biomass of fungi increased ($r^2 = 0.83$, $p < 0.02$). Any importance of such differences remains largely unknown.

A major remaining question about the effects of different species on soils is how long these effects will last after the original species is replaced by another.

TABLE 16.1 Some Proposed Generalizations Regarding Tree-Soil Interactions and the Extent of Supporting Evidence

Generalization	Evidence
Norway spruce acidifies soils by accumulating strongly acidic organic matter	Five common garden studies supported part or all of this generalization
Norway spruce degrades soils, particularly in comparison with beech	No evidence supports this view. Aboveground net primary production is higher in spruce forests than in beech forests; studies of adjacent stands show equal or greater net nitrogen mineralization and porosity under spruce
White pine may increase soil nitrogen availability	Three common garden experiments found notably higher net nitrogen mineralization under the white pine
Nitrogen-fixing trees increase soil carbon and rates of cycling of all nutrients	At least 12 studies have documented higher soil carbon and rates of nutrient cycling in litterfall under nitrogen fixers than under non-nitrogen fixers
Hardwoods promote soil nitrogen availability relative to conifers	Mixed evidence; at least some cases show equal or higher net nitrogen mineralization under conifers such as white pine (relative to green ash), larch, or Norway spruce (relative to beech)
Mull forest floors indicate more fertile conditions than mor forest floors	Probably supported regionally (mor forest floors are found on the poorest sites) but not locally (mull forest floors may indicate lower availability of phosphorus, for example)
Increases in soil carbon represent increased soil fertility	Some evidence, though untrue for Histosols. Common garden experiments do not show greater growth for species that develop greater soil carbon, but no "second-generation" reciprocal plantings have assessed the effect of increased carbon for soil fertility
Lignin: nitrogen ratio of litterfall is a good indicator of net nitrogen mineralization in soil	Moderately strong evidence

Source: Modified from Binkley (1996) and Binkley and Giardina (1998).

Will these changes have major legacies, or evaporate as merely ephemeral features that readily adjust to new species? Laboratory incubations of soils from the common garden trial in Wisconsin showed a fivefold difference in net nitrogen mineralization among species after 30 days (Scott, 1996). After 300 days, the differences among species had shrunk to less than twofold. The

TABLE 16.2 **Mechanisms by Which Species Differ in Their Effects on Soils**

Process	Common Size of Effect
Atmospheric deposition	Up to doubling of deposition of sulfur, nitrogen, and H^+ in polluted areas
Nitrogen fixation	Input of 50 to 150 kg/ha/yr of nitrogen for symbiotic nitrogen-fixing species
Mineral weathering by exudates, microflora	Mostly unknown, but some evidence of 0.3 to 1.5 $kmol_c$ ha^{-1} yr^{-1} greater input of base cations; higher phosphorus supply
Quantity, quality of carbon compounds added to soil	30 percent or more difference in quantity, large differences in quality, such as twofold range in lignin:nitrogen ratio
Soil communities	Populations, ratios of functional groups, and so on commonly differ by severalfold
Physical properties (temperature, water content, structure)	Moderate to large differences
Pedogenesis (such as podzolization)	Some notable case studies of horizon development (esp. E and Bs horizons)
Water loss through canopy interception	Up to twofold difference where conifer leaf area greatly exceeds leaf area of hardwoods

Source: Modified from Binkley and Giardina (1998).

laboratory incubations suggest that the apparent effect of species may be most pronounced on the relatively small, rapidly cycling pools of nitrogen, with little influence on the larger, more stable pools. Field tests are clearly needed to determine the robustness and longevity of tree species effects. The first installations of "second-generation" common garden experiments are now being done, so the first answers to this fundamental question will appear in coming decades.

HOW DOES REFORESTATION IMPROVE FORMER AGRICULTURAL SOILS?

Foresters and some soil scientists commonly believe that forests are the best type of vegetation for maintaining or improving soil conditions. Forest land uses are expected to protect soils better than agricultural land uses, and trees appear to protect soils better than pasture grasses. Such generalizations may not be supported by evidence for a variety of reasons. The first is that few generalizations apply to all forest soils, and even fewer apply to all agricultural situations. When forests replace row crops, land management practices such as

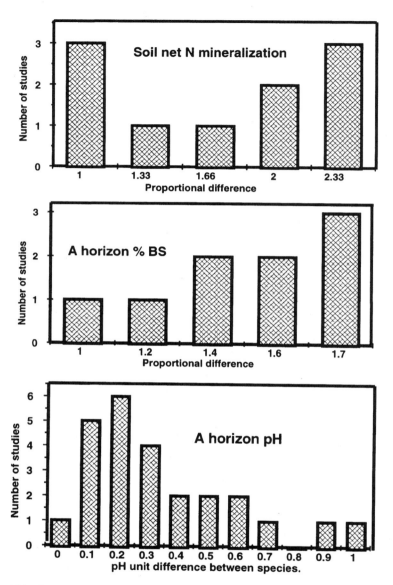

Figure 16.3 Lowest values reported within common gardens were set at 1.0, and values for other species were calculated as proportions of these lowest values. Rates of soil net nitrogen mineralization among species in common garden experiments (excluding nitrogen-fixing species) commonly differed by twofold or more, and base saturation often differed by 50 percent. Soil pH differences were commonly on the order of 0.2 unit among species, but differences of 0.5 unit or more have been reported (modified from Binkley and Giardina, 1998).

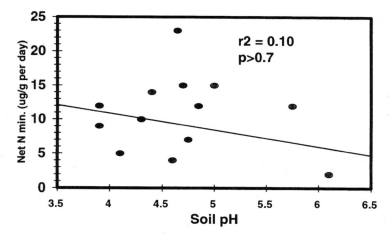

Figure 16.4 Soil net nitrogen mineralization from 11 species in a common garden species trail at Clonsast Bog, Ireland, showed no relationship with the soil pH that developed under the influence of each species (modified from Prescott et al., 1995).

annual fertilization stop; the change in vegetation and land management may also include a change in practices that favored sustained soil fertility at a high level. We will examine the effects of afforestation of grasslands and former agricultural fields to see what generalizations are supported by evidence.

A classic tree invasion story comes from grassland-aspen forest ecotones in Alberta, Canada. A case study presented by Jenny (1980) examined age sequences of grassland, invading aspen (45 years old), balsam poplar (85 years old) and Douglas-fir (150 years old). They found a different type of soil under each vegetation community. The grasslands were found on a black chernozem (Mollisol), and the old Douglas-fir site was a degraded (sic) brown-wooded soil (Alfisols). The age of the aspen stands supported the idea that tree invasion created a new O horizon and lowered the organic matter content of the A horizon. Exchangeable bases declined by two-thirds along the sequence, and total nitrogen dropped by almost half. The assumptions of a chronosequence should always be checked for reasonableness, and checking the implications for complete biogeochemical budgets is a good first step. If we assume that the soils all had a bulk density of 1.0, we can calculate the net change in soil base cations and nitrogen that would be necessary per year to produce the putative change over time. The sum of exchangeable bases differed by 450 $kmol_c$/ha, and soil nitrogen (0–80 cm) by 13,500 kg/ha. If such changes happened over 150 years, the rate of 3 $kmol_c$/ha annual loss of base cations would be fivefold the rate that would be expected to accumulate in trees and in the forest floor. The loss rate of nitrogen of 90 kg/ha annually is also severalfold the rate of accumulation in biomass. If these rates were true, then massive losses in soil leachates would need to be accounted for, which would be very difficult in such a dry climate. The most likely explanation of these rapid soil changes is that they didn't occur so rapidly or that the selection of sites spanned major differences in vegetation that predated the establishment of the trees.

When pastures are created in forested areas, it is common for the trees to return if grazing and other land management activities do not prevent tree reestablishment. Several chronosequence studies have looked at soil changes following reforestation of pastures, and most investigators expected to see substantial gains in soil carbon pools. Despite this expectation, soil carbon appears to accrue slowly, if at all (Table 16.3). A pair of chronosequences of forests developing after pasture abandonment in Puerto Rico showed no apparent carbon accretion across 60 years of forest development (Garcia-

TABLE 16.3 Rates of Carbon Accumulation Following Afforestation of Agricultural Fields

Location	Sampling Depth	Years Since Afforestation began	Carbon Accumulation (Mg/ha Annually)	Reference
South Carolina, USA	Forest floor	34	1.3	Richter et al.
	0–100 cm		0.07	(1995)
Karlsborg, Sweden	A horizon	60	0.02	Alriksson (1998)
Värnamo, Sweden	A horizon	50	0.0	
Hilo, Hawaii	0–50 cm	12	0.0	Bashkin and Binkley (1997)
Hilo, Hawaii	0–30 cm	3	0.0	Binkley and Resh (1999)
El Verde, Puerto Rico	0–15 cm	0–60+	0.0	Garcia-Montiel (1996)
Puerto Rico	0–15 cm	0–50	0.5	Brown and Lugo (1990)
La Selva, Costa Rica	0–10 cm	0–15	0.0–1.5, not significant	Reiners et al. (1994)
Broadbalk Wilderness, UK	0–23 cm	100	3.5, but the wilderness is tiny (see Figure 16.1)	Jenkinson et al. (1994)
New Hampshire, USA	Forest floor, A and B horizons	0–90	0.5, but resampling of same sites showed a rate closer to 0.0	Hamburg (1984), S. Hamburg (personal comm., 1997)

Montiel, 1996). Another three chronosequences from Puerto Rico found very high variation in two series, and some apparent carbon accretion in the third. The overall accretion was about 0.0 to 0.1 mg/ha of carbon annually (Lugo et al., 1986). In Costa Rica, a replicated chronosequence found that active pastures had 16 Mg/ha of carbon in soil depth of 0 to 10 cm, 5- to 10-year-old regenerating forests had 15 Mg/ha of carbon, and 10- to 15-year-old regenerating forests had 21 Mg/ha of carbon (Reiners et al., 1994). The rate of carbon accretion for that 5-year period would average 1.5 Mg/ha annually; this anomalously high rate may not be real given that the overall chronosequence trend was not significant.

Changes in soil chemistry are influenced by changes in soil biota, which may change soil chemistry. In the Puerto Rico reforestation example, the earthworm communities changed dramatically as pastures reverted to forests (Zou and Gonzalez, 1997). Pastures contained over 800 worms per square meter (1750 kg/ha fresh mass). Adjacent forests that were more than 60 years old had only 30 worms per square meter (20 kg/ha fresh mass).

We would not be surprised if future research found sites that accrued carbon after pastures were reforested, as we do not expect many generalizations to hold across all situations. However, currently available evidence does not support the idea of major soil carbon accretion following afforestation of pastures (Post and Kwon, 2000).

Carbon accretion also appears to be very slow when trees are planted on old plowed fields. The single best data set for soil changes following afforestation of agricultural soils comes from the work of D. Richter, C. Wells, and their colleagues on a site in South Carolina, (Richter et al., 1994, 1999). Wells designed a very thorough soil sampling scheme for a 5-year-old plantation of loblolly pine on a former agricultural field, and then repeated the sampling at 5-year intervals and archived the soils. After 30 yr, the forest floor had accumulated just 35 Mg/ha carbon (1.2 Mg/ha carbon annually), and the mineral soil showed almost no accumulation (2.1 Mg/ha carbon, or 0.07 Mg/ha carbon annually; Richter et al., 1995). The pH (in dilute salt) of the 0- to 15-cm-depth soil dropped from 4.8 in, 1962 to 3.9 in, 1990, driven largely by the drop in base saturation (from 65 percent to 15 percent). In this case, the rate of loss of base cations from the soil was about 1.2 $kmol_c$/ha, which was accounted for by an accumulation of base cations in the trees and forest floor of 0.6 $kmol_c$/ha annually and by net leaching losses (rain input/output) of about 0.6 $kmol_c$/ha. How sustainable is forest productivity with this change in soil fertility? Ongoing studies will examine the changes in soil chemistry into a second rotation of loblolly pine. Meanwhile, this highly weathered soil had accumulated substantial stores of exchangeable calcium in the "subsoil" down to 6 m. The pool of exchangeable calcium at a depth of 5 to 6 m was two to three times the size of the calcium pool removed by trees and lost in leaching over 30 years. The origin of the deep pool of available cations is not clear; did the residual cations come from long-term pedogenesis or from residual fertilizer?

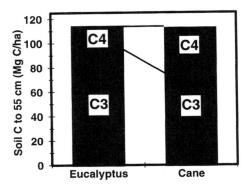

Figure 16.5   Old carbon derived from sugar cane disappeared from soils in *Eucalyptus* stands in Hawaii at the same rate that new carbon derived from *Eucalyptus* accumulated (modified from Bashkin and Binkley, 1997).

Reforestation of agricultural species sometimes follows cropping of species with the carbon-4 photosynthetic pathway, such as corn (maize) and sugarcane. These carbon-4 species differ in the natural abundance of ^{13}C (relative to ^{12}C; both are stable isotopes) from carbon-3 trees. The prior crops essentially "label" the soil organic matter with a ^{13}C signature, which allows researchers to follow the loss of old soil carbon and the gain of new soil carbon under the influence of trees. In Hawaii, reforestation with *Eucalyptus* trees showed no net increase in soil carbon after 10 to 15 years in both chronosequence and repeated-sampling studies (Bashkin and Binkley, 1998; Binkley and Resh, 1999). However, the lack of net change resulted from large offsetting losses of old cane-derived carbon (about 1500 kg/ha annually) and the gain of new *Eucalyptus*-derived carbon (Figure 16.5).

Nitrogen-fixing trees may produce greater rates of soil carbon accretion on former agricultural soils than do other tree species. For example, nitrogen-fixing species in Puerto Rico and Hawaii accumulated about 1 Mg/ha annually more soil carbon than *Eucalyptus* plantations, as a result of greater accumulation of new soil carbon, and lower rates of loss of soil carbon that accumulated from prior land use (Resh et al., 1999).

Other studies have also reported surprisingly little change in soil carbon with afforestation, and expectations of large amounts of carbon sequestration in soils in discussions of global change probably need to be revisited (Post and Kwon, 2000).

HOW DOES HARVESTING AFFECT SOIL FERTILITY?

Tree harvesting can dramatically affect forest soils, with major effects on soil temperature and moisture (typically increased), soil compaction/strength (especially where heavy equipments operates), and nutrient losses. The soil nutrient supply typically increases after harvesting, largely as a result of lowered microbial competition for nitrogen and somewhat from increased decomposition (Prescott, 1997). Forest productivity depends on soil nutrient supplies, and the removal of biomass from forests unavoidably removes nutrients. The quantity of nutrients removed generally increases with the biomass removed, as illustrated by the nitrogen and calcium contents of stems + bark for fast-growing plantations of species of *Eucalyptus* and *Pinus* in the tropics (Figure 16.6)

Nutrient removals also depend on the types of biomass removed; bolewood is low in nutrient concentration, but bark and foliage are quite high. Conventional harvests that remove only boles remove fewer nutrients than whole-tree harvests. For example, bole-only and whole-tree harvesting were compared with site capitals and input rates (Johnson et al., 1982). Before harvesting, tree ages ranged from 50 to 120 years, and total stand biomass was about 190 mg/ha (Table 16.4). Stem-only harvesting removed only 22 percent of the stand biomass and 0.1 to 7 percent of the ecosystem's nutrient capital. Whole-tree harvesting increased biomass by a factor of 2.6 and increased nutrient removal by a factor of 2.6 to 3.3.

Which nutrients are most likely to limit future productivity at this site? Based on comparisons with ecosystem capital, calcium appears most critical. The whole-tree harvest removed about 240 years' worth of calcium inputs from precipitation, compared to about 40 or 50 years' worth of nitrogen and phosphorus. These highly weathered soils (Typic Paleudults) are particularly low in both available and total mineral calcium. Although nitrogen and phosphorus are generally considered the primary limiting nutrients in the region, whole-tree harvesting removed only 9 and 2 percent of the total ecosystem content of these elements, respectively. The calcium removed in whole-tree harvests was more than double the calcium held on cation exchange sites, or 15 percent of the ecosystem capital to a soil depth of 45 cm. Tree roots

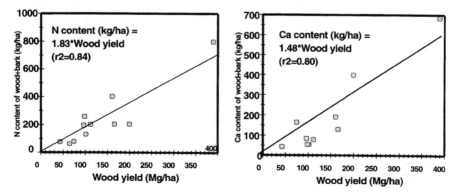

Figure 16.6   Wood yield and nutrient content of stemwood + bark in 7- to 14-year-old tropical plantations of *Eucalyptus* and *Pinus* (data from Gonçalves et al., 1997).

TABLE 16.4   Biomass and Nutrient Content (kg/ha) of Oak-History Watersheds Before Harvesting and Removal in Sawlog and Whole-tree Harvests

Component	Biomass	N	P	K	Ca
Tree					
Foliage	3,900	60	4	50	40
Branch	35,300	85	7	35	200
Stem	133,800	40	16	90	910
Stump	14,700	30	2	10	100
Total	187,700	415	29	185	1,250
Forest floor	13,700	150	12	20	160
Soil (0–45 cm)					
Extractable	—	15	40	280	500
Total	78,700	3,000	1,370	24,200	6,070
Total ecosystem	280,100	3,565	1,411	24,405	7,480
Harvest removal					
Stem-only	57,300	99	6	32	370
Whole-tree					
Branches	31,700	75	6	30	180
Stems	133,800	240	16	90	910
Total	165,530	315	22	120	1,090

Source: Calculated from Johnson et al. (1982).

extend deeper than 45 cm on these sites, and the roots could be mining deeper soil horizons to resupply the surface soil with calcium (Johnson et al., 1985).

This is one of the few whole-tree harvesting studies with very thorough tracking of the recovery of the vegetation and soils. Johnson and Todd (1998) found that most of the woody debris (80 percent) decomposed within 15 years after logging, and they noted that coarse woody debris could not play a major ecological role in such forests (because it disappears too quickly). Pools of exchangeable base cations were higher in the bole-only harvest areas than in the whole-tree harvest areas, but there were no signs of nutrient deficiencies in the foliage from any of the regenerating trees. Perhaps the biggest surprise of the study was a large increase in soil carbon (9 to 27 Mg/ha in 15 years) and nitrogen (1050 kg/ha average over 15 years). Transfers from decomposing organic matter (for carbon and nitrogen) and atmospheric deposition (for nitrogen) are too small to account for these increases. The authors speculated that perhaps preharvest soil samples that were stored (air dry) lost some carbon, overestimating the differences between the preharvest period and 15 years later. They could not account for the occult nitrogen increase of about 45 kg/ha annually (beyond estimated atmospheric deposition). Few studies can match this one for the massive amount of data collected, and the surprising lack of Ca^{2+} deficiency and the appearance of unexplainable quantities of nitrogen are sobering relative to conclusions from smaller studies.

Nutrient losses can also increase after logging as a result of increased water outflow, decreased plant uptake, and decreased input of labile carbon to fuel uptake by microbes. In most temperate forests, increases in leaching are small relative to the nutrient removals in biomass. Few tropical case studies have been examined, but the potential for larger increases in nutrient leaching appears even larger than the temperate-forest record-setting Hubbard Brook watersheds in New Hampshire (Table 16.5). The tropical rain forest shown in Figure 16.7 was on an Oxisol soil, and the increased concentrations of nitrate in soil solution after logging substantially increased the concentration of aluminum in soil solution. More studies of tropical rain forest situations are clearly warranted.

Nutrient losses in harvesting and leaching over the long run depend in part on rotation length. Short rotations produce small amounts of biomass that is relatively high in nutrients (especially with a high bark:wood ratio). Longer rotations produce more wood with lower nutrient concentrations. These features are illustrated for differing rotation periods for *Eucalyptus deglupta* in Figure 16.8 Three rotations of 7 yr would remove about twice the potassium, calcium, and magnesiun as one rotation of 21 years. The yield from the 21-year rotation is higher, of course, than the yield for a single 7-yr rotation, but rapid early growth of *Eucalyptus* actually produced more harvestable biomass over 3- to 7-year rotations than from one 21-year rotation. Higher nutrient outputs result from harvesting more biomass with higher nutrient concentrations and with three times the harvest-related soil leaching losses.

Does forest harvesting, site preparation, or fire lead to net losses of carbon from soils? Johnson (1992b) summarized available information from forest studies around the world, and found that some showed substantial losses after these treatments and others showed substantial gains (Figure 16.9). Studies that included revegetation by nitrogen-fixing species tended to increase soil

TABLE 16.5   Nutrient Losses (kg/ha) in Biomass Removal and 3-year Postharvest Soil Leaching (in Guyana) and Watershed Stream Loss (Malaysia and New Hampshire)

Location, Treatment, Reference	Component	N	K	Ca	Mg
Guyana, 0.3-ha clearing	Biomass removal	590	60	80	28
Oxisol, Brouwer	Leaching	180	40	24	31
(1996)	Biomass/leaching	3.3	1.5	3.3	0.9
Malaysia, 10-ha clearcut,	Biomass removal	120	100	200	40
Utisol, Malmer (1993)	Leaching	40	190	25	15
	Biomass/leaching	3.0	0.5	8	2.7
New Hampshire, USA,	Biomass removal	445	216	578	51
22-ha clearcut, whole-	Leaching	61	23	32	
tree removal, Yanai	Biomass/leaching	7.3	9	18	
(1998)					

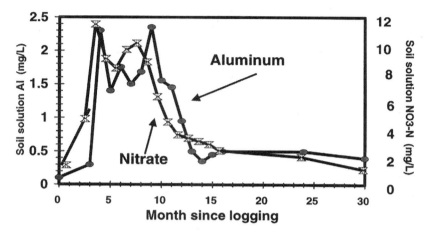

Figure 16.7   Clearing a 0.3-ha patch within a mature rain forest in Guyana dramatically increased soil solution concentrations of nitrate (at 1.2-m depth), which allowed the concentrations of aluminum and other cations to increase. Solutions from control plots remained below 1 mg/L of nitrate nitrogen and 0.6 mg/L for aluminum (data from Brouwer, 1996).

carbon by 20 percent or so. On a global scale, one might conclude that forest management activities have little net effect on soil carbon pools. The impacts on a local scale, however, may be substantial. At present, we do not have enough information to predict, for any given site (or forest type), the degree of gain or loss of soil carbon following major management practices.

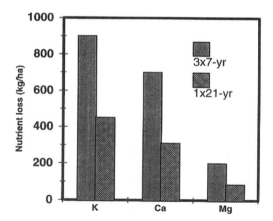

Figure 16.8   Nutrient losses from three 7-yr rotations of *Eucalyptus deglupta* compared with one 21-year rotation in Malaysia (data from Fölster and Khanna, 1997).

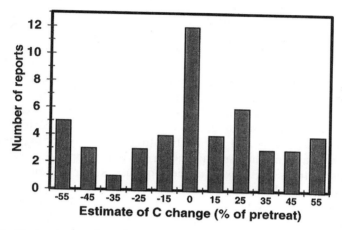

Figure 16.9 Best estimate of net loss or gain of soil carbon following harvesting, site preparation, or fire (data from Johnson, 1992b).

The complete effect of harvesting on the sustainability of forest soil productivity goes beyond the impacts on nutrient supply to include the effects on soil organic matter and bulk density. The combination of changes in nutrient supply, soil organic matter, and bulk density (both soil strength and aeration) determine the ability of the soil to support tree growth. The full suite of potential effects of harvesting on each of these features, and the interactions among the effects, provides an incredible challenge for forest soils research and management. This challenge is being addressed in a large, coordinated study of the effects of harvesting on soil productivity in the United States and Canada, with over 60 sites in the late, 1990s (Powers, 1999; Powers et al., 1999). The central treatments of this project include deliberate manipulation of soil organic matter (mostly forest floor) and soil compaction after harvesting, along with ameliorative practices such as fertilization. Forest practices of the future will be different from those of the present, but the net impacts of those practices will still depend on the overall effects on soil nutrients, organic matter, and bulk density.

WILL ACID DEPOSITION DEGRADE FOREST SOILS?

The answer to this question might seem to be deceptively simple: of course! The true answer is more complicated, with important factors including type of acid deposition, rate, soil type, and vegetation type. Much of the best work in this area has been in Europe, and we use case studies from Sweden, Norway, and Germany to illustrate the major points.

Acid deposition refers to rain or snow inputs with elevated levels of sulfuric and nitric acids. Rain in unpolluted areas is a dilute solution of carbonic acid, with a pH between 5 and 6 (depending on the presence of dust aerosols and other features). The burning of fossil fuels releases large amounts of sulfur dioxide to the atmosphere, and prompt oxidation provides sulfuric acid to rainwaters. Nitrogen oxides are created under high temperatures and pressures inside engines and other industrial systems, and complete oxidation provides nitric acid rainwater. The pH of rainfall in more polluted areas tends to be between 4.0 and 4.5 as a result of these "strong" or "mineral" acids.

Sulfur rarely limits forest growth, so sulfuric acid deposition probably cannot increase growth, and any acidification effect may (or may not) reduce growth. The major mechanism of concern has two components, both focusing on soil base cations. First, over sufficient periods, sulfur acid deposition could lead to sulfate leaching, and the sulfate anions may remove base cations, lowering the residual supply in the soil. Second, higher concentrations of sulfate in solution will lead to higher concentrations of all cations, and the concentration of aluminum increases faster than that of other ions (see Chapter 5). Further, the potential toxicity of aluminum may depend on the ratios of base cations to aluminum, not just on the aluminum concentration itself. The net effect of sulfate inputs could include both cation depletion and reductions in the ratios of base cations to aluminum.

How serious is the problem of sulfuric acid rain in forests? Around major point sources of sulfur dioxide, such as unregulated copper smelting plants, the devastation can be extensive. For example, the Copper Hills in Tennessee around a copper smelter were almost completely devegetated. The direct effects of sulfur dioxide and sulfuric acid deposition were not separate in this case, but the devastation was unarguable. Could the impacts of high-concentration problems be extrapolated to chronic low-concentration situations? This is a classic question in ecotoxicology and epidemiology. The empirical evidence shows substantial increases in sulfate adsorption in soils in high-deposition areas, as well as increased sulfate concentrations in soil solutions and leachates. There are no definitive cases of forest productivity decline as a result of sulfuric acid deposition, and this is a case where Type I, II, and III errors need to be considered.

Given the challenge of documenting changes in soils, and changes in the effects of soils on plants, only severe pollution problems can be documented with enough certainty to minimize the risk of a Type I error (believing in pollution problems that don't exist). Conversely, relatively small degradations in forest (and agricultural) soils could have large, economically costly effects on forests, crops, and aquatic ecosystems; the cost of a Type II error (failing to believe in real pollution impacts) could be very high. In the United States, legislation in the early, 1990s substantially reduced sulfur dioxide production and sulfuric acid deposition. The politicians weighted the potential risk of a Type II error (pollution impacts are probably expensive even if they cannot be proven) more highly than that of a Type I error (the cost of sulfur controls may exceed any neglible environmental effect). This is also a case where the risk of a Type III error may be important. It may be true that acid deposition hurts ecosystems, but if the effect was due to nitric acid rain, then controlling sulfuric acid rain would not help.

Nitric acid rain is important in part because most forests are (or were?) limited by low supplies of available nitrogen. Inputs of nitric acid may simply fertilize the forests. In fact, the acidity associated with nitric acid is consumed if the nitrate is retained within the forest because all of the major nitrate retention mechanisms involve reduction ($=$ consumption of H^+) to organic forms. Forests should not be harmed by nitric acid rain until they reach some state of nitrogen excess, where much of the added nitrate remains in the soil solution, leaching base cations and raising soil solution concentrations of aluminum.

How much nitrogen can be safely added to forests? Many politicians and scientists have worked to determine "critical loads," which if exceeded would lead to degradation of soils or aquatic ecosystems. Unfortunately, the current state of knowledge has confidence intervals that are as large as or larger than the estimated critical loads.

In Sweden, most forests retain most of the incoming nitrogen, with the same net retention apparent from both low-deposition and high-deposition areas (Figure 16.10). A similar trend has been reported from across Europe, even where deposition rates are higher than in Sweden.

Two major hypotheses have been advanced for how excessive nitrogen deposition could lower soil fertility and tree growth: changes in base ca- tion:aluminum ratios and nutritional imbalance in leaves. Many laboratory studies and a few field studies have shown that some tree species are impaired

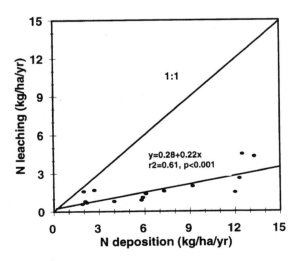

Figure 16.10 Net retention of nitrogen deposition in small watersheds in Sweden is about 80 percent (leaching $=$ 20 percent) across a wide gradient in deposition (modified from Nohrstedt, 1993).

by low concentrations of base cations in combination with high concentrations of aluminum (reviewed by Cronan and Grigal, 1995). H. Sverdrup and colleagues (1994) developed an elegant hypothesis to explain why forest growth should decline with excessive nitrogen. Overall site productivity is controlled by climatic conditions, and within this upper bound, forest growth is reduced by the limiting supplies of resources or the active impedance of toxic factors. According to the author's hypothesis, base cation leaching from the addition of excessive amounts of sulfate and nitrate would lead to lower base cation:aluminum ratios and lower growth. They plotted this hypothesis as a clear relationship between growth (as a percentage of optimal or unpolluted growth) and the base cations:aluminum ratio. Growth should be unrelated to the molar ratio of cations as long as the ratio was above about 1, but it would decline with further declines in the ratio. As with any good hypothesis, one could challenge the idea based on the information and ideas that led to its development. However, well-stated hypotheses such as this one can be tested most conclusively by experimentation. If a challenging experiment supports the idea, we can be more confident that we can accept it with a low Type I risk of believing untrue things. If the data clearly refute the hypothesis, then we cannot accept the hypothesis as true without an extemely high risk of a Type II error of believing in untrue things.

Forest fertilization often increases the ionic strength of soil solutions, and increased ionic strengths lower the base cation:aluminum ratio (see Chapter 5). Therefore, the hypothesis of Sverdrup and colleagues (1994) can be assessed by examining the growth response of trees to fertilization when that fertilization lowers the critical molar ratio below 1. Two chronic fertilization experiments in Sweden have sufficient data to test the hypothesis. At the lower productivity site at Stråsan, fertilization lowered the base cation:aluminum ratio and increased growth, but this is not a conclusive test because the ratio never dropped below 5 (Figure 16.11). A more rigorous test came from the more productive Skogaby site, where the molar ratio was less than 1 even without fertilization, and fertilization further lowered it to 0.2 to 0.3. The hypothesis predicted that growth should decline with fertilization, when in fact it increased. Therefore, the hypothesis must be rejected, or at least needs to be modified to include the effects of nitrogen on tree nutrition.

Nutrient deficiencies often limit forest growth, and alleviating the deficiency of one element may lead to opportunities for increasing growth by fertilization with other elements. This approach is embedded in some ideas about nutrient ratios; trees in a plantation with low nitrogen:phosphorus ratios may be more likely to increase their growth after phosphorus fertilization than after nitrogen fertilization. This doesn't mean that nitrogen fertilization would lower growth, but simply that the nitrogen supply is adequate. Can addition of an abundant element, such as nitrogen in this example, lead to such an imbalance in nutrient ratios that growth would actually decline? This has been observed in some cases in fertilization trials (see Chapter 12), but it is rare.

Could nitrogen deposition lead to imbalanced nutrition that would reduce

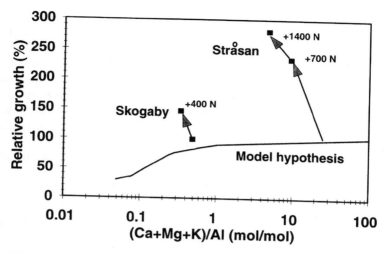

Figure 16.11 Forest growth responses to fertilization in relation to the base cation:aluminum ratio. The line representing the model hypothesis is from Sverdrup et al. (1994). (From Binkley and Hogberg, 1997, based on an idea developed with L.-O. Nilsson and unpublished data from J. Bergholm).

growth or put trees at risk? The most intensively studied case comes from the Fichtelgebirge in southern Germany. R. Oren and colleagues (1988) found that unhealthy, yellowish trees in a 30-year-old stand of Norway spruce had high concentrations of nitrogen and low concentrations of magnesium relative to healthy trees. They found that this pattern was most pronounced on older foliage, and hypothesized that when the nitrogen supply is in excess, magnesium is translocated from older foliage to new foliage to prevent problems from excessive nitrogen. This nutritional disharmony hypothesis was consistent with the observed data, but an experiment would be necessary to determine the level of confidence warranted.

These researchers chose an adjacent 10-year-old plantation of Norway spruce to test the hypothesis that excess nitrogen interfered with magnesium nutrition (Buchmann et al., 1995). Treatments included addition of nitrogen to make the problem worse; addition of magnesium to alleviate the problem; and addition of carbohydrates to the soil to alleviate the problem by tying up the excess nitrogen in the microbial biomass. Addition of nitrogen increased foliar concentrations slightly, and addition of carbohydrate lowered nitrogen concentrations slightly; addition of magnesium increased foliar magnesium levels.

Although the experiment achieved the desired changes in foliar chemistry, it produced no changes in tree growth. Even a magnesium deficiency could not be proven, aside from any role of nitrogen in inducing the magnesium deficiency. The researchers concluded that either the younger stand could not represent the nutrition of the older forest or that the nutritional disharmony hypothesis was not supported.

If forests that receive large amounts of nitrogen deposition are not currently at risk, should preventive measures of adding lime or nonnitrogen fertilizer be considered? Unfortunately, concerns about nitrogen saturation of forests are not allayed when soils are limed, as liming usually has either no effect on nitrogen leaching or increases nitrification and leaching (Kreutzer, 1995). Additions of nutrients other than nitrogen (sometimes called vitality fertilizers) generally do not improve forest growth. For example, vitality fertilization (48 kg/ha phosphorus, 43 kg/ha potassium, 218 kg/ha calcium, 46 kg/ha magnesium and 75 kg/ha sulfur) of a Norway spruce stand near Skogaby, Sweden produced only nonsignificant growth increases of 7 to 10 percent (Nilsson and Wiklund, 1992), and this marginal increase may have resulted from a higher nitrogen supply in vitality plots (nitrogen uptake increased by 50 percent in vitality plots, presumably from increases in turnover of soil nitrogen). Another vitality experiment in southern Sweden found that increases in relative basal area growth were 15 percent with vitalizer (25 kg/ha phosphorus, 60 kg/ha potassium, 40 kg/ha calcium, 8 kg/ha magnesium, and 50 kg/ha sulfur), compared with 25 percent for nitrogen (150 kg/ha) and 34 percent for nitrogen + calcium + phosphorus (Nohrstedt et al., 1993). Given the clear nitrogen limitation in these stands, the increased growth with the vitality ferilization treatment may have resulted an indirect increase in the nitrogen supply.

How much nitrogen can forest ecosystems hold? Over the past few decades, hundreds of studies have examined the rate at which nitrogen becomes available for plant uptake, but we know very little about the rate at which nitrogen is stabilized in unavailable pools. In the absence of much data, several teams have developed scenarios about nitrogen retentions (Johnson, 1992a; Aber et al., 1998; Nadelhoffer et al., 1999) and fertilization trials have been examined for clues to nitrogen retention by forests. The scenarios tell us that a variety of features are logically possible — for example, that nitrogen nitrogen retention will depend on plant uptake, or on microbial uptake, or on rates of accumulation in soil humus through microbial or chemical processes. The fertilization experiments have tended to use rates that greatly exceed rates of nitrogen deposition, so inferences about the fate of nitrogen deposited from the atmosphere are confounded by the excessive rates. Nonetheless, these experiments have often shown the remarkable capacity of forest soils to store nitrogen. The bottom line is that much more work will be needed to determine which of the descriptions of the nitrogen retention capacities of forests is well supported by good experimentation and which ones are not.

Does acid deposition increase weathering rates? Increases in the concentrations of H^+ should tend to increase rates of mineral weathering. A great deal of information is available on weathering processes in soils, but we still don't know the degree to which weathering can be increased by increasing concentrations of H^+. In an intriguing study in Norway, plots of Norway spruce were irrigated with pH 3 water (acidified with sulfuric acid) for 5 years (Stuanes et al., 1992). Acid irrigation doubled the rate of loss of base cations from the soil profile, but base saturation declined only slightly. Base saturation returned to the levels found in nonacidified plots within about 10 years. If the weathering rates among plots had been constant, base saturation should have remained depleted in the acidified plots. As noted in Chapter 9, rates of mineral weathering vary by an order of magnitude among mineral types, and the quantitative importance (and variability) of biotic process remains unknown.

What is the current state of knowledge about the effects of acid deposition on forests? This is a very large question, but a variety of generalizations appear to be supported. Nitrogen is being retained in almost all forests, but a significant number of forests do show saturation in which inputs are matched or exceeded by outputs. The majority of European forests remain responsive to nitrogen additions, although some are not, and some that were previously responsive no longer are (reviewed by Binkley and Högberg, 1997). Fertilization with vitalization fertilizers that aim to rebalance nutrition disrupted by nitrogen deposition generally do not increase growth as much as simply adding more nitrogen. In no case has tree growth been found to decline with nitrogen saturation. In fact, European forests in areas receiving high inputs of acids have experienced substantially *increased* growth in the past few decades, typically by 20 percent to 40 percent (see Elfving and Tegnhammar, 1996, and Elfving et al., 1996, for Sweden and Norway; Spiecker et al., 1996, for the rest of Europe). Many interesting questions remain. The current evidence for critical ratios of base cations and aluminum is weak, but viable hypotheses remain. For example, in a seedling study, Ericsson (1995) found that birch shifted allocation toward greater root production when supplies of nitrogen and phosphorus became limiting, which should help counter the effect of lower supplies. In contrast, proportional allocation to roots decreased when potassium or magnesium supplies became limiting, which would further restrict the seedlings' ability to obtain these elements. We need to know if this lack of root response to potassium and magnesium limitations pertains to field-grown trees. Although very few forest sites are nitrogen saturated, experimental additions of yet more nitrogen to these forests are needed to see if negative effects do in fact develop. The rates of mineral weathering in forests remain poorly known, especially under the influence of acid deposition and different species. More work is needed to determine if accelerated base cation losses will be replenished by mineral weathering.

During the, 1980s, when scientists, politicians, and the public were very concerned about acid rain and forest decline, the overall growth of the forests of Europe was increasing at unprecedented rates. How did our perceptions of forest decline deviate so strongly from reality? In part, normal dynamics in forests were suddenly thought to be very abnormal; thin canopies are common on suppressed trees, and Norway spruce trees apparently experience crown thinning as a result of drought or other factors and then recover

(Kandler, 1992, 1993). The quick belief in forest decline was a complex social phenomenon and may be explained in large part by a few key points (see Innes, 1993):

- Unhealthy trees exist, and it is often difficult to assign a cause.
- Greenhouse studies can demonstrate nutritional problems that resemble field symptoms.
- Assessments of forest canopy conditions are difficult to perform and interpret (most trees in a forest look poorer than the dominant trees, for example).
- Changes in soil chemistry over a period of decades could be demonstrated even if the causes could not.

Perhaps a major contributor to the mismatch between perception and reality was largely sociological. The risk of a Type II error (failing to believe the real impacts of pollution) was frightening; some scientists predicted massive growth reductions and even deforestation, and many people were very upset by such prospects. Another contributor is the simple, irreducible uncertainty about forests and forest soils; there are so few things that we can say with confidence that many alternative scenarios are difficult to rule out.

A final point about the acid rain beliefs of the, 1980s is that the purported decline of the forests needed to be addressed by a variety of disciplines; any one scientist may have felt that his or her own reservations about part of the puzzle would not be enough to discount the whole picture. When no one had a complete view of the emperor's new clothes, most were reluctant to proclaim that the clothes were not substantial. An important illustration of this problem comes from a, 1988 quotation from an eminent forest soil scientist (whom we respect):

> Researchers on the effects of acid atmospheric deposition on forests in Europe almost exclusively focus on the causal mechanisms involved, and pay little if any attention to quantifying the decrease in productivity of forest land.... [T]here can be little doubt that acid atmospheric deposition is a major factor limiting forest productivity in Europe, with little hope for dramatic improvement in decades to come.

This soil scientist noted that few quantitative data were available, yet he boldly concluded that the situation was bleak. He understood the changes that acidic deposition could create in soils and assumed that the stories he heard about the trees were true. When the growth data were synthesized in the, 1990s, the period of expected decline was actually found to be one of accelerating growth; there was no decline in need of any soils-based explanation. Perhaps this cautionary tale will lead researchers and forest managers to retain more than a little doubt when assessing such complex issues with so little data on actual effects.

IF THE CLIMATE WARMS, WILL FOREST SOILS CHANGE?

Rates of decomposition depend on temperature: higher temperatures increase rates of decomposition, just like other chemical reactions. Many incubation studies have shown that rates of carbon dioxide release from soils increase by about 1.5- to 2.0-fold when temperatures increase by 10°C (Verburg, 1998). This relationship is used in models of global ecosystem biogeochemistry that link rates of decomposition to temperature changes. As with many questions that address spatial and temporal scales beyond the range of easy experimentation, we don't know if these expectations warrant much confidence. For example, Verburg (1998) used a 400-m² greenhouse over a dwarf-shrub, pine, and birch ecosystem in southern Norway to examine the effects of increased temperature (5°C above ambient) and carbon dioxide concentration on decomposition of fresh litter. He found no effect of the greenhouse temperature on the mass loss from litter of birch, pine, or heather.

At least five factors may conspire to cloud the relationship between air temperature and soil carbon dynamics. Increases in air temperature do not translate well into increases in soil temperature. Soil temperature is moderated relative to air temperature on both daily and seasonal scales, and the energy budget of the soil includes radiation components that would not change much with air temperature. The future set of combinations of temperature, precipitation, and soil moisture levels remains unknown. Decomposition rates are sensitive to moisture as well as temperature, and rates may increase or decrease, depending on future water regimes. The current distributions of tree species (and understory species) show strong relationships with temperature and water supplies, so changes in these factors could alter species distributions, and species differ substantially in their effects on soil biogeochemistry. Forest management practices have major effects on soil carbon pools, and management changes over the decades even when climate does not.

Under elevated carbon dioxide concentrations and altered environmental conditions, it is not possible to predict changes in belowground tree production with any confidence. Some attempts have been made to examine the pattern of soil carbon dioxide evolution with seasonal temperature in the hope that this relationship could be used as a basis for modeling future responses to warmer air temperatures. This approach is probably simplistic because the overall ecophysiology of trees may also differ under new environmental conditions, including reduced transpiration in response to elevated carbon dioxide (Mooney et al., 1991). More important, the seasonal course of soil carbon dioxide evolution probably relates more to the seasonal course of incoming light and photosynthesis than to soil temperature per se (Giardina et al., 1999b), so any assumption that temperature patterns account for the carbon dioxide evolution patterns is unlikely to be directly useful.

The final reason why short-term incubations probably do not scale up well to long periods and large areas is a gap in logic. Incubations of over 30 days release carbon from very labile pools, and any changes in pools with rapid turnover do not result in large changes in soil carbon storage. The major effect of changing climate on soil carbon storage would entail changes in the less labile, semistabilized pool of carbon that comprises the large majority of soil carbon. This pool may or may not show the same sensitivity to temperature (or other factors) shown by the labile pool. For example, Giardina et al. (1999) examined the relationship between average carbon turnover period in 27 forest soils across a gradient of mean annual temperature from 5° to 25°C and found no trace of a relationship ($r^2 = 0.01$). These authors also compiled data on incubation studies that determined the temperature responsiveness of decomposition in both short-term (10- to 30-day) and long-term (1-yr) incubations. Although short-term incubations responded strongly to temperature, decomposition rates of more stabilized soil carbon showed only 10–20 percent increases when temperature increased by 10° to 15°C. We clearly have a long way to go before our predictions of climate effects on soil biogeochemistry can be extrapolated to global levels.

SUMMARY

The study and management of forest soils involve important questions with long-term implications for sustained and enhanced productivity of forests. Many questions about the sustainability of forest soils can be phrased clearly, but the available answers are typically fuzzy as a result of the great complexity of soils, the range of conditions, and the changes in conditions over time. A variety of cases studies evaluated some of the key issues of forest soils in recent decades, including rates of long-term changes in soils, the effects of individual tree species on soils, the effects of reforestation on soils, the effects of acid deposition on forests, and possible effects of changing climate on soils. In this chapter and book, we advocate an approach to forest soils that includes clear articulation of questions and evidence, explicit consideration of levels of uncertainty inherent in the answers to questions about forest soils, and discussion (and even debate) on issues that connect forest soils with policy, management, and science. In the long run, long-term monitoring and experimentation will remain the best option for validating ideas about how forest soils work and how they respond to management.

References

Aber, J., W. McDowell, K. Nadelhoffer, A. Magill, G. Berntson, M. Kamakea, S. McNulty, W. Currie, L. Rustad, and I. Fernandez. 1998. Nitrogen saturation in temperate forest ecosystems. *Bioscience* 48:921–934.

Adams, M. A., and P. M. Attiwill. 1986. Nutrient cycling and nitrogen mineralization in eucalypt forests of south-eastern Australia. II. Indices of nitrogen mineralization. *Plant and Soil* 92:341–362.

Agee, J. 1993. *Fire ecology of Pacific Northwest forests.* Island Press, Washington, DC.

Ahmed, R., D. Khan, and S. Ismail. 1985. Growth of *Azadiracta indica* and *Melia azederach* on costal sand using highly saline water for irrigation. *Pakistan Journal of Botany* 17:229–233.

Akkermans, A. D. L., and C. van Dijk. 1979. The formation and nitrogen fixing activity of the root nodules of *Alnus glutinosa* under field conditions. Pp. 511–520 in: *Symbiotic nitrogen fixation in plants*, P. S. Nutman (ed). Cambridge University Press, Cambridge.

Alban, D. H. 1972. The relationship of red pine site index to soil phosphorus extracted by several methods. *Soil Science Society of America Proceedings* 36:664–666.

Albaugh, T. J., H. L. Allen, P. M. Dougherty, L. Kress, and J. S. King. 1998. Leaf area and above- and belowground growth responses of loblolly pine to nutrient and water additions. *Forest Science* 44:1–12.

Alexander, L. T., and J. G. Cady. 1962. *Genisis and hardening of laterite in soils.* USDA Technical Bulletin 1282.

Allen, H. L. 1987. Forest fertilizers. *Journal of Forestry* 85:37–46.

Allen, H. L., and R. Ballard. 1982. Fertilization of loblolly pine. Pp. 163–181 in *Symposium on the loblolly pine ecosystem (east region).* School of Forest Resources, North Carolina State University, Raleigh.

Allen, M. F. 1991. *The ecology of mycorrhizae.* Cambridge University Press, Cambridge.

Allison, F. F. 1965. *Decomposition of wood and bark sawdusts in soil, nitrogen requirements and effects on plants.* USDA Technical Bulletin 1332.

Alriksson, A. 1998. Afforestation of farmland: Soil changes and the uptake of heavy metals and nutrients by trees. Ph.D. thesis, Swedish University of Agricultural Sciences, Uppsala.

Amaranthus, M. P., and J. M. Trappe. 1993. Effects of erosion on ecto- and VA-mycorrhizal inoculum potential of soil following forest fire in southwest Oregon. *Plant and Soil* 150:41–49.

Amer, F. D., D. Bouldin, C. Black, and F. Duke. 1955. Characterization of soil phosphorus by anion exchange resin adsorption and P-32 equilibration. *Plant and Soil* 6:391–408.

Anderson, H. W., and M. Hyatt. 1979. Feasibility of hand application of urea to forest land in Western Washington. Pp. 205–208 in: *Forest fertilization conference, contribution* #40. Institute of Forest Resources, University of Washington.

Andreason, O. 1988. Suiting forest management to a changed environment. Pp. 67–75 in: *Forest health and productivity, the Marcus Wallenberg Foundation symposia proceedings* #5. Falun, Sweden.

Archie, S. G., and M. Smith. Survival and growth of plantations in sludge-treated soils and older forest growth studies. Pp. 105–114 in: *Municipal sludge application to Pacific Northwest forest lands*, C. S. Bledsoe (ed.). Contribution #41, Institute of Forest Resources, University of Washington, Seattle.

Armson, K. A., and V. Sadreika. 1974. *Forest tree nursery soil management and related practices*. Publication of the Ontario Ministry of Natural Resources, Toronto.

Arseneault, D., and S. Payette. 1997. Landscape change following deforestation at the arctic tree line in Quebec, Canada. *Ecology* 78:693–706.

Aubert, G., and R. Travernier. 1972. Soil survey. Pp. 17–44 in: *Soils of the humid tropics*. U.S. National Academy of Sciences, Washington, D.C.

Auten, J. T. 1945. Prediction of site index for yellow-poplar from soil and topography. *Journal of Forestry* 43:662–668.

Ayers, A. S., M. Takahashi, and Y. Kanehiro. 1947. Conversion of non-exchangeable potassium to exchangeable forms in a Hawaiian soil. *Soil Science Society of America Proceedings* 11:175–181.

Baker, H. P. 1906. The holding and reclamation of sand dunes and sand wastes. *Forest Quarterly* 4:282–288.

Baker, J. B. 1973. Intensive cultural practices increase growth of juvenile slash pine in Florida sandhills. *Forest Science* 19:197–202.

Ballard, R. 1971. Interrelationships between site factors and productivity of radiata pine at Riverhead Forest, New Zealand. *Plant and Soil* 35:371–380.

Ballard, R. 1984. Fertilization of plantations. Pp. 327–360 in: *Nutrition of plantation forests*, G. D. Bowen and E. K. S. Nambiar (eds.). Academic Press, London.

Ballard, R., and J. G. A. Fiskell. 1974. Phosphorus retention in coastal plain forest soils: I. Relationships to soil properties. *Soil Science Society of America Proceedings* 38:250–255.

Baranov, I. Y. 1959. Geographical distribution of seasonally frozen ground and permafrost. Pp. 193–219. In V. A. Obruchev (ed.). *Institute of Permafrost Studies, Academy of Science*, Moscow. General Geocryology. Natural Resource Council of Canada Technical Translation No. 1121 (1964).

Barber, S. A. 1995. *Soil nutrient bioavailability: a mechanistic approach*. Wiley, New York.

Barclay, H. J., and H. Brix. 1985. Effects of urea and ammonium nitrate fertilizer on growth of a young thinned and unthinned Douglas-fir stand. *Canadian Journal of Forest Research* 14:952–955.

Barkman, J., and C. R. Schwintzer. 1998. Rapid N_2 fixation in pines? Results of a Maine field study. *Ecology* 79:1453–1457.

Barnes. B. V., K. S. Pregitzer, T. A. Spies, and V. H. Spooner. 1982. Ecological forest site classification. *Journal of Forestry* 80:493 498.

Barnes, B. V., D. R. Zak, S. R. Denton, and S. H. Spurr. 1998. *Forest ecology*, 4th ed. Wiley, New York.

Barnes, K. L., and C. W. Ralston. 1955. *Soil factors related to the growth and yield of slash pine plantations.* Florida Agricultural Experiment Station Bulletin. 559.

Barnes, R. 1980. An allocation and optimization approach to tree growth modeling: concepts and application to nitrogen economy. Unpublished manuscript.

Barney, C. W. 1951. Effect of soil temperature and light intensity on root growth of loblofly pine seedlings. *Plant Physiology* 26:146–163.

Barros, N. F., and R. F. Novais. 1996. Eucalypt nutrition and fertilizer regimes in Brazil. Pp. 335–355 in: *Nutrition of eucalypts.*, P. M. Attiwill and M. A. Adams (eds.). CSIRO, Collingwood, Australia.

Barros, N. F., R. F. Novais, and L. C. L. Neves. 1990. Fertilição e correção do solo para planntio de Eucalipto. Pp. 127–186 in: *Relação solo-Eucalipto*, N. F. Barros and R. F. Novais (eds.). Department of Forest Soils, Federal University of Viçosa, M. G., Brazil.

Bashkin, M. A., and D. Binkley. 1998. Changes in soil carbon following afforestation in Hawaii. *Ecology* 79:828–833.

Bates, C. G. 1918. Concerning site. *Journal of Forestry* 16:383–388.

Baule, H., and C. Fricker. 1970. *The fertilizer treatment of forest trees.* BLV Verlagsgesel-lshcaft, Munich.

Beare, M. H., D. C. Coleman, D. A. Crossley, Jr., P. F. Hendrix, and E. P. Odum. 1995. A hierarchal approach to evaluating the significance of soil biodiversity to biogeochemical cycling. *Plant and Soil* 170:5–22.

Beaufils, E. R. 1973. *Diagnosis and recommendation integrated system (DRIS).* Soil Science Bulletin #1, University of Natal, Pietermaritzburg, South Africa.

Beck, D. F. 1971. *Polymorphic site index curves for white pine in the southern Appalachians.* USDA Forest Service Research Paper SE-80.

Beck, D. F., and K. B. Trousdell. 1973. *Site index: accuracy of prediction.* USDA Forest Research Paper SE-108.

Beese, J. O., and K. J. Meiwes. 1995. 10 Jahre Waldkalkung—Stand und Perspektiven. *Allgemeine Forstzeitschrift* 50:946–949.

Beets, P. N., and D. S. Pollock. 1987. Uptake and accumulation of nitrogen in *Pinus radiata* stands as related to age and thinning. *New Zealand Journal of Forestry Science* 17:353–371.

Benckiser, G. 1997. *Fauna in soil ecosystems.* Marcel Dekker, New York.

Bengtson, G. (ed). 1968. *Forest fertilization: theory and practice.* Tennessee Valley Authority, Muscle Shoals, AL.

Bengtson, G. 1973. Fertilizer use in forestry: materials and methods of application. Pp. 97–153, in: *Proceedings of the International Symposium on Forest Fertilization.* FAO-IUFRO, Paris.

Bengtson, G. 1976. Fertilizers in use and under evaluation in silviculture: a status report. in: *Proceedings XVI IUFRO world congress, Working Group on Forest Fertilization.* Oslo, Norway.

Bengtson, G. W, 1981. Nutrient conservation in forestry: a perspective. *So. Journal of Applied Forestry* 5:50–58.

Benzian, B. 1965. *Experiments on nutrition problems in forest nurseries.* Forestry Commission Bulletin #37. Her Majesty's Stationery Office, London.

Berg, B., and E. Matzner. 1997. Effect of N deposition on decomposition of plant litter and soil organic matter in forest systems. *Environmental Reviews* 5:1–25.

Bergholm, J., F. Braekke, J. Frank, L. Hallbäcken, M. Ingerslev, and E. Mälkönen. 1997. Soil chemistry and weathering. Pp. 74–89 in: *Imbalanced forest nutrition—vitality measures.* Section of Ecology, Swedish University of Agricultural Sciences, SNS Project 1993–1996, Uppsala.

Bergkvist, B., and L. Folkeson. 1995. The influence of tree species on acid deposition, proton budgets and element fluxes in south Swedish forest ecosystems. *Ecology Bulletin* 44:90–99.

Berry, E. C. 1994. Earthworms and other fauna in the soil. Pp. 61–90 in:*Soil biology: effects on soil quality*, J. L. Hatfield and B. A. Stewart (eds.) Lewis Publications, Boca Raton, FL.

Beskow, G. 1935. Soil freezing and frost heaving. *Sveriges Geologiska Undersokning* 26:1–242.

Bigger, C. M., and D. W. Cole. 1983. Effects of harvesting intensity on nutrient losses and future productivity in high and low productivity red alder and Douglas-fir stands. Pp. 167–178 in: *IUFRO symposium on forest site and continuous productivity*, R. Ballard and S. P. Gessel (eds.). USDA Forest Service General Technical Report PNW-163, Portland, OR.

Bilan, M. V. 1971. Some aspects of tree root distribution. Pp. 69–80 in *Mycorrhizae*, F. Haeskaylo (ed.). USDA Forest Service Miscellaneous Publication 1189.

Binkley, D. 1981. Nodule biomass and acetylene reduction rates of red alder and Sitka alder on Vancouver Island, B. C. *Canadian Journal of Forest Research* 11:281–286.

Binkley, D. 1983. Interaction of site fertility and red alder on ecosystem production in Douglas-fir plantations. *Forest Ecology and Management* 5:215–227.

Binkley, D. 1984a. Douglas-fir stem growth per unit of leaf area increased by interplanted Sitka alder and red alder. *Forest Science* 30:259–263.

Binkley, D. 1984b. Ion exchange resin bags: factors affecting estimates of nitrogen availability. *Soil Science Society of America Journal* 48:1181–1184.

Binkley, D. 1986a. *Forest nutrition management.* Wiley, New York.

Binkley, D. 1986b. Soil acidity in loblolly pine stands with interval burning. *Soil Science Society of America Journal* 50:1590–1594.

Binkley, D. 1992. H^+ budgets. Pp. 450–466 in: *Atmospheric deposition and nutrient cycling*, D. W. Johnson and S. E. Lindberg (eds.). Springer-Verlag, New York.

Binkley, D. 1995. The influence of tree species on forest soils: processes and patterns. Pp. 1–34 in: *Proceedings of the trees and soils workshop, Lincoln University*, D. J. Mead and I. S. Cornforth (eds.). Agronomy Society of New Zealand Special Publication #10, Lincoln University Press, Canterbury.

Binkley, D. 1996. Bioassays of the influence of *Eucalyptus saligna* and *Albizia falcataria* on soil nutrient supply and limitation. *Forest Ecology and Management* 91:229–234.

Binkley, D. 1999. Disturbance in temperate forests of the northern hemisphere. Pp. 469–482 in: *Ecosystems of disturbed ground*, L. Walker (ed.). Elsevier Science, Amsterdam.

Binkley, D., J. Aber, J. Pastor, and K. Nadelhoffer. 1986. Nutrient availability patterns in some Wisconsin forests: ion exchange resin bags and on-site soil incubations. *Biology and Fertility of Soils* 2:77–82.

Binkley, D., C. Burnham, and H. L. Allen. 1999. Water quality impacts of forest fertilization with nitrogen and phosphorus. *Forest Ecology and Management* 121:191–213.

Binkley, D., K. Cromack, Jr., and D. Baker. 1994. Nitrogen fixation by red alder: biology, rates, and controls. Pp. 57–72 in: *The biology management of red alder*, D. Hibbs (ed.). Oregon State University Press, Corvallis.

Binkley, D., and C. Giardina. 1997. Biological nitrogen fixation in plantations. Pp. 297–337 in: *Management of soil, water, and nutrients in tropical plantation forests*, E.K.S. Nambiar and A. Brown (eds.). ACIAR Monograph 43, Canberra, Australia.

Binkley, D., and C. Giardina. 1998. Why do tree species affect soils? The warp and woof of tree-soil interactions. *Biogeochemistry* 42:89–106.

Binkley, D., and S. Greene. 1983. Production in mixtures of conifers and red alder: the importance of site fertility and stand age. Pp. 112–117 in: *IUFRO symposium on forest site and continuous productivity*, R. Ballard and S. P. Gessel (eds.). USDA Forest Service General Technical Report PNW-163, Portland, OR.

Binkley, D., and S. C. Hart. 1989. The components of nitrogen availability assessments in forest soils. *Advances in Soil Science* 10:57–116.

Binkley, D., and P. Högberg. 1997. Does atmospheric deposition of acidity and nitrogen threaten Swedish forests? *Forest Ecology and Management* 92:119–152.

Binkley, D., and L. Husted. 1983. Nitrogen accretion, soil fertility and Douglas-fir nutrition in association with redstem ceanothus. *Canadian Journal of Forest Research* 13:122–125.

Binkley, D., A. M. O'Connell, and K. V. Sankaran. 1997. Stand development and productivity. Pp. 419–442 in: *Management of soil, nutrients and water in tropical plantation forests*, E. K. S. Nambiar and A. G. Brown (eds.). Australian Centre for International Agricultural Research Monograph #43.

Binkley, D., and S. Resh. 1999. Rapid changes in soils following *Eucalyptus* afforestation in Hawaii. *Soil Science Society of America Journal* 63:222–225.

Binkley, D., and D. Richter. 1987. Nutrient cycles and H$^+$ budgets. *Advances in Ecological Research* 16:1–51.

Binkley, D., D. Richter, M. B. David, and B. Caldwell. 1992b. Soil chemistry in a loblolly/longleaf pine forest with interval burning. *Ecological Applications* 2:157–164.

Brinkley, D., F. W. Smith, and Y. Son. 1995. Nutrient supply and limitation in an age-sequence of lodgepole pine in southwestern Wyoming. *Canadian Journal of Forest Research* 25:621–628.

Binkley, D., and P. Sollins. 1990. Acidification of soils in mixtures of conifers and red alder. *Soil Science Society of America Journal* 54:1427–1433.

Binkley, D., P. Sollins, R. Bell, D. Sachs, and D. Myrold. 1992a. Biogeochemistry of adjacent conifer and alder/conifer ecosystems. *Ecology* 73:2022–2034.

Binkley, D., P. Sollins, and W. G. McGill. 1985. Natural abundance of N-15 as a tracer of alder-fixed nitrogen. *Soil Science Society of America Journal* 49:444–447.

Binkley, D., and D. W. Valentine. 1991. Fifty-year biogeochemical effects of Norway spruce, white pine, and green ash in a replicated experiment. *Forest Ecology and Management* 40:13–25.

Binkley, D., D. W. Valentine, C. Wells, and U. Valentine. 1989. An empirical analysis of the factors contributing to 20-year decrease in soil pH in an old-field plantation of loblolly pine. *Biogeochemistry* 7:39–54.

Birk, E., and P. Vitousek. 1984. Patterns of N retranslocation in loblolly pine stands: response to N availability. *Bulletin of the Ecological Society of America* 65:100.

Bissett, J., and D. Parkinson. 1980. Long-term effects of fire on the composition and activity of the soil microflora of a subalpine, coniferous forest. *Canadian Journal of Botany* 58:1704–1721.

Blake, J., H. N. Chappell, W. S. Bennett, S. R. Webster, and S. P. Gessel. 1990. Douglas fir growth and foliar nutrient responses to nitrogen and sulfur fertilization. *Soil Science Society of America Journal* 54:257–262.

Bledsoe, C. S., and R. J. Zasoski. 1981. Seedling physiology of eight tree species grown in sludge-amended soils. Pp. 93–100, in: *Municipal sludge application to Pacific Northwest forest lands*, C. S. Bledsoe (ed.). Contribution #41, Institute of Forest Resources, University of Washington, Seattle.

Bloomfield, C. 1954. A study of podzolization. III. The mobilization of iron and aluminum by *Rimu* (*Dacrydium cupressium*). *Journal of Soil Science* 5:39–45.

Bockheim, J. G., C. Tarnocai, J. M. Kimble and C. A. S. Smith. 1997. The concept of gelic materials in the new Gelisol order for permafrost-affected soils. *Soil Science* 162:927–939.

Boerner, R. 1982. Fire and nutrient cycling in temperate ecosystems. *BioScience* 32:187–192.

Boerner, R., and R. T. T. Forman. 1982. Hydrologic and mineral budgets of New Jersey Pine Barrens upland forests following two intensities of fire. *Canadian Journal of Forest Research* 12:503–510.

Boerner, R., T. Lord, and J. Peterson. 1988. Prescribed burning in the oak-pine forest of the New Jersey Pine Barrens: effects on growth and nutrient dynamics of two Quercus species. *American Midland Naturalist* 120:108–119.

Boggie, R. 1972. Effects of water-table height on root development of *Pinus contorta* on deep peat in Scotland. *Oikos* 23:304–312.

Bohn, H. L., B. L. McNeal, and G. A. O'Connor. 1985. *Soil chemistry*, 2nd ed. Wiley, New York.

Bonner, F. T. 1968. *Responses to soil moisture deficiency by seedlings of three hardwood species*. USDA Forest Service Research Note 50–70.

Boring, L., and W. Swank. 1984a. Symbiotic nitrogen fixation in regenerating black locust (*Robinia pseudoacacia* L.) stands. *Forest Science* 30:528–537

Bormann, B. T., F. H. Bormann, W. B. Bowden, R. S. Pierce, S. P. Hamburg, D. Wang, M. C. Snyder, C. Y. Li, and R. C. Ingersoll. 1993. Rapid N_2 fixation in pines, alder, and locust: evidence from the sandbox ecosystem study. *Ecology* 74:583–598.

Boul, S. W. 1973. Soil genesis, morphology, and classification. Pp. 1–37. In: *A review of soils research in tropical Latin America*, (P. A. Sanchez (ed.). North Carolina Agricultural Experiment Station Technical Bulletin 219.

Buol, S. W., F. D. Hole, R. J. McCracken, and R. J. Southard. 1997. *Soil genesis and classification*, 4th ed. Iowa State University Press, Ames.

Boul, S. W., F. D. Holeand, and R. J. McCracken. 1980. *Soil genesis and classification,* 2nd ed. Iowa State University Press, Ames.

Bowen, G. D. 1965. Mycorrhiza inoculation in nursery practice. *Australian Forestry* 29:231–237.

Bowen, G. D., and A. D. Rovira. 1969. The influence of microorganisms on growth and metabolism of plant roots. Pp. 170–199 in: *Root growth,* W. J. Whittington (ed.). Butterworths, London.

Boyer, W. D. 1987. Volume growth loss: a hidden cost of periodic prescribed burning in longleaf pine? *Southern Journal of Applied Forestry* 11:154–157.

Boyle, J. R., and G. K. Voigt. 1973. Biological weathering in silicate minerals. *Plant and Soil* 38:191–201.

Boyle, J. R., G. K. Voigt, and B. L. Saweney. 1974. Chemical weathering of biotite by organic acids. *Soil Science* 117:42–45.

Brady, N. C., and R. R. Weil. 1996. *The nature and properties of soils.* Prentice-Hall, Upper Saddle River, NJ.

Bray, J. R., and E. Gorham. 1964. Litter production in forests of the world.Pp. 101–157 in: *Advances in ecological research 2,* J. B. Cragg (ed.). Academic Press, New York.

Bredemeier, M., K. Blanck, N. Lamersdorf, and G. A. Wiedey, 1995. Response of soil water chemistry to experimental "clean rain" in the NITREX roof experiment at Solling, Germany. *Forest Ecology and Management* 71:31–44.

Brix, H. 1981. Effects of nitrogen fertilizer source and application rates on foliar nitrogen concentration, photosynthesis and growth of Douglas-fir. *Canadian Journal of Forest Research* 11:775–780.

Brix, H. 1983. Effects of thinning and nitrogen fertilization on growth of Douglas-fir: relative contribution of foliage quantity and efficiency. *Canadian Journal of Forest Research* 13:167–175.

Broadfoot, W. M. 1969. Problems in relating soil to site index for southern hardwoods. *Forest Science* 15:354–364.

Broadfoot, W. M. 1973. *Water table depth and growth of young cottonwood.* USDA Forestry Service Research Note SO-167.

Brouwer, L. C. 1996. *Nutrient cycling in pristine and logged tropical rain forest.* Tropenbos Guyana Series 1. Thesis, Utrecht University.

Brown, A. G., E. K. S. Nambiar, and C. Cossalter. 1997. Pp. 1–24 in: *Management of soil, nutrients, and water in tropical plantation forests,* E. K. S. Nambiar and A. G. Brown (eds.). ACIAR Monograph #43, Canberra.

Brown, B. A. 1995. Toward a theory of podzolization. Pp. 253–274 in: *Carbon forms and functions in forest soils,* (W. W. McFee and M. J. Kelly (eds.). Soil Science Society of America, Madison, WI.

Brown, R. J. E. 1967. Permafrost in Canada. Map 1246A. Geological Survey Canada, National Research Council of Canada, Ottawa.

Brown, R. J. E. 1970. *Permafrost in Canada; its influence on northern development.* University of Toronto Press, Toronto.

Bruijnzeel, L. A., 1982. Hydrological and biogeochemical aspects of man-made forests in South-Central Java, Indonesia. PhD thesis, Free University of Amsterdam, the Netherlands.

Bruijnzeel, L. A. 1991. Nutrient input-output budgets of tropical forest ecosystems: a review. *Journal of Tropical Ecology* 7:1–24.

Bryant, R. B., and R. W. Arnold. 1994. *Quantitative modeling of soil forming processes.* Soil Science Society of America, Madison, WI.

Buchmann, N., R. Oren, and R. Zimmermann, 1995. Response of magnesium-deficient saplings in a young, open stand of *Picea abies* (L.) Karst. to elevated soil magnesium, nitrogen, and carbon. *Environmental Pollution* 87:31–43.

Bunnell, F. L., D. E. N. Tait, P. W. Flanagan, and K. van Cleve. 1977. Microbial respiration and substrate weight loss. I. A general model of the influences of abiotic variables. *Soil Biology and Biochemistry* 9:33–40.

Burger, J. A., and and W. L. Pritchett. 1984. Effects of clearfelling and site preparation on nitrogen mineralization in a Southern pine stand. *Soil Science Society of America Journal* 48:1432–1437.

Burger, J. A., and J. L. Torbert. 1992. *Restoring forests on surface-mined land.* Virginia Tech Extension Publication Collection. Blacksberg, VA.

Burns, R. M., and E. A. Hebb. 1972. *Site preparation and reforestation of droughty, acid sands.* USDA Agricultural Handbook 426.

Campbell, R. E., M. B. Baker, Jr., and P. F. Ffolliott. 1977. Wildfire effects on a ponderosa pine ecosystem: an Arizona case study. USDA Forest Service Research Paper RM-191.

Carmean, W. H. 1961. Soil survey refinements needed for accurate classification of black oak site quality in southeastern Ohio. *Soil Science Society of America Proceedings* 25:394–397.

Carmean, W. H. 1975. Forest site quality evaluation in the United States. *Advances in Agronomy* 27:209–269.

Carter, M. R., D. A. Holmstrom, L. M. Cochrane, P. C. Brenton, J. A., Van Roestel, D. R. Langille, and W. G. Thomas. 1996. Persistence of deep loosening of naturally compacted subsoils in Nova Scotia. *Canadian Journal of Soil Science* 76:541–547.

Carter, R. E. 1992. Diagnosis and interpretation of forest stand nutrient status. Pp. 90–97 in: *Forest fertilization: sustaining and improving nutrition and growth of western forests,* H. N. Chappell, G. F. Weetman, and R. E. Miller (eds.). College of Forest Resources Contribution #73, University of Washington, Seattle.

Carter, R. E., J. Otchere-Boateng, and K. Klinka. 1983. Dieback of a 30-year-old Douglas-fir plantation in the Brittain River Valley, British Columbia: symptoms and diagnosis. *Forest Ecology and Management* 7:249–263.

Chandler, C., P. Cheney, P. Thomas, L. Trabaud, and D. Williams. 1983. *Fire in forestry, Volume 1: Forest fire behavior and effects.* Wiley, New York.

Chang, S. X. 1996. Fertilizer N efficiency and incorporation and soil N dynamics in forest ecosystems of northern Vancouver Island. Ph.D. thesis, Faculty of Forestry, University of British Columbia, Vancouver.

Chapin, F. S. H. III, and A. J. Bloom. 1976. Phosphate adsorption: adaptation of tundra graminoids to a low temperature, low phosphorus environment. *Oikos* 26:111–121.

Chapin, F. S. III, A. J. Bloom, C. B. Field, and R. H. Waring, 1987. Plant responses to multiple environmental factors. *Bioscience,* 37:49–57.

Chapin, F. S. H. III, and K. Van Cleve. 1981. Plant nutrient absorption and retention under differing fire regimes. Pp. 301–321 in: *Fire regimes and ecosystem properties,*

H. A. Mooney, T. M. Bonnicksen, N. L. Christensen, J. E. Lotan, and W. A. Reiners (eds.). USDA Forest Service General Technical Report WO-26.

Chappell, H. N., D. W. Cole, S. P. Gessel, and R. B. Walker. 1991. Forest fertilization research and practice in the Pacific Northwest. *Fertilizer Research* 27:129–140.

Charman, D. J., R. Aravema, and B. G. Warner. 1994. Carbon dynamics in a forested peatland in north-eastern Ontario, Canada. *Journal of Ecology* 82:55–62.

Chen, Q. 1987. Nitrogen transformations in adjacent cypress and loblolly pine ecosystems. M.S. thesis, Duke University, Durham.

Chijicke, E. O. 1980. *Impact on soils of fast-growing species in lowland humid tropics.* Food and Agricultural Organization Forestry Paper No. 21, Rome.

Chorover, J., P. M. Vitousek, D. A. Everson, A. M. Esperanza, and D. Turner. 1994. Solution chemistry profiles of mixed-conifer forests before and after fire. *Biogeochemistry* 26:115–144.

Christensen, N. L. 1987. The biogeochemical consequences of fire and their effects on the vegetation of the coastal plain of the southeastern United States. Pp. 1–21 in: *The role of fire in ecological systems*, L. Trabaud (ed.). SPB Academic Publishing, the Hague, the Netherlands.

Clayton, J. L. 1979. Nutrient supply to soil by rock weathering. Pp. 75–96 in: *Impact of intensive harvesting on forest nutrition*. State University of New York Press, Syracuse.

Clayton, J. L. 1984. A rational basis for estimating element supploy rate from weathering. Pp. 405–419 in: *Sixth North American forest soils conference*, E. L. Stone (ed.). University of Tennessee Press, Knoxville.

Cleaves, E. T., D. W. Fisher, and O. P. Bricker. 1974. Chemical weathering of serpentinite in the eastern Piedmont of Maryland. *Geological Society of America Bulletin* 85:437–444.

Cleaves, E. T., A. E. Godfrey, and O. P. Bricker. 1970. Geochemical balance of a small watershed and its geomorphic implications. *Geological Society of America Bulletin* 81:3015–3032.

Clements, F. E. 1916. *Plant succession.* Carnegie Institute Publication 242, Washington, DC.

Cochran, P. H. 1969. *Thermal properties and surface temperatures of seedbeds.* USDA Forest Service PNW Forest and Range Experiment Station, Portland, OR.

Coile T. S. 1935. Relation of site index for shortleaf pine to certain physical properties of the soil. *Journal of Forestry* 33:726–730.

Coile, T. S, 1952. Soil and the growth of forests. *Advances in Agronomy* 4:329–398.

Coile, T. S, and F. X. Schumacher. 1964. *Soil-site relations, stand structure, and yields of slash and loblolly pine plantations in the Southern United States.* T. S. Coil, Inc., Durham, NC.

Cole, D. W., and M. Rapp. 1981. Elemental cycling in forest ecosystems. Pp. 341–409. in: *Dynamic properties of forest ecosystems*, D. E. Reichle (ed.). Cambridge University Press, Cambridge.

Coleman, D. C., and D. A. Crossley, Jr. 1996. *Essentials of soil ecology.* Academic Press, San Diego, CA.

Coleman, D. C., J, Mende, and G. Uehara. 1989. *Dynamics of soil organic matter in tropical ecosystems.* University of Ottawa Press, Ottawa, Canada.

Colman, S. M., and D. P. Dethier (eds.). 1986. *Rates of chemical weathering of rocks and minerals.* Academic Press, Orlando, FL.

Comerford, N. B., and R. F. Fisher. 1982. Use of discriminant analysis for classification of fertilizer-responsive sites. *Soil Science Society of America Journal* 46:1093–1096.

Compton, J. E., and D. W. Cole. 1998. Phosphorus cycling and soil P fractions in Douglas-fir and red alder stands. *Forest Ecology and Management* 110:101–112.

Cornwell, S. M., and E. L. Stone. 1968. Availability of nitrogen to plants in acid coal mine spoils. *Nature* 217:768–769.

Cote, B., and C. Camire. 1984. Growth, nitrogen accumulation, and symbiotic dinitrogen fixation in pure and mixed plantings of hybrid poplar and black alder. *Plant and Soil* 78:209–220.

Cotta, B. 1852. *Praktische Geohnoise fur Land- und Forstwirte und Techniker.* Dresden.

Cotta, H. 1809. *Systematische Anleitung zur Taxoation der Waldungen.* Berlin.

Cotton, F. A., and G. Wilkinson. 1988. *Advanced inorganic chemistry,* 5th ed. Wiley, New York.

Courchesne, F., and G. R. Gobran. 1997. Mineralogic variations of bulk and rhizosphere soils from a Norway spruce stand. *Soil Science Society of America Journal* 61:1245–1249.

Covington, W. W., and S. Sackett. 1984. The effect of a prescribed fire in Southwestern ponderosa pine on organic matter and nutrients in woody debris and forest floor. *Forest Science* 30:183–192.

Cowles, H. C. 1899. The ecological relations of the vegetation. *Botanical Gazette* 27:95–116, 167–202, 281–308, 361–391.

Creasey, J., A. C. Edwards, J. M. Reid, D. A. MacLeod, and M. S. Cresser. 1986. The use of catchment studies for assessing chemical weathering rates in two contrasting upland areas in northeast Scotland. Pp. 468–502 in: *Rates of chemical weathering of rocks and minerals,* S. M. Colman and D. P. Dethier (eds.). Academic Press, Orlando.

Cronan, C. S., and D. F. Grigal. 1995. Use of calcium/aluminum ratios as indicators of stress in forest ecosystems. *Journal of Environmental Quality* 24:209–226.

Cross, A. F., and W. H. Schlesinger. 1995. A literature review and evaluation of the Hedley fractionation: applications to the biogeochemical cycle of soil phosphorus in natural ecosystems. *Geoderma* 64:197–214.

Crossley, D. A., B. R. Mueller, and J. C. Perdue. 1992. Biodiversity of microarthropods in agricultural soils: relations to processes. *Agricultural Ecosystems and Environment* 40:37–46.

Crouch, G. L. 1982. Pocket gophers and reforestation on western forests. *Journal of Forestry* 80:662–664.

Cstantini, A., M. R. Nestor, and M. Podberscek. 1995. Site preparation for *Pinus* establishment in southeastern Queensland. *Australian Journal of Experimental Agriculture* 35:1159–1164

Cuevas, E., and E. Medina. 1983. Root production and organic matter decomposition in a terra firme forest of the upper Rio Negro basin. Pp. 653–666 in: *International symposium on root ecology and its applications.* Gumpenstein, Irdning, Austria.

Czapowskyj, M. M. 1973. Establishing forest on surface-mined lands as related to fertility arid fertilization. Pp. 132–139 in: *Symposium on forest fertility*. USDA Forest Service General Technical Report NE-3.

Dalla-Tea, F., and M. A. Marco. 1996. Fertilizers and *Eucalypt* plantations in Argentina. Pp. 327–333 in: *Nutrition of Eucalypts*, P. M. Attiwill and M. A. Adams (eds.). CSIRO, Collingwood, Australia.

Dangerfield, J., and H. Brix. 1979. Comparative effects of ammonium nitrate and urea fertilizers on tree growth and soil processes. Pp. 133–139 in: *Forest fertilization conference*, Contribution #40, Institute of Forest Resources, University of Washington, Seattle.

Daniels, R. B., and R. D. Hammer. 1992. *Soil geomorphology*. Wiley, New York.

Darfus, G. H., and R. F. Fisher. 1984. Site relations of slash pine on dredge mine spoils. *Journal of Environmental Quality* 13:457–492.

Davey, C. B. 1984. Nursery soil organic matter: management and importance. Pp. 81–86 in: *Forest nursery manual*, M. L. Duryea and T. D. Landis (eds.). Martinus Nijnoff/Jung, the Hague, the Netherlands.

Davidson, E., D. D. Myrold, and P. M. Groffman. 1990. Denitrification in temperate forest ecosystems. Pp. 196–220 in: *Sustained productivity of forest soils*, S. P. Gessel, D. S. Lacate, G. F. Weetman, and R. F. Powers (eds.). University of British Columbia Faculty of Forestry Publication, Vancouver.

DeAngelis, D. L. 1992. *Dynamics of nutrient cycling and food webs*. Chapman & Hall, London.

DeBano, L. F., D. G. Neary, and P. F. Ffolliott. 1998. *Fire's effects on ecosystems*. Wiley, New York.

DeBano, L. F., S. M. Savage, and D. A. Hamilton. 1976. The transfer of heat and hydrophobic substances during burning. *Soil Science Society of America Proceedings* 40:779–782.

DeBell, D. S., T. G. Cole, and C. D. Whitesell. 1997. Growth, development, and yield of pure and mixed stands of *Eucalyptus* and *Albizia*. *Forest Science* 43:286–298.

DeBell, D. S., and M. A. Radwan. 1979. Growth and nitrogen relations of coppiced black cottonwood and red alder in pure and mixed plantings. *Botanical Gazette Supplement* 140:S97–S101.

Dell, B. 1996. Diagnosis of nutrient deficiencies in eucalypts. Pp. 417–440 in: *Nutrition of eucalypts*, P. M. Attiwill and M. A. Adams (eds.). CSIRO, Collingwood, Australia.

Degens, B. P. 1998. Decreases in microbial functional diversity do not result in corresponding changes in decomposition under different moisture conditions. *Soil Biology and Biochemistry* 20:1989–2000.

Dement, J. A. and E. L. Stone. 1968. Influence of soil and site on red pine plantations in New York. Cornell University Agriculture Experiment Station Btrll. 1020.

DeMontigny, L. E., C. M. Preston, P. G. Hatcher, and I. Kögel-Knaber. 1993. Comparison of humus horizons from two ecosystem phases on northern Vancouver Island using ^{13}C CPAMS NMR spectroscopy and CuO oxidation. *Canadian Journal of Soil Science* 73:9–25.

Denison, W. C. 1973. Life in tall trees. *Scientific American* 228:75–80.

Denison, W. C. 1979. *Lobaria oregana*, a nitrogen-fixing lichen in old-growth Douglas-fir forests. Pp. 266–275 in: *Symbiotic nitrogen fixation in the management of temperate forests*, J. C. Gordon, C. T. Wheeler and D. A. Perry (eds.). Forest Research Laboratory, Oregon State University, Corvallis.

Derome, J., and P til, A. 1989. The liming of forest soils in Finland. *Meddelser fran der Norsk Institut for Skogforskning* 42:147–155.

Derome, J., M. Kukkola, and Mlknen, E. 1986. *Forest liming on mineral soils. Results of Finnish experiments*. National Swedish Environmental Protection Board Report 3084, Solna, Sweden.

Derr, H. J. and W. F. Mann, Jr. 1970. *Site preparation improves growth of planted pines*. USDA Forest Service Research Note SO-106.

DeWitt, C. T. 1967. Photosynthesis: its relationship to overpopulation. Pp. 315–320 in: *Harvesting the sun*. F. A. San Pietro et al. (eds.). Academic Press, New York.

Dindal, D. L. 1990. *Soil biology guide*. Wiley, New York.

Dinkelaker, B., C. Hehgler, G. Neumann, L. Eltrop, and H. Marschner. 1997. Root exudates and mobilization of nutrients. Pp. 441–452 in: *Trees: contributions to modern tree physiology*. Backhuys Publishers, Leiden, the Netherlands.

Dissmeyer, G. E., and J. G. Greis. 1983. Sound soil arid water management good economics. Pp. 194–202 in: *The managed slash pine ecosystem*, E. L. Stone (ed.). School of Forest Resources and Conservation, University of Florida Press, Gainesville.

Doran, J. W., D. C. Colman, D. F. Bezdicek, and B. A. Stewart. 1994. *Defining soil quality for a sustainable environment*. Soil Science Society of America, Madison, WI.

Dorsser, J. C. van, and D. A. Rook. 1972. Conditioning of radiata pine seedlings by undercutting and wrenching: description of methods, equipment, and seedling response. *New Zealand Journal of Forestry* 17:61–73.

Dougherty, P. M., H. L. Allen, K. W. Kress, R. Murthy, C. A. Maier, T. J. Albaugh, and D. A. Sampson. 1998. An investigation of the impacts of elevated carbon dioxide, irrigation, and fertilization on the physiology and growth of loblolly pine. Pp 149–168 in: *The productivity and sustainability of southern forest ecosystems in a changing environment. Ecological studies*, Vol. 128. Springer-Verlag, New York.

Driscoll, C. T., N. van Breeman, and J. Mulder. 1984. Aluminum chemistry in a forested spodosol. *Soil Science Society of America Journal* 49:437–444.

Dunker, R. E., C. L. Hooks, S. L. Vance, and R. G. Darmody. 1995. Deep tillage effects on compacted surface-mine land. *Soil Science Society of America Journal* 59:192–199.

Duxbury, J. M., M. Scott Smith, and J. W. Doran. 1989. Soil organic matter as a source and a sink of plant nutrients. Pp. 33–67 in: *Dynamics of soil organic matter in tropical ecosystems*, D. Coleman, J. M. Oades, and G. Uehara (eds.). University of Hawaii, Honolulu.

Dyck, W. J., C. A. Mees, and P. D. Hodgkiss. 1987. Nitrogen availability and comparison to uptake in two New Zealand *Pinus radiata* forests. *New Zealand Journal of Forestry Science* 17:338–352.

Dyck, W. J., and M. F. Skinner. 1990. Potential for productivity decline in New Zealand radiata pine forests. Pp. 318–332 in: *Sustained productivity of forest soils*. S. P. Gessel et al. (eds.) University of British Columbia, Faculty of Forestry Publication, Vancouver.

Dyrness, C. T. 1969. *Hydrologic properties of soils on three small watersheds in the western Cascades*. USDA Forest Service Research Note PNW-111.

Dyrness, C. T. 1976. *Effect of wildfire on soil wettability in the high Cascades of Oregon.* USDA Forest Service Research Paper PNW-202, Portland.

Dyrness, C. T., and R. A. Norum. 1983. The effects of experimental fires on black spruce forest floors in interior Alaska. *Canadian Journal of Forest Research* 13:879–893.

Dyrness, C. T., K. Van Cleve, and J. D. Levison. 1989. The effect of wildfire on soil chemistry in four forest types in interior Alaska. *Canadian Journal of Forest Research* 19:1389–1396.

Ebermayer, E. 1876. *Die gesammte Lehre der Waldstreu mit Rueksicht auf die chemische Statik des Waldbaues.* J. Springer, Berlin.

Edmonds, R. L., and K. P. Mayer. 1981. Survival of sludge-associated pathogens and their movement into groundwater. Pp. 79–86 in: *Municipal sludge application to Pacific Northwest forest lands,* C. S. Bledsoe (ed.). Contribution #41, Institute of Forest Resources, University of Washington, Seattle.

Edwards, C. A. 1998. *Earthworm ecology.* St. Lucie Press. New York.

Edwards, C. A., D. E. Reichke, and A. Crossley. 1970. The role of soil invertebrates in turnover of organic matter and nutrients. Pp. 147–172 in: *Analysis of temperate forest ecosystems,* D. E. Reichle (ed.). Springer-Verlag, New York.

Ekblad, A., and K. Huss-Danell. 1995. Nitrogen fixation by *Alnus incana* and nitrogen transfer from *A. incana* to *Pinus sylvestris* influenced by macronutrients and ectomycorrhiza. *New Phytologist* 131:453–459.

Elfving, B., and L. Tegnhammer. 1996. Trends of tree growth in Swedish forests 1953–1992: an analysis based on sample trees from the National Forest Inventory. *Scandinavian Journal of Forest Research* 11:38–49.

Elfving, B., L. Tegnhammar, and B. Tveite. 1996. Studies on growth trends of forests in Sweden and Norway. Pp. 61–88 in: *Growth trends in European forests,* H. Spiecker, K. Mielikäinen, M. Köhl, and J. P. Skovsgaard (eds.). Springer-Verlag, Berlin.

Ellis, R. C., and A. M. Graley. 1983. Gains and losses in soil nutrients associated with harvesting and burning eucalyptus rainforest. *Plant and Soil* 74:437–450.

Emsley, J. 1984. The phosphorus cycle. Pp. 147–162 in: *The natural environment and the biogeochemical cycles,* Vol. 1, Part A, O. Hutzinger (ed.). Springer-Verlag, Berlin.

Ericsson, T. 1995. Growth and shoot:root ratio of seedlings in relation to nutrient availability. *Plant and Soil* 168:205–214.

Eriksson, H. M. 1996. Effects of tree species and nutrient application on distribution and budgets of base cations in Swedish forest ecosystems. Ph.D. thesis, Swedish University of Agricultural Sciences, Uppsala.

Evans, D. O., and L. T. Szott. 1995. *Nitrogen fixing trees for acid soils.* Nitrogen Fixing Tree Association, Morrilton, AR.

Evans, J. 1992. *Plantation forestry in the tropics,* 2nd ed. Clarendon Press, Oxford.

Ewel, J., C. Berish, B. Brown, N. Price, and J. Raich. 1981. Slash and burn impacts on a Costa Rican wet forest site. *Ecology* 62:816–829.

Ewers, B., D. Binkley, and M. Bashkin. 1996. Influence of adjacent stand on spatial patterns of carbon and nitrogen in *Eucalyptus* and *Albizia* plantations. *Canadian Journal of Forest Research* 26:1501–1503.

Eyk, J. J. Van der. 1957. Reconnaissance soil survey in northern Suririam. Ph.D. dissertation, Wageningen.

Eyre, S. R. 1963. *Vegetation and soils.* Aldine, Chicago.

Fahey, V. J. 1983. Nutrient dynamics of aboveground detrites in lodgepole pine (*Pinus contarta* ssp. *latifolia*) ecosystems, southeastern Wyoming. *Ecology Monographs* 53:51–72.

Fahey, T. J., J. W. Hughes, M. Pu, and M. A. Arthur. 1988. Root decomposition and nutrient flux following whole-tree harvest of northern hardwood forest. *Forest Science* 34:744–768.

Fayle, D. C. F. 1975. Extension and longitudinal growth during the development of red pine root systems. *Canadian Journal of Forest Research* 5:109–121.

Feagley, S. E., M. S. Valdez, and W. H. Hudnall. 1994. Papermill sludge, phosphorus, potassium, and lime effect on clover grown on a mine soil. *Journal of Environmental Quality* 23:759–765.

Feller, M. C. 1981. Catchment nutrient budgets and geological weathering in *Eucalyptus regnans* ecosystems in Victoria. *Australian Forestry* 44:502–510.

Feller, M. C. 1983. Impacts of prescribed fire (slashburning) on forest productivity, soil erosion, and water quality on the coast. Pp. 57–91 in: *Prescribed fire-forest soils symposium proceedings.* Land Management Report #16, Ministry of Forests, British Columbia, Victoria.

Fellin, D. G., and P. C. Kennedy. 1972. *Abundance of arthropods inhabiting duff and soil after prescribed burning on forest clearcuts in northern Idaho.* USDA Forest Service Research Note INT-162.

Fernow, B. E. 1907. *History of forestry.* University Press, Toronto.

Ferrians, O. 1965. Permafrost map of Alaska. U.S. Geological Survey, Miscellaneous Map 1–445.

Fife, D. N., and E. K. S. Nambiar. 1995. Effect of nitrogen on growth and water relations of radiata pine families. *Plant and Soil* 168:279–285.

Fink, S. 1992. Physiologische und strukturelle Veränderungen an Bäumen unter Magnesiummangel. Pp. 16–26 in: *Magnesiummangel in Mitteleuropäischen Waldökosystemen,* G. Glatzel, R. Jandel, M. Sieghardt, and H. Hager (eds.) Forstliche Schriftenreihe Band 5, Universität für Bodenkultur, Vienna.

Fisher, R. F. 1972. Spodosol development and nutrient distribution under Hydnaceae fungal mats. *Soil Science Society of America Proceedings* 30:492–495.

Fisher, R. F. 1979. Allelopathy. Pp. 313–332 in: *Plant disease,* Vol. IV, (J. G. Horsfall and F. B. Cowling (eds.). Academic Press, New York.

Fisher, R. F. 1980. Soils interpretions for silviculture in the Southeastern coastal plain. Pp. 323–300 in: *Proceedings first biennial southern silvicultural research conference,* J. P. Barnett (ed.). USDA Forest Service General Techical Report SO-34.

Fisher, R. F. 1984. Predicting tree and stand response to cultural practices. Pp. 53–66. in: *Forest soils and treatment impacts,* E. L. Stone (ed.). University of Tennessee Press, Knoxville.

Fisher, R. F. 1995. Soil organic matter: clue or conundrum? Pp. 1–12 in: *Carbon forms and functions in forest spoils,* W. W. McFee and J. M. Kelly (eds.). Soil Science Society of America. Madison, WI.

Fisher, R. F., and F. W. Adrian. 1980. Bahiagrass reduces slash pine seedling survival and growth. *Tree Plantation Notes* 32(2):19–21.

Fisher, R. F., and R. P. Eastburn. 1974. Afforestation alters prairie soil nitrogen status. *Soil Science Society of America Proceedings* 38:366–368.

Fisher, R. F., and A. S. R. Juo. 1995. Mechanisms of tree growth in acid soils. Pp. 1–18 in *Nitrogen fixing trees for acid soils*, D. O. Evans and L. T. Szott (eds.). Nitrogen Fixing Tree Research Reports (Special Issue). Winrock International and NFTA, Morrilton, AR.

Fisher, R. F., and R. J. Miller. 1980. Description of soils having disturbed or discontinuous horizons. *Soil Science Society of America Journal* 44:1279–1281.

Fisher, R. F., and W. L. Pritchett. 1982. Slash pine response to different nitrogen fertilizers. *Soil Science Society of America Journal* 46:113–136.

Fisher, R. F., and E. L. Stone. 1969. Increased availability of nitrogen and phosphorus in the root zone of conifers. *Soil Science Society of America Proceedings* 33:955–961.

Fogel, R., and G. Hunt. 1979. Fungal and arboreal biomass in a western Oregon Douglas fir ecosystem: distribution patterns and turnover. *Canadian Journal of Forest Research* 9:245–256.

Fölster, H., and P. K. Khanna. 1997. Dynamics of nutrient supply in plantation soils. Pp. 339–378 in: *Management of soil, nutrients, and water in tropical plantation forests* E. K. S. Nambiar and A. G. Brown (eds.). ACIAR Monograph #43, Canberra.

Fox, T. 1995. Forest soils and low-molecular-weight organic acids. Pp. 53–62 in: *Carbon forms and functions in forest soils*, W. W. McFee and J. M. Kelly (eds.). Soil Science Society of America, Madison, WI.

Fox, T. R., and N. Comerford. 1990. Low-molecular-weight organic acids in selected forest soils of the southeastern USA. *Soil Science Society of America Journal* 54:1139–1144.

Fox, T. R., and N. Comerford. 1992a. Rhizosphere phosphatase activity and phosphatase hydrolyzable organic phosphorus in two forested Spodosols. *Soil Biology and Biochemistry* 24:579–583.

Fox, T. R., and N. B. Comerford. 1992b. Influence of oxalate loading on phosphorus and aluminum solubility in Spodosols. *Soil Science Society of America Journal* 56:290–294.

Franklin, J. F., C. T. Dyrness, D. G. Moore, and R. F. Tarrant. 1968. Chemical soil properties under coastal Oregon stands of alder and conifers. Pp. 157–172 in: *Biology of alder*, J. M. Trappe, J. F. Franklin, R. F. Tarrant and G. H. Hansen (eds.). USDA Forest Service, Portland.

French, H. M. 1976. *The periglacial environment*. Longman, London.

Froehlich, H. A., and D. H. McNabb. 1984. Minimizing soil compaction in Pacific Northwest forests. Pp 159–192 in: *Forest soils and treatment impacts*, E. L. Stone (ed.). University of Tennesse Press, Knoxville.

Frothingham, E. H. 1918. Height growth as a key to site. *Journal of Forestry* 16:754–760.

Gadgil, R. L. 1971a. The nutritional role of *Lupinus arboreus* in coastal sand dune forestry: 1. The potential influence of undamaged lupin plants on nitrogen uptake by *Pinus radiata*. *Plant and Soil* 34:357–367.

Gadgil, R. L. 1971b. The nutritional role of *Lupinus arboreus* in coastal sand dune forestry: III. Nitrogen distribution in the ecosystem before tree planting. *Plant and Soil* 35:114–126.

Gadgil, R. L. 1976. Nitrogen distribution in stands of *Pinus radiata* with and without lupin in the understorey. *New Zealand Journal of Forestry Science* 6:33–39.

Gadgil, R. L. 1983. Biological nitrogen fixation in forestry: research and practice in Australia and New Zealand. Pp. 317–332 in: *Biological nitrogen fixation in forest ecosystems: foundations and applications*, J. C. Gordon and C. T. Wheeler (eds.). Martinus Nijhoff/Junk, the Hague, the Netherlands.

Gahoonia, T. S., and N. E. Nielsen. 1992. The effects of root-induced changes on the depletion of inorganic and organic phosphorus in the rhizosphere. *Plant and Soil* 143:185–191.

Garcia-Montiel, D. 1996. Changes in nutrient cycling during tropical reforestation. PhD thesis, Colorado State University, Ft. Collins.

Garcia-Montiel, D., and D. Binkley. 1998. Effect of *Eucalyptus saligna* and *Albizia facaltaria* on soil processes and nitrogen supply in Hawaii. *Oecologia* 113:547–556.

Gedroiz, K. K. 1912. Colloidal chemistry in relation to the problems of soil science. *Zhur. Opit. Agr.* 13.

Gent, J. A., Jr., H. L. Allen, R. G. Campbell, and C. G. Wells. 1984. Magnitude, duration, and economic analysis of loblolly pine growth response following bedding and phosphorus fertilization. Pp. 1–18 in: *Phosphorus fertilization in young loblolly pine stands*. Report #17, North Carolina State Forest Fertilization Cooperative, Raleigh.

Gerdemann, J. M. 1968. Vesicular-arbuseular mycorrhizae and plant growth. *Annual Review of Phytopathology* 6:397–418.

Gerdemann, J. W., and J. M. Trappe. 1975. Taxonomy of Endagonaceae. Pp. 35–51 in: *Endomycorrhizas*, F. F. Sanders, B. Mosse, and P. B. Tinker (eds.). Academic Press, New York.

Gessel, S. P., and A. N. Balci. 1965. Amount and composition of forest floors under Washington coniferous forests. Pp. 11–23 in: *Forest-soil relationships in North America*. C. T. Youngberg (ed.). Oregon State University Press, Corvallis.

Gholz, H. L., and R. F. Fisher. 1982. Organic matter production and distribution in slash pine plantation ecosystems. *Ecology* 63:1827–1839.

Gholz, H. L., R. F. Fisher, and W. L. Pritchett. 1985. Nutrient dynamics in slash pine plantation ecosystems. *Ecology* 63:1827–1839.

Gholz, H. L., C. S. Perry, W. P. Cropper, Jr., and L. C. Hendy. 1985. Litterfall, decomposition, and nitrogen and phosphorus dynamics in a chronosequence of slash pine (*Pinus elliottii*) plantations. *Forest Science* 31:463–485.

Giardina, C., S. Huffman, D. Binkley, and B. Caldwell. 1995. Alders increase phosphorus supply in a Douglas-fir plantation. *Canadian Journal of Forest Research* 25:1652–1657.

Giardina, C., M. G. Ryan, and R. M. Hubbard. 1999b. Do carbon turnover rates in mineral soil vary with temperature and texture? Manuscript submitted to *Nature*.

Giardina, C. P., R. L. Sanford, Jr., I. C. Dockersmith, and V. J. Jaramillo. 1999a. The effects of slash burning on soil nutrient availability during the land preparation phase of shifting cultivation. *Plant and Soil*, in press.

Giesler, R., M. Högberg, and P. Högberg. 1998. Soil chemistry and plants in fennoscandian boreal forest as exemplified by a local gradient. *Ecology* 79:119–137.

Giesler, R., U. S. Lundstrom, and H. Grip. 1996. Comparison of soil solution chemistry assessment using zero-tension lysimeters or centrifugation. *European Journal of Soil Science* 47:63–74.

Gilmore, A. R., W. A. Geyer, and W. R. Boggess. 1968. Microsite and height growth of yellow-poplar. *Forest Science* 14:420–426.

Gobran, G. R., and S. Clegg. 1996. A conceptual model for nutrient availability in the mineral soil-root system. *Canadian Journal of Soil Science* 76:124–131.

Goldammer, J. G. 1993. Historical biogeography of fire: tropical and subtropical. Pp. 298–314 in: *Fire in the environment: the ecological, atmospheric, and climatic importance of vegetation fires*, P. J. Crutzen and J. G. Goldammer (eds.). Wiley, Chichester, UK.

Goldich, S. S. 1938. A study of rock weathering. *Journal of Geology* 46:17–58.

Gonçalves, J. L. M., N. F. Barros, E. K. S. Nambiar, and R. F. Novais. 1997. Soil and stand management for short-rotation plantations. Pp. 379–417 in: *Management of soil, nutrients, and water in tropical plantation forests*, E. K. S. Nambiar and A. G. Brown (eds.). ACIAR Monograph #43, Canberra.

Gonçalves, J. L. M., B. Raij, J. L. M. Gonçalves, and J. C. Florestais. 1996. Pp. 219–232 in: *Recomendações de adubação e calagem para o Estado de São Paulo*, 2nd ed. Instituto Agronômico de Campinas, Campinas, Brazil.

Gosz, J. K., G. E. Likens, and F. H. Bormann. 1972. Nutrient content of litter fall on the Hubbard Brook Experimental Forest, New Hampshire. *Ecology* 53:769–784.

Gould, A. B., J. W. Hendrix, and R. S. Ferriss. 1996. Relationship of mycorrhizal activity to time following reclamation of surface mine land in western Kentucky. I. Propagule and spore population densities. *Canadian Journal of Botany* 74:247–261.

Gracen, E. L., and R. Sands. 1980. Compaction of forest soils: a review. *Australian Journal of Soil Research* 18:163–189.

Graham, R. C., and H. B. Wood. 1991. Morphologic development and clay redistribution in lysimeter soils under chaparral and pine. *Soil Science Society of America Journal* 55:1638–1646.

Grant, D., and D. Binkley. 1987. Rates of free-living nitrogen fixation in some Piedmont forest types. *Forest Science* 33:548–551.

Gray, L. E. 1971. Physiology of vesicular-arbuscular mycoorrhizae. Pp. 145–150 in *Mycorrhizae*, E. Haeskaylo (ed.). USDA Forest Service Miscellaneous Publication 1189.

Gray, T. R. G., and S. T. Williams. 1971. *Soil micro-organisms*. Longman, London.

Grayson, S. J., D. Vaughan, and D. Jones. 1997. Rhizosphere carbon flow in trees, in comparison with annual plants: the importance of root exudation and its impact on microbial activity and nutrient availability. *Applied Soil Ecology* 5:29–56.

Grebe, C. 1852. *Forstliche Gebirgskunde, Bodenkunde und Klimalehre*. Vienna.

Green, R. N., R. L. Trowbridge, and K. Klinka. 1993. *Towards a taxonomic classification of humus forms*. Forest Science Monograph 29. Society of American Foresters, Bethesda, MD.

Greenwood, N., and A. Earnshaw. 1984. *Chemistry of the elements*. Pergamon Press, Oxford.

Grier, C. 1975. Wildfire effects on nutrient distribution and leaching in a coniferous ecosystem. *Canadian Journal of Forest Research* 5:599–607.

Grier, C. C. 1978. A *Tsuga heterophylia–Picea sitchensis* ecosystem of coastal Oregon: decomposition and nutrient balances of fallen logs. *Canadian Journal of Forest Research* 8:198–206.

Griffith, B. G., F. W. Haretwell, and T. E. Shaw. 1930. *The evolution of soil as affected by the old field white pine-mixed hardwood succession in central New England.* Harvard Forestry Bulletin 15.

Grigal, D. F. 1984. Shortcomings of soil surveys for forest management in: *Forest land classification: Experience, problems, perspectives*, NCR-102 J. G. Bockheim (ed.), North Central Forest Soils Committee, Madison, WI.

Grigal, D. F., and J. McColl. 1977. Litter decomposition following forest fire in northeastern Minnesota. *Journal of Applied Ecology* 14:531–538.

Grubb, P. J. 1995. Mineral nutrition and soil fertility in tropical rain forests. Pp. 308–330 in: *Tropical forests: management and ecology*, A. E. Lugo and C. Lowe (eds.). Springer-Verlag, New York.

Guerrini, I. A., J. L. M. Gonçalves, and R. L. V. Bôas. 1999. Uses of industrial residues in Brazilian plantation forestry. *The forest alternative — Principles and practice residual of uses*, C. Henry, R. Harrison, and B. Bastian (eds.). College of Forest Resources, University of Washington, Seattle.

Hacskaylo, E. 1971. Metabolite exchanges in ectomyconhizae. Pp. 175–182 in: *Mycorrhizae*, E. Hacskaylo (ed.). USDA Forest Service Miscellaneous Publication 1189.

Haig, I. T. 1929. *Colloidal content and related soil factors as indicators of site quality.* Yale University School of Forestry Bulletin 24.

Haines, B. L., J. B. Waide, and R. L. Todd. 1982. Soil solution nutrient concentrations sampled with tension and zero-tension lysimeters: report of discrepancies. *Soil Science Society of America Journal* 46:658–661.

Haines, L. W., and W. L. Pritchett. 1964. The effects of site preparation on the growth of slash pine. *Soil Crop Science Society of Florida Proceedings* 24:27–34.

Haines, L. W., and W. L. Pritchett. 1965. The effects of site preparation on the availability of soil nutrients and on slash pine growth. *Soil Crop Science Society of Florida Proceedings* 25:356–364.

Haines, L. W., T. F. Maki, and S. G. Sanderford. 1975. The effects of mechanical site preparation treatments on soil productivity and tree (*Pinus taeda* L. and *P. elliottii* Engelm. var. *elliottii*) growth. Pp. 379–395 in: *Forest Soils and Forest Land Management*, B. Bernier and C. Winget (eds.), Laval University Press, Quebec.

Hall, S. J., and D. G. Raffaelli. 1993. Food webs: Theory and reality. *Advances in Ecological Research* 24:18239.

Hallbäcken, L., and J. Bergholm. 1997. Nutrient dynamics. Pp. 112–136 in: *Imbalanced forest nutrition — vitality measures.* Section of Ecology, Swedish University of Agricultural Sciences, SNS Project 1993–1996, Uppsala.

Hamburg, S. 1984. Effects of forest growth on soil nitrogen and organic matter pools following release from subsistence agriculture. Pp. 145–158 in: *Forest soils and treatment impacts*, E. L. Stone (ed.). University of Tennessee Press, Knoxville.

Hamilton, W. J., Jr., and D. B. Cook. 1940. Small mammals and the forest. *Journal of Forestry* 38:468–473.

Handley, W. R. C. 1954. *Mull and mor formation in relation to forest soils.* United Kingdom Forestry Commission, *Bulletin* 23. London.

Hannah, P. R. 1968. *Topography and soil relations for white and black oak*. USDA Forest Service Research Paper NC-25.

Hanson, E. A., and J. O. Dawson. 1982. Effect of *Alnus glutinosa* on hybrid *Populus* height growth in a short-rotation intensively cultured plantation. *Forest Science* 28:49–59.

Harley, J. L. 1969. *The biology of mycorrhizae*, 2nd ed. Leonard Hill, London.

Harrington, C. A., and D. S. DeBell. 1995. Effects of irrigation, spacing and fertilization on flowering and growth in young *Alnus rubra*. *Tree Physiology* 15:427–432.

Harrington, T. B., and M. B. Edwards. 1996. Structure of mixed pine and hardwood stands 12 years after various methods and intensities of site preparation in the Georgia Piedmont. Canadian Journal of Forest Research 26:1490–1500.

Harrison, R. B., C. L. Henry, D. W. Cole, and D. Xue. 1995. Long-term changes in organic matter in soils receiving applications of municipal biosolids. Pp. 139–153 in: *Carbon forms and functions in forest spoils*, W. W. McFee and J. M. Kelly (eds.). Soil Science Society of America, Madison, WI.

Harrison, R. B., and D. W. Johnson. 1992. Inorganic sulfate dynamics. Pp. 104–118 in: *Atmospheric deposition and nutrient cycling*, D. W. Johnson and S. E. Lindberg (eds.). Springer-Verlag, New York.

Hart, S. C. 1999. Nitrogen transformations in fallen tree boles and mineral soil of an old-growth forest. *Ecology* 80:1385–1394.

Hart, S. C., D. Binkley, and R. Campbell. 1986. Predicting loblolly pine current growth and growth response to fertilization. *Soil Science Society of America Journal* 50:230–233.

Hart, S. C., D. Binkley, and D. A. Perry. 1997. Influence of red alder on soil nitrogen transformations in two conifer forests of contrasting productivity. *Soil Biology and Biochemistry* 29:111–1123.

Hart, S. C., and M. K. Firestone. 1989. Evaluation of three in situ soil nitrogen availability assays. *Canadian Journal of Forest Research* 19:185–191.

Hart, S. C., and M. K. Firestone. 1991. Forest floor-mineral soil interactions in the internal nitrogen cycle of an old-growth forest. *Biogeochemistry* 12:103–127.

Hart, S. C., G. E. Nason, D. D. Myrold, and D. A. Perry. 1994. Dynamics of gross nitrogen transformations in an old-growth forest: the carbon connection. *Ecology* 75:880–891.

Harwood, C., and W. Jackson. 1975. Atmospheric losses of four plant nutrients during a forest fire. *Australian Forestry* 38:92–99.

Hatch, A. B. 1937. *The physical basis for mycotrophy in Pinus*. Black Rock Forestry Bulletin 6.

Heath, B. 1985. Levels of asymbiotic nitrogen fixation in leaf litter in Northwest forests. M.S. thesis, Oregon State University, Corvallis.

Hebb, E. A., and R. M. Burns. 1975. *Slash pine productivity and site preparation on Florida sandhill sites*. USDA Forest Service Research Paper SE-135.

Hedderwick, G. W., and G. M. Will. 1982. *Advances in the aerial application of fertilizer to New Zealand Forests: use of an electronic guidance system and dust-free fertilizer*. Forest Research Institute Bulletin #34, Rotorua.

Hedley, M. J., J. W. B. Stewart, and B. S. Chauhan. 1982. Changes in inorganic and organic soil phosphorus fractions by cultivation practices and by laboratory incubations. *Soil Science Society of America Journal* 46:970–976.

Heiberg, S. O., and R. F. Chandler. 1941. A revised nomenclature of forest humus layers for the northeastern United States. *Soil Science* 52:87–99.

Heilman, P. E. 1981. Root penetration of Douglas-fir seedlings into compacted soil. *Forest Science* 27:660–666.

Heilman, P. E., and G. Ekuan. 1983. Nodulation and nitrogen fixation by red alder and Sitka alder on coal mine spoils. *Canadian Journal of Forest Research* 12:992–997.

Helvey. I. D. 1971. A summary of rainfall interception by certain conifers of North America. Pp. 103–113, in *Biological effects on the hydrological cycle*. Purdue University Press, West Lafayette, MD.

Helvey, J. D., and J. H. Patric. 1988. Research on interception loss. Pp. 130–137 in: *Forest hydrology and ecology at Coweeta*, W. T. Swank and D. A. Crossley, Jr. (eds.). Springer-Verlag, New York.

Hendricks, J. LJ., and L. R. Boring. 1999. N2–fixation by native herbaceous legumes in burned pine systems of the southeastern United States. *Forest Ecology and Management* 113:167–177.

Hendrickson, O., and D. Burgess. 1989. Nitrogen-fixing plants in a cut-over lodgepole pine stand of southern British Columbia. *Canadian Journal of Forest Research* 19:936–939.

Hendrix, P. 1995. *Earthworm ecology and biogeography in North America*. CRC Press, Boca Raton, FL.

Herbauts, J. 1982. Chemical and mineralogical properties of sandy and loamy-sandy ochreous brown earths in relation to incipient podzolization in a brown earth-podzol evolutive sequence *Journal of Soil Science* 33:743–762.

Herbert, M. A. 1996. Fertilizers and *Eucalyptus* plantations in South Africa. Pp. 303–325 in: *Nutrition of eucalypts*, P. M. Attiwill and M. A. Adams (eds.). CSIRO, Collingwood, Australia.

Herbert, M. A., and A. P. G. Schönau. 1989. Fertilising commercial forest species in southern Africa: research progress and problems. Southern Africa Forestry Journal 151:58–70.

Hesselman, H. 1917. *Studier over salteterbildningen i naturliga jordmaner och des betydelse i vaxtekologiskt avseende*. Medd Skogsforskosanst, Stockholm #13–14.

Hesselman, H. 1926. Studier over batrskogens humustache, dess egenskaper och beroende av skogsvarden. *Statens Skogsforsoksant Meddel*. 22:169–552.

Hewlett, J. D. 1972. *An analysis of forest water problems in Georgia*. Georgia Forest Research Council Report 30.

Heyward, F., and R. M. Barnette. 1934. *Effect of frequent fires on chemical composition of forest soils in the longleaf region*. Florida Agricultural Experiment Station Bulletin 265.

Heyward, F., and R. M. Barnette. 1936. *Field characteristics and partial chemical analyses of the humus layer of longleaf pine forest soils*. Florida Agricultural Experiment Station Bulletin 302.

Hilgard, E. W. 1906. *Soils*. Macmillan, New York.

Hills, G. A. 1952. *The classification and evaluation of site for forestry*. Ontario Department of Lands Forest Research Report 24.

Hills, G. A. 1961. *The ecological basis for land-use planning.* Ontario Department of Lands Forest Research Report 46.

Hockman, J. N., and H. L. Allen. 1990. Nutritional diagnoses in loblolly pine stands using a DRIS approach. Pp. 500–514 in: *Sustained productivity of forest soils,* S. P. Gessel, D. S. Lacate, G. F. Weetman, and R. F. Powers (eds.). University of British Columbia Faculty of Forestry Publication, Vancouver.

Hoffman, R. J. and R. F. Ferreira. 1976. *A reconnaissance of effects of a forest fire on water quality in Kings Canyon National Park.* USDI Geological Survey Open File Report 76–497.

Högberg, P. 1997. ^{15}N natural abundance in soil-plant systems. *New Phytologist* 137:179–203.

Hole, F. D. 1981. Effects of animals on soils. *Geoderma* 25:75–112.

Holmen, H. 1969. Afforestation of peatlands. Skogs-o. Lantbr.-akad. Tidskr. 108:216–235.

Holmen, H. 1971. Forest fertilization in Sweden. Skogs-o. Lantbr.-akad. Tidskr. 110:156–162.

Holstener-Jorgensen, H. 1983. Forest fertilization research in Denmark: results and perspectives. Pp. 339–345, in: *IUFRO symposium on forest site and continuous productivity,* R. Ballard and S. P. Gessel (eds.). USDA Forest Service General Technical Report PNW-163, Portland, OR.

Hoopmans, P., D. W. Flinn, P. W. Geary, and I. B. Tomkins. 1993. Sustained growth response of *Pinus radiata* on podzolised sands to site management practices. *Australian Forestry* 56:27–33.

Hoover, M. D. 1949. Hydrologic characteristics of South Carolina piedmont forest soils. *Soil Science Society of America Proceedings* 14:353–358.

Hoover, M. D., and H. A. Lunt. 1952. A key for the classification of forest humus types. *Soil Science Society of America Proceedings* 16:368–370.

Houston, M., and T. Smith. 1987. Plant succession: life history and competition. *American Naturalist* 130:168–198.

Hovmand, M. F. 1999. Cumulated deposition of strong acid and sulphur compounds to a spruce forest. *Forest Ecology and Management* 114:19–30.

Hue, N. V., G. R. Craddock, and F. Adams. 1986. Effect of organic acids on aluminum toxicity in subsoils. *Soil Science Society of America Journal* 50:28–34.

Hoyle, M. C. 1971, Effects of the chemical environment on yellow birch, root development and top growth. *Plant and Soil* 35:623–633.

Huikari, O. 1973. *Results of fertilization experiments on peatlands drained for forestry.* Finnish Forest Research Institute Report 1.

Humphreys, F. R., and F. G. Craig. 1981. Effects of fire on soil chemical, structural, and hydrological properties. Pp. 177–200 in: *Fire and the Australian biota,* A. M. Gill, R. Groves, and I. Nobel (eds.). Australian Academy of Sciences Press, Canberra.

Humphreys, F. R., and M. J. Lambert. 1965. An examination of a forest site which has exhibited the ashbed effect. *Australian Journal of Soil Research* 3:81–94.

Humphreys, F. R., and W. L. Pritchett. 1971. Phosphorus adsorption and movement in some sandy forest soils. *Soil Science Society of America Proceedings* 35:495–500.

Hungerford, R. 1980. Microenvironmental response to harvesting and residue management. Pp. 37–73 in: *Environmental consequences of timber harvesting in Rocky Mountain coniferous forests.* USDA Forest Service General Technical Report INT-90, Ogden, Utah.

Hunter, I. R., and J. D. Graham. 1983. Three-year response of *Pinus radiata* to several types and rates of phosphorus fertiliser on soils of contrasting phosphorus retention. *New Zealand Journal of Forestry Science* 13:229–238.

Hüttl, R. F., and W. Schaaf (eds.) 1997. *Magnesium deficiency in forest ecosystems.* Kluwer Academic Press, Dordrecht, the Netherlands.

Ingestad, T. 1979. Mineral nutrient requirements of *Pinus sylvestris* and *Picea abies* seedlings. *Physiologia Plantarum* 45:373–380.

Ingestad, T. 1981. Growth, nutrition, and nitrogen fixation in grey alder at varied rates of nitrogen addition. *Physiologia Plantarum* 50:353–364.

Ingestad, T. 1982. Relative addition rate and external concentration: driving variables used in plant nutrition research. *Plant, Cell and Environment* 5:443–453.

Innes, J. 1993. *Forest health: its assessment and status.* CAB International, Wallingford, UK.

Isaac, L. A., and H. G. Hopkins. 1937. The forest soil of the Douglas-fir region and changes wrought upon it by logging and slashburning. *Ecology* 18:264–279.

Jackson, D. S. 1965. Species siting: climate, soil, and productivity. *New Zealand Journal of Forestry* 10:90–102.

Jackson, D. S., H. H. Gifford, and J. D. Graham. 1983. Lupin, fertiliser, and thinning effects on early productivity of *Pinus radiata* growing on deep Pinaki sands. *New Zealand Journal of Forestry Science* 13:159–182.

Jackson, J. K. 1977. Irrigated plantations. Pp. 277–285 in: *Savanna afforestation in Africa.* Food Agriculture Organization, Rome.

Jencks, E. M., E. H. Tyron, and M. Contri. 1982. Accumulation of nitrogen in minesoils seeded to black locust. *Soil Science Society of America Journal* 46:1290–1293.

Jenkinson, D. S., N. J. Bradbury, and K. Coleman. 1994. How the Rothamsted classical experiments have been used to develop and test models for the turnover of carbon and soil nitrogen. Pp. 117–138 in: *Long-term experiments in agricultural and ecological studies,* R. A. Leigh and A. E. Johnston (eds.). CAB International, Oxon.

Jenny, H. 1941. *Factors of soil formation.* McGraw-Hill, New York.

Jenny, H. 1980. *The soil resource. Origen and behavior.* Ecology Studies 37. Springer-Verlag, New York.

Johnson, C. E., A. H. Johnson, and T. G. Siccama. 1991. Whole-tree clear-cutting effects on exchangeable cations and soil acidity. *Soil Science Society of America Journal* 55:502–508.

Johnson, C. M., and P. R. Needham. 1966. Ionic composition of Sagehen Creek, California, following an adjacent fire. *Ecology* 47:636–639.

Johnson, D. W. 1992a. Nitrogen retention in forest soils. *Journal of Environmental Quality* 21:1–12.

Johnson, D. W. 1992b. Effects of forest management on soil carbon storage. *Water, Air, and Soil Pollution* 64:83–120.

Johnson, D. W. 1995. Soil properties beneath ceanothus and pine stands in the eastern Sierra Nevada. *Soil Science Society of America Journal* 59:918–924.

Johnson, D. W., G. S. Henderson, and D. Todd. 1981. Evidence of modern accumulation of sulfate in an east Tennessee forested Ultisol. *Soil Science* 132:422–446.

Johnson, D. W., G. S. Henderson, D. D. Huff, S. E. Lindberg, D. D. Richter, D. S. Shriner, D. E. Todd, and J. Turner. 1982. Cycling of organic and inorganic sulphur in a chestnut oak forest. *Oecologia* 54:141–148.

Johnson, D. W., and S. E. Lindberg. 1992. *Atmospheric deposition and forest nutrient cycling.* Springer-Verlag, New York.

Johnson, D. W., R. B. Susfalk, R. A. Dahlgren, and J. M. Klopatek. 1998. Fire is more important than water for nitrogen fluxes in semi-arid forests. *Environmental Science and Policy* 1:79–86.

Johnson, D. W., and D. Todd. 1985. Nitrogen availability and conservation in young yellow-poplar and loblolly pine plantations fertilized with urea. *Agonomy Abstracts* 77:220.

Johnson, D. W., and D. Todd. 1998. Harvesting effects on long-term changes in nutrient pools of mixed oak forest. *Soil Science Society of America Journal* 62:1725–1735.

Johnson, N. E., and H. D. Smith. 1983. Forest productivity: economic factors involved. Pp. 101–113, in: *Maintaining forest site productivity.* Appalachian Society of American Foresters, Clemson.

Johnsrud, S. C. 1979. Nitrogen fixation by root nodules of *Alnus incana* in a Norwegian forest ecosystem. *Oikos* 30:475–479.

Jones, H. E., P. Högberg, and H. Ohlsson. 1994. Nutritional assessment of a forest fertilization experiment in northern Sweden by root bioassays. *Forest Ecology and Management* 64:59–69.

Jurgensen, M. F., R. T. Graham, M. J. Larsen, and A. E Harvey. 1992. Clearcutting, woody residue removal, and nonsymbiotic nitrogen fixation in forest soils of the Inland Northwest. *Canadian Journal of Forest Research* 22:1172–1178.

Jurgensen, M. F., M. J. Larsen, S. D. Spano, A. F. Harvey, and M. R. Gale. 1984. Nitrogen fixation associated with increased wood decay in Douglas fir residue. *Forest Science* 30:1038–1044.

Kadeba, O., and J. R. Boyle. 1978. Evaluation of phosphorus in forest soils: comparison of phosphorus uptake, extraction method and soil properties. *Plant and Soil* 49:285–297.

Kandler, O. 1992. Historical declines and diebacks of central European forest and present conditions. *Environmental Toxicology and Chemistry* 11:1077–1093.

Kandler, O. 1993. The air pollution/forest decline connection: the "Waldsterben" theory refuted. *Unasylva* 44:39–49.

Kane, M. B. 1981. Fertilization of juvenile loblolly pine plantation: impacts on fusiform rust incidence. M.S. thesis, North Carolina State University, Raleigh.

Katznelson, H., J. W. Rowatt, and E. A. Peterson. 1962. The rhizosphere effect of mycorrhizal and non-mycorrhizal roots of yellow birch seedlings. *Canadian Journal of Botany* 40:378–382.

Kaufman, C. M. 1968. Growth of horizontal roots, height, and diameter of planted slash pine. *Forest Science* 14:265–274.

Kaufman, C. M., W. L. Pritchett, and R. E. Choate. 1977. *Growth of slash pine (Pinus elliottii Engel. var. elliottii) on drained flatwoods.* Florida Agricultural Experiment Station Bulletin 792.

Kaupenjohann, M. 1997. Tree nutrition. Pp. 275–308 in: *Magnesium deficiency in forest ecosystems*, R. F. Hüttl and W. Schaaf (eds.). Kluwer Academic Press, Dordrecht, the Netherlands.

Kay, B. D. 1997. Soil structure and organic carbon: a review. Pp. 169–197 in: *Soil processes and the carbon cycle*, R. Lal, J. M. Kimble, R. F. Follett and B. A. Stewart (eds.) CRC Press, Boca Raton, FL.

Kaye, J., and S. C. Hart. 1997. Competition for nitrogen between plants and soil microorganisms. *Tree* 12:139–143.

Kaye, J., and S. C. Hart. 1998. Ecological restoration alters nitrogen transformations in a ponderosa pine–bunchgrass ecosystem. *Ecological Applications* 8:1052–1060.

Kaye, J., S. Resh, M. Kaye, and R. Chimner. 1999. Nutrient and carbon dynamics in a replacement series of *Eucalyptus* and *Albizia* trees. Manuscript submitted to *Ecology*.

Keith, H. 1997. Nutrient cycling in eucalypt ecosystems. Pp. 197–226 in: *Eucalypt ecology*, J. Williams and J. Woinarski (eds.). Cambridge University Press, Cambridge.

Kellman, M., and M. Kading. 1992. Facilitation of tree seedling establishment in a sand dune succession. *Journal of Vegetation Science* 3:679–688.

Keresztesi, B. 1988. Black locust: the tree of agriculture. *Outlook on Agriculture* 17:77–85.

Kessell, S. L. 1927. Soil organisms: the dependence of certain pine species on a biological soil factor. *Empire Forestry Journal* 6. 70–74.

Kevin, D. K. McE. 1962. *Soil animals*. Witherby, London.

Khanna, P. K., and R. J. Raison. 1986. Effect of fire intensity on solution chemistry of surface soil under a *Eucalyptus pauciflora* forest. *Australian Journal of Soil Research* 24:423–434.

Klawitter, R. A. 1970a. *Does bedding promote pine survival and growth on ditched wet sands?* USDA Forest Service Research Note NE-109.

Klawitter, R. A. 1970b. Water regulation on forest land. *Journal of Forestry* 68:338–342.

Klinge, H. 1976. Nahrstoffe, Wasser, and Durchwurzelung von Podsolen und Latosolen unter tropischem Regenwald bei Manaus/Amazonien. *Biogeographica*. 7:45–58.

Knight, D. H., T. J. Fahey, and S. W. Running. 1985. Water and nutrient outflow from contrasting lodgepole pine forests in Wyoming. *Ecological Monographs* 55:29–48.

Knight, H. 1966. Loss of nitrogen from the forest floor by burning. *Forestry Chronicle* 42:149–152.

Knight, P. J., and I. D. Nicholas. 1996. Eucalypt nutrition: New Zealand experience. Pp. 275–302 in: *Nutrition of eucalypts*. P. M. Attiwill, and M. A. Adams (eds.). CSIRO, Collingwood, Australia.

Kodama, H. F., and D. H. Van Lear. 1980. Prescribed burning and nutrient cycling relationships in young loblolly pine plantations. *Southern Journal of Applied Forestry* 4:118–121.

Kohmann, K. 1972. Root ecological investigations on pine (*Pinus silvestris*) II. The root system's reaction to fertilization. Meddr. Norske Skogfors Ves. 30:392–396.

Kormanik, P. P., W. C. Bryan, and R. C. Schultz. 1977. The role of mycorrhizae in plant growth and development. Pp. 1–10 in *Physiology of root-Microorganisms Associ-*

ations, H. M. Vines (ed.). Proceedings of the Symposium of the Southern Section of the American Society of Plant Physiology, Atlanta.

Kozlowski, T. T. 1968. Soil water and tree growth. Pp. 30–57 in: *Seventeenth annual forestry symposium on "The ecology of southern forest,"* N. E. Linnartz (ed.). Louisiana State University Press, Baton Rouge.

Kraemer, J. F., and R. K. Hermann. 1979. Broadcast burning: 25-year effects on forest soils in the western flanks of the Cascade Mountains. *Forest Science* 25:427–439.

Kramer, P. J. 1969. *Plant and soil-water relationships: a modern synthesis.* McGraw-Hill, New York.

Krammes, J. S., and L. F. DeBano. 1965. Soil wettability: a neglected factor in watershed management. *Water Resources Research* 1:283–286.

Kreutzer, K. 1995. Effects of forest liming on soil processes. *Plant and Soil* 168–169:447–470

Krugman, S. L., and E. C. Stone. 1966. The effect of cold nights on the root regeneration potential of ponderosa pine seedlings. *Forest Science* 12:451–459.

Kubiena, W. L. 1953. *The Soils of Europe.* Thomas Morby, London.

Kuhry, P., B. J. Nicholson, L. D. Gignac, D. H. Vitt, and S. E. Bayley. 1993. Development of sphagnum-dominated peatlands in boreal continental Canada. *Canadian Journal of Botany* 71:10–22.

Kushla, J. D., and R. F. Fisher. 1980. Predicting slash pine response to nitrogen and phosphorus fertilization. *Soil Science Society of America Journal* 44:1301–1306.

Laine, J., K. Minkkinen, J. Sinisalo, I. Savolainen, and P. J. Martikainen. 1997. Greenhouse impact of a mire after drainage for forestry. Pp. 437–447 in: *Northern forested wetlands: ecology and management*, C. C. Trettin, M. F. Jurgensen, D. F. Grigal, M. R. Gale and J. K. Jeglum (eds.). Lewis Publishers, Boca Raton, FL.

Lal, R., J. M. Kimble, R. F. Follett, and B. A. Stewart. 1998a. *Soil processes and the carbon cycle.* CRC Press, Boca Raton, FL.

Lal, R., J. M. Kimble, R. F. Follett, and B. A. Stewart. 1998b. *Management of carbon sequestration in soil.* CRC Press, Boca Raton, FL.

Landsburg, J. D., and P. H. Cochran. 1980. Prescribed burning effects on foliar nitrogen content in ponderosa pine. Pp 209–213 in: *Proceedings of the sixth conference on fire and forest meteorology, April* 22–24. Society of American Foresters, Washington.

Larson, M. M., S. H. Patel, and J. P. Vimmerstedt. 1995. Allelopathic interactions between herbaceous species and trees grown in topsoil and spoil media. *J. Sustain. Forestry* 3:39–52.

Lavelle, P. 1997. Faunal activities and soil processes: adaptive strategies that determine ecosystem function. *Advance in Ecological Research* 27:93–133.

Lea, R., and R. Ballard. 1982. Predicting loblolly pine growth response from N fertilizer using soil-N availability indices. *Soil Science Society of America Journal* 46:1096–1099.

Leaf, A. L., R. E. Leonard, J. V. Berglund, A. R. Eschner, P. H. Cochran, J. B. Hart, G. M. Marion, and R. A. Cunnigham. 1970. Growth and development of *Pinus resinosa* plantations subjected to irrigation-fertilization treatments. Pp. 97–118 in: *Tree growth and forest soils*, C. T. Youngberg and C. B. Davey (eds.). Oregon State University Press, Corvallis.

Lehotsky, K. 1972. Sand dune fixation in Michigan—thirty years later. *Journal of Forestry* 70:155–160.

Levy, G. 1972. Premiers resultats concernant deux experiences d'assainissement du sol sur plantations de resineux. *Annual Science of Forestry* 29:427–450.

Lewis, W. M., Jr. 1974. Effects of fire on nutrient movement in a South Carolina pine forest. *Ecology* 55:1120–1127.

Likens, G., F. H. Bormann, R. S. Pierce, J. S. Eaton, and N. Johnson. 1977. *Biogeochemistry of a forested ecosystem.* Springer-Verlag, New York.

Linder, S. 1995. Foliar analysis for detecting and correcting nutrient imbalances in Norway spruce. *Ecological Bulletins* 44:178–190.

Lindsay, W. 1979. *Chemical equilbria in soils.* Wiley, New York.

Lindsay, W., and P. L. G. Vlek. 1977. Phosphate minerals. Pp. 639–672 in: *Minerals in soil environments,* J. B. Dixon and S. B. Weed (eds.). Soil Science Society of America, Madison, WI.

Little, S., and J. Ohmann. 1988. Estimating nitrogen lost from forest floor during prescribed fires in Douglas-fir/western hemlock clearcuts. *Forest Science* 34:152–164.

Löfvenius, M. O. 1993. Temperature and radiation regimes in pine shelterwood and clear-cut area. Ph.D. thesis, Swedish University of Agricultural Sciences, Umeå.

Lorio, P. L., V. K. Howe, and C. N. Martin. 1972. Loblolly pine rooting varies with microrelief on wet sites. *Ecology* 53:1134–1140.

Lugo, A. E., M. J. Sanchez, and S. Brown. 1986. Land use and organic carbon content of some subtropical soils. *Plant and Soil* 96:185–196.

Lundgren, B. 1978. *Soil conditions and nutrient cycling under natural and plantation forests in Tanzanian highlands.* Swedish University of Agricultural Science, Uppsala.

Lutz, H. J., and R. F. Chandler. 1946. *Forest soils.* Wiley, New York.

Lyford, W. H. 1943. The palatability of freshly fallen forest tree leaves to millipedes. *Ecology* 24:252–261.

Lyford, W. H. 1952. Characteristics of some podzolic soils of the northeastern United States. *Soil Science Society of America Proceedings* 16:231-234.

Lyford, W. H. 1963. *Importance of ants to brown podzolic soil genesis in New England.* Harvard Forestry Paper 7.

Lyford, W. H. 1973. Forst soil microtopography. Pp. 47–58, in *Proceedings of the first soil microcommunities conference.* U.S. Atomic Energy Commission, Syracuse, NY.

Lyford, W. H., and D. W. MacLean. 1966. *Mound and pit microrelief in relation to soil disturbance and tree distribution in New Brunswick, Canada.* Harvard Forestry Paper 15.

Lyford, W. H., and B. F. Wilson. 1966. *Controlled growth of forest tree roots: techniques and application.* Harvard Forestry Paper 16.

Lynch, J. M. 1990. *The rhizosphere.* Wiley Interscience, Chichester, UK.

Lyr, H., and U. Hoffman. 1967. Growth rates and growth periodicity of tree roots. Pp. 181–236 in: *International review of forestry research,* Vol. 2. J. A. A. Romberger and P. Mikola (eds.). Academic Press, New York.

MacCarthy, P., C. E. Clapp, R. L. Malcolm, and P. R. Bloom. 1990. *Humic substances in soil and crop sciences: selected readings.* American Society of Agronomy, Madison, WI.

MacFarlane, I. C. 1969. *Maskey engineering handbook.* University of Toronto Press, Toronto.

Mader, D. L. 1953. Physical and chemical characteristics of the major types of forest humus found in the United States and Canada. *Soil Science Society of America Proceedings* 17:155–158.

Mader, D. L. 1964. Where are we in soil-site classification? Pp. 23–32 in: *Applications of soils information in forestry.* New York State College of Agricultural Miscellaneous Publication.

Magnusson, T. 1992. Temporal and spatial variation of the soil atmosphere in forest soils of northern Sweden. Ph.D. thesis, Swedish University of Agricultural Sciences, Umeå.

Mailly, D., P. Ndiaye, H. A. Margolis, and M. Pineau. 1994. Dune stabilization and reforestation with filao (*Casuarina equisetifolia*) in the northern coast zone of Senegal. *Forestry Chronicle* 70:282–290.

Makkonen-Spiecker, K., and H. Spiecker. 1997. Influence of magnesium supply on tree growth. Pp. 215–254 in: *Magnesium deficiency in forest ecosystems,* R. F. Hüttl and W. Schaaf (eds.). Kluwer Academic Press, Dordrecht, the Netherlands.

Malloch, D. W., K. A. Pirozymski, and P. H. Raven. 1980. Ecological and evolutionary significance of mycorrhizal symbiosis in vascular plants. *Proceedings of the National Academy of Sciences* 77:2113–2118.

Malm, D., and G. Moller. 1975. Skillnader i volymtillvaxtokning efter godsling med urea resp ammoniumnitrat. Pp. 46–63 in: *Foreningen Skogstradsforadling, 1974 arsbok.* Institutet for Skogsfofbattring.

Malmer, A. 1993. Dynamics of hydrology and nutrient losses as response to establishment of forest plantation. A case study on tropical rain forest land in Sabah, Malaysia. Ph.D. thesis, Swedish University of Agricultural Sciences, Umeå.

Mann, W. F., and J. M. McGilvray. 1974. *Response of slash pine to bedding and phosphorus application in southeastern flatwoods.* USDA Forest Service Research Paper SO-99.

Marchand, D. E. 1971. *Chemical weathering, soil development, and geochemical fractionation in a part of the White Mountains, Mono and Inyo Counties, California.* Geological Survey Professional Papers (U.S.) #352J.

Marks, G. C., and T. T. Koslowski. 1973. *Ectomycorrhizae: their ecology and physiology.* Academic Press, New York.

Marschner, H. 1995. *Mineral nutrition of higher plants.* Academic Press, London.

Martinez Velazquez, A., and D. A. Perry. 1997. Factors influencing the availability of nitrogen in thinned and unthinned Douglas-fir stands in the central Oregon Cascades. *Forest Ecology and Management* 93:195–203.

Marx, D. H. 1972. Ectomycorrhizae as biological deterrents to pathogenic root infections. *Annual Review of Phytopathology* 10:429–454.

Marx, D. H. 1977. The role of mycorrhizae in forest production. Pp. 151–161 in: *TAPPI conference papers.* Atlanta, GA.

Marx, D. H., W. C. Bryan, and C. B. Davey. 1970. Influence of temperature on aseptic synthesis of ectomycorrhizae by *Thelephora terrestris* and *Pisolithus tinctorius* on loblolly pine. *Forest Science* 16:424–431.

Mast, A. 1989. A laboratory and field study of chemical weathering with special reference to acid deposition. Ph.D. thesis, University of Wyoming.

Matson, P. A., and R. H. Waring. 1984. Effects of nutrient and light limitation on mountain hemlock: susceptibility to laminated root rot. *Ecology* 65:1517–1524.

Matthews, S. W. 1973. This changing earth. *National Geographic Magazine* 143:1–37.

Matziris, D., and B. Zobel. 1976. Effects of fertilization on growth and quality characteristics of loblolly pine. *Forest Ecology and Management* 1:21–30.

May, J. T., C. C. Parks, and H. F. Perkins. 1969. Establishment of grasses and tree vegetation on spoil from kaolin strip mining. Pp. 137–147 in: *Ecology and revegetation of devastated land* Vol. 2. Gordon and Breach, New York.

McClaugherty, C. A., J. D. Aber, and J. M. Melillo. 1982. The role of fine roots in the organic matter and nitrogen budgets of two forested ecosystems. *Ecology* 63:1481-1490.

McColl, J. G., and N. Gressel. 1995. Forest soil organic matter: characterization and modern methods of analysis. Pp. 13–32 in: *Carbon forms and functions in forest spoils*, W. W. McFee and J. M. Kelly (eds.). Soil Science Society of America, Madison, WI.

McColl, J. G., and R. F. Powers. 1998. Decomposition of small diameter woody debris of red fir determined by nuclear magnetic resonance. *Communications in Soil Science and Plant Analysis* 29:2691–2704.

McFee, W W., and E. L. Stone. 1965. Quantity, distribution, and variability of organic matter and nutrients in a forest podzol in New York. *Soil Science Society of America Proceedings* 29:432–436.

McNabb, D. H., F. Gaweda, and H. A. Froehlich. 1989. Infiltration, water repellency, and soil moisture content after broadcast burning a forest site in southwest Oregon. *Journal of Soil and Water Conservation* 44:87–90.

McQuilkin, W. E. 1935. Root development of pitch pine, with some comparative observations on shortleaf pine. *Journal of Agricultural Research* 51:983–1016.

Mead, D. J., and W. L. Pritchett. 1971. A comparison of tree responses to fertilizers in field and pot experiments. *Soil Science Society of America Proceedings* 35:346–349.

Meetemeyer, V. 1978. Macroclimate and lignin control of decomposition. *Ecology* 59:465–472.

Mehrotra, V. S. 1998. Arbuscular mycorrhizal associations of plants colonizing coal mine spoil in India. *Journal of Agricultural Science* 130(2):125–133.

Melillo, J. M., J. D. Aber, A. E. Linkins, A. Ricca, B. Fry, and K. J. Nadelhoffer. 1989. Carbon and nitrogen dynamics along the decay continuum: plant litter to soil organic matter. Pp. 53–62 in: *Ecology of arable land*, M. Clarholm and L. Bergström (eds.). Kluwer Academic Press, Amsterdam.

Melin, F. 1963. Some effects of forest tree roots on mycorrhizal *Basidiomycetes*. Pp. 125–145 in: *Symbiotic associations. Proceedings of the 13th symposium of the Society of General Microbiology*. Cambridge University Press, London.

Melin, J., H. Nommik, U. Lohm, and J. Flower-Ellis. 1983. Fertilizer nitrogen budget in a Scots pine ecosystem attained by using root-isolated plots and ^{15}N technique. *Plant and Soil* 74:249–263.

Merriam, C. H. 1898. *Life zones and crop zones of the United States*. USDA Biological Survey Bulletin 10. Washington, DC.

Messina, M. G., and W. H. Conner. 1998. *Southern forested wetlands: ecology and management.* Lewis Publishers, Boca Raton, FL.

Metz, L. J., C. G. Wells, and P. P. Kormanik. 1970. *Comparing the forest floor and surface soil beneath four pine species in the Virginia Piedmont.* USDA Forest Service Research Paper SF-55.

Mika, P. G., J. A. Moore, R. P. Brockley, and R. F. Powers. 1992. Fertilization response by interior forests: when, where, and how much? Pp. 127–142 in: *Forest fertilization: sustaining and improving nutrition and growth of western forests,* H. N. Chappell, G. F. Weetman, and R. E. Miller (eds.). College of Forest Resources, University of Washington, Seattle.

Mikola, P. 1973. Application of mycolrhizal symbiosis in forestry practice. Pp. 383–411 in: *Ectomycorrhizae: their ecology and physiology,* G. C. Marks and T. T. Kozlowski (eds.). Academic Press, New York.

Mikola, P. 1980. *Tropical mycorrhiza research.* Clarendon Press, Oxford.

Mikola, P., P. Uomala, and E. Malkonen. 1983. Application of biological nitrogen fixation in European silviculture. Pp. 279–294 in: *Biological nitrogen fixation in forest ecosystems: foundations and applications,* J. C. Gordon and C. T. Wheeler (eds.). Martinus Nijhoff/Junk, the Hague, the Netherlands.

Miller, H. G. 1981. Forest fertilization: some guiding concepts. *Forestry* 54:157–167.

Miller, R. E., and R. E. Bigley. 1990. Effects of burning Douglas-fir logging slash on stand development and site productivity. Pp. 362–376 in: *Sustained productivity of forest soils,* S. P. Gessel, D. S. Lacate, G. F. Weetman, and R. F. Powers (eds.). University of British Columbia Faculty of Forestry Publication, Vancouver.

Miller, R. E., and M. D. Murray. 1978. The effects of red alder on growth of Douglas-fir. Pp. 286–306 in: *Utilization and management of alder,* D. G. Briggs, D. S. DeBell, and W. A. Atkinson (eds.). USDA Forest Service General Technical Report PNW-70, Portland OR.

Miller, R. E., and M. D. Murray. 1979. Fertilizer versus red alder for adding nitrogen to Douglas-fir forests of the Pacific Northwest. Pp. 356–373 in: *Symbiotic nitrogen fixation in the management of temperate forests,* J. C. Gordon, C. T. Wheeler and D. A. Perry (eds.). Forest Research Laboratory, Oregon State University, Corvallis.

Minderman, G. 1968. Addition, decomposition and accumulation of organic matter in forests. *Journal of Ecology* 56:355–362.

Minkkinen, K., J. Laine, H. Nykanen, and P. J. Martikainen. 1997. Importance of drainage ditches in emissions of methane from mires drained for forestry. *Canadian Journal of Forest Research* 27:949–952.

Minore, D., and C. E. Smith. 1971. *Occurrence and growth of four northwestern tree species over shallow water tables.* USDA Forest Service Research Note PNW. 160.

Minore, D., C. E. Smith, and R. F. Wollard. 1969. *Effects of high-soil density on seedling root growth of seven northwestern tree species.* USDA Forest Service Research Note PNW-112.

Montes, R. A., and N. L. Christensen. 1979. Nitrification and succession in the Piedmont of North Carolina. *Forest Science* 25:287–297.

Mooney, H. A., B. G. Drake, R. J. Luxmoore, W. G. Oechel, and L. F. Pitelka. 1991. Predicting ecosystem response to elevated CO_2 concentrations. *Bioscience* 41:96–104.

Moore, D. G. 1975. *Effects of forest fertilization with urea on stream quality — Quilcene Ranger District, Washington.* USDA Forest Service Research Note PNW-241.

Moore, J. A., P. G. Mika, J. W. Schwandt, and T. M. Shaw. 1994. Nutrition and forest health. Pp. 173–176 in: *Interior cedar-hemlock-white pine forests: ecology and management.* Department of Natural Resource Sciences, Washington State University, Pullman.

Morby, F. E. 1982. Nursery site selection, layout and development. Pp. 9–16 in: *Forest nursery manual,* M. L. Duryea and T. D. Landis (eds.). Martinus Nijhoff/Junk, the Hague, the Netherlands.

Moro, L., and J. L. M. Gonçalves. 1995. Efeito da "cinza" de biomassa florestal sobre a produtividade de povoamentos puros de *Eucalyptus grandis* e avaliação financeira. *IPEF, Piracicaba* (48/49):18–27.

Morris, L. A., W. L. Pritchett, and B. F. Swindell. 1983. Displacement of nutrients into windrows during site preparation of a flatwoods forest. *Soil Science Society of America Journal* 47:591–594.

Moser, K. M. 1985. Stem growth and leaf area of loblolly pine mixed with nitrogen-fixing *Lespedeza.* Master's project, School of Forestry and Environmental Studies, Duke University, Durham, NC.

Moser, M. 1967. Ectotrophic nutrition at timberline. Midd. d. Forst. Bundesversuchsanst. *Wien.* 75:357–380.

Motavalli, P. P., C. A. Palm, E. T. Elliott, S. D. Frey, and P. C. Smithson. 1995. Nitrogen mineralization in humid tropical forest soils: mineralogy, texture, and measured nitrogen fractions. *Soil Science Society of America Journal* 59:1168–1175.

Mroz, G. D., M. F. Jurgensen, A. E. Harvey, and M. J. Larsen. 1980. Effects of fire on nitrogen in forest floor horizons. *Soil Science Society of America Journal* 44:395–400.

Muller, P. E. 1879. Studier over skovjord. som bidrag til skovdyrkningens theori. Tidsskr. Skovbr. 3:1–124.

Myers, E. A. 1979. Design and operational criteria for forest irrigation systems. Pp. 265–272 in: *Utilization of municipal sewage effluent and sludge on forest and disturbed land,* W. E. Sopper and S. N. Kerr (eds.). Pennsylvania State University Press, University Park.

Myrold, D. D., and K. Huss-Danell. 1994. Population dynamics of *Alnus*-infective *Frankia* in a forest soil with and without host trees. *Soil Biology and Biochemistry* 26:533–540.

Nadelhoffer, K. J., J. D. Aber, and J. M. Melillo. 1983. Leaf-litter production and soil organic matter dynamics along a nitrogen-availability gradient in southern Wisconsin (U.S.A.). *Canadian Journal of Forest Research* 13:12–21.

Nadelhoffer, K. J., B. A. Emmett, P. Gundersen, O. J. Kjønaas, C. J. Koopmans, P. Schleppi, A. Tietema, and R. F. Wright. 1999. Nitrogen deposition makes a minor contribution to carbon sequestration in temperate forests. *Nature* 398:145–148.

NADP. 1999. National Atmospheric Deposition Program (NRSP-3)/National Trends Network. NADP Program Office, Champaign, IL.

Nambiar, E. K. S., and A. G. Brown. 1997. *Management of soil, nutrients and water in tropical plantation forests.* ACIAR, Canberra, Australia.

Nambiar, E. K. S., and D. N. Fife. 1991. Nutrient retranslocation in temperate conifers. *Tree Physiology* 9:185–207.

Nambiar, E. K. S., and R. Sands. 1992. Effects of compaction and simulater root channels in the subsoil on root development, water-uptake and growth of radiata pine. *Tree Physiology* 10:297–306.

Näsholm, T. 1994. Removal of nitrogen during needles senescence in Scots pine (*Pinus sylvestris* L.). *Oecologia*, 99:290–296.

National Academy of Sciences. 1977. *Leucaena; promising forage and tree crop for the tropics.* National Academy of Sciences, Washington, DC.

Neal, J., E. Wright, and W. B. Bollen. 1965. *Burning Douglas-fir slash: physical, chemical, and microbial effects in the soil.* Oregon State University, Forest Research Laboratory, Corvallis.

Negi, J. D. S. 1984. Biological productivity and cycling of nutrients in mmanaged and manmade ecosystems. PhD thesis, Garhwal University, Srinagar (U.P.), India.

Nelson, E. E., E. M. Hansen, C. Y. Li, and J. M. Trappe. 1978. The role of red alder in reducing losses from laminated root rot. Pp. 273–306 in: *Utilization and management of alder,* D. G. Briggs, D. S. DeBell, and W. A. Atkinson (eds.). USDA Forest Service General Technical Report PNW-70, Portland, OR.

Newman, H. C., and W. C. Schmidt. 1980. Silviculture and residue treatments affect water use by a larch/fir forest. Pp. 75–110 in: *Environmental consequences of timber harvesting in Rocky Mountain coniferous forests.* USDA Forest Service General Technical Report INT-90, Ogden, UT.

Nicolas, J. R. J., and K. M. Hinkel. 1996. Concurrent permafrost aggradation and degradation induced by forest clearing, central Alaska, U.S.A. Arctic Alpine Research 28:294–299.

Nihlgård, B. 1971. Pedological influences of spruce planted on former beech forest soils in Scania, south Sweden. *Oikos* 22:302–314.

Nihlgård, B., and B. Popovic. 1984. *Effekter av olika kalkningsmedel I skogsmark — en litteratur versikt.* Statens Naturvardsverk PM 1851, Solna.

Nilsson, L.-O., and K. Wiklund. 1994. Nitrogen uptake in a Norway spruce stand following ammonium sulphate application, fertigation, irrigation, drought, and nitrogen-free-fertilization. *Plant and Soil* 164:221–229.

Nohrstedt, H. -?. 1990. Effects of repeated nitrogen fertilization with different doses on soil properties in a *Pinus sylvestris* stand. *Scandinavian Journal of Forest Research* 5:3–15.

Nohrstedt, H. -?. 1993. *Nitrogen status of the Swedish forest.* Forestry Research Institute of Sweden Redog relse No. 8.

Nohrstedt, H. -?, Sikstr m, U., and E. Ring. 1993. *Experiments with vitality fertilisation in Norway spruce stands in southern Sweden.* SkogForsk Report 2/93, Kista, Sweden.

Nohrstedt, H. -?, E. Ring, L. Klemedtsson, and ?. Nilsson. 1994. Nitrogen losses and soil water acidity after clearfelling of fertilized experimental plots in a *Pinus sylvestris* stand. *Forest Ecology and Management* 66:69–86.

North Carolina State Forest Nutrition Cooperative. 1997. *Ten-year growth and foliar responses of mid-rotation loblolly pine plantations to nitrogen and phosphorus fertilization.* North Carolina State Forest Nutrition Cooperative Report #39, Raleigh.

Nykvist, N., and Rosén, K. 1985. Effect of clear-felling and slash removal on the acidity of northern coniferous soils. *Forest Ecology and Management* 11:157–169.

Oades, J. M., G. P. Gillman, and G. Uehara. 1989. Interactions of soil organic matter and variable-charge clays. Pp. 69–95 in: *Dynamics of soil organic matter in tropical ecosystems*, D. Coleman, J. M. Oades, and G. Uehara (eds.). University of Hawaii, Honolulu.

O'Connell, A. M., and K. V. Sankaran. 1997. Organic matter accretion, decomposition and mineralisation. Pp. 443–480 in: *Management of soil, nutrients, and water in tropical plantation forests*, E. K. S. Nambiar and A. G. Brown (eds.). ACIAR Monograph #43, Canberra.

O'Loughlin, C. L., L. K. Rowe, and A. J. Pearce. 1982. *Exceptional storm influences on slope erosion and sediment yield in small forest catchments, North Westland, New Zealand.* Institute of Engineers, Australia. National Symposium on Forest Hydrology, Melbourne.

O'Loughlin, C. L. and A. Watson. 1981. Note on root-wood strength deterioration in *Nothofagus fusca* and *N. truncata* after clearfelling. *New Zealand Journal of Forest Science* 11:183–185.

Olsen, S. R., and L. E. Sommers. 1982. Phosphorus. Pp. 403–430 in: *Methods of soil analysis, part 2: chemical and microbiological properties*, A. L. Page (ed.). American Society of Agronomy, Madison, WI.

Olson, J. 1981. Carbon balance in relation to fire regimes. Pp. 327–378 in: *Fire regimes and ecosystem properties.* USDA Forest Service GTR-WO-26, Washington, DC.

Oren, R., E.-D. Schulze, K. S. Werk, and J. Meyer. 1988. Performance of two *Picea abies* (L.) Karst. stands at different stages of decline. VII. Nutrient relations and growth. *Oecologia* 77:163–173.

Outcalt, K. W. 1994. *Evaluations of a restoration system for sandhills longleaf pine communities.* Rocky Mountain Forest and Range Experiment Station, Forest Service, USDA General Technical Report 247.

Paces, T. 1986. Rates of weathering and erosion derived from mass balance in small drainage basins. Pp. 531–551 in: *Rates of chemical weathering of rocks and minerals*, S. M. Colman and D. P. Dethier (eds.). Academic Press, Orlando, FL.

Paré, D., and B. Bernier. 1989a. Origin of phosphorus deficiency observed in declining sugar maple stands in the Quebec Appalachians. *Canadian Journal of Forest Research* 19:24–34.

Paré, D., and B. Bernier. 1989b. Phosphorus-fixing potential of Ah and H horizons subjected to acidification. *Canadian Journal of Forest Research* 19:132–134.

Parrotta, J. A., D. D. Baker, and M. Fried. 1996. Changes in dinitrogen fixation in maturing stands of *Casuarina equisetifolia* and *Leucaena leucocephala*. *Canadian Journal of Forest Research* 26:1684–1691.

Pankhurst, C., B. M. Doube, and V. V. S. R. Gupta. 1997. *Biological indicators of soil health.* CAB International. Wallingford, UK.

Paton, T. R., G. S. Humphreys, and P. B. Mitchell. 1995. *Soils, a new global view.* Yale University Press, New Haven, CT.

Paul, E. A., and F. E. Clark. 1996. *Soil microbiology and biochemistry*, 2nd ed. Academic Press, New York.

Pavich, M. J. 1986. Processes and rates of saprolite production and erosion on a foliated granitic rock of the Virginia Piedmont. Pp. 552–590 in: *Rates of chemical weathering of rocks and minerals*, S. M. Colman and D. P. Dethier (eds.). Academic Press, Orlando, FL.

Pearson, G. A. 1931. *Forest types in the Southwest as determined by climate and soil.* USDA Technical Bulletin 247.

Pereira, A. R. 1990. Biomass e ciclagem de nutrientes minerais em povomantos jovens de *Eucalyptus grandis* e *Eucalyptus urophylla* em regiao de cerrado. M.S. thesis, University of Viçosa, Viçosa.

Perkins, P. V., and A. R. Vann. 1997. The bulk density amelioration of minespoil with pulverised fuel ash. *Soil Technology* 10:111–114.

Persson, T., H. Lundkvist, A. Wir?n, R. Hyv?nen, and B. Wess?n. 1989. Effects of acidification and liming on carbon and nitrogen mineralization and soil organisms in mor humus. *Water, Air and Soil Pollution* 45:77–96.

Pessotii, J. E. S., L. T. B. Rizzo, and P. F. M. Vaillant. 1983. Caracteristicas do meio fisico do distrito florestal norte. *Salvador, INTEC, Copener Florestal Ltda* 1(3):1–5.

Peterson, C. 1982. Regional growth and response analysis for unthinned Douglas-fir. Pp. 3–25 in: *Regional forest nutrition research project biennial report 1980–1982.* Contribution #46, Institute of Forest Resources, University of Washington, Seattle.

Peterson, C., P. J. Ryan, and S. P. Gessel. 1984. Response of Northwest Douglas-fir stands to urea: correlations wtih forest soil properties. *Soil Science Society of America Journal* 48:162–169.

Pettersson, F. 1994. *Predictive functions for impact of nitrogen fertilization on growth over five years.* SkogForsk Report #3/1994, Uppsala, Sweden.

Pfeil, W. 1860. *Die deutsche Holzzucht Begrundet auf der Eigentumlichkeit der Forstholzer unf ihr Verhalten zu den verschiedenen Standorten.* H. Laupp, Tubingen.

Piccolo, A. 1996. *Humic substances in terrestrial ecosystems.* Elsevier, Amsterdam.

Pichtel, J. R., W. A. Dick, and P. Sutton. 1994. Comparison of amendments and management practices for long-term reclamation of abandoned mine lands. *Journal of Environmental Quality* 23:766–772.

Pimentel, D., and M. Pimentel. 1979. *Food, energy and society.* Edward Arnold, London.

Ponomareva, V. V. 1964. *Theory of podzolization*, trans. A. Gourevitch. Israel Program for Science Translations, Jerusalem.

Popovic, B., and F. Andersson. 1984. *Markkalkning och skogsproduktion—litteratur versikt och revision av svenska kalkningsf rs k.* Statens Naturv rdsverk PM 1792, Solna, Sweden.

Post, M. W., W. Emanuel, P. Zinke, and A. Stangenberger. 1982. Soil carbon pools and world life zones. *Nature* 298:156–159.

Powers, R. F. 1980. Mineralizable nitrogen as an index of nitrogen availability to forest trees. *Soil Science Society of America Proceedings* 44:1314–1320.

Powers, R. F. 1992. Fertilization response of subalpine Abies forests in California. Pp. 114–126 in: *Forest fertilization: sustaining and improving nutrition and growth of western forests*, H. N. Chappell, G. F. Weetman, and R. E. Miller (eds.). College of Forest Resources Contribution #73, University of Washington, Seattle.

Powers, R. F. 1999. If you build it, will they come? Survival skills for silvicultural studies. *Forestry Chronicle*, in press.

Powers, R. F., D. H. Alban, R. E. Miller, A. E. Tiarks, C. G. Wells, P. E. Avers, R. G. Cline, R. O. Fitzgerald, and N. S. Loftus, Jr. 1990. Sustaining site productivity in North American forests: Problems and prospects. Pp. 49–70 in: *Sustained productiv-*

ity of forest soils. S. P Gessel et al. (eds.). University of British Columbia, Faculty of Forestry Publication, Vancouver.

Powers, R. F., D. J. Mead, J. A. Burger, and M. W. Ritchie. 1994. Designing long-term site productivity experiments. Pp. 247–286 in: *Impacts of forest harvesting on long-term site productivity*, W. J. Dyck and D. W. Cole (eds.). Chapman & Hall, London.

Powers, R. F., and G. T. Ferrell. 1996. Moisture, nutrient, and insect constraints on plantation growth: the "Garden of Eden" study. *New Zealand Journal of Forestry Science* 26:126–144.

Powers, R. F., A. E. Tiarks, and J. R. Boyle. 1998. Assessing soil quality: practicable standards for sustainable forest productivity in the United States of America. Pp. 53–80 in: *The contribution of soil science to the development of the implementation of criteria and indicators of sustainable forest management.* SSSA Special Publication Number 53. Soil Science Society of America, Madison, WI.

Prescott, C. E. 1996. Influence of forest floor type on rates of litter decomposition in microcosms. *Soil Biology and Biochemistry* 10:1319–1325.

Prescott, C. E. 1997. Effects of clearcutting and alternative silvicultural systems on rates of decomposition and nitrogen mineralization in a coastal montane coniferous forest. *Forest Ecology and Management* 95:253–260.

Prescott, C. E., L. P. Coward, G. F. Weetman, and S. P. Gessel. 1993. Effects of repeated nitrogen fertilization on the ericaceous shrub, salal (*Gaultheria shallon*), in two coastal Douglas-fir forests. *Forest Ecology and Management* 61:45–60.

Prescott, C. E., K. D. Thomas, and G. F. Weetman. 1995. The influence of tree species on nitrogen mineralisation in the forest floor: lessons from three retrospective studies. Pp. 59–58 in *Proceedings of the trees and soils workshop, Lincoln University*, D. J. Mead and I. S. Cornforth (eds.). Agronomy Society of New Zealand Special Publication # 10, Lincoln University Press, Canterbury.

Priester, D. S., and W. R. Harmes. 1971. *Microbial populations in two swamp soils of South Carolina.* USDA Forest Service Research Note SE-150.

Pritchett, W. L. 1972. The effect of nitrogen and phosphorus fertilizers on the growth and composition of loblolly and slash pine seedlings in pots. *Soil Crop Science Society of Florida Proceedings* 32:161–165.

Pritchett, W. L., and W. H. Smith. 1972. Fertilizer responses in young pine plantations. *Soil Science Society of America Proceedings* 36:660–663.

Pritchett, W. L., and W. H. Smith. 1974. *Management of wet savanna soils for pine production.* Florida Agricultural Experiment Station Technical Bulletin 762.

Pyatt, D. G. 1970. *Soil groups of upland forests.* For. Comm. For. Rec. (London) 71.

Quesnel, H. J., and L. M. Lavkulich. 1981. Distinguishing the forest floors of three ecosystems. *Soil Science Society of American Journal* 45:624–628.

Radwan, M. A., and J. Shumway. 1983. Soil nitrogen, sulfur, and phosphorus in relation to growth response of western hemlock to nitrogen fertilization. *Forest Science* 29:469–477.

Raich, J. W., R. H. Riley, and P. M. Vitousek. 1994. Use of root-ingrowth cores to assess nutrient limitations in forest ecosystems. *Canadian Journal of Forest Research* 24:2135–2138.

Raison, R. J. 1979. Modification of the soil environment by vegetation fires, with particular reference to nitrogen transformations: a review. *Plant and Soil* 51:73–108.

Raison, R. J., P. K. Khanna, M. J. Connell, and R. A. Falkiner. 1990. Effects of water availability and fertilization on nitrogen cycling in a stand of *Pinus radiata*. *Forest Ecology and Management* 30:31–43.

Raison, R. J., P. Khanna, and P. Woods. 1985. Mechanisms of element transfer to the atmosphere during vegetation fires. *Canadian Journal of Forest Research* 15:132–140.

Ralston, C. W. 1964. Evaluation of forest site productivity. Pp. 171–201 in: *International review of forest research*, Vol. 1. J. A. Romberger and P. Mikola (eds.). Academic Press, New York.

Ramann, E. 1893. *Forstliche Bodenkunde und Standortslehre*. Julius Springer, Berlin.

Raupach, M. 1967. Soil and fertilizer requirements for forest of *Pinus radiata*. *Advances in Agronomy* 19:307–353.

Rehfeuss, K. E. 1979. Underplanting of pines with legumes in Germany. Pp. 374–387 in: *Symbiotic nitrogen fixation in the management of temperate forests*, J. C. Gordon, C. T. Wheeler, and D. A. Perry (eds.). Forest Research Laboratory, Oregon State University, Corvallis.

Rehfuess, K. E., and A. Schmidt. 1971. Effects of lupine establishment and nitrochalk on state of nutrition and increment in older pine stands of the Oberpfalz. Forstw. Cbl. 90:237–259.

Rehfeuss, K. E., F. Makeschin, and J. Volkl. 1984. Amelioration of degraded pine sites (*Pinus sylvestris* L.) in southern Germany. Pp. 933–946 in: *IUFRO symposium on site and productivity of fast growing plantations*, D. C. Grey, A. P. G. Schonau, C. J. Schutz, and A. Van Laar (eds.). South African Forest Research Institute, Pretoria.

Reich, P., D. Griegal, J. Aber, and S. Gower. 1997. Nitrogen mineralization and productivity in 50 hardwood and conifer stands on diverse soils. *Ecology* 78:335–347.

Reiners, W. A., A. F. Bouwman, W. F. J. Parsons, and M. Keller. 1994. Tropical rain forest conversion to pasture: changes in vegetation and soil properties. *Ecological Applications* 4:363–377.

Remezov, N. P., and P. S. Progrebnyak. 1965. *Forest soil science* (English translation). U. S. Department of Commerce, Clearinghouse for Federal Scientific and Technical Information, Springfield, VA.

Remezov, N. P., and P. S. Pogebnyak. 1969. *Forest soil science*. USDA National Science Foundation, Washington, DC.

Resh, S., D. Binkley, and J. Parrotta. 1999. Retention of old soil carbon under N-fixing trees. Manuscript in review.

Reuss, J. O. 1989. Soil-solution equilibria in lysimeter leachates under red alder. Pp. 547–559 in: *Effects of air pollution on western forests*, R. K. Olson and A. S. Lefohn (eds.). Air and Waste Mangement Association, Pittsburgh.

Reuss, J. O., and D. W. Johnson. 1986. *Acid deposition and the acidification of soils and waters*. Springer-Verlag, New York.

Rhoades, C. 1997. Single-tree influences on soil properties in agroforestry: lessons from natural forest and savanna ecosystems. *Agroforestry Systems* 35:71–94.

Rhoades, C., and D. Binkley. 1995. Factors influencing decline in soil pH in Hawaiian *Eucalyptus* and *Albizia* plantations. *Forest Ecology and Management* 80:47–56.

Rhodes, C., G. E. Eckert, and D. C. Coleman. 1998. Effect of pasture trees on soil nitrogen and organic matter: implications for tropical montane forest restoration. *Restoration Ecology* 6:262–270.

Richards, B. N. 1961. Soil pH and mycunhiza development in *Pinus*. *Nature* (London) 190:105.

Richards, B. N., and G. K. Voight. 1965. Nitrogen accretion in coniferous forest ecosystems. Pp. 105–116 in: *Forest-soil Relationships in North America*. C. T. Youngberg (ed.). Oregon State University Press, Corvallis.

Richter, D. D., and L. I. Babbar. 1991. Soil diversity in the tropics. *Advances in Ecological Research* 21:315–389.

Richter, D. D., P. J. Comer, K. S. King, H. A. Sawin, and D. S. Wright. 1988. Effects of low ionic strength solutions on pH of acid forested soils. *Soil Science Society of America Journal* 52:261–264.

Richter D. D., D. Markewitz, C. Wells, H. L. Allen, R. April, P. Heine, and B. Urrego. 1994. Soil chemical changes during three decades in an old-field loblolly pine (*Pinus taeda* L.) ecosystem. *Ecology* 75:1463–1473.

Richter, D. D., D. Markewitz, S. E. Trumbore, and C. G. Wells. 1999. Rapid accumulation and turnover of soil carbon in a re-establishing forest. *Nature* 400:56–58.

Richter, D. D., C. W. Ralston, and W. Harms. 1982. Prescribed fire: effects on water quality and forest nutrient cycling. *Science* 215:661–663.

Ring, E. 1995. Nitrogen leaching before and after clear-felling of fertilised experimental plots in a *Pinus sylvestris* stand in central Sweden. *Forest Ecology and Management* 72:151–166.

Robertson, W K., W. H. Smith, and D. M. Post. 1975. Effect of nitrogen and placed phosphorus and dolomitic limestone in an Aeric Haplaquod on slash pine growth and composition. *Soil Crop Science Society of Florida Proceedings* 34:58–60.

Rochelle, J. A. 1979. The effects of forest fertilization on wildlife. Pp. 164–167 in: *Proceedings of the forest fertilization conference*. Contribution #40, Institute of Forest Resources, University of Washington, Seattle.

Rockwood, D. L., C. L. Windsor, and J. F. Hodges. 1985. Response of slash pine progenies to fertilization. *Southern Journal of Applied Forestry* 9:37–40.

Romell, L. G. 1922a. Luftvaxlingen I marken som ekologisk faktor. *Statens Skogsforsoksanst. Meddel.* 19:125–359.

Romell, L. G. 1922b. Die Bodenventilation als okologiseher Faktor. *Meddel. Statens Skogsforsoksanstalt* 19:281–359.

Romell, L. G. 1935. *Ecological problems of the humus layer in the forest*. Cornell University Mem. 170.

Romell, L. G., and S. O. Heiberg. 1931. Types of humus layer in the forests of the northeasten U. S. *Ecology* 12:567–608.

Rose, R., D. L. Haase, and D. Boyer. 1995. *Organic matter management in forest tree nurseries: theory and practice*. Nursery technology Cooperative, Oregon State University, Corvallis.

Roulet, N. T., and T. R. Moore. 1995. The effect of forestry drainage practices on the emission of methane from northern peatlands. *Canadian Journal of Forest Research* 25:491–499.

Ryan, M. G., D. Binkley, and J. H. Fownes. 1996. Age-related decline in forest productivity: pattern and process. *Advances in Ecological Research* 27:213–262.

Sabey, B. R., R. L. Pendleton, and B. L. Webb. 1990. Effect of municipal sewage sludge application on growth of two reclamation shrub species in copper mine spoils. *Journal of Environmental Quality* 19:580–586.

Sallenave, H. 1969. The cultivation of maritime pine in southwest France. *Phosphorus Agricultural* 54:17–26.

Sanchez, P. A. 1976. *Properties and management of soils in the tropics.* Wiley, New York.

Sanchez, P., M. P. Gichuru, and L. B. Katz. 1982. Organic matter in major soils of the tropical and temperate regions. *Transactions of the 12th International Congress of Soil Science (New Delhi)* 1:99–114.

Sanford, R. L., J. Saldarriaga, K. E. Clark, C. Uhl, and R. Herrera. 1985. Amazon rain forest fires. *Science* 227:53–55.

Saric, M. R., and B. C. Loughman (eds.). 1983. *Genetic aspects of plant nutrition.* Martinus Nijhoff/Junk, the Hague, the Netherlands.

Sartz, R. S. 1973. *Snow and frost depths on north and south slopes.* USDA Forest Service Research Note NC-157.

Sayn-Wittgenstein, L. 1969. The northern forest. *Canadian Pulp and Paper Industrial* 22:77–78.

Schenck, N. C. 1982. *Methods and principals of mycorrhizae research.* American Phytopathology Society, St. Paul, MN.

Schenck, N. C. 1983. Can mycorthizae control root disease? *Plant Disease* 65:230–234.

Schiess, P., and D. W. Cole. 1981. Renovation of wastewater by forest stands. Pp. 131–148 in: *Municipal sludge application to Pacific Northwest forest lands*, C. S. Bledsoe (ed.). Contribution #41, Institute of Forest Resources, University of Washington, Seattle.

Schimel, J. P., L. E. Jackson, and M. K. Firestone. 1989. Spatial and temporal effects on plant-microbial competition for inorganic nitrogen in a California grassland. *Soil Biology and Biochemistry* 21:1059–1066.

Schlenker, G. 1964. Entwicklung des Sudwestdeutschlund angewandten Verfahrens der forstuchen Standortskunde. Pp. 5–26 in: *Standort. Wald und Waldwirtschaft in Obersebwahen, "Oberschwabishe Fiehtenreviere."* Stuttgart.

Schlesinger, W. H. 1995. An overview of carbon cycle in: *Soils and global change.* R. Lal, J. M. Kimble, E. Levine and B. A. Stewart (eds.). CRC Press, Boca Raton, FL.

Schlesinger, W. 1997. *Biogeochemistry: an analysis of global change.* Academic Press, San Diego, CA.

Schnitzer, M. 1991. Soil organic matter — the next 75 years. *Soil Science* 151:41–48.

Schoch, P., and D. Binkley. 1986. Prescribed burning increased nitrogen availability in a mature loblolly pine stand. *Forest Ecology and Management* 14:13–22.

Schoenholtz, S. H., J. A. Burger, and R. E. Kreh. 1992. Fertilizer and organic amendment effects on mine soil properties and revegetation success. *Soil Science Society of America Journal* 56:1177–1184.

Schubert, K. R. 1982. *The energetics of biological nitrogen fixation.* American Society of Plant Physiologists, Rockville, MD.

Schultz, R. P. 1971. *Stimulation of flower and seed production in a young slash pine orchard.* USDA Forest Service Research Paper SE-91.

Schultz, R. P. 1972. Root development of intensively cultivated slash pine. *Soil Science Society of America Proceedings* 36:158–162.

Schultz, R. P., and L. P. Wilhite. 1969. *Differential response of slash pine families to drought.* USDA Forest Service Research Note SE-104.

Schultz, R. P., and L. P. Wilhite. 1974. Changes in a flatwood site following intensive preparation. *Forest Science* 20:230–237.

Schumacher, F. X., and T. S. Coile. 1960. *Growth and yields of natural stands of the southern pines.* T. S. Coile, Inc., Durham, NC.

Scott, N. 1996. Plant species effects on soil organic matter turnover and nutrient release in forests and grasslands. Ph.D. thesis, Colorado State University, Ft. Collins.

Scott, N., and D. Binkley. 1997. Litter quality and annual net N mineralization: comparisons across sites and species. *Oecologia* 111:151–159.

Scott, W. 1970. Effect of snowbrush on the establishment and growth of Douglas-fir seedlings. M.S. thesis, Oregon State University, Corvallis.

Seastedt, T. R., and A. K. Knapp. 1993. Consequences of nonequilibrium resource availability across multiple time scales: the transient maxima hypothesis. *American Naturalist* 41:621–633.

Sharma, E. 1988. Altitudinal variation in nitrogenase activity of the Himalayan alder naturally regenerating on landslide-affected sites. *New Phytologist* 108:411–416.

Sharma, E. 1993. Nutrient dynamics in Himalayan alder plantations. *Annals of Botany* 72:329–336.

Sharma, E., R. S. Ambasht, and M. P. Singh. 1985. Chemical soil properties under five age series of *Alnus nepalensis* plantations in the eastern Himalayas. *Plant and Soil* 84:105–113.

Sheikh, M. I. 1986. *Afforestation of arid and semi-arid areas in Pakistan.* Food and Agriculture Organization and Pakistan Forest Institute, Peshawar.

Shetty, K. G., B. A. D. Hetrick, D. A. H. Figge, and A. P. Schwab. 1994. Effects of mycorrhizae and other soil microbes on revegetation of heavy metal contaminated mine spoil. *Environmental Pollution* 86:181–188.

Shoulders, E., and A. E. Tiarks. 1984. Response of pines and native forage to fertilizer. Pp. 105–126 in: *Agroforestry in the Southern United States*, N. E. Linnartz and M. K. Johnson (eds.). Louisiana Agricultural Experiment Station, Baton Rouge.

Shumway, J., and W. A. Atkinson. 1978. Predicting nitrogen fertilizer response in unthinned stands of Douglas-fir. *Communications in Soil Science and Plant Analysis* 9:529–539.

Sibanda, H. M., and S. D. Young. 1989. The effect of humus acids and soil heating on the availability of phosphate in oxide-rich tropical soils. Pp. 71–83 in: *Mineral nutrients in tropical forests and savannas*, J. Proctor (ed.). Blackwell Scientific, Oxford.

Silberbursh, M., and S. A. Barber. 1983. Sensitivity of simulated phosphorus uptake to parameters used by mechanistic-mathematical model. *Plant and Soil* 74:93–100.

Silkworth, D. R., and D. F. Grigal. 1981. Field comparison of soil solution samplers. *Soil Science Society of America Journal* 45:440–442.

Silvester, W. B. 1977. Dinitrogen fixation by plant associations excluding legumes in: *A treatise on dinitrogen fixation*, Wiley, W. F. Hardy (ed.). New York.

Silvester, W. B., P. Sollins, T. Verhoeven, and S. P. Cline. 1982. Nitrogen fixation and acetylene reduction in decaying conifer boles: effects of incubation time, aeration and moisture content. *Canadian Journal of Forest Research* 12:646–652.

Simmons, G. L., and P. E. Pope. 1988. Influence of soil water potential and mycorrhizal colonization on root growth of yellow poplar and sweet gum seedlings grown in compacted soil. *Canadian Journal of Forest Research* 18:1392–1396.

Smalley, G. W. 1964. Topography, soils, and the height of planted yellow-poplar. *Journal of the Alabama Academy of Science* 35:39–44.

Smethurst, P. J. 1999. Nutrient concentrations in soil solution as indicators of nutrient-supply limitations to plant growth: a review. *Forest Ecology and Management*, in review.

Smith, D. W. 1970. Concentrations of soil nutrients before and after fire. *Canadian Journal of Soil Science* :17–29.

Smith, S. S. F. 1980. Mycorrhizas of autotrophic higher plants. *Biological Reviews* 55:475–510.

Smith, W. H., C. A. Hollis, and J. W. Gooding III. 1977. Influence of soil factors on fusiform rust incidence. Pp. 81–88 in: *Management of fusiform rust in Southern pines*, R. J. Dinus and R. A. Schmidt (eds.). University of Florida, Gainesville.

Snowdon, P., and M. L. Benson. 1992. Effects of combinations of irrigation and fertilisation on the growth and biomass production of *Pinus radiata*. *Forest Ecology and Management* 52:87–116.

Snowdon, P., and H. D. Waring. 1984. Long-term nature of growth responses obtained to fertilizer and weed control applied at planting and their consequences for forest management. Pp. 701–712 in: *IUFRO symposium on site and productivity of fast growing plantations*. South African Forest Research Institute, Pretoria.

Snyder, G. G., H. F. Haupt, and G. H. Belt, Jr. 1975. *Clearcutting and burning slash alter quality of streamwater in northern Idaho*. USDA Forest Service Research Paper INT-168.

Soil Survey Staff. 1975. *Soil taxonomy—a basic system of soil classification for making and interpreting soil surveys*. USDA Handbook 436.

Soil Survey Staff. 1993. *Soil survey manual*. U.S. Department of Agriculture, U.S Government Printing Office, Washington, DC.

Sollins, P. 1982. Input and decay of coarse woody debris in coniferous stands in western Oregon and Washington. *Com. Journal of Forest Research* 12:18–28.

Sollins, P., S. P. Cline, T. Verhoeven, D. Sachs, and G. Spycher. 1987. Patterns of log decay in old-growth Douglas-fir forests. *Canadian Journal of Forest Research* 17:1585–1595.

Sollins, P., K. Cromack, Jr., and C. Y. Li. 1981. Role of low-molecular-weight organic acids in the inorganic nutrition of fungi and higher plants. Pp. 607–619 in: *The fungal community*, D. T. Wicklow and G. C. Carroll (eds.). Marcel Dekker, New York.

Sollins, P., C. C. Grier, F. M. McCorison, K. Cromack, Jr., R. Fogel, and R. L. Fredriksen. 1980. The internal element cycles of an old-growth Douglas-fir ecosystem in western Oregon. *Ecological Monographs* 50:261–285.

Sommers, L. E., C. M. Gilmour, R. E. Wildung, and S. M. Beck. 1980. The effect of water potential on decomposition processes in soils. Pp. 97–117 in: *Water potential relations in soil microbiology*. Soil Science Society of America, Madison, WI.

Sopper, W. E., and S. N. Kerr (eds.). 1979. *Utilization of municipal sewage effluent and sludge on forest and disturbed land*. Pennsylvania State University Press, University Park.

Sort, X., and J. M. Alcaniz. 1996. Contribution of sewage sludge to erosion control in the rehabilitation of limestone quarries. *Land Degeneration and Development* 7:69–76.

Spiecker, H. 1995. Growth dynamics in a changing environment — long-term observations. *Plant and Soil* 168–169:555–561.

Spiecker, H., K. Mielikäinen, M. Köhl, and J. P. Skovsgaard (eds.). 1996. *Growth trends in European forests*. Springer-Verlag, Berlin.

Sposito, G. 1989. *The chemistry of soils*. Oxford University Press, New York.

Spurr, S. H. 1952. *Forest inventory*. Ronald Press, New York.

Stanford, G., and S. J. Smith. 1972. Nitrogen mineralization potentials of soils. *Soil Science Society of America Proceedings* 36:465–472.

Stape, J. L., A. N. Gomes, and T. F. Assis. 1997. Estimativa da produtividade de povoamentos monoclonais de *Eucalyptus grandis* × *urophylla* no Nordeste do Estado da Bahia-Brazil em função das variabilidades pluviométrica e edáfica. Pp. 192–198 in: *Proceedings of the IUFRO conference on silviculture and improvement of eucalypts*. EMBRAPA/CNPF, El Salvador.

Stark, J., and S. C. Hart. 1997. High rates of nitrification and nitrate turnover in undisturbed coniferous forests. *Nature* 385:61–64.

Stark, L. R., W. R. Wenerick, F. M. Williams, S. E. Stevens, Jr., and P. J. Wuest. 1994. Restoring the capacity of spent mushroom compost to treat coal mine drainage by reducing the inflow rate: a microcosm experiment. *Water, Air, and Soil Pollution* 75:405–420.

Stark, N. M. 1977. Fire and nutrient cycling in a Douglas-fir/larch forest. *Ecology* 58:16–30.

Steinbrenner, E. C. 1975. Mapping forest soils on Weyerhaeuser lands in the Pacific Northwest. Pp. 513–525 in: *Forest soils and forest land management*, B. Bernier and C. H. Winget (eds.). Laval University Press, Quebec.

Steinbrenner, E. C., and J. H. Rediske. 1964. *Growth of ponderosa pine and Douglas-fir in controlled environment*. Weyerhaeuser Forestry Paper 1.

Stephens, F. R. 1965. Relation of Douglas-fir productivity to some zonal soils in the north-western Cascades of Oregon. Pp. 245–260 in: *Forest-soil relationships in North America*, C. T. Youngberg (ed.). Oregon State University Press, Corvallis.

Stevenson, F. J. 1994. *Humus chemistry*, 2nd ed. Wiley, New York.

Steward, F. C. 1964. *Plants at work*. Addison-Wesley, Reading, MA.

Stobbe, P. C., and J. R. Wright. 1959. Modern concepts of the genesis of Podzols. *Soil Science Society of America Proceedings* 23:161–164.

Stoeckeler, J. H. 1960. *Soil factors affecting the growth of quaking aspen forests in the Lake States*. Minnesota Agricultural Experiment Station Technical Bulletin 323.

Stone, E. L. 1968. Microelement nutrition of forest trees: a review. Pp. 132–179 in:

Forest fertilization: theory and practice, G. W. Bengtson (ed.). Tennessee Valley Authority, Muscle Shoals, AL.

Stone, E. L. 1975. Effects of species on nutrient cycles and soil change. *Philosophical Transactions of the Royal Society, London (B)* 271:149–162.

Stone, E. L., and P. J. Kalisz. 1991. On the maximum extent of tree roots. *Forest Ecology and Management* 46:59–102.

Stork, N. E., and P. Eggleton. 1992. Invertebrates as determinates of soil quality. *American Journal of Alternatives in Agriculture* 7:38–47.

Stottlemyer, R., D. Toczydlowski, and R. Herrmann. 1998. *Biogeochemistry of a mature boreal ecosystem: Isle Royale National Park, Michigan.* USDI National Park Service Scientific Monograph NPS/NRUSGS/NRSM-98/01, Ft. Collins, CO.

Stout, B. B. 1956. *Studies on the root systems of deciduous trees.* Black Rock Forestry Bulletin 15.

Strader, R., D. Binkley, and C. Wells. 1989. Nitrogen mineralization in high elevation forests of the Appalachians. I. Regional patterns in spruce-fir forests. *Biogeochemistry* 7:131–145.

Strand, R. F., and L. C. Promnitz. 1979. Growth response falldown associated with operational fertilization. Pp. 209–213 in: *Proceedings of the forest fertilization conference.* Contribution #40, Institute of Forest Resources, University of Washington, Seattle.

Stuanes, A. O., H. Van Miegroet, D. W. Cole, and G. Abrahamsen. 1992. Recovery from acidification. Pp. 467–494 in: *Atmospheric deposition and forest nutrient cycling,* D. W. Johnson and S. E. Lindberg (eds.). Springer-Verlag, New York.

Stump, L. M., and D. Binkley. 1993. Relationships between litter quality and nitrogen availability in Rocky Mountain forests. *Canadian Journal of Forest Research* 23:492–502.

Sutton, R. F. 1980. Root system morphogenesis. *New Zealand Journal of Forestry* 10:264–292.

Sutton, R. F. 1991. *Soil properties and root development in forest trees: A review.* Forestry Canada. Information Report O-X-413.

Sutton, R. F., and T. P. Weldon. 1993. Jack pine establishment in Ontario: 5–year comparison of stock types with and without bracke scarification, mounding, and chemical site preparation. *Forestry Chronicle* 69:545–553.

Sutton, R. F., and T. P. Weldon. 1995. White spruce establishment in boreal Ontario mixedwood: 5–year results. *Forestry Chronicle* 71:633–638.

Sverdrup, H., P. Warfvinge, and B. Nihlg rd, B. 1994. Assessment of soil acidification effects on forest growth in Sweden. *Water, Air and Soil Pollution* 78:1–36.

Swanson, D. K. 1996. Susceptibility of permafrost soils to deep thaw after forest fires in Interior Alaska, U.S.A., and some ecologic implications. *Arctic and Alpine Research* 28:217–227.

Swanson, F. J. 1981. Fire and geomorphic processes. Pp. 421–444 in: *Fire regimes and ecosystem properties,* H. A. Mooney, T. M. Bonnicksen, N. L. Christensen, J. E. Lotan, and W. A. Reiners (eds.). USDA Forest Service General Technical Report WO-26.

Swanston, D. N. and C. T. Dyrness. 1974. Stability of steep land. *Journal of Forestry* 71:264–269.

Swift, M. J., O. W. Heal, and J. M. Anderson. 1979. *Decomposition in terrestrial ecosystems.* University of California Press, Berkeley.

Switzer, G. L., L. E. Nelson, and W. H. Smith. 1968. The mineral cycle in forest stands. Pp. 1–9 in: *Forest fertilization—theory and practice,* G. Bengtson (ed.). Tennessee Valley Authority, Knoxville.

Switzer, M. 1979. Energy relations in forest fertilization. Pp. 243–246 in: *Proceedings of the forest fertilization conference.* Contribution #40, Institute of Forest Resources, University of Washington, Seattle.

Sykes, D. J. 1971. Effects of fire and fire control on soil and water relations in northern forests. Pp. ??. In: *Fire in the northern environment,* C. W. Slaughter et al. (eds.). USDA Forest Service, Pacific Northwest Forest and Range Experiment Station, Portland, OR.

Sylvia, D. M., J. J. Fuhrmann, P. G. Hartel, and D. A. Zuberer. 1998. *Principles and applications of soil microbiology.* Prentice-Hall, Upper Saddle River, NJ.

Talibudeen, O., J. D. Beasley, P. Lane, and N. Rajendran. 1978. Assessment of soil potassium reserves available to plant roots. *Journal of Soil Science* 29:207–218.

Tamm, C. O., and L. Hällbacken. 1985. Changes in soil pH over a 50-year period under different forest canopies in Southwest Sweden. *Water, Air and Soil Pollution.*

Tamm, C. O., H. Holmen, B. Popovic, and G. Wiklander. 1974. Leaching of plant nutrients from soils as a consequence of forestry operations. *Ambio* 3:211–221.

Tan, K. H. 1993. *Principles of soil chemistry,* 2nd ed. Marcel Dekker, New York.

Taylor, C. M. A. 1991. *Forest fertilization in Britain.* Bulletin 95, Forestry Commission, Her Majesty's Stationery Office, London.

Teng, Y., and V. R. Timmer. 1995. Rhizosphere phosphorus depletion induced by heavy nitrogen fertilization in forest nursery soils. *Soil Science Society of America Journal* 59:227–233.

Theodorou, C., and G. D. Bowen. 1969. The influence of pH and nitrate on mycorrhizal associations of *Pinus radiata* D. Don. Australia Journal of Botany 17:59–67.

Theodorou, C., and G. D. Bowen. 1971. Influence of temperature on the mycorrhizal associations of *Pinus radiata* D. Don. Australia Journal of Botany 19:13–20.

Tiedmann, A. R., C. E. Conrad, J. H. Dieterich, J. W. Hornbeck, W. F. Megahan, L. A. Viereck, and D. D. Wade. 1979. *Effects of fire on water: a state-of-knowledge review.* USDA Forest Service General Technical Report WO-10.

Tiedemann, A. R., J. D. Helvey, and T. D. Anderson. 1978. Stream chemistry and watershed nutrient economy following wildfire and fertilization in eastern Washington. *Journal of Environmental Quality* 7:580–588.

Tilki, F., and R. F. Fisher. 1998. Tropical leguminous species for acid soils: studies on plant form and growth in Costa Rica. *Forest Ecology and Management* 108:175–192.

Timmer, V. R., and L. D. Morrow. 1984. Predicting fertilizer growth response and nutrient status of jack pine by foliar diagnosis. Pp. 335–351 in: *Forest soils and treatment impacts,* E. L. Stone (ed.). University of Tennessee, Knoxville.

Timmer, V. R., and Y. Teng. 1999. Foliar nutrient analysis of sugar maple decline: retrospective vector diagnosis. *Proceedings of the Maple Dieback Symposium, USDA Forest Service.* In press.

Torbert, J. L., and J. A. Burger. 1984. Long-term availability of applied phosphorus to loblolly pine on a Piedmont soil. *Soil Science Society of America Journal* 48:1174–1178.

Torn, M., S. Tgrumbore, O. Chadwick, P. Vitousek, and D. Hendricks. 1997. Mineral control of soil organic carbon storage and turnover. *Nature* 389:170–173.

Torry, J. G. 1978. Nitrogen fixation by actinomycete-nodulated angiosperms. *Bioscience* 28:586–592.

Toumey, J. W. 1916. *Foundations of silviculture upon an ecological basis.* Yale School of Forestry, New Haven, CT.

Trettin, C. C., M. F. Jurgensen, D. F. Grigal, M. R. Gale, and J. K. Jeglum. 1997. *Northern forested wetlands: ecology and management.* Lewis Publishers. Boca Raton, FL.

Troedsson, T., and W. H. Lyford. 1973. *Biological disturbance and small-scale spatial variations in a forested soil near Garpenberg, Sweden.* Studia Forestalia Auccica 109.

Trout, L. C., and T. A. Leege. 1971. Are the northern Idaho elk herds doomed? *Idaho Wildlife Review* 24:3–6.

Truax, B., D. Gagnon, F. Lambert, and N. Chevrier. 1994. Nitrate assimilation of raspberry and pin cherry in a recent clearcut. *Canadian Journal of Botany* 72:1343–1348.

Tubbs, C. H. 1973. Allelopathic relationships between yellow birch and sugar maple seedlines. *Forest Science* 19:139–145.

Turner, J., D. W. Cole, and S. P. Gessel. 1976. Mineral nutrient accumulation and cycling in a stand of red alder (*Alnus rubra*). *Journal of Ecology* 64:965–974.

Turner, J., and S. P. Gessel. 1990. Forest productivity in the southern hemisphere with particular emphasis on managed forests. Pp. 23–39 in: *Sustained productivity of forest soils,* S. P Gessel et al. (ed.). University of British Columbia, Faculty of Forestry Publications, Vancouver.

Turner, J., and M. J. Lambert. 1996. Nutrient cycling and forest management. Pp. 229–248 in: *Nutrition of eucalypts.* P. M. Attiwill, and M. A. Adams (eds.). CSIRO, Collingwood, Australia.

Turner, J., M. J. Lambert, and S. P. Gessel. 1979. Sulfur requirements of nitrogen fertilized Douglas-fir. *Forest Science* 25:461–467.

Ugolini, F. C., H. Dawson, and J. Zachara. 1977. Direct evidence of particle migration in the soil solution of a podzol. *Science* 195:603–605.

Ulrey, A. L., R. C. Graham, and L. H. Bowen. 1996. Forest fire effects on soil phyllosilicates in California. *Soil Science Society of America Journal* 60:309–315.

Ulrich, B. 1983. Interaction of forest canopies with atmospheric constituents. Pp. 33–45 in: *Effects of accumulation of air pollutants in forest ecosystems,* B. Ulrich and J. Pankrath (eds.). D. Reidel, Boston.

Ursic, S. J. 1970. *Hydrologic effects of prescribed burning and deadening upland hardwoods in northern Mississippi.* USDA Forest Service Research Report Paper SO-54.

Valentine, D., and H. L. Allen. 1990. Foliar responses to fertilization identify nutrient limitation in loblolly pine. *Canadian Journal of Forest Research* 20:144–151.

Valentine, D., and D. Binkley. 1992. Topography and soil acidity in an Arctic landscape. *Soil Science Society of America Journal* 56:1553–1559.

Valentine, K. W. G. 1986. *Soil resource surveys for forestry.* Clarendon Press, Oxford.

Van Cleve, K., and C. T. Dyrness. 1985. The effect of the Rosie Creek Fire on soil fertility. Pp. 7–11 in: *Early results of the Rosie Creek Fire Research Project 1984*, G. Juday and C. T. Cyrness (eds.). Agricultural and Forestry Experiment Station Miscellaneous Publication #85-2, University of Alaska, Fairbanks.

van den Driessche, R. 1954. Soil fertility in forest nurseries. Pp. 63–74 in: *Forest Nursery Manual*, M. L. Duryea and T. D. Landis (eds.). Martinus Nijhoff/Junk, the Hague, the Netherlands.

van den Driessche, R. 1979. Estimating potential response to fertilizer based on tree tissue and litter analysis. Pp. 214–220 in: *Proceedings of the forest fertilization conference.* Contribution #40, Institute of Forest Resources, University of Washington, Seattle.

van den Driessche, R., and D. Ponsford. 1995. Nitrogen induced potassium deficiency in white spruce (*Picea glauca*) and Engelmann spruce (*Picea engelmannii*) seedlings. *Canadian Journal of Forest Research* 25:1445–1454.

van Kessel, C., R. E. Farrell, J. P. Roskoski, and K. M. Keane. 1994. Recycling of the naturally-occurring ^{15}N in an established stand of *Leucaena leucocephala. Soil Biology and Biochemistry*, 26:757–762.

Van Lear, D. H., and J. F. Hosner. 1967. Correlation of site index and soil mapping units. *Journal of Forestry* 65:22–24.

Van Miegroet, H., and D. W. Cole. 1984. The impact of nitrification on soil acidification and cation leaching in a red alder ecosystem. *Journal of Environmental Quality* 13:586–590.

Van Miegroet, H., D. W. Cole, D. Binkley, and P. Sollins. 1989. The effect of nitrogen accumulation and nitrification on soil chemical properties in alder forests. Pp. 515–528 in: *Effects of air pollution on western forests*, R. K. Olson and A. S. LeFohn (eds.). Air and Waste Management Association, Pittsburgh.

Van Rees, K. C. J., and N. B. Comerford. 1990. The role of woody roots of slash pine seedlings in water and potassium absorption. *Canadian Journal of Forest Research* 20:1183–1191.

Verburg, P. 1998. Organic matter dynamics in a forest soil as affected by climate change. Ph.D. thesis, University of Wageningen, the Netherlands.

Verbyla, D. L., and R. F. Fisher. 1989. An alternative approach to conventional soil-site regression modeling. *Canadian Journal of Forest Research* 19:179–184

Verstraten, J. 1977. Chemical erosion in a forested watershed in the Oesling, Luxembourg. *Earth Surface Processes* 2:175–184.

Vihnaneck, R., and T. Ballard. 1988. Slashburning effects on stocking, growth, and nutrition of young Douglas-fir plantations in salal-dominated ecosystems of eastern Vancouver Island. *Canadian Journal of Forest Research* 18:718–722.

Vincent, A. B. 1965. *Black spruce: a review of its silvics, ecology and silviculture.* Department of Forestry Canada, Publication 1100. Ottawa.

Vitousek, P. M., L. R. Walker, L. D. Whiteaker, and P. A. Matson. 1993. Nutrient limitations to plant growth during primary succession in Hawaii Volcanoes National Park. *Biogeochemistry* 23:197–215.

Vogt, K., R. L. Edmonds, and D. J. Vogt. 1981. Nitrate leaching in soils after sludge application. Pp. 59–66 in: *Municipal sludge application to Pacific Northwest forest lands*, C. S. Bledsoe (ed.). Contribution #41, Institute of Forest Resources, University of Washington, Seattle.

Vogt, K. A., C. C. Grier, C. E. Meier, and R. L. Edmonds. 1982. Mycorrhizal role in net primary production and nutrient cycling in *Abies amabilis* ecosystems in western Washington. *Ecology* 63:370–380.

Voigt, G. K. 1965. Nitrogen recovery from decomposing tree leaf tissue and forest humus. *Soil Science Society of America Proceedings* 29:756–759.

Voigt, G. K. 1971. Mycorrhizae and nutrient mobilization. Pp. 122–131 in: *Mycorrhizae*, E. Haeskaylo (ed.). USDA Forest Service Miscellaneous Publication 1189.

Voss, R. D., and W. D. Shrader. 1984. *Crop rotations: effect on yields and response to nitrogen*. Cooperative Extension Service, Iowa State University, Ames.

Vosso, J. A. 1971. Field inoculation with mycorrhizae fungi. Pp. 187–196 in: *Mycorrhizae*, E. Hacskaylo (ed.). USDA Forest Service Miscellaneous Publication 1189.

Waisel, Y., A. Eshel, and U. Kafkafi. 1996. *Plant roots: The hidden half*, 2nd ed. Marcel Decher, New York.

Waksman, S. A. 1936. *Humus: origin, chemical composition and importance in nature*. Williams & Wilkins, Baltimore.

Waksman, S. A. 1952. *Soil microbiology*. Wiley, New York.

Walker, J., R. J: Raison, and P. K. Khanna. 1986. Fire. Pp. 185–216 in: *Australian soils: the human impact*, J. Russell and R. Isbell (eds.). University of Queensland Press, Queensland, Australia.

Wallwork, J. A. 1970. *Ecology of soil animals*. McGraw-Hill, New York.

Waring, H. D., and P. Snowdon. 1977. Genotype-fertilizer interaction. Pp. 19–22 in: *Annual report*. CSIRO, Division of Forest Research, Canberra, 1976–1977.

Waring, R. H., and G. B. Pitman. 1983. Physiological stress in lodgepole pine as a precursor for mountain pine beetle attack. *Zeitschrift fur angewandte Entomologie* 96:266–270.

Waring, R. H., and G. B. Pitman. 1985. Modifying lodgepole pine stands to change susceptibility to mountain pine beetle attack. *Ecology* 66:889–897.

Watson, R. 1917. Site determinations, classification, and application. *Journal of Forestry* 15:553–565.

Weast, R. 1982. *CRC handbook of chemistry and physics*. CRC Press, Boca Raton, FL.

Weber, M. G. 1987. Decomposition, litterfall, and forest floor nutrient dynamics in relation to fire in eastern Ontario jack pine ecosystems. *Canadian Journal of Forest Research* 17:1496–1506.

Weetman, G. F., and R. M. Fournier. 1982. Graphical diagnoses of lodgepole pine responses to fertilization. *Soil Science Society of America Journal* 46:381–398.

Weetman, G. F., and R. M. Fournier. 1984. Ten-year growth results of nitrogen source and interprovincial experiments on jack pine. *Canadian Journal of Forest Research* 14:424–430.

Weetman, G. F., and B. Webber. 1972. The influence of wood harvesting on the nutrient status of two spruce stands. *Canadian Journal of Forest Research* 2:351–369.

Wei, X., and J. P. Kimmins. 1998. Asymbiotic nitrogen fixation in harvested and wildfire-killed lodgepole pine forests in the central interior of British Columbia. *Forest Ecology and Management* 109:343–353.

Wells, C. G. 1971. Effects of prescribed burning on soil chemical properties and nutrient availability. Pp. 86–99 in: *Proceedings of the prescribed burning symposium.* USDA Forest Service, Southeast Forestry Experiment Station, Ashville, NC.

Wells, C., R. Campbell, L. DeBano, D. Lewis, R. Fredriksen, E. Franklin, R. Froclich, and P. Dunn. 1979. *Effects of fire on soil. A state-of-knowledge review.* USDA Forest Service General Technical Report WO-7, Washington, DC.

Wells, S. G. 1987. The effects of fire on the generation of debris flows in southern California. *Geological Society of America, Reviews in Engineering Geology* 7:105–114.

West, S. D., R. D. Taber, and D. A. Anderson. 1981. Wildlife in sludge-treated plantations. Pp. 115–122 in: *Municipal sludge application to Pacific Northwest forest lands,* C. S. Bledsoe (ed.). Contribution #41, Institute of Forest Resources, University of Washington, Seattle.

White, E. H., and W. L. Pritchett. 1970. *Water-table control and fertilization for pine production in the flatwoods.* Florida Agricultural Experiment Station Technical Bulletin 743.

White, E. H., W. L. Pritchett, and W. K. Robertson. 1971. Slash pine root biomass and nutrient concentrations. Pp. 165–176 in: *Forest Biomass Studies,* H. E. Young (ed.). Proceedings of the XV IUFRO Congress, Gainesville, FL.

White, W. D., and S. G. Wells. 1981. Geomorphic effects of the La Mesa. Pp. 73–90 in: *The La Mesa fire symposium,* LA-9236-NERP. Los Alamos National Laboratory.

Whittaker, R. H., and G. M. Woodwell. 1972. Evolution of natural communities. Pp. 137–159 in: *Ecosystem structure and function,* J. A. Weins (ed.). Proceedings of the 31st annual biological colloquium. Oregon State University Press. Corvallis.

Wilde, S. A. 1946. *Forest soils and forest growth.* Chronica Botanica, Waltham, MA.

Wilde, S. A. 1958. *Forest soils.* Ronald Press, New York.

Wilde, S. A. 1966. A new systematic terminology of forest humus layers. *Soil Science* 101:403–407.

Wilde, S. A., and G. K. Voigt. 1967. The effects of different methods of tree planting on survival and growth of pine plantations on clay soils. *Journal of Forestry* 65:99–101.

Wilkinson, S. R., and A. J. Ohlrogge. 1964. Mechanism for nitrogen-increased shoot/root ratios. *Nature* 204:902–904.

Will, G. M. 1966. *Root growth and dry-matter production in a high-producing stand of Pinus radiata.* New Zealand Forestry Research Note 44.

Will, G. M. 1972. Copper deficiency in radiata pine planted on sands at Mangawhai Forest. *New Zealand Journal of Forestry Science* 2:217–221.

Will, G. M. 1985. *Nutrient deficiencies and fertiliser use in New Zealand exotic forests.* Forest Research Institute Bulletin #97, Rotorua.

Williams, R., and J. Fraústo da Silva. 1996. *The natural selection of the chemical elements.* Oxford University Press, New York.

Wollum, A. G. 1973. Characterization of the forest floor in stands along a moisture gradient in southern New Mexico. *Soil Science Society of America Proceedings* 37:637–640.

Wolt, J. D 1994. *Soil solution chemistry: applications to environmental science and agriculture.* Wiley, New York.

Wood, H. B. 1977. Hydrologic differences between selected forested and agricultural soils in Hawaii. *Soil Science Society of America Journal* 41:132–136.

Wood, P. J., A. F. Willens, and G. A. Willens. 1975. An irrigated plantation project in Abu Dhabi. *Commonwealth Forestry Review* 54:139–146.

Wood, T., and F. H. Bormann. 1984. Phosphorus cycling in a Northern Hardwood forest: biological and chemical control. *Science* 223:391–393.

Woodmansee, R. G., and L. S. Wallach. 1981. Effects of fire regimes on biogeochemical cycles. Pp. 379–400 in: *Fire regimes and ecosystem properties*, H. A. Mooney, T. M. Bonnicksen, N. L. Christensen, J. E. Lotan, and W. A. Reiners (eds.). USDA Forest Service General Technical Report WO-26.

Woods, R. V. 1976. *Early silviculture for upgrading productivity on marginal Pinus radiata sites in the southeast region of South Australia.* Wood Forestry Department Bulletin 24.

Woodward, J. 1699. Thoughts and experiments on vegetation. *Philosophical Transactions* 21:382–398.

Woodwell, G. M., and R. H. Whittaker. 1967. Primary production and the cation budget of the Brookhaven Forest. Pp. 151–166 in: *Symposium on primary productivity and mineral cycling in natural ecosystems*, H. E. Young (ed.). University of Maine Press, Orono.

Wooldridge, D. D. 1970. Chemical and physical properties of forest litter layers in central Washington. Pp. 327–337 in: *Tree growth and forest soils*, C. T. Youngberg and C. B. Davey (eds.). Oregon State University Press, Corvallis.

Worst, R. H. 1964. A study of the effects of site preparation and spacing on planted slash pine in the coastal plains of southeast Georgia. *Journal of Forestry* 62:556–557.

Wright, H. A., and A. W. Bailey. 1982. *Fire ecology: United States and southern Canada.* Wiley, New York.

Wright, R. F. 1976. The impact of forest fire on the nutrient influxes to small lakes in northeastern Minnesota. *Ecology* 57:549–663.

Wright, R. J., and S. C. Hart 1997. Nitrogen and phosphorus status in a ponderosa pine forest after 20 years of interval burning. *Écoscience* 4:526–533.

Wu, G., K. Haibara, T. Koike, and Y. Aiba. 1994. Available cation dynamics in forest soil measured by a combined in situ IER method. *Proceedings of the 105th Annual Japanese Forestry Society meeting, Tokyo.*

Yanai, R. D. 1992. Phosphorus budget of a 70-year-old northern hardwood forest. *Biogeochemistry* 17:1–22.

Yanai, R. D. 1998. The effect of whole-tree harvest on phosphorus cycling in a northern hardwood forest. *Forest Ecology and Management* 104:281–295.

Yanai, R. D., T. G. Siccama, M. A. Arthur, C. A. Federer, and A. J. Friedland. 1999. Accumulation and depletion of base cations in forest floors in the northeastern US.

Young, C. E., Jr. 1979. Cost analysis of land application of municipal effluent on forest land. Pp. 273–284 in: *Utilization of municipal sewage effluent and sludge on forest and disturbed land*, W. E. Sopper and S. N. Kerr (eds.). Pennsylvania State University Press, University Park.

Young, C. E., Jr., and R. H. Brendemuehl. 1973. *Response of slash pine to drainage and rainfall.* USDA Forest Service Research Note SE-186.

Youngberg, C. T., and A. G. Wollum II. 1970. Nonleguminous symbiotic nitrogen fixation. Pp. 383–395 in: *Tree growth and forest soils*, C. T. Youngberg and C. B. Davey (ed.). Oregon State University Press, Corvallis.

Youngberg, C. T., and A. G. Wollum II. 1976. Nitrogen accretion in developing *Ceanothus velutinus* stands. *Soil Science Society of America Journal* 40:109–112.

Youngberg, C. T., A. G. Wollum II, and W. Scott. 1979. *Ceanothus in Douglas-fir clear-cuts: nitrogen accretion and impact on regeneration.* Pp. 224–233 in: *Symbiotic nitrogen fixation in the management of temperate forests*, J. C. Gordon, C. T. Wheeler, and D. A. Perry (eds.). Forest Research Laboratory, Oregon State University, Corvallis.

Youssef, R. A., and M. Chino. 1987. Studies on the behavior of nutrients in the rhizosphere I: Establishment of a new rhizobox system to study nutrient status in the rhizosphere. *Journal of Plant Nutrition* 10:1185–1196.

Zabowski, D., and F. C. Ugolini. 1990. Lysimeter and centrifuge soil solutions: seasonal differences between methods. *Soil Science Society of America Journal* 54:1130–1135

Zahner, R. 1968. Water deficits and growth of trees. Pp. 191–254 in: *Water deficits and plant growth*, Vol. 2. Academic Press, New York.

Zak, B. 1964. Role of mycorrhizae in root disease. *Annual Review of Phytopathology* 2:377–393.

Zasoski, R. J. 1981. Effects of sludge on soil chemical properties. Pp. 45–48 in: *Municipal sludge application to Pacific Northwest forest lands*, C. S. Bledsoe, (ed.). Contribution #41, Institute of Forest Resources, University of Washington, Seattle.

Zavitkovski, J., and M. Newton. 1968. Ecological importance of snowbrush *Ceanothus velutinus* in the Oregon Cascades. *Ecology* 49:1134–1145.

Zeman, L. J., and O. Slaymaker 1978. Mass balance model for calculation of ionic output loads in atmospheric fallout and discharge from a mountainous basin. *Hydrology Science Bulletin* 23:103–117.

Zimmerman, M. H., and C. L. Brown. 1971. *Tree structure and function.* Springer-Verlag, New York.

Zinke, P. J. 1967. Forest interception studies in the United States. Pp. 137–161 in: *Proceedings of the international symposium on forest hydrology*, W. E. Sopper and H. W. Hull (eds.). Pergamon Press, London.

Zobel, B. J., and J. T. Talbert. 1984. *Applied forest tree improvement.* Wiley, New York.

Zon, R. 1913. Quality classes and forest types. *Society of American Forestry Proceedings* 8:100–104.

Zöttl, H. W. 1960. Dyrnamik der stickstoffmineralisation im organischen waldboden-material: I. Beziehung zwischcn Bruttomineralisation und nettomineralisation. *Plant and Soil* 13:166–182.

Zöttl, H. W. 1973. Diagnosis of nutritional disturbances in forest stands. Pp. 75–95 in: *FAO-IUFRO international symposium on forest fertilization,* Paris.

Zou, X., D. Binkley, and K. Doxtader. 1992. A new method for estimating gross phosphorus mineralization and immobilization rates in soils. *Plant and Soil* 147:243–250.

Zou, X., and G. Gonzalez. 1997. Changes in earthworm density and community structure during secondary succession in abandoned tropical pastures. *Soil Biology and Biochemistry* 29:627–629.